Histochemistry in Pathology

EDITED BY

M. Isabel Filipe
MB BS, PhD, MRCPath
Senior Lecturer and Honorary Consultant, Department
of Histopathology, Guy's Hospital Medical School, London

Brian D. Lake
BSc, PhD, FRCPath
Reader in Histochemistry, Institute of Child Health,
The Hospital for Sick Children, London

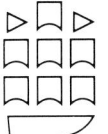

CHURCHILL LIVINGSTONE
EDINBURGH LONDON MELBOURNE AND NEW YORK 1983

CHURCHILL LIVINGSTONE
Medical Division of Longman Group Limited

Distributed in the United States of America by
Churchill Livingstone Inc., 1560 Broadway, New York,
N.Y. 10036, and by associated companies,
branches and representatives throughout the world.

First published 1983

ISBN 0 443 02429 4

British Library Cataloguing in Publication Data
Histochemistry in pathology.
  1. Pathology   2. Histochemistry
  I. Filipe, M. Isabel   II. Lake, Brian D.
  616.07'583   RB113

Library of Congress Cataloging in Publication Data
Main entry under title:
Histochemistry in pathology
  Included bibliographical references and index.
  1. Histology, Pathological.   2.Histochemistry.
  I. Filipe, M. Isabel. II. Lake, Brian D. [DNLM.
1. Histochemistry. 2. Histological technics.
QS 525 H673]
RB27.H57 1983   616.07'583   82-17706

Printed and bound in Great Britain by
William Clowes (Beccles) Limited, Beccles and London

Preface

Over the past few years we have become increasingly aware of the need for a practical book on the use of histochemistry in human pathology. The medical literature contains much information on histochemical methods and their interpretation in pathological material, but the references are widely scattered and there is no book which collects this information together. Professor A. G. E. Pearse has contributed extensively to the subject over the years and it was his suggestion that we should undertake the task of compiling a practical manual in which useful methods of known value in diagnosis could be described.

Histochemistry has now penetrated all fields of biological science, has contributed to the understanding of the relationship between structure and function in cells and tissues and has acquired an important role in diagnosis.

Its value in localising events at the histological level is unique and is an area not accessible by any other means. It provides a direct visual correlation between structure and function, so that individual cellular metabolic activities can be readily distiguished. In this and other aspects it has advantages over biochemical analysis. For example, a minimal amount of tissue is required; the functional heterogeneity at individual cell levels rather than the average activity can be visualised and a permanent record for interpretation is provided which can always be reviewed and often allows retrospective study.

Histopathology has moved in the last few decades from *macro* to *micro*, and the pathologist is now facing a new era with smaller and smaller biopsies to interpret. It is particularly in this respect that the more specific techniques of histochemistry will greatly reduce the number of equivocal and inconclusive diagnoses with consequent benefit to the patient.

Histochemical techniques are now an essential part of the histopathological diagnosis of muscle disease and storage disorders, in the classification and diagnosis of leukaemias, lymphomas, and tumours of the neuroendocrine system, and in determining the site of origin of a variety of malignancies. Histochemistry increases the accuracy of the histological diagnosis in the assessment of malignant transformation and has a role in monitoring response to therapy, and in prognosis. Its use, however, has not yet gained wide acceptance.

This book is therefore aimed at the pathologist. It is not a pathology text. We assume that the reader is familiar with general pathological principles and has access to standard pathological texts. In the same way the book is not a histochemistry text. It includes histochemical techniques of established value in particular diagnostic problems. There is no pretence to cover every possible pathological condition, but we hope to have covered most disorders that are encountered in routine pathology.

We have excluded histological methods and subjects for which histochemistry as a means of diagnosis is well covered in pathology textbooks.

Finally, good co-operation between clinician, surgeon and pathologist should be encouraged, so that maximum information can be extracted from a biopsy specimen. However, histochemical techniques should not be used 'blind' as a sterile academic exercise; they should be used intelligently in relation to the clinical history. The methods included are simple and well-tried and all can be carried out in any routine laboratory.

We hope that many pathologists and technicians will find this book useful and thus be persuaded to adopt histochemistry as an integral part of histopathology.

London,   M.I.F.
1983   B.D.L.

Acknowledgements

This book could not have been written without the help, advice and encouragement of our many friends and colleagues, too numerous to mention by name. Some, however, deserve especial acknowledgement of their particular suport. This book in part fulfils the vision of the late Dr Martin Bodian, whose encouragement led to the establishment of a diagnostic histochemistry laboratory at the Hospital for Sick Children. His successor, Professor Albert Claireaux, has provided continued support since 1965. On the technical side, Mrs Virpi Smith, FIMLS (Hospital for Sick Children) and the technical staff at the Westminster Medical School, in particular Miss Phyl Brock and Mr J. Patel, have given expert assistance for many years. We thank Miss Barbara Archer for her comments on copper-associated protein. Our thanks are also due to Mrs K. Matharu for invaluable secretarial support.

We would like to thank our Publishers, Mr Andrew Stevenson and his colleagues at Churchill Livingstone, for their help in putting this book together.

Contributors

Olga B. Bayliss High FIMLS MPhil PhD
Chief Medical Laboratory Scientific Officer,
Histopathology Department, Guy's Hospital
Medical School, London, UK

M. Isabel Filipe MB BS PhD MRCPath
Senior Lecturer and Honorary Consultant,
Department of Histopathology, Guy's Hospital
Medical School, London, UK

E. Heyderman MB BS BDs MRCPath
Senior Lecturer and Honorary Consultant,
Department of Histopathology, St Thomas'
Hospital Medical School, London, UK

E.J. Holborow MA MD FRCP FRCPath
Professor of Immunopathology and Head,
Immunology Section, Bone & Joint Research Unit,
The London Hospital Medical College, London,
UK

David Hopwood MD BSc PhD MRCPath
Senior Lecturer in Pathology, Dundee Medical
School and Honorary Consultant, Tayside Health
Board, Ninewells Hospital, Dundee, UK

O.A.N. Husain MD FRCPath FRCOG
Consultant Pathologist, Cytology Department,
Charing Cross and St Stephen's Hospitals, London,
UK

Nasser Javadpour MD
Urologist in charge and Senior Investigator,
National Cancer Institute, NIH; Consultant,
National Naval Hospital and Walter Reed Army
Hospital, Washington DC, USA

A.C. Jöbsis MB BS MD
Consultant Pathologist and Senior Lecturer in
Pathology, University Hospital, Amsterdam, The
Netherlands

Margaret A. Johnson BSc PhD
Lecturer in Experimental Neurology, University of
Newcastle upon Tyne, UK

Brian D. Lake BSc PhD FRCPath
Reader in Histochemistry, Institute of Child
Health, The Hospital for Sick Children, London,
UK

P.L. Lantos MD PhD MRCPath
Professor of Neuropathology, Institute of
Psychiatry; Honorary Consultant Neuropathologist,
The Bethlem Royal Hospital and The Maudsley
Hospital, London, UK

Micheline Levy MD
Maître de Recherches, INSERM, Hôpital Necker
Enfants-Malades, Paris, France

Z. Lojda MD
Professor, Laboratory of Histochemistry, Faculty of
Medicine, Prague, Czechoslovakia

D.H. Mackenzie MA MD FRCPath
Professor of Histopathology, Westminster Medical
School; Consultant Pathologist, Westminster
Hospital, London, UK

A. McQueen MB ChB FRCPath
Senior Lecturer in Dermatopathology, University of
Glasgow, Glasgow, UK

D.Y. Mason BM BCh DM MRCPath
University Lecturer in Haematology, University of
Oxford, Oxford, UK

Liliane Morel-Maroger MD
Maître de Recherches, INSERM, Hôpital Tenon,
Paris, France

M.G. Ormerod PhD
Scientist, Institute of Cancer Research, Royal
Cancer Hospital, The Haddow Laboratories,
Sutton, Surrey, UK

A.G.E. Pearse MA MD FRCP FRCPath
Professor of Histochemistry, University of London;
Consultant Pathologist, Hammersmith Hospital,
London, UK

J.P. Sloane MB BS MRCPath
Consultant Histopathologist, Royal Marsden
Hospital, Sutton, Surrey, UK

H. Smith BSc MD (Melb.) FRCPA
Consultant Haematologist, Royal Brisbane Hospital,
Brisbane, Queensland, Australia

Peter J. Stoward MA MSc DPhil FRSE
Reader in Histology, Department of Anatomy,
University of Dundee, Dundee, UK

C.R. Taylor MA MD DPhil
Professor of Pathology, University of Southern
California; Chief of Immunopathology, Los Angeles
County-USC Medical Center, Los Angeles,
California, USA

Nancy E. Warner SB MD
Professor and Chairman, Department of Pathology,
University of Southern California School of
Medicine, Los Angeles, California, USA

Moshe Wolman MD
Professor and Chairman, Department of Pathology,
Tel Aviv University Sackler School of Medicine,
Tel Aviv, Israel

Contents

General principles of fixation

D. Hopwood

FIXATION AND THE CLINICIAN

For the pathologist there is a dilemma presented by fixation of tissues. On the one hand, with the development of more histochemical methods which have become useful diagnostically, a variety of fixative procedures from an increased armamentarium are found to be necessary. In some cases, no fixation is required. On the other hand the clinician needs a simple uniform routine for dealing with biopsies and surgical specimens.

The problem is fundamentally related to two factors, namely the multiplicity of reactions both chemical and physical involved in tissue fixation and the specific needs of the techniques. Placing the specimen in formaldehyde solution of one kind or another immediately limits the number of investigations the pathologist can pursue unless there has been prior consultation with the clinician.

The processes involved in fixation from theoretical and practical aspects have been reviewed. Earlier work is well summarised by Baker (1960). More recent work has been covered by the author (Hopwood 1969, 1972) and Pearse (1980).

At the most informed level, there is discussion between the clinician and pathologist over handling of the biopsy specimen. Here, the diagnosis is usually suspected and the histochemical options available are maximal. Examples of this are seen in storage diseases and secreting APUD tumours. Other situations where the pathologist receives unfixed material regularly include muscle and kidney biopsies where this routine has now become well-established. Also, material from theatre for frozen section is received fresh and if some tissue remains, then appropriate histochemistry is possible. This also includes lymph nodes where culture may be initiated (Table 1.1).

Table 1.1 Fixation and the clinician

1. Prior discussion, diagnosis suspected, pathologist to collect
   a. metabolic diseases, APUD tumours
   b. muscle, kidney, jejunum
   c. liver
2. Suspected by pathologist at cut-up (minus one)
   a. lipids, EM, some enzymes
   b. post fixation required e.g. dichromate
3. After paraffin wax sections
   tumours — mucosubstances
   pigments, inorganic substances — X-ray
   dispersive microscopy
   immunoperoxidase

A less favourable, although more common, receipt of tissue is at the cut-up. Here, the biopsy or surgical specimen will have been placed, usually unopened, in the routine fixative of the department — most often formaldehyde. Fixation will have occurred at room temperature for varying periods up to 24 hours or more if the specimen has come from another hospital. The histochemical possibilities are much more severely limited here and depend upon the diagnostic acumen of the pathologist, often in the light of clinical information given. At this stage, it is possible to perform lipid and mucin stains and investigate some of the enzymes which will withstand formaldehyde fixation. At this point in processing, some specific form of postfixation may be indicated such as chromate for the chromaffin reaction. Electron microscopy is also possible from this material. Success depends a great deal on the time the tissue has already spent in formaldehyde. Some institutes do not process all their material, but retain a small portion in fixative whilst the remainder goes forward for routine sectioning and staining.

The most common way specimens are closely examined is as a paraffin section, stained with haematoxylin and eosin. At this stage, there are severe limitations on the types of histochemical investigations possible. Commonly, this is limited to the nature of the mucosubstances in tumours and virtually no enzyme studies are attempted. This probably explains why the most common histochemical interest of pathologists is limited to mucosubstances and the nature of various pigments investigated. A recent advent in this area merits attention. Crocker et al (1980) have shown that it is possible to determine the nature of particulate and crystalline material in paraffin wax-embedded tissue by a combination of electron microscopy and X-ray energy dispersive microscopy (see also Ch. 28). Also there has been recent interest in the application of immunoperoxidase techniques for paraffin wax-embedded material. This allows retrospective study of material already in departmental files.

ADDITIVES

A number of substances have been added to the fixative solution used in histochemical techniques. These are largely related to electron histochemistry where they fall into two areas. In the ultrastructural demonstration of mucosubstances, both ruthenium red and alcian blue have been added to the primary and secondary fixative solutions (Behnke & Zelander 1970). The non-ionic detergent, Triton X-100, has been used as an agent to facilitate the penetration of ruthenium red during fixation. It should also be mentioned that other non-ionic detergents have been used to enhance the penetration of antibodies and lectins into cells (Laurila et al 1978).

A variety of substances have been used as cryoprotectants to prevent damage during freezing and thawing. These include polyvinylpyrrolidone (Novikoff 1956), sucrose (Holt & Withers 1958), glycerol (Melnick 1967) and demethylsulphoxide (Cope 1968). It should be remembered that these may produce artefacts which may be relevant ultrastructurally. These additives may also inhibit enzyme activity.

ARTEFACTS

There are a number of artefacts which are associated with fixation and the pathologist should be aware of these.

The physical loss of some tissue components occurs during fixation. These are, for obvious reasons, mostly small molecules. Important in the present context are various co-factors for enzymes (Hopwood 1968), and polypeptide hormones. Elsewhere in this chapter, the loss of mucosubstances, lipids and nucleic acids during fixation have been mentioned. Similarly, inorganic substances are lost and gradients they may have in life disappear.

The rearrangement of substances, e.g. glycogen and iron, within cells and tissues is the well-known phenomenon of false localisation. The diffusion of enzymes may also occur, even in fixed material incubated with buffer at 37°C (Hardonk et al 1977).

Loss of enzyme activity usually follows fixation, although sometimes this may be reversed subsequently by washing the tissue in buffer, usually to a histochemically acceptable level. Different fixatives affect enzyme activity to a different extent (Lake & Ellis 1976). The same fixative reduces activity to varying levels when reacting with individual enzymes (Hopwood 1972). Arborgh and his colleagues (1976) showed that most enzyme fixation took place rapidly and that from a histochemical viewpoint, any time between 5 minutes and 24 hours was satisfactory provided morphological preservation was adequate. It is important to differentiate between histochemical preservation of enzyme activity and enzyme activity measured in fixed tissues biochemically (Christie & Stoward 1974).

Besides the loss of enzyme activity, fixation also alters other protein-dependent reactions, e.g. antigen-antibody reactions. The dilemma is the same as with the enzymes, that is, morphological preservation versus biological activity.

Further points to bear in mind are:
1. potential reaction between various components in a fixation mixture
2. false-negative effects of fixation
3. false-positive results.
Nonspecific binding occurs with glutaraldehyde and with formaldehyde in line with their general reactivity towards tissues (Hopwood 1969). Glutaraldehyde also has the effect of producing aldehyde groups in tissues due to its bifunctional nature and will give false-positive PAS reactions.

Further, Pentilla et al (1975) have shown that injured cells behave differently to non-injured in respect to their volume in various fixatives. 4% formaldehyde should be used with buffers of 310 mmol and glutaraldehyde-formaldehyde mixtures with 100–150 mmol buffers. The osmolality of two commonly used fixative solutions are: glutaraldehyde 480 mmol, paraformaldehyde 1280 mmol (Hayat 1973). The problems of cell volume change have been pursued further by Collins et al (1977) who showed changes which were more obvious with SEM and time-lapse cinematography than on electron microscopy in transmission mode.

FIXATIVES AS HISTOCHEMICAL REAGENTS

Fixation may be used to differentiate various cellular features histochemically. Some years ago, we showed that glutaraldehyde fixation discriminated between adrenalin and noradrenalin-containing cells by light and electron microscopy in the adrenal medulla (Coupland & Hopwood 1966), the noradrenalin only forming an electron-dense deposit with the glutaraldehyde.

Simionescu et al (1972) introduced a lead-containing fixative which demonstrates glycogen and dextrans in tissues by electron microscopy.

Karnovsky's osmium tetroxide potassium ferrocyanide fixative also enhances glycogen and components of the cell membranes and fuzzy coat (1971). More recently, Ainsworth (1977) has described the use of osmium tetroxide-potassium ferrocyanide to demonstrate polyvinylpyrrolidone and dextrans used in tissues as tracers.

NO FIXATION

In certain circumstances, it is essential that the tissue is not fixed before it is studied. From a diagnostic viewpoint, probably the most important are the metabolic disorders and most enzyme techniques, in particular, dehydrogenases where fixation must be avoided. It is often also useful to make dabs from fresh unfixed spleen, lymph nodes, lung and brain biopsies, tumours, etc. Remember porphyria cutanea tarda and the brick-red autofluorescence in cryostat sections.

MUCOSUBSTANCES

The pathologist's chief interests in mucosubstances are directed towards glycogen and acid mucosubstances as diagnostic aids. Glycogen is best demonstrated in unfixed cryostat sections covered with celloidin and stained by the PAS technique (Lake 1970). Rossman's fluid retains 70% of glycogen (Smitherman et al 1972), whilst formaldehyde retains about 12%. Nonetheless, in the majority of surgical specimens, this is enough to be demonstrated. Polarisation of glycogen within the cells is less after aqueous fixatives than after alcoholic fixatives. The mechanism whereby glycogen is fixed remains uncertain, but Smitherman and his colleagues (1972) think that physical entrapment is important. Mercuric chloride reduces the amount of glycogen which may be demonstrated.

Glycosaminoglycans (GAGS) and proteoglycans (e.g. mucins) remain soluble after tissue has been fixed in formaldehyde or glutaraldehyde where losses may mount during processing to 70%. Although mucins can be demonstrated in routinely fixed surgical material, successful attempts have been made to improve their fixation. Various cationic dyes and cetylpyridiniums have been added to the fixative with increased retention of proteoglycans and GAGS.

A more recently introduced technique to investigate mucosubstances involves the use of lectins, a group of substances derived largely from plants which bind to specific sugar residue on cell surfaces. They have also been used to demonstrate differences in sugars in cell membrane development (Jamieson et al 1979) and we have employed concanavalin A to study the distribution of glycosyl and mannosyl residues in human oesophagus (Hopwood et al 1978). Evidence to date suggests that the fixative used is not critical and routinely fixed sections can be used with satisfactory results.

NUCLEIC ACIDS AND NUCLEOPROTEINS

Formaldehyde does not react with nucleic acids under normal fixation conditions (Hopwood 1975).

Although some nucleic acids are lost during fixation, they are mostly trapped in the cells. The associated histones with the nucleic acids do react with formaldehyde (Brutlag et al 1969). This reaction with formaldehyde remains reversible for some time (McGhee & Hippez 1975).

The interest of the pathologist in nucleic acids is chiefly directed towards recognising atypia and RNA-rich plasma cells. Formaldehyde is not the optimum fixative and usually Carnoy is the one recommended, especially if any quantitative work is envisaged. Other than haematoxylin and eosin, the commonly used histochemical techniques for demonstrating nucleic acids are methyl green pyronin and Feulgen. In the latter, which is used quantitatively, an acid hydrolysis is required and this varies in time with the fixative first employed.

LIPIDS

The general pathologist's interest in lipids is limited. He notes the hole left by extracted lipids in steatosis of the liver which, in many ways, denotes the attitude to this important class of substance. And yet lipids are present unobserved in many tissues in the surgical routine, for example, gall bladder epithelium, other than in cholesterolosis (Boyd 1922) and oesophageal epithelium (Hopwood et al 1977).

A modest repertoire of methods for the histochemical demonstration of a variety of lipids and lipoproteins exists. These are outlined by Pearse (1968). In general the techniques for their demonstration call for frozen sections of unfixed or formaldehyde-fixed tissues. The methods for the fixation of lipids, making them insoluble, are few. The roles of osmium tetroxide and potassium dichromate are well known, although both of these alter the chemical reactivity. The standard fixatives for phospholipids (Baker's and Elftman's) are given in Appendix 4. Formaldehyde has also been shown to react with various lipids. The reactions with those containing amine groups, such as phosphatidyl ethanolamine, would be expected. Formaldehyde has been shown to react with ethylene double bonds to form 1,2 glycols (Jones 1973).

ENZYMES

The ability of enzymes to withstand fixation varies enormously. The dehydrogenases are sensitive even to brief fixation in dilute aldehydes, whereas the peroxidase activity of haemoglobin can persist in red blood cells for several weeks (Torack 1974) in buffered formaldehyde. Biochemical and morphological studies, largely of aldehyde fixation of enzymes, have been pursued and these have been summarised by Hopwood (1972) and Pearse (1980). From the detail, some generalities appear.

Some form of fixation is desirable in general, although in the case of the dehydrogenases this may have to be limited to a short time with low fixative concentrations. Unbound enzymes are liable to be lost from the specimen. To overcome this difficulty, a variety of protective agents, such as polyvinyl-pyrrolidone, have been introduced into the incubation medium.

In general with aldehyde fixation, it was found that the longer the fixation period and the higher the temperature at which it was carried out, the less enzyme activity remained. At the same time it was found that washing fixed tissue in buffer restored enzyme activity considerably, at the same time retaining good morphological preservation.

Phosphate-buffered 4% formaldehyde (with added sucrose) has been found to be a good general fixative for preservation of hydrolase activity as Holt & Hicks reported (1961) with subsequent storage in cold gum sucrose. We have found that glutaraldehyde-paraformaldehyde mixtures introduced by Karnovsky (1965) are useful for the ultrastructural demonstration of enzymes. One should remember that if the commonly used methods of fixation do not work, there are other worthwhile alternatives which should be tried, for example acrolein (Saito & Keino 1976). The fixation requirements for the specific enzymes should be sought in the appropriate chapter and Appendix 5.

IMMUNOCYTOCHEMISTRY

Fixation and processing of tissues for subsequent immunological techniques produces some problems of artefact and interpretation and no single method is applicable to all problems. Basically, there are the usual two horns of a dilemma: on the one hand, preservation of morphology and on the other, preservation of antigenicity without diffusion artefacts.

In our department, the routine practice for the investigation of patients' autoantibodies is to use fresh unfixed cryostat sections of animal (rat) tissue and an indirect immunofluorescent technique. Similarly, skin and kidney biopsies are investigated as unfixed cryostat sections, but directly labelled fluorescent antibodies are used. On lymphoid tissue, we often give brief fixation with ethanol which has two functions: to intensify the fluorescence and to reduce the background. Fixation can be successfully employed for immunofluorescent studies, but much depends on the structure of the antigen. Interestingly, osmium tetroxide has been shown to preserve growth hormone and prolactin in pituitary (Baskin et al 1979). Leatham & Atkins (1980) have ranked fixatives in their effects on morphological preservation of lymphoid tissue along with the preservation of immunoglobulin antigenicity. Susa was found to be the best fixative for demonstrating specific staining and to produce the minimum background staining. This was followed by Bouin and formol sublimate. Formalin fixation generally gave poor specificity and enhanced background staining.

Immunoperoxidase techniques have established themselves in routine pathological investigations. They have various advantages over the fluorescent technique, namely, they provide permanent preparations and some authors feel useful detail is shown (Curran & Gregory 1978). Immunoperoxidase techniques on paraffin section give similar results to fluorescent techniques, although the latter is less time-consuming. For further views see Chapters 3 and 20, and Bullock & Petrusz (1982).

The effect of various proteases on tissue digestion has been examined in fixed sections prior to immunohistochemical staining (Mepham et al 1979). There seems to be general agreement that trypsin is the most useful, the time of digestion needed being related to the mode and duration of fixation. The need for freshly prepared trypsin solutions should be stressed. This technique cannot be applied to all antigens, particular problems being encountered with complement (Huang et al 1976). Heyderman (1979) has counselled against the use of digestion in immunoperoxidase techniques on reasonable theoretical grounds. We use the digestion technique in various of our routine investigations and to date have had no problems, in line with other pathology departments.

Piris & Thomas (1980), in a quantitative study on mucosal plasma cells, compared formaldehyde saline and formaldehyde mercuric chloride fixatives and reported a better staining and higher cell counts with the latter. These techniques offer the advantage of retrospective studies on filed material.

Many of the antigens investigated up to now have been relatively stable and 'easy'. Smaller molecules are more labile and difficult to investigate. Such include the smaller peptides in the nervous system and the diffuse endocrine system in the gut and elsewhere. These have been successfully studied by Pearse & Polak (1975) using benzoquinone fixation.

A further area of difficulty for immunohistochemical methods has been pointed out by Matthew (1981). He showed that xylene decreased immunoreactivity in frozen sections and in nontrypsinised material fixed in formaldehyde-calcium, formaldehyde-saline or Bouin, whereas chloroform or Inhibisol© gave better results. For electron immunocytochemistry, the problem of morphological and antigenic preservation are even more acute (McLean & Nakane 1974, Laurila et al 1978, Willingham & Yamada 1979).

BIOGENIC AMINES

Over the past 15 or so years, methods have been elaborated which use formaldehyde vapour both to fix tissue and produce a fluorescent product with biogenic amines. These have characteristic wavelengths for their fixation and emission spectra and are fully detailed by Falk & Owman (1965). Greater sensitivity is reported by including magnesium ions (Lorén et al 1977). A simple rapid method for producing monoamine fluorescence has been described by de la Torre & Surgeon (1976) who used cryostat sections with glyoxylic acid-induced fluorescence. The complete process from obtaining the tissue to examining for fluorescence under the microscope takes under 20 min. The details are given in Appendix 2.

POLYPEPTIDE HORMONES

Tissue in which polypeptide-containing cells are to be demonstrated must be dealt with in a specific manner. Ideally, the specimen should be prepared for the following procedures, depending on the tissue.

1. Immunocytochemistry — benzoquinone fixation
2. Radioimmunoassay — frozen
3. Electron microscopy — glutaraldehyde
4. Fluorogenic amine content — formaldehyde vapour and glyoxylic acid
5. Nonspecific esterase or cholinesterase — cold formaldehyde calcium or frozen
6. Lead haematoxylin, argyrophil — paraffin sections. Bouin masked metachromasia
7. Recently, neuron-specific enolase has been found to be a better general marker for these cells.

Other methods may be used in the investigation of these tissues, but this is not the appropriate place to discuss them.

INORGANIC MATERIALS

The investigation and fixation of these substances has received scant attention. Some substances such as iron, calcium and copper are commonly looked for in routine formaldehyde-fixed material. Walker and his colleagues (1971) have shown that there was little correlation between histological grading of iron in biopsies and the concentration measured by atomic absorption spectroscopy. Part of the problem was iron loss into the fixative. In a similar way Renaud (1959) reported loss of calcium into aqueous formaldehyde but not into ethanol. Again, copper may be lost unless the biopsy is fixed in the presence of rubeanic acid. A possible way round this last problem is to demonstrate the presence of copper-binding protein, using the prolonged orcein stain on formaldehyde fixed material (Shikata et al 1974).

Relatively simple techniques have recently been described for the analysis of inorganic material in routine surgical material, even after H & E sections have been prepared. Crocker et al (1980) used micro-incineration of paraffin sections followed by scanning electron microscopy and X-ray energy spectroscopy. They have since shown that the micro-incineration

stage may be omitted and the back-scattered electrons used for analysis (Crocker et al 1981).

HAEMATOLOGY

A number of histochemical techniques have been applied to the study of white and red cells. Those commonly used in this department are PAS, Sudan black B, acid and alkaline phosphatase and nonspecific esterase. A variety of fixative methods have been proposed. Hayhoe & Quaglino (1980) suggest the use of formaldehyde vapour alone, but aqueous fixatives also work well.

CYTOLOGY

Cytological techniques have been subsequently applied for a number of years to gynaecological and other problems. The repertoire has been extended recently by aspiration cytology (Melcher & Linehan-Smith 1981). Various attempts have been made to quantify and automate the cell scanning. In the case of the cervix, this has been with the Feulgen reaction (Millett & Hussain 1979) and naphthylamidase activity (Hussain & Millett 1979). Alternatively, cells may be collected on a Millipore© filter for histochemical analysis.

Similar preparations of cells may be made from touch preparations or dabs from unfixed material such as lymph nodes and tumours where architectural relations may still remain. Smear preparations of brain biopsies have been used for many years and their use has recently been reviewed (Adam et al 1981).

PREPARATIVE TECHNIQUES

Having received the specimen and fixed or frozen it where this is appropriate, the next problem is to produce sections for incubation. There are basically two methods. The older is the freezing microtome where the specimen and knife are cooled by a carbon dioxide jet or dry ice. More recently, controllable water-cooled thermo-elements have been introduced to replace the carbon dioxide. We have found that these work well for fixed human and animal material. Fresh tissue sectioning is best accomplished using a cryostat. In routine histopathology practice the fresh tissue is usually frozen to a chuck using an acetone dry ice mixture. This method is unsatisfactory for most histochemical procedures, and other methods are available and preferable. For quenching with minimum artefact, the tissue should be placed in isopentane cooled in liquid nitrogen after dusting with starch (see also Appendix 1). For small pieces, blocks of 12% gelatin provide a good medium for proper orientation and support, for sectioning. Sections are then mounted either on slides or coverslips and allowed to dry at room temperature. Alternatively, sections may be moved, possibly through a buffer, to the incubation medium. Valuable specimens on chucks may be conserved for 1–2 weeks at $-20°C$ by wrapping them and the chuck with Parafilm© or similar material. More complete details of the practical procedures are given by Pearse (1980) and Bancroft & Stevens (1982) where freeze drying and freeze substitutions are also described.

For some histochemical techniques in use at ultrastructural level, 20–40 μm sections are used for incubation. Unfixed tissue sections may be prepared either with a Vibratome© (Oxford Instruments) or a Sorvall TC-2 tissue chopper. For fixed tissues, a freezing microtome may be used.

PLASTIC EMBEDDING

The use of plastic embedding materials has helped certain areas of pathological practice considerably, especially renal, lymph node, lung, needle and bone marrow biopsies. Early attempts to apply histochemical immunoperoxidase methods met with limited success. Using glycol methacrylate, we found it is possible to demonstrate alkaline phosphatase and nonspecific esterase activity in thin mounted sections. Mucosubstance histochemistry is also possible. Great difficulty has been encountered in demonstrating immunoglobulins. More recently, Rodning et al (1980) have demonstrated IgA and lysozyme in immunocytes and Paneth cells in epoxy resin sections. IgA was very sensitive to glutaraldehyde. Sections were successfully etched with sodium ethoxide or methoxide. Oxidizing etching agents were inimical to immunoglobulin reactivity.

HAZARDS

Biological

Unfixed surgical specimens sent for frozen section pose a potential for infection to all those who handle the tissue. There are two agents which have caused most concern to pathologists in recent years, namely the hepatitis B virus and the tubercle bacillus.

The risk of acquiring hepatitis B infection has been shown to be highest to workers in haematology and chemical pathology laboratories (Grist 1980). Although most hospitals have some system for identifying specimens which are positive for HBsAg, workers remain exposed to two groups of material. These obviously include patients who are unscreened asymptomatic carriers. Known HBsAg-positive patients in whom the information is not passed on are a potential source for this hazard, clinical notes being kept by separate hospitals and departments.

Laver and his colleagues (1979) have shown that the primary mode of transmission in the laboratory is for the hepatitis B virus to enter through trivial cuts and abrasions on the hands from contaminated material. Protection will be afforded by the use of waterproof dressings and gloves. Care should also be taken in the disposal of sharp objects. A further discussion of this topic is given in the Bulletin of the RCP (1981).

Active tuberculosis remains a problem in unfixed surgical and necropsy specimens. The diagnosis may be unsuspected in up to one-third to one-half of the patients (Edlin 1978, Cameron & McCoogan 1981). Clearly, this raises the problem of contamination of cryostats and other instruments. Methods for this disinfection are detailed by Howie (1978) who also deals with the more general problem of the prevention of infection in laboratories.

Chemical

The safety of various chemicals used in preparative techniques has been questioned. Problems may be considered in particular in relation to fixatives, buffers and embedding materials for electron microscopy.

Attention has been drawn by Nature (1980) to two toxicological studies still running on the effects of formaldehyde vapour. All aldehydes and their vapours should be regarded as toxic.

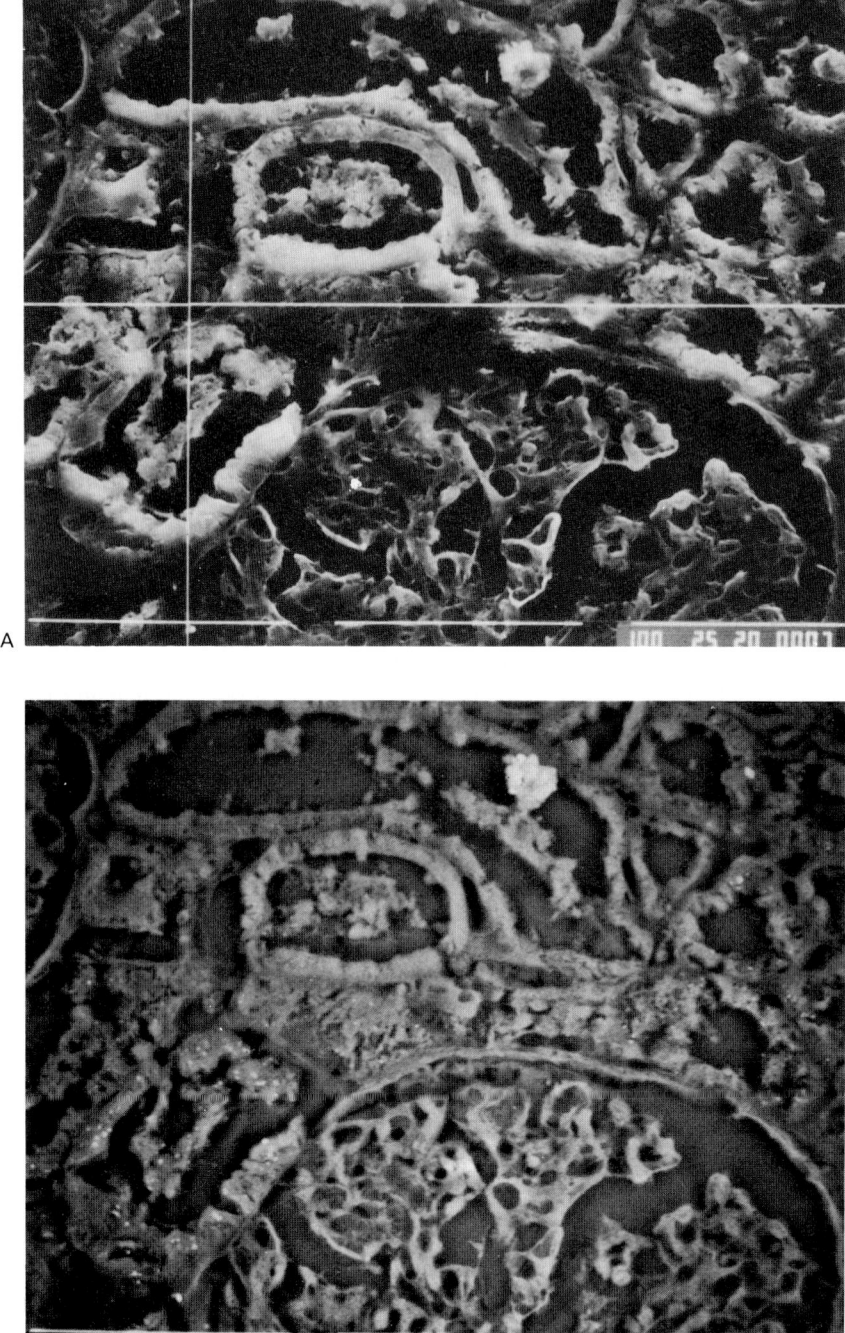

Fig. 1.1 A renal biopsy from a patient with rheumatoid arthritis treated with gold, who developed nephrotic syndrome and was found to have membranous glomerulonephritis.

A. Paraffin wax section (deslipped H & E), lightly coated with carbon. ×350

B. Same field, same magnification, back-scatter electron image. Note tiny bright granule in tubular epithelium, especially just to left of glomerulus. ×350

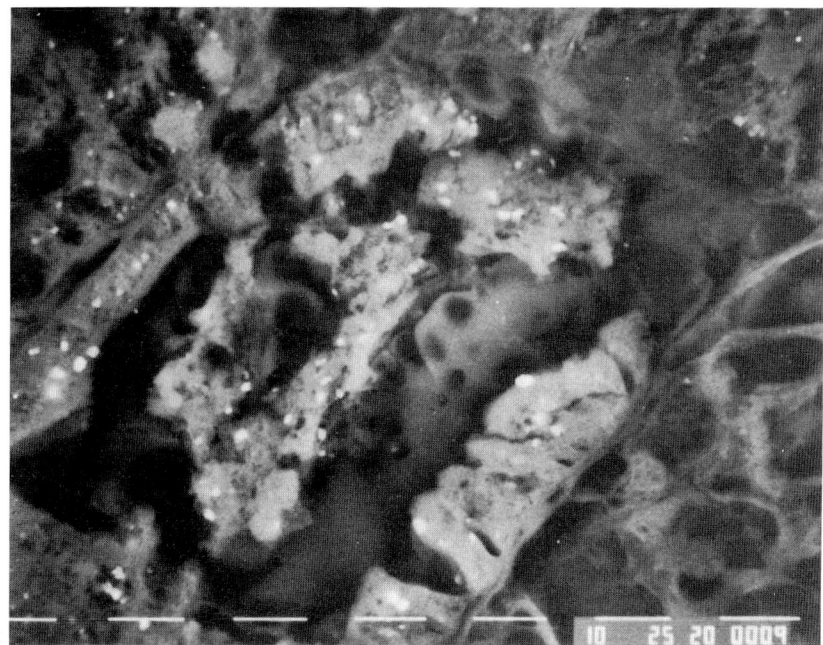

Fig. 1.1 C. Higher-power picture of tubule with bright granules showing more clearly. On X-ray analysis, these granules proved to be gold. × 1000
(Photographs kindly supplied by D.A. Levison and P.R. Crocker, Department of Histopathology, St Bartholomew's Hospital)

Attention should also be drawn to the buffer system often used in electron microscopy, namely cacodylate. This contains arsenic which may be absorbed through the skin and the dust from the powder through the respiratory tract.

The epoxy resin embedding media, their hardener and catalysts are all toxic (especially Spurr's). In susceptible individuals, they cause dermatitis on the hands and the use of gloves for their handling is obviously necessary. The methacrylates, which are enjoying a new popularity for 1 μm sections, also produce toxic vapours.

Substances used in the staining reactions, histochemical and electron microscopic, also have their toxicities and carcinogenicities. Some of these such as DAB and the dianisidines, are well known. The dangers of enzyme inhibitors should be self-evident and they should be handled with appropriate care.

Obviously, care must be taken in the storage of dangerous materials, that is, both the inflammable substances used preparatively, and toxic chemicals. Attention should also be drawn to the hazard of storing inflammable materials in an ordinary domestic refrigerator with an internal thermostat which can spark.

Although some of the hazards of the more commonly used procedures have been mentioned here, this is obviously not the place for a complete and detailed description of chemical and biological hazards. These are dealt with more fully, along with good laboratory practice, general precautions and legislative aspects, in the following: Howie (1978), Bretherick (1980) and Weakley (1981).

CONCLUSION

It seems that we are still some distance from finding a magic juice which will give both perfect or even acceptable morphological preservation and at the same time, good biochemical or immunological activity. Such a fixative would obviously facilitate our investigations on biopsies sent by the clinicians. Until such a fixative is evolved, we, as pathologists, must inform our clinical colleagues of histochemical possibilities for diagnosis and the fixation procedures required. At the same time, in spite of all the ideal situations described for various techniques, it is often worth trying a method from tissue fixed under non-ideal conditions. It may well succeed. (See Fig. 1.1).

REFERENCES

Adam J H, Graham D I, Doyle D 1981 Brain biopsy — the smear technique for neurosurgical biopsies. Chapman and Hall, London

Ainsworth S K 1977 An ultrastructural method for the use of polyvinylpyrrolidone and dextrans as electron opaque tracers. Journal of Histochemistry and Cytochemistry 25: 1254–1259

Arborgh B, Bell P, Brunk U, Collins V P 1976 The osmotic effect of glutaraldehyde during fixation. Journal of Ultrastructural Research 56: 339–350

Baker J R 1960 Principles of biological microtechnique. Methuen, London

Bancroft J D, Stevens A 1982 Theory and practice of histological techniques, 2nd edn. Churchill Livingstone, Edinburgh

Baskin D G, Erlandsen S L, Parsons J A 1979 Immunocytochemistry with osmium-fixed tissue. I. Light microscopic localization of growth hormone and prolactin with the unlabelled antibody enzyme method. Journal of Histochemistry and Cytochemistry 27: 867–872

Behnke O, Zelander T 1970 Preservation of intercellular substances by the cationic dye Alcian Blue in preparative procedures for electron microscopy. Journal of Ultrastructural Research 31: 424–438

Boyd W 1922 Studies in gall bladder pathology. British Journal of Surgery 10: 337–356

Bretherick L (ed) 1980 Hazards in the chemical laboratory, 3rd edn. Chemical Society, London

Brutlag D, Schlehuber C, Bonner J 1969 Properties of formaldehyde-treated nucleohistone. Biochemistry 8: 3214–3216

Bulletin of the Royal College of Pathologists 1981 34: 3–7

Bullock G R, Petrusz P 1982 Techniques in immunocytochemistry. Academic Press, London, vol 1

Cameron H M, McGoogan E 1981 A prospective study of 1152 hospital autopsies. II. Analysis of inaccuracies in clinical diagnoses and their significance. Journal of Pathology 133: 285–300

Christie K N, Stoward P J 1974 A quantitative study of the fixation of acid phosphatase by formaldehyde and its relevance to histochemistry. Proceedings of Royal Society B 186: 137–164

Collins V P, Arborgh D, Brunk U 1977 A comparison of the effects of three widely-used glutaraldehyde fixatives on cellular volume and structure. Acta pathologica et microbiologica scandinavica 85A: 157–168

Cope G H 1968 Low-temperature embedding in matter-miscible methacrylates after treatment with antifreezes. Journal of the Royal Microscopical Society 88: 235–257

Coupland R E, Hopwood D 1966 The mechanism of the differential staining reaction for adrenalin and nonadrenalin storing granules in tissues fixed in glutaraldehyde. Journal of Anatomy 100: 227–243

Crocker P R, Doyle D V, Levison D A 1980 A practical method for the identification of particulate and crystalline material in paraffin-embedded tissue specimens. Journal of Pathology 131: 165–173

Crocker P R, Toulson E, Levison D A 1981 The back-scattered electron image (BEI) for identification of particles in paraffin sections. Jeol News 19E: 10–14

Curran R C, Gregory J 1978 Demonstration of immunoglobulin in cryostat and paraffin sections of human tonsils by immunofluorescence and immunoperoxidase techniques. Journal of Clinical Pathology 31: 974–983

De La Torre J C, Surgeon J W 1976 A methodological approach to rapid and sensitive monoamine histofluorescence using a modified glyoxylic acid technique: the SPG method. Histochemistry 49: 81–93

Edlin G P 1978 Active tuberculosis unrecognised until necropsy. Lancet 1: 650–652

Falk B, Owman C 1965 A detailed methodological description of fluorescence method for the cellular demonstration of biogenic amines. Acta Universitatis Lundensis II (7): 1

Grist N R 1980 Hepatitis in clinical laboratories 1977–78. Journal of Clinical Pathology 33: 471–473

Hardonk M J, Haarsma T J, Kijkhuis F W J, Poel M, Koudstaal J 1977 Influence of fixation and buffer treatment on the release of enzymes from the plasma membrane. Histochemistry 54: 57–66

Hayat M A 1973 Specimen preparation. In: Hayat M A (ed) Electron microscopy of enzymes. Van Nostrand Reinhold, London, vol 1

Hayhoe F G J, Quaglino D 1980 Haematological cytochemistry. Churchill Livingstone, Edinburgh

Heyderman E 1979 Immunoperoxidase techniques in histopathology: applications, methods and controls. Journal of Clinical Pathology 32: 971–978

Holt S J, Hicks R M 1961 Journal of Biophysical and Biochemical Cytology 11: 31

Holt S J, Withers R F J 1958 Studies in enzyme histochemistry. V. An appraisal of indigogenic reactions for esterase localization. Proceedings of Royal Society B 148: 520–532

Holt S J, Hobbiger E L, Pawan G L S 1960 Journal of Biophysical and Biochemical Cytology 7: 383

Hopwood D 1969 Fixatives and fixation: a review. Histochemical Journal 1: 323–360

Hopwood D 1972 Theoretical and practical aspects of glutaraldehyde fixation. Histochemical Journal 4: 267–303

Hopwood D 1975 The reactions of glutaraldehyde with nucleic acids. Histochemical Journal 7: 267–276

Hopwood D, Logan K R, Coghill G, Bouchier I A D 1977 Histochemical studies of mucosubstances and lipids in normal human oesophageal epithelium. Histochemical Journal 9: 153–161

Hopwood D, Logan K R, Milne G, Bouchier I A D 1978 Concanavalin A receptors in normal and inflamed oesophageal epithelium. Histochemistry 57: 255–263

Howie J 1978 Code of practice for the prevention of infection in clinical laboratories and post mortem rooms. HMSO, London

Huang S-N, Minassian H, More J D 1976 Application of immunofluorescent staining on paraffin sections improved by trypsin digestion. Laboratory Investigation 35: 303–390

Hussain O A N, Millett J A 1979 The detection of malignancy in the cervix. In: Pattison J R, Bitensky L, Chayen J (eds) Quantitative cytochemistry and its applications. Academic Press, London

Jamieson J D, Hull B E, Galardy R E, Maylié-Pfenninger M F 1979 Acinar cells: relationship to secretagogue action in secretory mechanisms. Cambridge University Press, SEB Symposia 33

Jones D 1973 Reactions of aldehydes with unsaturated fatty acids during histological fixation. In: P J Stoward (ed) Fixation in histochemistry. Chapman and Hall, London, p 1–45

Karnovsky M J 1965 A formaldehyde-glutaraldehyde fixative of high osmolarity for use in electron microscopy. Journal of Cell Biology 27: 137A

Karnovsky M J 1971 Use of ferrocyanide-reduced osmium tetroxide in electron microscopy. 11th Annual Meeting of the American Society of Cell Biology, p 146

Lake B D 1970 The histochemical evaluation of the glycogen storage diseases. A review of techniques and their limitations. Histochemical Journal 2: 441–450

Lake B D, Ellis R B 1976 What do you think you are quantifying? An appraisal of histochemical methods in the measurement of the activities of lysosomal enzymes. Histochemical Journal 8: 357–366

Laurila P, Virtanen I, Wartiovaara J, Stenman S 1978 Fluorescent antibodies and lectins stain intracellular structures in fixed cells treated with non-ionic detergent. Journal of Histochemistry and Cytochemistry 26: 251–257

Laver J L, Van Drunen N A, Washburn J W, Balfour H H 1979 Transmission of hepatitis B virus in clinical laboratory areas. Journal of Infectious Diseases 140: 513–576

Leatham A, Atkins N 1980 Fixation and immunohistochemistry of lymphoid tissue. Journal of Clinical Pathology 33: 1010–1012

Lorén I, Björklund A, Lindvall O 1977 Magnesium ions in catecholamine fluorescence histochemistry. Histochemistry 52: 223–239

McGhee J D, Hippez P H Von 1975 Formaldehyde as a probe of DNA structure. Biochemistry 14: 1281–1296

McLean I W, Nakane P K 1974 Periodate-lysine-paraformaldehyde fixative. Journal of Histochemistry and Cytochemistry 22: 1077–1083

Matthew J B 1981 Influence of clearing agent on immunohistochemical staining of paraffin-embedded tissue. Journal of Clinical Pathology 34: 103–105

Melcher D H, Linehan Smith R S 1981 Fine needle aspiration cytology. In: Anthony P P, MacSween R N W (eds) Recent advances in histopathology 11. Churchill Livingstone, Edinburgh

Melnick P J 1967 Histochemical enzyme procedures in variability assay of tissues and organs. Cytobiology 4: 136

Mepham B L, Frater W, Mitchell B S 1979 The use of proteolytic enzymes to improve immunoglobulin staining by the PAP technique. Histochemical Journal 11: 345–357

Millett J A, Hussain O A N 1979 Analysis of chromatin in carcinoma in situ. In: Pattison J R, Bitensky L, Chayen J (eds) Quantitative cytochemistry and its applications. Academic Press, London, p 37–42

Nature 1980 Academy recommends cut in formaldehyde exposure. 284: 587

Novikoff A B 1956 Preservation of fine structure of isolated liver cell particulates with polyvinylpyrrolidone. Journal of Biophysical and Biochemical Cytology 2 (supplement): 65–66

Pearse A G E 1968 Histochemistry, theoretical and applied, 3rd edn. Churchill, London, vol 1

Pearse A G E 1980 Histochemistry, theoretical and applied, 4th edn. Churchill Livingstone, Edinburgh, vol 1

Pearse A G E, Polak J M 1975 Bifunctional reagents as vapour and liquid phase fixatives for immune histochemistry. Histochemical Journal 7: 179–186

Pentilla A, McDowell E M, Trump B F 1975 Effects of fixation and post-fixation treatments on volume of injured cells. Journal of Histochemistry and Cytochemistry 23: 251–270

Piris J, Thomas N D 1980 A quantitative study of the influence of fixation on immunoperoxidaae staining of rectal mucosal plasma cells. Journal of Clinical Pathology 33: 361–364

Renaud S 1959 Superiority of alcoholic over aqueous fixation in the histochemical detection of calcium. Stain Technology 34: 267–271

Rodning C B, Erlandsen S L, Coulter H D, Wilson I D 1980 Immunohistochemical localization of IgA antigens in sections embedded in epoxy resins. Journal of Histochemistry and Cytochemistry 28: 199–205

Saito T, Keino H 1976 Acrolein as a fixative for electron cytochemistry. Journal of Histochemistry and Cytochemistry 24: 1258–1269

Shikata T, Uzawa T, Yoshiwara A, Akatsuka T, Yamazaki S 1974 Staining methods of Australian antigen in paraffin section. Detection of cytoplasmic inclusion bodies. Japanese Journal of Experimental Medicine 44: 25–36

Simionescu N, Simionescu M, Palade G E 1972 Permeability of intestinal capillaries: pathway followed by dextrans and glycogen. Journal of Cell Biology 58: 365–392

Smitherman M L, Lazarow A, Sorenson R L 1972 The effect of light microscopic fixatives on the retention of glycogen in protein matrices and the particulate state of native glycogen. Journal of Histochemistry and Cytochemistry 20: 463–471

Torack R M 1974 Peroxidase activity in autopsy material. Archives of Pathology 98: 233–236

Walker R J, Miller J P G, Dymock I W, Shilkin K B, Williams R 1971 Relationship of hepatic iron concentration to histochemical grading and to total chelatable body iron in conditions associated with iron overload. Gut 12: 1011–1014

Weakley B S 1981 A beginner's handbook in biological transmission electron microscopy, 2nd edn. Churchill Livingstone, Edinburgh

Willingham M C, Yamada S S 1979 Development of a new primary fixative for electron microscopic immunocytochemical localization of intracellular antigens in cultured cells. Journal of Histochemistry and Cytochemistry 27: 947–960

Substances identified by histochemical methods

Peter J. Stoward

INTRODUCTION

The methods available to the histopathologist for visualising, and in some cases quantifying, the different kinds of substances and cells present in sections of tissue fall roughly into six groups of increasing complexity:

1. Simple routine histological methods, such as haematoxylin and eosin

2. A more extended array of histological and quasi-histochemical methods for staining particular cells more selectively than is possible with the older and simpler methods of the previous group

3. Routine histochemical methods

4. 'Special' histochemical techniques for either visualising or confirming rarer pathological lesions or for use in automated cytological analyses

5. Immunocytochemical methods

6. Cytochemical and quantitative histochemical techniques for investigating the molecular mechanisms underlying pathological processes.

Most pathologists prefer haematoxylin and eosin, and perhaps a trichrome method in addition, for the routine staining of specimens of tissue presented for histological examination. Not only are the methods simple to perform but they stain many cells and tissue elements clearly. They have also stood the test of time and are adequate for recognising the vast majority of common histological lesions. However, they do not, and cannot, reveal all diagnostically important features. Consequently, several histopathology laboratories now routinely employ a wider range of histological methods, such as those listed in Table 2.1.

Unfortunately, even these methods, numerous as they are, are inadequate for investigating some disorders of skeletal muscle, the neuro-endocrine system, the gastro-intestinal tract, liver, kidney and peripheral blood. For such so-called 'special' cases, a systematic histochemical analysis is appropriate. In this chapter, some reliable and well-tried histochemical methods for such an analysis are brought together in a logical order, together with comments on their underlying rationale.

SUBSTANCES IDENTIFIED IN ROUTINE HISTOCHEMICAL DIAGNOSES

Most specialised cells in a section of tissue contain a preponderance of one particular macromolecular substance. The aim of a histochemical analysis is to visualise this substance relatively specifically so that the cell in which it is present can be recognised and distinguished from morphologically similar cells. Five types of chemical substances can be found in varying proportions in a mammalian cell:

    a. nucleic acids

    b. proteins (and peptides)

    c. mucosubstances

    d. lipids

    e. inorganic salts.

Various names are used in the histochemical literature to describe individual mucosubstances. Examples are mucin, mucoid, mucopolysaccharide, mucoprotein, sialoprotein, and sulphomucin. The term 'mucosubstance', although not officially recommended, has passed into common parlance for embracing all carbohydrate-containing macromolecular substances and is, therefore, used here. There are three kinds of mucosubstance:

1. polysaccharides (composed entirely of carbohydrate)

Table 2.1 Histological and quasi-histochemical methods used routinely for the visualisation of specific tissue components and histopathological features. The simplest and more informative ones are marked*

Histological feature visualised	Methods[1]
Blood cells (in smears and tissues)	Wright*, Giemsa or Leishman[2]
Brush borders of e.g. enterocytes and renal proximal tubules	Alkaline phosphatase
Calcification	Alkaline phosphatase *Von Kossa
Calcium/urate deposits	*Harris's haematoxylin Von Kossa Gomori's methenamine silver
Central nervous system: a. Nerve cells and axons	Holmes's silver Bielschowsky's silver *Nissl's cresyl fast violet
b. Glia	Mallory's phosphotungstic acid-haematoxylin (PTAH) Holzer's crystal violet Cajal's gold-mercuric chloride
c. Myelin	Solochrome cyanine *Luxol fast blue Weigert-Pal haematoxylin (Kultschitsky's modification) Osmium tetroxide-α-naphthylamine (OTAN)
Cell proliferation	*Methyl green-pyronin (to reveal DNA and RNA) Feulgen-Schiff

Connective tissue:
  a. Macrophages
  b. Mast cells
	*Acid phosphatase-haematoxylin
	*Metachromasia (towards e.g. azure A)
	Csaba's alcian blue-safranin
	Chloroacetate esterase

c. Fibres (collagen, elastin, reticulin)	*Trichrome stains (e.g. Mallory, Masson) Weigert-French or Verhoeff (for elastin) Allochrome (for reticulin)[3] Silver impregnation (for reticulin)[4]
Endocrine cells (especially in gastro-intestinal tract)	Diazonium salt coupling[5] Masson-Fontana alkaline silver Grimelius's silver Gomori's aldehyde fuchsin Solcia's lead haematoxylin
Glycogen accumulation	*Diastase (or saliva)-PAS Best's or Southgate's mucicarmine
Lipid deposits (including cholesterol)	*Oil red O Sudan black or Sudan IV Osmium tetroxide (for unsaturated lipids) *Schultz's iron-sulphuric acid (for cholesterol) Acid haematin (for phospholipids)
Melanogenesis	*DOPA oxidase
Micro-organisms	Gram Ziehl-Neelsen Gomori's methenamine silver *PAS

Histological feature visualised	Methods[1]
Mucus (especially in or on hypersecreting cells in respiratory and gastro-intestinal tracts)	*Diastase-PAS *Alcian blue (pH 2.5 and 1.0) – PAS High iron diamine-alcian blue pH 2.5 or PAS Chromic acid-Schiff or -methenamine silver
Muscle fibre typing	*Myofibrillar ATPase pH 9.5 after preincubation at pH 4.6 or 4.2 *NADH tetrazolium reductase
Phagocytic cells	*Acid phosphatase
Pigments: melanin	Masson-Fontana ammoniacal silver *DOPA oxidase
lipofuscin/ceroid	*Autofluorescence
Protein deposits, granules and inclusions (e.g. viruses): a. general	*Eosin (preceded by haematoxylin) *Phloxine-tartrazine
b. amyloid	Methyl violet *Congo red Thioflavine T (preferably preceded by alcian blue)
c. copper-binding protein (and hepatitis B antigen)	*Shikata's prolonged orcein[6]
d. fibrin	*MSB (Martuis-scarlet-blue) – Obadiah[7]
Resorption (e.g. in bone)	*Acid phosphatase
Vascular system	Alkaline phosphatase

Notes
1. Except where indicated by a footnote, practical details of all these methods, original references and some background explanation may be found in Drury & Wallington (1980) or Bancroft & Stevens (1982). Lillie & Fullmer (1976) may be consulted for critical discussions of their relative specificity.
2. Lillie & Fullmer (1976), pages 747–748. Wright's stain is faster than the other methods, and is suitable for automated staining of smears for routine haematology. Chloroacetate esterase and alkaline phosphatase are also very useful for the routine identification of certain blood cells (see Table 2.3).
3. Lillie & Fullmer (1976), pages 704–705
4. Slidders et al (1958)
5. Often referred to, incorrectly, as alkaline diazo methods
6. Shikata et al (1974)
7. Lendrum et al (1962)

2. proteoglycans or glycosaminoglycans (consisting of long polysaccharide chains covalently attached to a relatively small protein core) and

3. glycoproteins (proteins bearing numerous covalently-linked short oligosaccharide side chains).

Glycoproteins are divided further into neutral and acid glycoproteins, depending on whether or not they contain sialic acid or a sulphated sugar (or both) as a component. Nearly all glycosaminoglycans possess uronic acid and sulphate ester groups in various proportions, and thus are known as acid glycosaminoglycans (or acid mucopolysaccharides in older terminologies). Glycosaminoglycans bearing sulphate ester groups are commonly said to be sulphated.

Lipids are also classifiable into chemically well-defined substances (Bayliss High 1982) but in routine histopathology it is normally sufficient to divide them into two classes, acidic and unsaturated lipids as one class, and cholesterol and its esters as the other.

Each of the substances (a), (b), (c) and (d), all complex macromolecules, possesses chemical residues and end-groups often not present to any significant extent in the other three. Sometimes such groups, e.g. hydroxyl ($-OH$) or sulphate ester ($-OSO_3H$), are referred to as radicals in the histochemical and histological literature. This is wrong. A radical has an unpaired electron (for example, $\cdot OH$) and is highly injurious towards cell membranes. A chemical residue is a monomeric component such as an amino acid within a protein or a sugar within a complex saccharide.

Table 2.2 lists a selection of reliable histochemical techniques which may be used routinely for the

Table 2.2 Principal chemical groups and residues identifiable in the substances commonly present in mammalian tissues, and recommended histochemical methods for visualising them. Experimental details of the routine methods may be found in Bancroft & Stevens (1982) and those marked* in the Appendices

Substance	Identifiable groups and residues	Method	Routine or special
Nucleic acids	1. $-PO_3H$	Haematoxylin (as in H & E)	Routine
		Methyl green-pyronin	Routine
		Acridine orange, Hoechst 33342	Special*
	2. Deoxyribose	Feulgen-Schiff	Special*
Proteins	1. $-NH_2$	Eosin (as in H & E)	Routine
		Phloxine-tartrazine	Routine
	2. $-SH$	Ferric ferricyanide	Routine
	$-SH$ & $-SS-$	Performic acid-alcian blue or -azure A	Routine
		Shikata's prolonged orcein	Special*
	3. Tryptophan	DMAB-nitrite	Routine*
Mucosubstances	1. *Vic*-glycols	PAS	Routine
		Diastase-PAS (for glycogen)	Routine
		Chromic acid-methenamine silver	Routine
	2. $-SO_3H$	Azure A metachromasia	Routine
		High iron diamine	Special*
	3. Uronic acid	Alcian blue (pH 2.5 and 1.0) $-PAS$	Routine
	Sialic acid	High iron diamine-alcian blue pH 2.5	Special*
		High iron diamine-PAS	Special
		Critical electrolyte concentration techniques (alcian blue + $MgCl_2$)	Special*
		Hyaluronidase and neuraminidase digestion	Special*
Lipids	1. Long aliphatic chains	Oil red O (for all lipids)	Routine*
	2. Free fatty acids, olefin bonds, and phospho-proteins	Nile blue sulphate	Routine*
	3. Cholesterol & esters	Schultz's method	Routine*
		Perchloric acid-naphthoquinone (PAN)	Routine*
Metals	1. Iron	Perls' method	Routine
	2. Calcium	Von Kossa	Routine
		Alizarin	Routine
		Glyoxal-*bis*(2-hydroxyanil)	Special*
	3. Copper	Rubeanic acid	Special*
		Rhodamine	Special
	4. Aluminium	Solochrome Azurine	Special*

detection of the principal residues and end-groups of the five types of substances referred to above. Most proteins are also antigenic and some have catalytic (enzymic) properties which can be exploited for their visualization as well.

GENERAL PRINCIPLES OF ROUTINE HISTOCHEMICAL METHODS

The histochemical methods listed in Table 2.2 are nearly all based on one of the four following principles:

1. Simple ionic interactions of either positively-charged basic dyes (B^+) or negatively-charged acid dyes (A^-) with groups of opposite charge in tissue macromolecules.

2. Reactions of aldehydes with Schiff's reagent, or occasionally methenamine silver, to form a coloured product of unknown structure.

3. Coupling of aromatic diazonium salts ($Ar.N^+_2$) with electron-rich centres in the aromatic residues (e.g. tyrosine) of cellular proteins and hormones.

4. If the macromolecule being detected is an enzyme, conversion to an insoluble coloured precipitate of the primary reaction products released by the catalytic activity of the enzyme acting on a suitable substrate.

When a particular chemical group or residue cannot take part in any of the first three of these reactions, it is first converted to a form in which it can. Glycogen, for example, cannot be visualised directly (except with mucicarmine) since its only chemically reactive groups, primary and *vicinal* hydroxyl groups, neither carry an electrical charge nor react with either Schiff's reagent or diazonium salts. However, the *vicinal* hydroxyl groups (also known as 1,2-glycols) can be made to react with Schiff's reagent by first oxidising them to aldehydes with periodic acid.

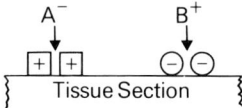

Fig. 2.1 Reactions of acid (A^-) and basic (B^+) dyes with, respectively, cationic and anionic macromolecular substances in tissue sections

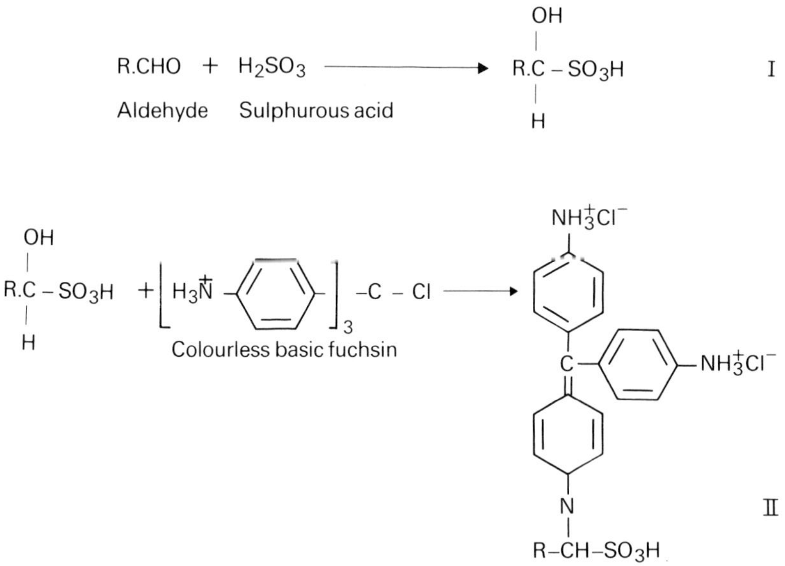

Fig. 2.2 Possible course of reaction of tissue aldehydes (R.CHO) with Schiff's reagent. Aldehydes react first with the sulphurous acid present in the reagent to form a sulphonic acid intermediate (I) which then combines with the colourless analinium form of basic fuchsin to give the coloured derivative (II)

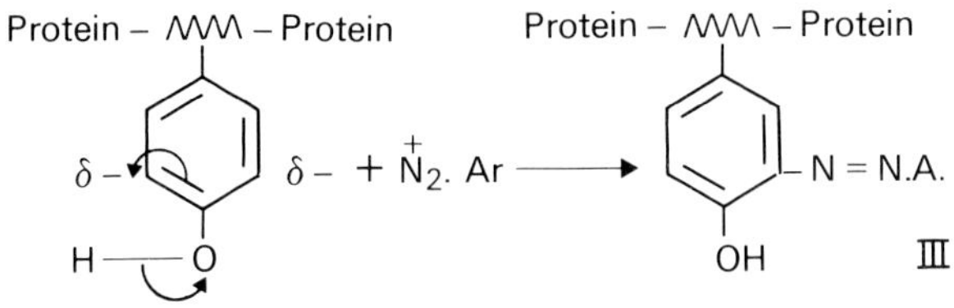

Fig. 2.3 Coupling of aromatic diazonium salts with the electron-rich centres of aromatic amino-acid residues of proteins to form coloured azo derivatives (III)

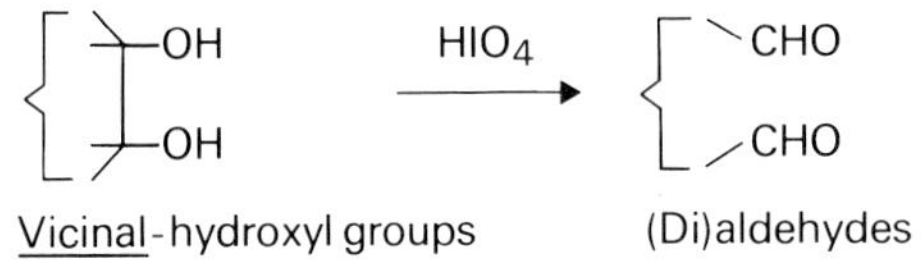

Vicinal-hydroxyl groups   (Di)aldehydes

Fig. 2.4 Selective cleavage of *vicinal*-hydroxyl groups (in the sugar residues of glycogen and glycoproteins) by periodic acid (HIO_4) to yield dialdehydes

In some instances, reactions (1)–(3) can be rendered relatively specific for the substance one wishes to localise by controlling the reaction conditions (e.g. pH).

SUBSTANCES LOCALISABLE

Nucleic acids

Both RNA and DNA can be localised in cells by the affinity of their negatively-charged phosphate ester groups for almost any basic dye, but particularly haematoxylin or methyl green and pyronin. Haematoxylin imparts a bluish-black colour to nuclei (containing DNA), but its staining of RNA is usually only apparent in cells whose cytoplasm is particularly rich in this nucleic acid (e.g. serous acinar cells). Haematoxylin is also taken up by other basophilic substances such as sulphated glycosaminoglycans. Therefore, it is not specific for nucleic acids exclusively.

Methyl green and pyronin, on the other hand, are more selective, staining DNA and RNA respectively. Because of the favourable molecular geometry of its molecule, methyl green intercalates between the stacked base-pairs of DNA and binds to its nucleotide phosphate groups. However, it is unable to fit easily onto single-stranded polynucleotides such as RNA and, therefore, stains this nucleic acid poorly. In contrast, pyronin, which has a flatter molecular shape, is taken up readily by RNA.

The specificity of the methyl green-pyronin technique (or any method involving basic dyes) is customarily confirmed by showing that in a control section which has been treated with ribonuclease before exposure to the dyes, the staining of the presumed cytoplasmic RNA is lost.

Fluorescent basic dyes are being increasingly used for revealing cellular DNA and RNA, particularly in automated searches for aberrant cells in, for example, cervical smears using fluorescence-activated cell sorting systems (Melamed & Darzynkiewicz 1981). The most widely-used dye for this purpose is acridine orange, which emits a green fluorescence when bound to undenatured DNA in nuclei, and a red fluorescence when taken up by cytoplasmic RNA. Other anionic constituents of cells, such as acid glycosaminoglycans, also bind this dye and therefore, it does not stain nucleic acids specifically. Fortunately, more selective fluorescent dyes are now available for DNA, such as ethidium bromide, propidium iodide and Hoechst 33342.

Nuclear DNA can also be visualised selectively, and quantified if desired, with the Feulgen-Schiff technique, in which sections of fixed tissue are initially treated with 5M HCl at room temperature for about 10 min. This treatment is called Feulgen hydrolysis. The deoxyribose component of DNA is selectively hydrolysed by the acid to form an aldehyde, which is then converted to a magenta-coloured derivative by reaction with Schiff's reagent. Basic fluorescent dyes, such as acriflavine, can be substituted for basic fuchsin in Schiff's reagent if greater sensitivity is required.

Proteins

Amine groups

The simplest way of revealing basic proteins in sections of tissue is to stain them with solutions of an acid dye, such as eosin, at a pH below 6. As the pH is lowered, more protein terminal amino groups become protonated to form $-NH_3^+$ groups and consequently take up an increasing amount of dye. However, this may lead to so many tissue proteins becoming stained that cellular detail is lost. The staining can be made relatively more specific for densely packed proteins, such as viruses and Paneth cell granules, by extracting loosely bound dye with a suitable solvent (a procedure known as differentiation). The extraction can be controlled more reliably with phloxine-tartrazine than with eosin. The phloxine method is thus preferred for the routine diagnosis of protein inclusions.

For reasons that are not fully understood, a few acid dyes stain some proteins relatively more selectively than others. For example, orcein, when employed in Shikata's 'prolonged' method, seems to be taken up specifically by hepatitis B virus and copper-binding protein (in diseases of liver) and, therefore, this dye is useful for the diagnostic recognition of these cellular inclusions.

Thiol and disulphide groups

Most methods for demonstrating proteins rich in the sulphur-containing amino acid residues cystine, cysteine and methionine are based on reactions of their thiol groups ($-SH$). If such groups are initially absent (as in cystine residues), they are produced by treating sections with an alkaline solution of sodium thioglycollate, which chemically reduces the disulphide bonds to free thiol groups.

Both the revealed and endogenous thiol groups can be detected by reactions with either an organomercurial (such as mercury orange) or a maleimide, followed by coupling with a diazonium salt. The latter appears to be the most specific. Thiol groups can be visualised less specifically, but more easily, by their ability to reduce ferric ferricyanide to give ferrous ions which then react with unreduced ferricyanide to form an insoluble blue pigment (Turnbull's blue). Alternatively, thiol and disulphide-containing proteins can be revealed collectively by oxidising them with a peracid, such as performic or peracetic acid, followed by staining of the sulphonic or sulphinic acids thus produced with a basic dye. This forms the basis of the performic acid-azure and -alcian blue methods listed in Table 2.2.

Tryptophan residues

Of all the other chemical end-groups and amino acid residues that it is possible to detect histochemically in proteins, only tryptophan and related indole residues are regularly investigated. They can be visualised specifically by reacting them with dimethy-aminobenzaldehyde (DMAB) followed by nitrous acid to form a blue-coloured product.

Mucosubstances

Complex carbohydrate-containing substances are stored intracellularly in mammalian cells either as the polysaccharide glycogen or in combination with proteins as glycosoaminoglycans and glycoproteins. Carbohydrate residues (e.g. fucose, mannose and N-acetyglucosamine) are also essential components of most cellular membranes. Their detection in situ is based on reactions of their (1) *vicinal* glycol groups, (2) anionic groups (sulphate ester, uronic acid), (3) anionic residues (e.g. sialic acid), or (4) specific monosaccharide entities. In some cells, the relative amount and reactivity of these groups and individual sugars allow one to infer the predominant carbohydrate-containing substances present.

Vicinal glycol groups

Any substance containing these groups can be readily localised with the periodic acid-Schiff (PAS) reaction, which depends on the unique susceptibility of the groups to become oxidised to dialdehydes by periodic acid. The dialdehydes, like the aldehydes exposed in nuclear DNA after a Feulgen hydrolysis, are converted to strongly coloured products by treating them with Schiff's reagent, or alternatively, with methenamine silver (the PAMS technique).

The beauty of the PAS technique is its simplicity and reliability. It works equally well with glycogen and glycoproteins, but the presence of glycogen in a tissue can be distinguished by showing that the PAS reaction no longer occurs in sections treated beforehand with diastase (or saliva if a rapid confirmation is required). However, it is doubtful whether acid glycosaminoglycans are revealed with the PAS technique (as normally practised), even though they contain free *vicinal*-glycols. Fortunately, they are identified more easily with the basic dyes described below.

Chromic acid is also employed routinely in some laboratories for oxidising carbohydrate *vicinal*-glycols; it oxidises the dialdehydes produced initially to carboxylic acid groups, which still react, however, with both Schiff's reagent and methenamine silver.

The chromic acid variant, especially when used with methenamine silver, is particularly useful in the diagnosis of renal diseases and fungal infections (Mowry 1981).

Anionic groups and residues

All acid glycosaminoglycans, in principle at least, can be detected by their affinity at an acid pH for the basic dye reagents azure A (or the virtually identical toluidine blue), alcian blue, high iron diamine and colloidal ferric iron.

At pH 3.5–4.0, azure A binds to most acid glycosaminoglycans and sialic acid-containing glycoproteins via their anionic groups. The bound dye is red (i.e. the dye exhibits metachromasia) in contrast to the orthochromatically-coloured blue dye taken up by other basophilic components of cells (e.g. nucleic acids). This enables acid mucosubstances to be identified rapidly in situ, but unfortunately the stained sections normally have to be mounted in water for examination. The usual dehydrating solvents destroy the metachromasia.

Alcian blue, or a comparable phthalocyanine dye, is perhaps the most useful basic dye reagent available at present for localising acid glycosaminoglycans and glycoproteins because by altering the staining conditions, it can be used to identify individual glycosaminoglycans selectively. This is usually accomplished by either adjusting the pH or the concentration of an inorganic salt ($MgCl_2$) in the dye solution. At pH 2.5, alcian blue stains both sulphated and carboxyl-containing acid glycosaminoglycans and glycoproteins, but at pH 1.0 only the more strongly ionised sulphated mucosubstances are stained.

Alcian blue is even more discriminating when it is dissolved in solutions of magnesium chloride of various concentrations (0.1–2.0M), and use is made of the 'critical electrolyte concentration' principle. The magnesium ions compete with the dye molecules for the anionic groups of the macromolecular carbohydrate complex and under the reaction conditions normally employed in histopathology laboratories (e.g. short staining times), this results in alcian blue 'staining' (i.e. being bound to) highly sulphated glycosaminoglycans selectively at high magnesium chloride concentrations ($\geqslant$ 1.0M), whereas the staining of hyaluronic acid- and sialic

acid-containing complexes is extinguished at low magnesium chloride concentrations (e.g. 0.2M).

Alcian blue (at either pH 2.5 or 1.0) can be preceded by the high iron diamine reagent for demonstrating sulphated and carboxylated glycoproteins simultaneously in two or three different colours, or followed by the PAS technique for visualising acid and neutral glycoproteins separately in the same preparation.

Sometimes not all the carbohydrate-containing substances present in a tissue react with any of these techniques because they are bound to, or are protected by, a protein masking their reactive end-groups. However, they usually react when treated briefly with mild alkali (e.g. potassium hydroxide dissolved in 75% ethanol), a procedure generally described as saponification.

The identity of some glycosaminoglycans, as inferred from any of the techniques outlined so far (particularly the critical electrolyte concentration method using alcian blue and magnesium chloride), can be confirmed by incubating sections of tissue before staining with a weakly buffered solution of a purified glycosidase, such as hyaluronidase, or one of the chondroitinases. This is particularly important for checking the identification of sulphated and hyaluronic acid-rich substances in situ. Sialic acid-rich glycoproteins can be confirmed by pretreatment of tissue sections with neuraminidase. Prior saponification enhances the susceptibility of some sialomucins towards neuraminidase digestion, and also renders previously unreactive mucosubstances PAS-positive, thus enabling them to be differentiated further. This is useful in the diagnosis of those disorders of the gastro-intestinal tract in which the distribution of different types of glycoprotein changes (Filipe 1979, Dawson 1981).

Specific monosaccharide residues

Specific sugars incorporated into plasma and other membrane proteins can be visualised selectively by their affinity for certain plant lectins (Kiernan 1981). Generally the lectin is conjugated to either horseradish peroxidase (which can be readily visualised with diaminobenzidine and hydrogen peroxide) or, less preferably, a fluorescent dye. At present, lectins are not used to a significant extent in diagnostic histopathology, partly because of their high cost. Nevertheless, they may become a common tool in the near future, as some lectins (e.g. peanut lectin) show promise of revealing possible pathological determinants, for example in cancer cells (for references, see Stoward et al 1980), not easily demonstrated by other means.

Lipids

Many types of lipid can be identified histochemically in sections of unfixed fresh tissue (Bayliss High 1982) but in routine histopathology, it is sufficient to concentrate on detecting three broad classes.

All lipids

Oil red O stains most lipids, but not those in the solid state, and is thus the best reagent for their general screening. Two major classes of lipid can be distinguished if sections stained with this dye are viewed with polarised light: unstained (crystalline) lipids appear birefringent (anisotropic) whereas the stained liquid lipids are non-refringent (isotropic). The birefringent lipid is usually cholesterol or one of its solid esters.

Sudan black also stains most lipids and is commonly employed as an alternative to oil red O. However, it does not stain sections as cleanly. The Sudan dyes also do not reveal free fatty acids, phosphoglycerides, certain protein-bound lipids and solid lipids. The first two named components are extracted by the dye solvent. These disadvantages can be largely avoided by treating tissues with bromine water first, which presumably brominates the olefinic bonds of unsaturated lipids, thus rendering them less soluble.

Acidic and unsaturated lipids

In water, the basic dye nile blue sulphate selectively binds to, or dissolves in, unsaturated hydrophobic lipids, free fatty acids and phospholipids, and is a good supplementary reagent to oil red O.

Cholesterol and its esters

Schultz's modification of the Liebermann-Burchardt reaction is widely used in histopathology for revealing the presence of cholesterol and its esters. The chemistry of the reaction has not been clarified.

Because of the harshness of the reagent needed, gentler tests are performed in some laboratories. Of these, the perchloric acid-naphthoquinone (PAN) method seems to be the most popular.

Metals

The only metals, or rather their salts, that are commonly sought histologically are iron and calcium. Occasionally, there is a need to look for deposits of copper, aluminium or barium.

Ferric iron (usually in the form of haemosiderin deposits) is easily demonstrable with Perls' method, first described in 1867, in which sections are treated with a fresh solution of potassium ferrocyanide in dilute hydrochloric acid. The acid releases the ferric ions from their protein attachments, whence they immediately react with ferrocyanide ions to form insoluble Prussian blue.

Several tests exist for visualising calcium salts. One of the oldest, and until recently perhaps the most frequently employed, is von Kossa's in which sections are placed in solutions of silver nitrate and exposed to light. Calcium phosphate deposits become coloured brown or red. Unfortunately, many other tissue components reduce silver salts, and thus the method is capricious and unspecific for calcium. As a result, there is a growing trend to apply more specific methods, particularly when the role of calcium in a disease process is being investigated. At present, the best seems to be alizarin and glyoxal bis-(2-hydroxyanil), which form coloured complexes with calcium.

Rubeanic acid is commonly employed for detecting copper (producing greenish-black complexes), and rhozodinate for lead and barium (to give intensely red deposits). Aluminium deposits are demonstrable with several dyes of the solochrome series, such as solochrome azurine.

Protein antigens

In principle, almost any protein or peptide can be localised specifically with immunocytochemical techniques if an antiserum against it is available. Such techniques are the only ones possible if standard histochemical techniques do not exist. Many peptide hormones and enzymic proteins have been localised by such means, and the list is growing. However, for the moment, the immunohistochemical localisation of hormone and enzyme antigens should only be attempted for confirming certain rare syndromes, particularly those involving the diffuse neuro-endocrine system (e.g. APUDomas), or in research investigations of the pathogenesis of a disease at the molecular level. On the other hand, the increasing availability of monoclonal antisera may result in immunohistochemical techniques replacing many existing histochemical and histological techniques in the near future.

Since immunocytochemical techniques and their applications are described in Chapter 3, they are not discussed further here.

Enzymes

Some enzymes exhibit a particularly high activity in certain specialised cells and organelles, and in these situations they are commonly either exploited for identifying the cells in sections of tissue or regarded as markers or indicators of their function, or if the activity is absent or substantially reduced, of cellular dysfunction. So far, about 30 such enzymes have been found valuable for the differential diagnosis of muscle diseases (Johnson & Walton 1981), certain neoplasias, storage diseases, disorders of the gastro-intestinal tract (Lojda 1981) and leukaemias (Catovsky et al 1981, Lojda 1981). These enzymes are listed in Table 2.3, together with the cell type and function they are assumed to signify. About 10 of the enzymes are worth identifying routinely.

Table 2.3 Enzymes used as markers of cell type and function (or dysfunction) in diagnostic histopathology. Particularly useful ones are indicated by an asterisk (*)

Enzyme	Cell type or function indicated	Diagnostic applications[1]
*Acetylcholinesterase	Nerve fibre tracts	Hirschsprung's disease (↑)
Acid esterase (E600-resistant)	T-lymphocytes (helper and suppressor subsets)	Lymphomas Wolman's disease (↓)
Acid β-galactosidase	Degradation of mucosubstances within lysosomes	G_{M1}-gangliosidosis (↓), atheroma
* Acid phosphatase	Active phagocytic cells (i.e. those containing numerous lysosomes, e.g. macrophages) Autolysis and necrosis Resorption (of bone)	Muscle fibre damage (↑) Metastatic adenocarcinomas of prostate origin (if activity inhibited by tartrate) Identification of histiocytes, particularly osteoclasts
Alkaline phosphatase	Epithelial cells with functioning brush borders Mature neutrophils of peripheral blood Osteoblasts, calcification	Renal tubule dysfunction (↓) Chronic myeloid leukaemia (↓) Myelofibrosis, leukaemoid reactions
Amine oxidase[2]	Sites of biogenic amine destruction	
Aminopeptidase M[3]	Activated macrophages Epithelial cells with functioning brush borders	Metastases originating from stomach, bile ducts, urinary bladder or kidney
ATPase (Mg²⁺-activated)	Intra-epithelial lymphocytes Functioning bile canaliculi	Liver disease
*ATPase (myofibrillar)[4]	Fast contracting Type II muscle fibres	Muscle fibre typing for myopathies and neuropathies Rhabdomyosarcoma
Catalase	Cells containing many peroxisomes (e.g. granular polymorphonuclear granulocytes)	Zellweger syndrome (↓)
Chloroacetate esterase	Neutrophils and mast cells	
Diaminopeptidyl peptidase IV (DPP IV) *DOPA oxidase[5]	T-lymphocyte subsets Capillary endothelial cells Melanocytes	Differential diagnosis of leukaemias Malignant lymphomas Amelanotic melanoma
Elastase[6]	Neutrophils (but not eosinophils)	
Endopeptidase		Coeliac disease
Enterokinase		Malabsorption syndromes
Glucose-6-phosphatase	Hepatocytes, renal proximal tubular cells and enterocytes	Type I glycogenosis deficiency
Glucose-6-phosphate dehydrogenase	Cells underoing rapid proliferation (i.e. comparatively rapid DNA and lipid synthesis)	Some carcinomas (↑) (e.g. basal cell carcinoma)[7]

Enzyme	Cell type or function indicated	Diagnostic applications[1]
Glutamate dehydrogenase	Cells with damaged mitochondria	Recent tissue damage (↑)
γ-Glutamyltransferase		Carcinomas[8]
3-Hydroxybutyrate dehydrogenase	β-Oxidation of fatty acids (i.e. lipid metabolism)	
3α- and β-Hydroxysteroid dehydrogenases	Steroid synthesis	
*Lactase[9]	Absorption and digestion of disaccarides in brush borders of e.g. enterocytes	Malabsorption syndrome (↓)
Maltase[9,10]	As lactase	Malabsorption syndrome (↓)
Malate dehydrogenase		Infarction (↓)
α-Mannosidase		Melanoblastomas, histiocytosis X (↑)
*NADH dehydrogenase[11]	Type I aerobic muscle fibres Aerobic metabolism	Muscle fibre typing for myopathies and neuropathies
NADPH dehydrogenase	Cytochrome electron transport Detoxification	
*Non-specific esterase[12]	Type I muscle fibres Cells containing lysosomes, monocytic markers and histocytes	Differential diagnosis of leukaemias Recent tissue injury
Peroxidase: eosinophil myelo* platelet	Eosinophils Intracellular cidosis, neutrophils, Platelets, megakaryocytes	Different diagnosis of leukaemias Zellweger syndrome
*Phosphorylase	Anaerobic metabolism	Ischaemia (↓) Type V glycogenosis deficiency
Phosphofructokinase	Glycolysis	Type VII glycogenosis deficiency
*Succinate dehydrogenase	Cells containing many mitochondria Type I muscle fibres Aerobic metabolism	Ischaemia (↓) Liver damage (↓) Reye's syndrome
Sucrase[9]	As lactase	Malabsorption syndrome (↓)
Trehalase[9]	As lactase	Malabsorption syndrome (↓)

Notes

1. The signs ↑ ↓ in parenthesis indicate that the activity is substantially increased or decreased respectively

2. Usually known in the histochemical literature as monoamine oxidase. Its EC recommended name is amine oxidase (flavin-containing)

3. Normally referred to simply as aminopeptidase. However, several aminopeptidases, with different subcellular locations, are now known. This one, also called membrane aminopeptidase, is situated on brush borders of e.g. enterocytes and renal proximal tubules and on the plasma membrane of activated macrophages

4. Myofibrillar calcium-activated ATPase demonstrable at pH 9.4 (Appendix 5). Usually demonstrated by loss of activity after preincubation of sections of unfixed muscle at pH 4.6 or 4.2 (see Appendix 5). However, Mg, Ca-activated actomyosin ATPase is probably a better enzyme for muscle typing. Mabuchi & Sréter's (1980) method for this enzyme enables the typing to be accomplished in one step without preincubation

5. EC recommended name is catechol oxidase

6. Normally referred to as elastase-like activity

7. Diagnostic reliability is questionable

8. Reviewed by Vanderlaan & Phares (1981)

9. Lojda (1981) recommends that in investigations of the malabsorption syndrome, the histochemical localisation of lactase, sucrase and trehalase are all obligatory. Other brush border enzymes should also be investigated if possible (e.g. maltase, alkaline phosphatase)

10. EC recommended name for maltase is α-D-glucosidase (3.2.1.20). Not the same as acid maltase (EC 3.2.1.3), a lysosomal enzyme

11. Normally called NADH-tetrazolium reductase in the histochemical literature. The name used here is the EC recommended one

12. There are many types of non-specific esterase. The enzyme referred to in this table is probably C-esterase because α-naphthyl acetate is the substrate usually employed for its histochemical localisation.

Lojda et al (1979) and Appendix 5 should be consulted for reliable histochemical methods for localising the enzymes. Purists sometimes worry that these methods may not be specific. In my view, it is sufficient if the methods give results (i.e. staining patterns) which can be correlated empirically either with the presence (or absence) of particular types of cell or with a defined clinical stage of a disease. Further, the purists' worries often turn out to be groundless or can be overcome by carrying out appropriate control experiments.

One area of pathology where enzyme histochemical techniques could be usefully applied more often than at present is in difficult forensic cases, particularly for estimating the time at which wounds occur before death. The usefulness is based on the observation that certain enzymes, which normally have a low histochemical activity, appear in a regular chronological sequence in the peripheral wound zone of injured tissues such as a myocardial infarct (Raekallio 1970).

Monoamine oxidase, non-specific esterase and ATPase are the first to appear, becoming evident 1h after injury, whereas acid phosphatase and alkaline phosphatase do not become histochemically demonstrable until 4 h after injury (Fig. 2.5). This enzyme 'clock' is a little slower in older subjects, and faster in younger ones (Raekallio & Mäkinen 1974).

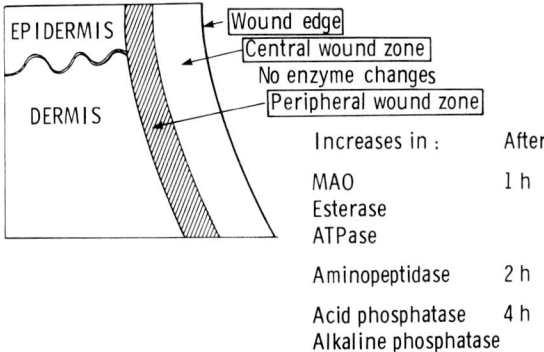

Fig. 2.5 Schematic representation of the time-dependent appearance of certain enzymes in the peripheral wound region of an injured tissue, e.g. cut skin (after Raekallio 1970) MAO = monoamine oxidase

LIMITATIONS OF HISTOCHEMISTRY IN ROUTINE HISTOPATHOLOGY

At best, the histochemical techniques suggested so far in this chapter will, if applied systematically, only tell the histopathologist what substances are present in a tissue and where they are situated. Unless certain criteria are met, they do not allow one to infer how much of the substance is present, or what the level of activity is if the substance is an enzyme. Even though one can estimate the intensity of 'staining' produced by a histochemical technique on a 1 to 5+ scale in different histological sites, it cannot be assumed that the intensity is necessarily related to the concentration of the reacting substance, especially if it is an enzyme. Before one can do so, it must be shown that the histochemical technique is specific and quantitatively valid in the senses which have been defined rigorously elsewhere (Stoward 1980).

A consequence of this caveat is that one must be cautious in interpreting histochemical observations in detailed biochemical and physiological terms. This particularly applies to observations on the apparent changes occurring in the activity of marker enzymes in diseased tissues. Qualitative histochemical techniques, as practised in routine laboratories, should not be relied on for unravelling the pathogenesis of a disease at the molecular level. The techniques were not designed for this.

In order to investigate a pathological mechanism at this level, a completely different approach is required. First, a hypothesis should be formulated in biochemical terms about the cellular changes taking place in the early stages of a disease. From this hypothesis, it usually becomes apparent what substances and enzymes it would be desirable to identify and perhaps quantify in situ. Unfortunately it often turns out that suitable cyto- and histochemical techniques do not exist for their localisation. New methods must, therefore, be developed rather than subverting existing routine ones. It is also necessary to validate the devised methods before applying them to the investigation in hand.

An example where this fundamental approach has been followed with conspicuous success is the work of the Karnovsky brothers and their colleagues on phagocytosis (Karnovsky et al 1981). They reasoned that polymorphonuclear leucocytes normally generate hydrogen peroxide on their plasma membranes when engulfing, say, bacteria. If this hypothesis is correct, one or more enzymes capable of producing hydrogen peroxide must be present on the plasma membranes of leucocytes. Further, the enzymes should be quiescent (i.e. relatively inactive) in unchallenged or resting leucocytes, and only become markedly active (i.e. stimulated) at the moment of phagocytosis during what is now known as the respiratory burst. This led the Karnovskys to devise methods for localising possible H_2O_2-generating enzymes such as NADH oxidase and D-amino acid oxidase. This is just the beginning of their story. Methods for the localization of other molecular components thought to be involved in the early stages of phagocytosis such as superoxide ions, are in the process of development at the time of writing (Karnovsky & Robinson 1981).

This approach to the investigation of a pathological mechanism contrasts sharply with that which is commonly followed. It is often assumed, for example, that if acid phosphatase can be demonstrated histochemically in a cell, then the cell is undergoing phagocytosis because acid phosphatase indicates the presence of lysosomes (as implied in Tables 2.1 and 2.3), which in turn are assumed to contain a range of enzymes capable of degrading phagocytosed material. Although this may be true, intralysosomal digestion is the last step in phagocytosis, and the presence of lysosomes does not prove that the cells in which they are situated are actually capable of phagocytosis. Some diseases are known in which lysosome-rich cells lose their ability to endocytose foreign materials. This again illustrates the danger of inferring too much about the physiological capacities of specialised cells from just one or two routine histochemical experiments.

QUANTITATIVE HISTOCHEMISTY

At present, quantitative histochemical techniques have no place in routine histopathology. On the other hand, I believe they will become commonplace in a few years' time for certain large-scale screening programmes, for monitoring the pharmacological and toxic effects of drugs, and for basic studies of the aetiology of many chronic diseases (Stoward 1981). These applications will depend on the development of completely new histochemical techniques and computer-assisted instrumentation (Ploem 1980) which are outside the scope of this chapter.

REFERENCES

Bancroft J D, Stevens A (eds) 1982 Theory and practice of histological techniques, 2nd ed. Churchill Livingstone, Edinburgh

Bayliss High O 1982 Lipids. In: Bancroft J D, Stevens A (eds) Theory and practice of histological techniques, 2nd ed. Churchill Livingstone, Edinburgh, ch 12, p 217

Catovsky D, Crockard A D, Matutes E, O'Brien M 1981 Cytochemistry of leukaemic cells. In: Stoward P J, Polak J M (eds) Histochemistry: the widening horizons of its applications in the biomedical sciences. John Wiley, London, ch 6, p 67

Dawson I M P 1981 The value of histochemistry in the diagnosis and prognosis of gastro-intestinal diseases. In: Stoward P J, Polak J M (eds) Histochemistry: the widening horizons of its applications in the biomedical sciences. John Wiley, London, ch 9, p 127

Drury R A B, Wallington E A 1980 Carleton's histological technique, 5th ed. Oxford University Press, Oxford

Filipe M I 1979 Mucins in the human gastro-intestinal epithelium: a review. Investigative and Cell Pathology 2: 195–216

Johnson M A, Walton J N 1981 Histochemistry — its contribution to the study of normal and diseased muscle. In: Stoward P J, Polak J M (eds) Histochemistry: the widening horizons of its applications in the biomedical sciences. John Wiley, London, ch 11, p 183

Karnovsky M J, Robinson J M 1981 Contribution of oxidative cytochemistry to our understanding of the phagocytic process. In: Stoward P J, Polak J M (eds) Histochemistry: the widening horizons of its applications in the biomedical sciences. John Wiley, London, ch 5, p 47

Karnovsky M J, Robinson J M, Briggs R T, Karnovsky M L 1981 Oxidative cytochemistry in phagocytosis: the interface between structure and function. Histochemical Journal 13: 1–22

Kiernan J A 1981 Histological and histochemical methods: theory and practice, 1st edn. Pergamon Press, Oxford, ch 11, p 157

Lendrum A C, Fraser D S, Slidders W, Henderson R 1962 Studies on the character and staining of fibrin. Journal of Clinical Pathology 15: 401–413

Lillie R D, Fullmer H M 1976 Histopathologic technic and practical histochemistry, 4th edn. McGraw-Hill, New York

Lojda Z 1981 The applications of enzyme histochemistry in diagnostic pathology. In: Stoward P J, Polak J M (eds) Histochemistry: the widening horizons of its applications in the biomedical sciences. John Wiley, London, ch 12, p 205

Lojda Z, Gossrau R, Schiebler T H 1979 Enzyme histochemistry, 1st ed. Springer-Verlag, Berlin

Mabuchi A, Sréter F A 1980 Actomyosin ATPase. II. Fiber typing by histochemical ATPase reaction. Muscle Nerve 3: 233–239

Melamed M R, Darzynkiewicz Z 1981 Acridine orange as a quantitative cytochemical probe for flow cytometry. In: Stoward P J, Polak J M (eds) Histochemistry: the widening horizons of its applications in the biomedical Sciences. John Wiley, London, ch 14, p 237

Mowry R W 1981 Contributions of practical carbohydrate histochemistry to the histopathological diagnosis of renal diseases, fungal infections, and some types of cancer. In: Stoward P J, Polak J M (eds) Histochemistry: the widening horizons of its applications in the biomedical sciences. John Wiley, London, ch 8, p 109

Ploem J S 1980 Appropriate technology for the quantitative assessment of the final reaction product of histochemical techniques. In: Trends in enzyme histochemistry and cytochemistry (Ciba Foundation Symposium 73). Excerpta Medica, Amsterdam, p 275

Raekallio J 1970 Enzyme histochemistry of wound healing. Progress in Histochemistry and Cytochemistry 1: no 2

Raekallio J, Mäkinen P L 1974 Effect of ageing on enzyme histochemical vital reactions. Zeitschrift für Rechtsmedizin 75: 105–111

Shikata T, Uzawa T, Yoshiwara N, Akatsuka T, Yamazaki S 1974 Staining methods of Australian antigen in paraffin sections — detection of cytoplasmic inclusion bodies. Japanese Journal of Experimental Medicine 44: 25–36

Slidders W, Fraser D S, Lendrum A C 1958 Silver impregnation of reticulin. Journal of Pathology and Bacteriology 75: 478–481

Stoward P J 1980 Criteria for the validation of quantitative histochemical enzyme techniques. In: Trends in enzyme histochemistry and cytochemistry (Ciba Foundation Symposium 73). Excerpta Medica, Amsterdam, p 11

Stoward P J 1981 The past, present and future of quantitative histochemistry. In: Stoward P J, Polak J M (eds) Histochemistry. The widening horizons of its applications in the biomedical sciences. John Wiley, London, ch 15, p 263

Stoward P J, Spicer S S, Miller R L 1980 Histochemical reactivity of peanut lectin — horseradish peroxidase conjugate. Journal of Histochemistry and Cytochemistry 28: 979–990

Vanderlaan M, Phares W 1981 γ-Glutamyltranspeptidase: a tumour cell marker with a pharmacological function. Histochemical Journal 13: 865–877

3

The value of immunohistochemistry in diagnosis

E. J. Holborow

Covalent linking of antibody molecules with other molecules was first attempted in the 1930s in the course of early investigations of the nature and reactive properties of antibody protein by J. Marrack and others. What emerged from these early experiments was the discovery that antibody could be labelled chemically in this way without destroying its specific reactivity with antigen. Albert Coons has left us a description of how in 1942 he first thought of using labelled antibody to trace streptococcal antigens in sections of pathological tissues from rheumatic fever patients (Coons 1976). As every schoolboy knows, in 1942 he took the memorable step of choosing the most fluorescent compound then known, fluorescein, as label. Having chosen fluorescein, he persuaded his colleagues in the Harvard Medical School Chemistry Department to prepare the isocyanate and couple it, in the first instance, to an anti-pneumococcal polysaccharide-specific antiserum. When sections from picric acid-alcohol-formalin fixed pieces of the livers of mice fatally infected with pneumococci were treated with the resulting dialysed soluble product and viewed under a microscope illuminated with a suitably filtered carbon-arc, individual bacteria and free antigen in the Kupffer cells fluoresced bright green (Kaplan et al 1950).

IN THE BEGINNING

This was an auspiciously chosen test system, since by using fixed tissues in which the bacterial polysaccharide still retained good antigenic reactivity, Coons and his co-workers did not at this initial stage encounter the troublesome non-specific staining that later bedevilled the use of unfixed frozen sections. By judicious absorption of fluorescein-conjugated antisera with acetone-dried mouse liver powder, Coons and his colleagues reduced their tendency to give this unwanted staining sufficiently to use labelled specific antibodies successfully to demonstrate rickettsial and mumps antigens respectively in unfixed tissue sections, as well as in smears and exudates from infected rats, human body lice, and monkeys (Coons et al 1950).

The fate of injected protein antigens in experimental animals became the next object of immunohistochemical attention in Coons' laboratory as a preliminary to his early studies of antibody production. Here the need to use frozen sections of unfixed tissue cut in a cryostat [Linderstrom-Lang & Morgensen's (1938) technique] appeared incontestable, since the usual histological fixatives were generally assumed to destroy the immunological reactivity of protein antigens, including antibody proteins. Despite this, the first specific immunohistochemical demonstration of the presence of antibody in the cytoplasm of plasma cells was with frozen sections of spleen and lymph nodes from rabbits hyperimmune to human gammaglobulin or ovalbumin which had been treated with 1/2000 antigen *after* having been fixed in 95% ethanol for 15 min at 37°C. They were then treated with fluorescein-labelled antibody with specific precipitating activity against human gammaglobulin or ovalbumin, when it was abundantly evident by fluorescence microscopy that the islands of cells in the spleen and other lymphoid organs that fluoresced bright green were collections of plasma cells, and that the antibody in their cytoplasm had retained its ability to bind antigen. This methodology, with antigen as the middle layer, Coons called the 'sandwich' technique (Coons et al 1955) and he used it to study cellular antibody production in specific primary and secondary responses (Leduc et al 1955).

This result foreshadowed the potentiality of labelled antibody as a specific and sensitive reagent for disclosing the presence of and differentiating antigenic constituents of human tissues, especially those of an exiguous character, such as specific cell products and markers. Among the earliest successes in the use of labelled antibodies in this way to detect tissue-specific antigens was Hill & Cruickshank's (1953) demonstration that fluorescein-labelled rabbit antiserum raised against a preparation of isolated rat glomeruli selectively stained glomerular and tubular basement membranes in sections of rat kidney. Following a related line of investigation, Ortega & Mellors (1956) broke new ground by using fluorescein-labelled goat anti-rabbit globulin antibodies to demonstrate the glomerular localization of rabbit γ-globulin in kidney sections from rats injected with rabbit anti-rat nephrotoxic serum. They embroidered the theme by likewise demonstrating that in the later and severer stage of such nephrotoxic nephritis, rabbit anti-rat γ-globulin antibodies labelled with fluorescein revealed deposits of autologous antibody on the affected glomerular basement membranes. The obvious next step was to use labelled anti-human globulin antibodies from immunized goats or rabbits to examine lymphoid tissues for antibody-forming cells by direct staining of γ-globulin as antigen, instead of the 'sandwich' method which detected it only as specific antibody.

Using air-dried sections of human lymph nodes and spleens, Ortega & Mellors (1957) showed that the Russell bodies in 'mature' plasma cells, and the cytoplasm of 'immature' (Marschalko) plasma cells, both stained specifically, and concluded that maturation does not precede the formation of gamma-globulin in these cells but accompanies it. They also confirmed a passing observation of Coons, noting a reticulated pattern of fluorescence often present in germinal centres in lymph nodes, which I shall refer to again below.

THE CLINICAL HARVEST

Mellors et al (1957) next moved immuno-histochemistry into the field of human immuno-pathology by using fluorescent anti-immunoglobulin

antibodies to demonstrate immunoglobulin deposits in the kidneys, chiefly in the glomeruli, in autopsy material from patients with lupus nephritis. This laid the groundwork for the use of immunofluorescence in the examination of biopsy material for deposits of immunoglobulin and complement resulting from immunopathological events. The most enduring benefits of this development have accrued in nephrology and dermatology, where, in the hands of the experienced fluorescence microscopist, it provides confirmatory evidence for the diagnosis in renal diseases (McGiven 1980 and Ch. 17) and in skin diseases (Fry 1981 and Ch. 11).

With the appearance on the scene of labelled anti-immunoglobulin as tracer, the scope of immunohistochemical methodology in pathology was immeasurably widened. Its reactivity with immunoglobulin protein as antigen, irrespective of its combining specificity as antibody, opened the way to the development of two major present uses of immunofluorescence, viz. the examination of biopsy material for immune deposits, already mentioned, and the even more widely practised technique of indirect immunofluorescence for the detection of anti-tissue antibodies in patients' sera. The first use of labelled antiglobulin to detect circulating antibodies in this way was to reveal the presence of antinuclear factors in systemic lupus erythematosus sera (Friou et al 1957, Holborow et al 1957). Since then diagnostic tests for a continually increasing range of tissue auto-antibodies reacting with different organ-specific and non-organ-specific antigens have become part of the routine service of the clinical immunology laboratory, often accounting for a good proportion of the total work-load undertaken.

A third use of labelled antibodies which is rapidly expanding with the advent of monoclonal antibodies is for identification of cell membrane antigenic markers on lymphocytes and other cells of the immune system involved in lymphoproliferative and lymphoneoplastic diseases (see Ch. 20). An example of how the use of immunohistochemical techniques for direct cell membrane antigens has progressed from investigative to diagnostic applications in this field begins with the germinal centre observations of Ortega & Mellors (1957) mentioned above. Since the latter workers saw positive staining with anti-γ-globulin conjugates in germinal centres, they concluded firstly that such centres are 'germinal' in that they are sites of origin of cells that form gammaglobulin, a conclusion amply confirmed since by direct immunohistochemical demonstration that centroblasts and centrocytes mostly belong to the B cell lineage (Stein et al 1980). Secondly, Ortega and Mellors concluded that germinal centres are miniature organs of internal secretion of gammaglobulin. Further immunohistochemical studies have shown this last conclusion to be incorrect. With the exception of the occasional plasma cell which may be present, germinal centre cells do not actively secrete Ig. The dendritic pattern of Ig staining revealed by immunohistochemical techniques in germinal centres in spleen or lymph nodes is the antibody component of immune complexes with antigen (White et al 1970) which have been formed elsewhere and have diffused or have been transported, in soluble form, into the germinal centres, where they become deposited extracellularly on the surfaces of the fine processes of the argyrophilic cells of Marshall (1956) which are now termed dendritic reticulum cells.

Localization of immune complexes in germinal centres in this way is complement-dependent (Papamichail et al 1975). Furthermore, C3 receptors have now been shown to be present on dendritic reticulum cells in both germinal centres of normal lymphoid tissue and in the neoplastic follicles of centroblastic-centrocytic lymphomas. Gerdes et al (1981) have demonstrated them immuno-histochemically by means of a rabbit antiserum against human C3 receptors isolated from human tonsil cells (B lymphocytes). This finding seems to fit very well the hypothesis that germinal centres constitute the machinery for imprinting immunological memory of specific antigen, a fundamental feature of adaptive immunity. Thus, with the primary antibody response to initial encounter with antigen, soluble antigen-antibody complexes are formed, and a (probably minute) proportion of these, through their complement-fixing property, become localized on the C3 receptors of germinal centre dendritic cells, where their extracellular location protects them from enzymic removal. The antigen thus 'trapped' acts as an oligoclonal stimulator of such B cells entering the lymphoid follicle as carry the matching surface Ig receptors, thus selecting and expanding the specific clones which, finally settling elsewhere in the traffic areas of the lymphoid system, implant specific memory cells ready to respond with enhanced antibody production to a further specific antigenic stimulus.

The most recent findings, however, cast some doubt on the correctness of detail of this interpretation. Although in reactive lymphoid follicles both immunoglobulins and complement are prominently disposed in the classical reticular, dendritic pattern and surrounded by large numbers of active B cells, in centroblastic-centrocytic lymphomas in contrast, while dendritic cells are also prominent and have numerous C3 receptors demonstrably intact, no immunoglobulin or C3 deposits are seen, despite the plentiful surrounding accumulation of lymphomatous B cells. It thus appears that it may not be antigen, but an as yet undefined receptor on dendritic cells that accounts for the initial aggregation of B cells around them in lymphoid follicles. Gerdes et al (1981) have in fact proposed that lack of an immunohistochemically demonstrable dendritic network of immunoglobulin and C3 is a useful diagnostic feature differentiating centroblastic-centrocytic lymphoma from benign reactive hyperplasia.

Rhodamine was early introduced (Chadwick et al 1958) as an alternative labelling fluorochrome, and for easier conjugation was later modified to the tetramethyl isothiocyanate form (Bergqvist & Nilsson 1974). Its main role in immunohistological work is to provide a contrasting colour in double staining fluorescence work, e.g. that of Brandtzaeg (1974) on the differential localization of free and bound secretory component in secretory epithelial cells, or the refined triple layer staining with the two fluorochromes, in conjunction with hapten-sandwich labelling of monoclonal antibodies (Wofsy et al 1978), recently used by Janossy et al (1981) to demonstrate the different distributions of helper and suppressor T lymphocytes in the synovial membrane of rheumatoid joints.

The first serious non-fluor competitor to establish itself as an immunohistochemical label was horseradish peroxidase (HRP), giving a useful insoluble dark-brown reaction product with diamino-benzidine. The advantages over fluorescein claimed for HRP were several, those most vigorously argued being avoidance of the trouble and expense of dark field ultraviolet illumination of the viewing microscope, the obscuring effect of autofluorescence of tissues, and the tendency of fluorescein staining of tissues and cells to fade (Nakane & Pierce 1967). In the context of present fluorochrome-label methodology, these advantages can hardly now be sustained. Ultraviolet light is unnecessary for fluorescein excitation, at least by transmitted light, since fluorescein isothiocyanate absorbs light more efficiently at 495 nm, a wavelength readily available from a 12-Volt iodine quartz lamp and appropriate interference filters. For rhodamine the corresponding exciting wavelength is 550 nm, and here the high energy peak at 546 nm provided by a mercury burner of the HB 200 type probably does have an advantage, though not by reason of its ultraviolet output. With epi-illumination, and especially where double-fluorochrome staining is used, the HB 200 lamp is for this reason usually supplied. Fading of fluorescence, however, now presents little problem, since it was found in my laboratory (Johnson & Araujo 1981) that addition of para-phenylene diamine to the glycol mounting fluid has a marked retarding effect on fading so that the same fields may be subjected to prolonged and repeated inspection.

HRP (or other enzymes) conjugated with antiglobulin antibody may be used for direct or indirect staining in the same way as fluorescein, or alternatively with the antibody bridge technique, where the primary antibody bound to the tissue used as substrate is linked to soluble immune complexes of peroxidase with anti-peroxidase (PAP) raised in the same species as the primary antibody. The link is an anti-immunoglobulin antibody raised against the species in question. This PAP method, introduced by Sternberger et al (1970), is claimed to have the advantage that the combining specificities of the antibody molecules involved do not run the risk of impairment as a result of chemical conjugation. It has become a popular method and adequate reagents are available commercially. However, the classical indirect staining procedure analogous to indirect immunofluorescence, with HRP conjugated to affinity purified antiglobulin by the carbohydrate oxidation method of Nakane & Kawaoi (1974) is reliably reported to give excellent results for both sensitivity and specificity (Heyderman 1979), and has the time saving advantage of being a 2-step rather than a 3-step procedure.

By the beginning of the 1970s, most of the essential techniques for immunofluorescence and immunoenzyme labelling had been established, including the use of the latter for ultrastructural location of antigens in tissues. A general realization of their potential in diagnosis stimulated an increasing demand for improved and simplified fluorescence microscopy, and optimally labelled reagents of defined specificity and potency. An additional spur to improved microscopy was the rapid growth of interest in membrane determinants of lymphocytes and tissue culture cells and their functional behaviour. This encouraged the production of a range of moderate and higher power objectives of high numerical aperture (NA 1.30–1.40) in combination with wide-angle low-power eyepieces, allowing detection of much smaller quantities of fluorochrome than conventional systems. With the introduction of vertical illuminators capable of providing epi-illumination, reproducible conditions of illumination were achieved, the use of oil was avoided, and the advantage of increasing intensity with increasing objective power attained. Another development in microscopy was the production of a wide range of interference filters of different short-pass and narrow-band transmittances, capable of providing filter combinations to deal with staining by more than one fluorochrome, in combination with phase-contrast microscopy (Ploem 1975). Modern microscopes with epi-illumination have easily interchangeable blocks containing complete sets of primary and secondary filters so that the required illumination may be instantly assembled by the user.

In the last decade also quality control of immunological reagents used in labelled antibody techniques has received long-overdue attention. The reproducibility of coupling of fluorochromes and enzymes to antibodies has been enhanced and the potency and specificity of the resulting conjugates markedly increased by the use of pure IgG or affinity purified antibody as the starting material, and by paying careful attention to fluorochrome:protein ratios and conditions of coupling (Goding 1976). As a result, the quality of commercial conjugates has undoubtedly improved. An essential step still required of the user, however, is to titrate any given conjugate to determine its optimal titre in direct or indirect staining. Potent conjugates, optimally labelled, may be used with excellent results at dilutions far beyond that giving non-specific staining due to the presence of even a small proportion of 'sticky' over-labelled molecules. The World Health Organization, in collaboration with the International Union of Immunological Societies, issues the range of reference conjugates shown in Table 3.1, and further

antigens, it was expected, might retain a degree of reactivity after conventional fixation.

Since the early uses of direct immunofluorescence were mostly in research, aimed at unravelling the complexities of antibody production in the immune

Table 3.1 Immunological comparison materials

Material	Intended use	available for immunofluorescence
WHO internat standards	Calibration of local materials for control of quality and for potency	FITC anti-human Ig conjugates: Sheep anti-human Ig Sheep anti-human IgM (anti-μ-chain) Sheep anti-human IgG (anti-γ-chain)
WHO internat reference preparations	Provided as a service reagent; also suitable for calibrating local material	Anti-nuclear factor (homogeneous) 66/233 (100 IU/ampoule) IgM class anti-nuclear antibody HL
WHO internat reference reagents	Qualitative use in identification (high specificity)	(in preparation)

candidate preparations relevant to quality assessment are undergoing collaborative studies to check their usefulness and suitability. WHO reference conjugates have been checked for total protein, antibody protein and fluorescein:protein weight and molar ratios. Their specificity and potency in indirect immunofluorescence have been determined in tests of a routine nature with sera containing antibody of a single defined class (Holborow et al 1981). It should be noted, however, that, as shown by Schuit et al (1981), conjugates against heavy and light immunoglobulin chains which display adequate specificity at optimal dilution in indirect tests, may give unwanted cross-reactions with other immunoglobulin epitopes in direct tests for membrane Ig of normal or leukaemic lymphocytes, or even of cytoplasmic Ig in polyclonal plasma cells in normal bone marrow preparations. Again, it is necessary for the user himself to conduct quality control tests in his chosen system in order to satisfy himself regarding specificity.

For at least two decades after immunofluorescent protein tracing in histology was introduced, the use of unfixed snap-frozen tissue sections was regarded as de rigueur, since, as already noted above, fixation in formal saline was thought to denature irremediably the antigenic structures at any rate of proteins, although the specific epitopes of polysaccharide

system, the requirement for frozen sections was initially of little logistical handicap. The extension of immunohistological methods to include examination of human biopsies of kidney, skin and, less often, other organs was likewise based on the use of cryostat sections, although bone marrow smears were often fixed in acetic acid-ethanol at $-20°C$ before staining (van Furth et al 1966). A first step towards remedying this restriction on fixation had been taken by Sainte-Marie (1962) who introduced a method for fixation of small tissue pieces in cold ethanol, followed by dehydration and clearing in pre-cooled alcohol and xylene, embedding in paraffin wax, storage of blocks in the cold, and sectioning under specified cool conditions. Although useful in fluorescence, the Sainte-Marie technique has not found general favour as a routine pathology method, requiring as it does special handling of biopsy material from the moment of taking, and resulting nevertheless in poor preservation of morphological detail.

The success of Taylor & Burns (1974) in demonstrating plasma cells and other immuno-globulin-containing cells in formalin-fixed, paraffin wax-embedded tissues, using peroxidase-labelled antibody, was an important step towards bringing immunohistochemistry within the scope of conventional histopathological methodology. They used the indirect method on 3–4 μm sections, dewaxed and taken through the alcohols to water, having first demonstrated and blocked endogenous peroxidase by brief treatment with alpha-naphthol pyronin. For the first layer, they used rabbit anti-human immunoglobulin antisera against Ig isotypes, and for the second swine anti-rabbit IgG peroxidase conjugate. In this way they were able to stain cytoplasmic Ig of different classes in the cytoplasm of plasma cells in a variety of clinical specimens containing lymphoid tissues, and judged the method especially useful in the diagnosis of multiple myeloma in bone biopsies. Although intracellular Ig remains antigenically reactive after formalin fixation, at least in part, it has been generally found that extracellular Ig, or surface Ig (sIg) on lymphocyte membranes is not demonstrable in sections of formaldehyde-fixed tissue, unless they have first been treated with a proteolytic enzyme to reveal the antigenic activity of Ig 'masked' by fixation. Comparing an FITC technique and one based on PAP in studying the distribution of Ig in cryostat and paraffin wax sections of human tonsil, Curran & Gregory (1978) concluded that treatment of paraffin wax sections with 0.1% buffered trypsin for a measured period both improves the staining of cytoplasmic plasma cell Ig and reveals the presence of sIg on B lymphocytes. Under their respective optimal staining conditions, the performances of FITC and PAP methods in detecting these two forms of Ig were judged equal, whether with paraffin wax or cryostat sections.

The masking effect of formaldehyde fixation seems to be due to extensive cross-linking of tissue fluid proteins which prevents immunoglobulin reagents reaching their antigens (Hed & Enestrom 1981), and is hence mostly seen with extracellular but not intracellular Ig. Masking is thus also a hindrance to immunohistochemical demonstration of immune deposits. Using protease VII as an alternative to trypsin, Sinclair et al (1981) have reported good overall concordance for detection of Ig and C3 deposits in paraffin wax sections of renal biopsies by FITC, PAP and indirect HRP methods. Immune deposits in the skin of patients with pemphigus, pemphigoid and lupus erythematosus have likewise been demonstrated by direct immunofluorescence in formalin-fixed tissue sections treated with trypsin (Mera et al 1980). Enzyme treatment thus deals quite effectively with 'masking'. Renewed examination of

fixation methods, however, has shown that the need to use it can be avoided by the use of formol sublimate (a 9:1 mixture of saturated mercuric chloride and formalin) as fixative before paraffin wax embedding (Bosman et al 1977). Piris & Thomas (1980) report that with this fixative cytoplasmic staining of plasma cells in the rectal mucosa by the indirect HRP method was much improved, and Mason et al (1980) confirm that the use of trypsin in the study of lymphomas may be avoided, and interpretation of staining classified by the use of mercury-based fixatives. Mason & Biberfeld (1980) have published a useful discussion of technical aspects of lymphoma immunohistology (see also Ch. 20).

THE FUTURE

The results of several studies in the last few years have gone far towards closing the gap that has separated exponents of immunoenzyme techniques from those who favour immunofluorescence, and it is generally agreed that expediency rather than prejudice should govern the choice of methodology for diagnostic purposes.

Recent work has also, as noted above, removed much force from the objection that fixed tissues are unsuitable for immunohistology. The challenge here is to persuade surgeons and their assistants to heed the advice of Heyderman (1979) and Mason et al (1980) and consider substituting Bouin's, Zenker's or other mercury-based fixatives for formol-saline.

The advent of monoclonal antibodies against immunoglobulin and tissue antigen epitopes is the development most likely to radically alter attitudes to and interpretations of immunohistology. Conventional antisera recognize multiple antigenic determinants, while monoclonal antibodies recognize individual epitopes. For this reason they should make a powerful contribution to standardized differential diagnosis in lymphoma, leukaemia and lympho-proliferative disease, as well as in the more precise differentiation of enzymes, hormones, oncofetal and membrane antigens in tumour specimens. A period of cautious appraisal will be necessary, however, to identify what monoclonal specificities are diagnostically reliable, and what fixation methods are compatible with their use.

REFERENCES

Bergqvist N R, Nilsson P 1974 The conjugation of immunoglobulins with tetramethyl rhodamine isothiocyanate by utilization of dimethylsulfoxide (DMSO) as a solvent. Journal of Immunological Methods 5: 189–198

Bosman F T, Lindeman J, Kuiper G, van der Wal A, Kerunig J 1977 The influence of fixation of immunoperoxidase staining of plasma cells in paraffin sections of intestinal biopsy specimens. Histochemistry 53: 57–62

Brandtzaeg P 1974 Mucosal and glandular distribution of immunoglobulin components: immunohistochemistry with a cold ethanol-fixation technique. Immunology 26: 1011–1114

Chadwick C P, McEntigart M G, Nairn R C 1958 Fluorescent protein tracers: a simple alternative to fluorescein. Lancet 1: 412–414

Coons A H 1976 The development of immunohistochemistry. Annals of the New York Academy of Sciences 177: 5

Coons A H, Kaplan M H 1950 Localization of antigen in tissue cells. II. Improvements in a method for the detection of antigen by means of a fluorescent antibody. Journal of Experimental Medicine 91: 1

Coons A H, Leduc E H, Connolly J M 1955 Studies on antibody production. I. A method for the histochemical demonstration of specific antibody and its application to a study of the hyperimmune rabbit. Journal of Experimental Medicine 102: 49

Coons A H, Snyder J E, Cheever F S, Murray E J 1950 Localization of antigen in tissue cells. IV. Antigens of rickettsiae and mumps virus. Journal of Experimental Medicine 91: 31

Curran R C, Gregory J 1978 Demonstration of immunoglobulin in cryostat and paraffin sections of human tonsil by immunofluorescence and immunoperoxidase techniques. Journal of Clinical Pathology 31: 974–983

Friou G J, Finch S C, Detre K D 1957 Nuclear localization of a factor from disseminated lupus serum. Federation Proceedings 16: 413

Fry L 1981 Autoimmunity in skin diseases. In: Holborow E J (ed) Clinics in immunology and allergy. W B Saunders, London

Gerdes J, Stein H, Bonk A, Lennert K 1981 Immunohistologic demonstration of C3 receptors in normal lymphoid tissue and malignant lymphomas. In: Knapp W (ed) Leukaemia markers. Academic Press, London, p 121–124

Goding J W 1976 Conjugation of antibodies with flurochromes; modification to standard methods. Journal of Immunological Methods 13: 215–226

Hed J, Enestrom S 1981 Detection of immune deposits in glomeruli: the masking effect on antigenicity of formalin in the presence of proteins. Journal of Immunological Methods 41: 57–62

Heyderman E 1979 Immunoperoxidase technique in histopathology: application, methods and controls. Journal of Clinical Pathology 32: 971–978

Hill A G S, Cruickshank B 1953 Study of antigenic components of kidney tissue. British Journal of Experimental Pathology 34: 27–34

Holborow E J, Johnson G D, Chantler S 1981 Use of intermediate reference preparations for immunofluorescence. In: Wick G, Schauenstein (eds) Selected practical topics of immunofluorescence. Elsevier/North Holland, Amsterdam

Holborow E J, Weir D M, Johnson G D 11957 A serum factor in lupus erythematosus with affinity for tissue nuclei. British Medical Journal 2: 732–734

Janossy G, Panayi G, Duke D, Bofil M, Poulter L W, Goldstein G 1981 Rheumatoid arthritis: a disease of T lymphocyte/macrophage immunoregulation. Lancet 2: 839–841

Johnson G D, Araujo G M de C N 1981 A simple method of reducing the fading of immunofluorescence during microscopy. Journal of Immunological Methods 43: 349–357

Kaplan M H, Coons A H, Deane H W 1950 Localization of antigen in tissue cells. III. Cellular distribution of pneumococcal polysaccharided type II and III in the mouse. Journal of Experimental Medicine 91: 15

Leduc E H, Coons A H, Connolly J M 1955 Studies on antibody production. II. The primary and secondary responses in tbe popliteal lymph node of the rabbit. Journal of Experimental Medicine 102: 61

Linderstrom-Lang K, Morgensen K B 1938 Comptes rendus des travaux du Laboratoire Carlsberg, Sér. Chim. 23 (4): 27

McGiven A R (ed) 1980 Immunological investigation of renal disease. Churchill Livingstone, Edinburgh

Marrack J 1934 Nature of antibodies. Nature 133: 292–293

Marshall A H E 1956 An outline of the reticular tissue. Livingstone, Edinburgh, p 23 et seq

Mason D Y, Bell J I, Christensson B, Biberfeld P 1980 An immunohistological study of human lymphoma. Clinical and Experimental Immunology 41: 235–248

Mason D Y, Biberfeld P 1980 Technical aspects of lymphoma immunohistology. Journal of Histochemistry and Cytochemistry 28: 731–745

Mellors R C, Ortega L G, Holman H R 1957 Role of gammaglobulins in pathogenesis of renal lesions in systemic lupus erythematosus and chronic membranous glomerulonephritis with an observation on the lupus erythematosus cell reaction. Journal of Experimental Medicine 106: 191–202

Mera S L, Young E W, Bradfield J W B 1980 Direct immunofluorescence of skin using formalin-fixed paraffin-embedded sections. Journal of Clinical Pathology 33: 365–369

Nakane P K, Pierce G B 1967 Enzyme-labelled antibodies: preparation and application for the localization of antigens. Journal of Histochemistry and Cytochemistry 14: 929–930

Nakane P K, Kawaoi A 1974 Peroxidase-labelled antibody. A new method of conjugation. Journal of Histochemistry and Cytochemistry 23: 1084–1091

Ortega L G, Mellors R C 1956 Analytical pathology. IV. The role of localized antibodies in the pathogenesis of nephrotic nephritis in the rat. Journal of Experimental Medicine 104: 151–170

Ortega L G, Mellors R C 1957 Cellular sites of formation of gamma globulin. Journal of Experimental Medicine 106: 627–640

Papamichail M, Gutierrez C, Embling P, Johnson G D, Holborow E J, Pepys M B 1975 Complement dependence of localization of aggregated IgG in germinal centres. Scandinavian Journal of Immunology 4: 343–347

Piris J, Thomas N D 1980 A quantitative study of the influence of fixation in immunoperoxidase staining of rectal mucosal plasma cells. Journal of Clinical Pathology 33: 361–364

Ploem J S 1975 General introduction to 5th International Conference on Immunofluorescence and Related Staining Techniques. Annals of the New York Academy of Sciences 254: 4–20

Sainte-Marie G 1962 A paraffin embedding technique for studies employing immunofluorescence. Journal of Histochemistry and Cytochemistry 10: 250

Schuit H R E, van der Linde P C M, Hijmans W 1981 The specificity of commercial conjugates to human immunoglobulins: results of performance tests on monoclonal and polyclonal plasma cells and lymphocytes.

Sinclair R A, Burns J, Dunnill M S 1981 Immunoperoxidase staining of formalin-fixed, paraffin-embedded, human renal biopsies with a comparison of the peroxidase-anti-peroxidase (PAP) and indirect methods. Journal of Clinical Pathology 34: 859–865

Stein H, Bonk A, Tolksdord G, Lennert K, Rodt H, Gerdes J 1981 Immunohistologic analysis of the organization of normal lymphoid tissue and non-Hodgkin's lymphomas. Journal of Histochemistry and Cytochemistry 28: 746–760

Sternberger L A, Hardy P H, Cuculis J J, Meyer H G 1970 The unlabelled antibody enzyme method of immunohistochemistry. Preparation and properties of soluble antigen-antibody complex (horseradish peroxidase-anti-horseradish peroxidase) and its use in the identification of spirochaetes. Journal of Histochemistry and Cytochemistry 18: 315–333

Taylor C R, Burns J 1974 The demonstration of plasma cells and other immunoglobulin-containing cells in formalin-fixed paraffin embedded tissues using peroxidase-labelled antibody. Journal of Clinical Pathology 27: 14

Van Furth R, Schuit H R E, Hijmans W 1966 The formation of immunoglobulins by human tissues in vitro. I. Methods and their specificity. Immunology 11: 1

White R G, French V I, Stark J M 1970 A study of the localization of a protein antigen in the chicken spleen and its relation to the formation of germinal centres. Journal of Medical Microbiology 3: 65–84

Wofsy L, Henry C, Cammisuli S 1978 Hapten-sandwich labelling of cell surface antigens. Contemporary Topics in Molecular Immunology 7: 215

4

Metabolic disorders — general view

B. D. Lake

COMMUNICATION

Metabolic disorders are generally rare and their investigation is usually carried out in centres specialising in such conditions. It may be possible to refer appropriately prepared tissue but unless there is an adequate degree of communication between the physician whose patient is under investigation, the surgeon who will take the biopsy and the pathologist who has to ensure that the biopsy is handled correctly, the appropriate preparative techniques cannot be anticipated. Figure 4.1 illustrates the triangle of communication.

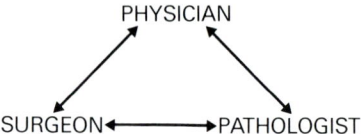

Fig. 4.1 The triangle of communication

Each should be aware of what is to be done, when it is to be done and what can be expected of the procedure. Without the necessary communication valuable tissue will be wasted, perhaps resulting in another operation with the attendant risks

FREEZE SOME–FIX SOME

In the investigation of a biopsy from a patient with a suspected metabolic disorder, histology is probably the least important aspect. However some portion of the tissue should be fixed in formalin for routine sections, to provide some morphological assessment in case the suspected diagnosis does not materialise. For example a needle biopsy of liver taken for the diagnosis of glycogen storage disease should be

divided so that at least 50% of the tissue (weighing a minimum of 8 mg) is frozen for biochemical assay, 25% is taken for histochemistry and frozen for cryostat sections, and the remaining 25% is divided between electron microscopy and histology. Whatever the diagnosis might be, it is valuable to take a portion of the biopsy and freeze some for preparation of cryostat sections. This is true for all specimens, whether or not a metabolic disorder is considered.

Many pathologists feel that cryostat sections are often of inferior quality. However, provided that the tissue has been frozen properly, the knife in the cryostat is sharp, and the anti-roll plate correctly adjusted there is no reason why perfect sections cannot be obtained. With the almost unlimited selection of techniques available for demonstration of practically everything within the cell, cryostat sections offer the most potent means of microscopic diagnosis in the majority of metabolic disorders.

A sample of fresh tissue will allow not only most histochemical techniques but is also available for biochemical assay and for extraction of components for detection by thin-layer chromatography. Although fresh tissue is important for most thin-layer procedures, sphingomyelin, cerebrosides, sulphatides, cholesterol and its esters, and triglycerides may be adequately detected in a semi-quantitative fashion in formalin-fixed tissue. By semiquantitative, I mean that the amount of the compound in question can be compared with that in normal tissue fixed in the same way.

The variety of substances encountered in the metabolic disorders means that a variety of methods are necessary for their preservation and subsequent detection. Cystine may be relatively insoluble, but cystinotic tissues fixed in aqueous media, processed and sectioned by routine methods will have had all cystine extracted because of the large volumes of aqueous solutions washing over the few micrograms of cystine present in a section. Cryostat sections of snap-frozen tissue stained in an alcoholic dye solution, or sections from alcohol-fixed tissue floated out on alcohol are necessary for demonstration of cystine. There are many methods for mucosubstances applicable to routine sections but the mucopolysaccharide deposited in the several mucopolysaccharidoses is not protein-bound (i.e. not a proteoglycan) and is extremely water-soluble. Attempts to preserve the soluble mucopolysaccharides by fixing tissue blocks in Lindsay's fixative (picric-dioxan) are only partially successful and extremely wasteful in valuable tissue. Snap-frozen tissue offers the best preservation of mucopolysaccharides, but detection through metachromasia with toluidine blue in one guise or another, although sensitive, does not give good localisation in tissue sections.

In the lipid storage diseases the deposited lipid survives formalin fixation and can be demonstrated in standard frozen sections. Cryostat sections of snap-frozen tissue can also be used. However the routine processing schedules of dehydration, clearing and impregnation with wax followed by sectioning, dewaxing etc. effectively removes most of the storage product. Thus the diagnosis of metachromatic leucodystrophy will be thwarted if routine sections of nerve are used. In some of the lipid storage diseases different cell types accumulate the substance in different forms, some of which resist standard procedures. The glial storage in Tay-Sachs disease is thus PAS-positive while the neuronal storage has been extracted to give a negative reaction (Fig. 4.2). The interstitial foamy cells in the kidney in Fabry's disease are PAS-positive in routine sections in contrast with the ballooned foamy glomerular epithelial cells which have had their stored hexosides extracted during processing. Smooth muscle deposits in Fabry's disease are retained in processing not only to show PAS and Sudan black positivity but also to exhibit birefringence. It is clear that wherever possible, sections of snap-frozen tissue are to be preferred, but in many instances formalin-fixed frozen sections can be used for diagnosis and in isolated cases routine paraffin wax-embedded tissue is adequate.

BIOPSY SITE

Metabolic disorders manifest their presence in many ways and it is not always necessary to biopsy the main target organ. In G_{M1}-gangliosidosis the brain stores

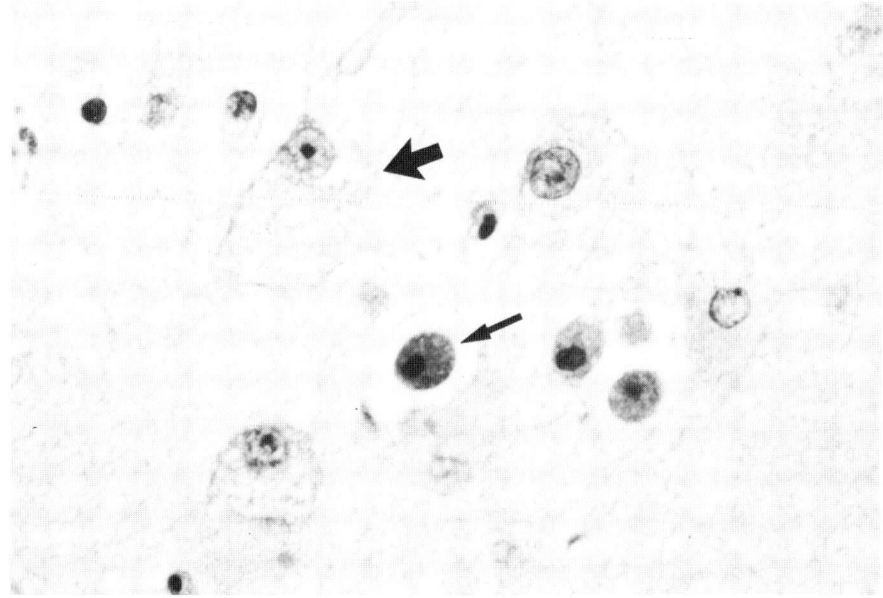

Fig. 4.2 Tay-Sachs disease. Routine section of brain stained by the PAS technique. Large ballooned neurons (large arrows) are unstained because the stored ganglioside has been extracted during processing. The glial cell storage (small arrows) resists extraction. ×600

the ganglioside within the neurons which are also accessible for diagnosis by suction rectal biopsy performed as an outpatient procedure. However a bone marrow aspirate will confirm the generalised nature of the disorder (if needed) or blood films can be examined and the diagnosis made by finding vacuolated lymphocytes and an absence of β-galactosidase activity. For other disorders urine sediment, a skin biopsy, or jejunal biopsy might be the site of choice.

DETECTION OF ENZYME ACTIVITY

Many of the metabolic disorders are a result of a deficient lysosomal enzyme activity. In order to detect the activity of most lysosomal enzymes it is necessary to use tissue fixed in formal-calcium (at $+4°C$ for up to 18 h). The tissue is rinsed before freezing or can be washed and soaked in gum-sucrose prior to freezing. Free-floating frozen or cryostat sections can then be prepared and incubated in the appropriate medium with sections of a normal control tissue, so that a valid comparison may be made. To be able to diagnose a deficiency state there should be an adequate reaction in the control normal tissue and absence in the case under investigation. Some lysosomal enzyme activities are amenable to demonstration in sections of snap-frozen tissue but acid esterase activity in particular must be detected in formal-calcium/gum sucrose-treated tissue.

Occasionally tissue will show an apparent increase in lysosomal enzyme activity. This increase may not be a quantitative increase but reflects a change in the state of the cell and indicates either greater catabolic activity or that the cell is affected by a lysosomal storage disease. In the latter instance the increase may be gross, even though the routine sections do not appear to show involvement. For example, an acid phosphatase reaction on cryostat sections is particularly helpful to show that the liver in Niemann-Pick group C (ophthalmoplegic lipidosis) and Farber's disease is involved in the disorder. Endothelial cell involvement in Hunter's disease (Fig. 4.3), smooth muscle cell involvement in Fabry's disease and the presence of storage cells in a bone marrow aspirate can also be easily detected by application of an acid phosphatase reaction.

In the detection of enzyme activity some background knowledge of the biochemistry of the enzymes involved is advisable. Many enzymes are present in tissues in several isoenzyme forms and although, for example, sulphatase A may be absent in metachromatic leucodystrophy the presence of sulphatase B makes it impossible to use histochemical

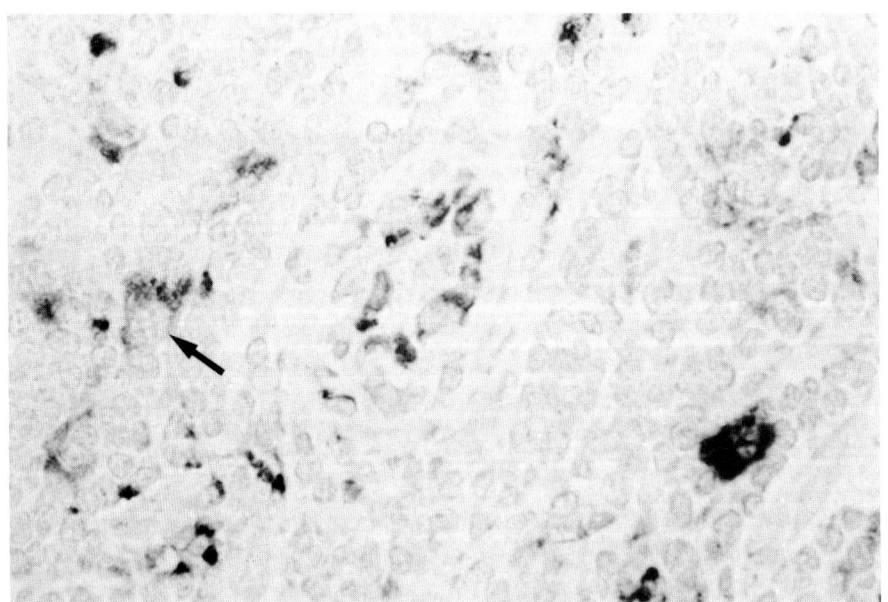

Fig. 4.3   Hunter's disease (mucopolysaccharidosis type 2). Cryostat section of adenoid stained to demonstrate acid phosphatase activity. Endothelial cells, normally negative, show activity indicating their involvement in the disease. A large foamy storage cell is present (arrow). ×600

techniques for the diagnosis of sulphatase A deficiency. Similar considerations apply to α-mannosidase and hexosaminidase for which adequate methods are otherwise available. In any test to detect a deficiency of some particular enzyme activity it is most important to know that the test is capable of detecting the deficient activity and a known deficient tissue should have been examined to validate the method. A positive control should also be included.

The specificity of the method should be examined, not only with a deficient case wherever possible, but to test that the reaction is showing what was intended. In studies of the reaction to demonstrate ornithine carbamylphosphate transferase (OCT) activity two complicating factors are apparent. Firstly carbamylphosphate spontaneously hydrolyses, releasing phosphate ions, and secondly carbamyl phosphate appears to act as a substrate for glucose-6-phosphatase. So attempts to demonstrate OCT deficiency or even OCT activity by a lead-phosphate capture reaction are unlikely to be successsful in unfixed tissue. However, the method does appear valid in lightly fixed tissue.

FUNCTIONAL MARKERS

The metabolic disorders affect the whole range of cell organelles and it may be useful to use methods to detect particular organelles, either to show their presence or to indicate their activity. In Zellweger's cerebro-hepato-renal syndrome there is an absence of peroxisomes in liver and kidney. This absence can be demonstrated in tissue prepared for peroxidase activity. Although animal tissue gives a good reaction, human liver or kidney requires a rather more careful preparation to ensure that peroxisomes are reliably demonstrated (see Ch. 16). Defects in the urea cycle, causing hyperammonaemia, can be mistaken for Reye's syndrome. A liver biopsy taken for assay of one or other of the urea cycle enzymes can be used to exclude Reye's syndrome; the succinate dehydrogenase reaction is normal in the urea cycle defects, but in Reye's syndrome no activity can be detected in an extremely fatty liver. A variety of toxins can also damage mitochondrial function and lead to absent succinate dehydrogenase. In other situations where mitochondrial function is abnormal, and numbers of mitochondria are increased and their structure is altered, an increased succinate dehydrogenase activity may be apparent. For example in the mitochondrial cytopathies (Egger et al 1981) the accumulation of abnormally structured mitochondria in ragged-red muscle fibres is readily detected after short incubation times, before the normal fibres show any reaction product (Fig. 4.4).

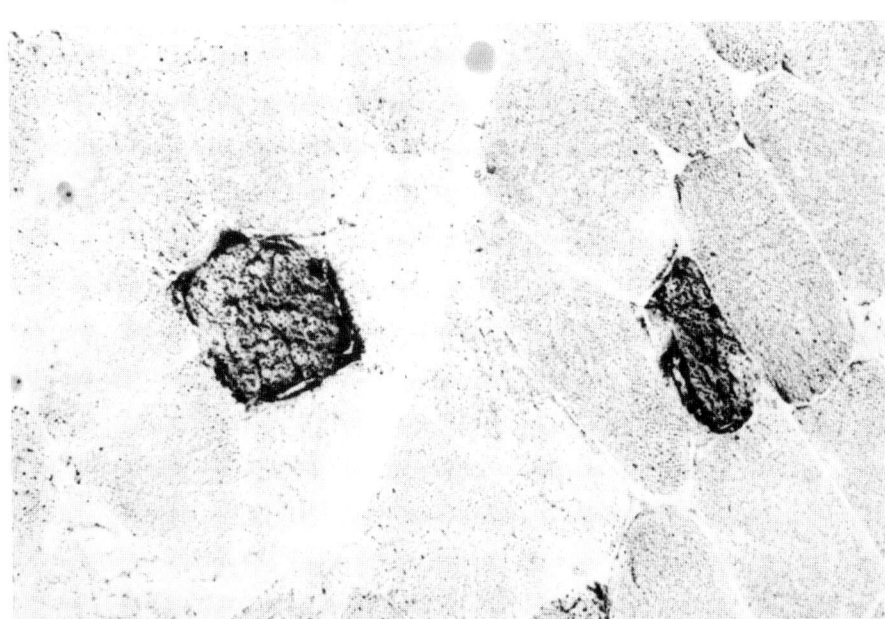

Fig. 4.4 Mitochondrial cytopathy. Cryostat section of a muscle biopsy stained to demonstrate succinate dehydrogenase activity. Two muscle fibres show excess activity indicating collections of abnormal mitochondria. These fibres correspond to the 'ragged-red' fibres seen in trichrome preparations. ×370

MULTIPLE STORAGE PRODUCTS

It used to be considered that in a defect of one particular enzyme only one storage product should accumulate, but it has become clear that many products are stored in most of the lysosomal enzyme deficiencies. There are two or possibly three main reasons for this phenomenon. The first is that the cell may find it difficult to accommodate a compound with a strongly polar end group or one with an awkward shape. To overcome the difficulty the cell may find it convenient to store the substance in association with other substances which may either neutralise the charge and/or help to pack the compound in an acceptable shape. Examples here are of the gangliosidoses where the gangliosides are stored as membranous cytoplasmic bodies (MCBs) with cholesterol and phospholipid. In Niemann-Pick disease the marked cholesterol content led some to believe at one time that the disorder should be considered as a cholesterol disorder as well as a sphingomyelin disorder. Subsequent studies have shown that deficient activity of sphingomyelinase is the cause of the disorder and that the cholesterol is there to package the sphingomyelin. Since the substances involved in the majority of lysosomal storage disorders are lipids, and other lipids are soluble in the lipid complexes, it is not surprising that several lipid components may be isolated from the storage cell.

The second reason for multiple storage products is that most of the enzymes are not substrate-specific and only recognize a particular end group. Thus in β-galactosidase deficiency any compound with a terminal β-galactose link will tend to accumulate. This will include G_{M1}-ganglioside and its asialo derivative G_{A1} in the brain, haematoside from red blood cell membranes in the liver and spleen and the hexasaccharide unit derived from the molecule shown in Figure 4.5. The deficiency of acid esterase in Wolman's disease leads to storage of cholesteryl esters and triglycerides (both are esters) and a variety of other minor esters (Fig. 4.6). Because of the variety of stored products no one special stain is the right one for the diagnosis but rather a range of special stains, coupled with careful assessment of the different cell types affected and their morphologies, is required for accurate diagnosis. It should be pointed out at this stage that a set of staining properties coupled with a morphological appearance of, say neurons, does not necessarily mean that the diagnosis is the same as in another case with apparently similar characteristics. In Tay-Sachs disease (G_{M2}-gangliosidosis) the staining reactions in the brain are virtually indistinguishable from those in G_{M1}-gangliosidosis. Inspection of the structure of the two gangliosides will show the reasons for this. The two disorders can be distinguished by either thin-layer chromatography of solvent extracts of brain (Lake & Goodwin 1976) or by staining to demonstrate β-galactosidase activity which will be absent in G_{M1}-gangliosidosis.

FETAL PATHOLOGY AND THE PLACENTA

In the examination of fetal tissue for confirmation in

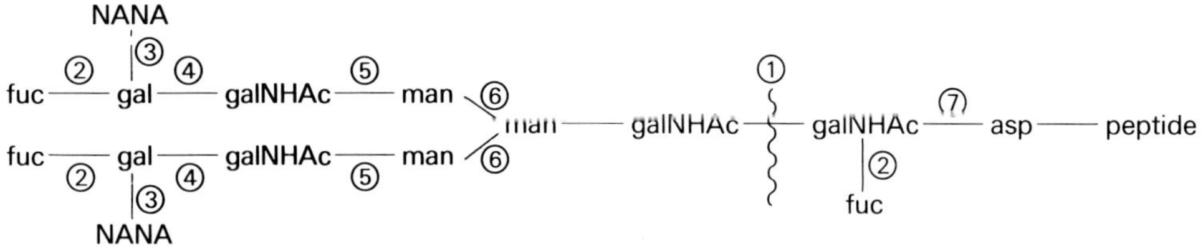

Fig. 4.5 A sialyloligosaccharide unit common to immunoglobulins and several plasma proteins, hormones etc. Cleavage by an endo β-N-acetylglucosaminidase as shown at 1 leads to the 12-membered compound which is susceptible to hydrolysis by a number of glycosidases.

2 α-fucosidase   3 sialidase   4 β-galactosidase
5 hexosaminidase B   6 α-mannosidase
7 aspartylglycosaminamidohydrolase

fuc = fucose   gal = galactose
galNHAc = N-acetylgalactosamine
man = mannose   asp = asparagine
NANA = N-acetylneuraminic acid

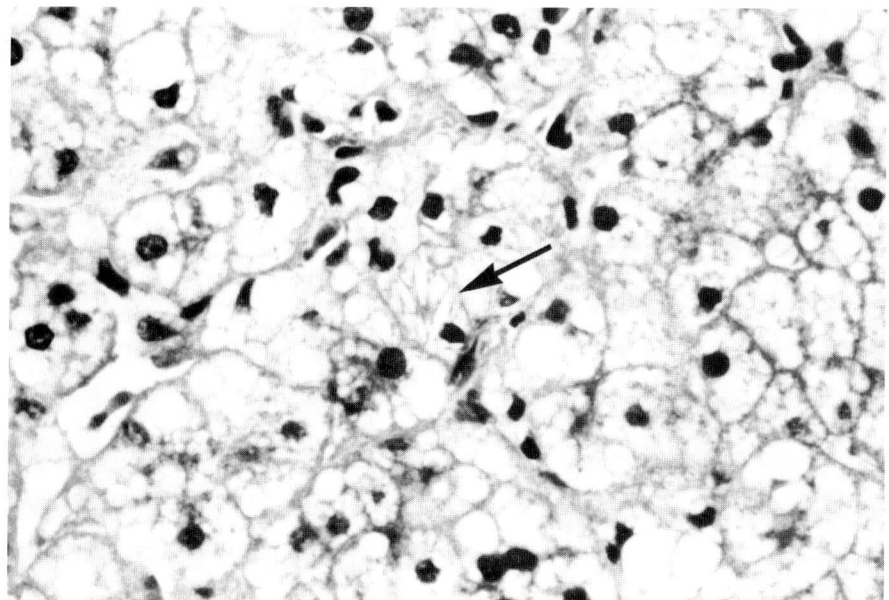

Fig. 4.6 Wolman's disease. Routine section of a liver biopsy stained with H & E showing foamy lipid-laden hepatocytes and storage cells. One of the storage cells (arrow) contains cholesteryl ester crystal shapes. ×600

an abortus following detection of an enzyme defect by amniocentesis and cell culture, it may be difficult to be certain that the storage is present. This is particularly true in the mucopolysaccharidoses where, unless electron microscopy is undertaken, confirmation of the diagnosis by microscopy is rarely possible. In these instances, sections of the placenta afford plentiful evidence of the disease (Fig. 4.7). Although in other storage disorders fetal tissue involvement may be relatively easy to establish (Lake 1976), the placenta should provide important additional evidence in most storage disorders.

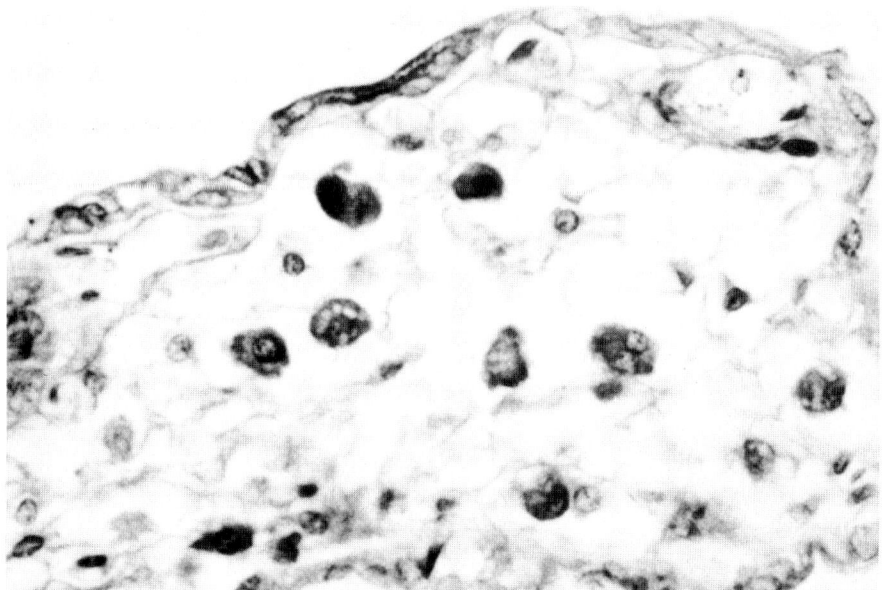

Fig. 4.7 Hurler's disease (mucopolysaccharidosis type 1). Routine section of placenta from an affected fetus of 20 weeks, stained with the colloidal iron technique. Large foamy storage cells are present in the stroma of a villus. ×650

BIOCHEMICAL CONFIRMATION REQUIRED

In all metabolic disorders it is mandatory to have the diagnosis confirmed by appropriate enzyme assay. This can be performed on the frozen tissue from which the cryostat sections have been taken. Why reach a diagnosis by microscopy and biochemical assay? Tissue biopsies are often very small and allow only one or two assays. Histochemical assessment is valuable in establishing the identity of the tissue and in providing a working diagnosis which then only requires minimal biochemistry for confirmation.

REFERENCES

Egger J, Lake B D, Wilson J 1981 Mitochondrial cytopathy. A multisystem disorder with ragged-red fibres in muscle. Archives of Disease in Childhood 56: 741–752
Lake B D 1977 Histochemical and ultrastructural studies in the diagnosis of inborn errors of metabolism. Records of the Adelaide Children's Hospital 1: 337–345
Lake B D, Goodwin H J 1976 Lipids. In: Smith I, Seakins J W T (eds) Chromatographic and electrophoretic techniques, 4th edn. Heinemann, London, vol 1, ch 14

GENERAL BACKGROUND READING

Hers H G, VanHoof F (eds) 1973 Lysosomes and storage diseases. Academic Press, New York
Stanbury J B, Wyngaarden J B, Fredrickson D S (eds) 1978 The metabolic basis of inherited disease, 4th edn. McGraw Hill, New York

5

The role of histochemistry in tumour pathology

J. P. Sloane & M. G. Ormerod

INTRODUCTION

Histochemical stains — both immunohistochemical and the more conventional stains — may assist in the diagnosis of tumours in several different ways. By staining cell products they may add functional data to the existing structural information. On the other hand by identifying structures normally invisible with conventional stains, they may add a new dimension to the morphological appearances. In the classification of neoplasms, such stains are of value if they exhibit specificity for some cell or tissue type; the histogenesis of the lesion can thus be inferred and the tumour appropriately classified. Sub-classification within histogenetic groups according to certain structural or functional parameters may also be of value especially if related to prognosis. Histochemical indicators of the neoplastic process itself are, of course, immensely valuable in the distinction of tumours from hyperplastic and dysplastic conditions but, unfortunately, these are all too rare.

Conventional histochemical stains have been available to the histopathologist for decades and considerable experience has been gained with them. While they have the advantages of being generally applicable to standard paraffin wax-embedded sections and not requiring any special equipment, they lack scope and specificity and in recent years other methods such as enzyme histochemistry, immunohistochemistry and various fluorescent techniques have been developed to increase diagnostic accuracy.

IMMUNOHISTOCHEMISTRY

The use of immunohistochemistry in diagnostic tumour pathology has expanded considerably in recent years. One of the main reasons, apart from the potential specificity of the technique, has been the introduction of stable conjugates of antibodies and enzymes which allows staining of specific tissue components as well as counterstaining by conventional dyes. Thus the product of interest can be visualised at the same time as the histological structure of the tissue; furthermore, the preparations are permanent. Immunohistological techniques using enzyme conjugates have thus largely superseded immunofluorescence in tumour pathology.

Other major factors which have contributed to the increase in the use of immunohistochemistry are the improvement in the quality, range and availability of polyclonal antisera and the introduction of monoclonal antibodies.

For detailed discussion of the techniques themselves, see Taylor (1978), Heyderman (1979) and Chapters 3, 20 and 23. In this Introduction discussion is limited to a few salient points. The detailed methods used in our laboratory are described in Appendix 6.

The most important factor in producing a first-class immunohistochemical stain is the quality of the reagents used. It is essential to ensure that the primary antibody is specific.

Using either polyclonal or monoclonal antibodies, cross-reactions may be observed whereby an antibody to one molecule reacts with a related determinant on a different molecule. For example, part of the carcino-embryonic antigen (CEA) is identical with a smaller molecule designated normal cross-reacting antigen (NCA) which is expressed strongly by polymorphonuclear leucocytes. Antisera to CEA frequently react with NCA and these cross-reacting antibodies need to be removed by absorption with purified NCA. In a monoclonal preparation all of the antibodies react with the same determinant so that it is necessary to select a monoclonal antibody which does not react with shared determinants. Another, more serious problem encountered with polyclonal antisera is the presence of unwanted antibodies. Immunocytochemical staining frequently reveals reactants which are not shown by other techniques. In a radioimmunoassay the cross-reacting antibodies will reduce the specificity of the assay but impurity antibodies will have no effect as long as the labelled antigen is reasonably pure since only those antibodies reacting with the label are of importance. However, in an immunohistological stain, any antibody present in the serum used which reacts with the tissue section will give a reaction product. It is important that antisera absorbed with purified antigen should be used to stain a section to check for irrelevant antibodies and, if demonstrated, they must be removed by absorption. Some effort may be needed in order to discover an appropriate absorbant.

Clearly there are many problems in producing high-quality immunocytochemical reagents and the amount of time and level of immunochemical expertise required made it difficult for histopathologists to incorporate the technique into a routine diagnostic service, before a wide range of highly characterised antisera became commercially available (see Ch. 3).

Another problem is the lack of reproducibility of conventional antisera which are only available in finite quantities and limited by the size of the animal immunised. Different animals injected with the same immunogen produce slightly different antisera directed against different determinants on the molecule. When a technique is established, it is not possible to distribute precisely the same reagent to all laboratories. In contrast, the recently developed myeloma hybrids producing monoclonal antibodies are immortal. Once a particular antibody has been fully characterised and its value established, the same antibody should be available world-wide indefinitely. As monoclonal antibodies are virtually pure solutions of specific antibody, further purifications and absorptions in the histopathology laboratory are superfluous. It should be noted, however, that as monoclonal antibodies react with only one epitope on the molecule they may have less rather than greater specificity by virtue of more cross-reactions.

Originally, immunohistochemical stains used conjugates made with horseradish peroxidase which gives a brown reaction product with diaminobenzidine. This is an inconvenient colour as many endogenous pigments such as melanin, haemosiderin and lipofuscin are also brown and may be misinterpreted as a positive reaction. Recently more use has been made of alkaline phosphatase with which red or blue reaction products may be generated. When choosing an enzyme conjugate regard should be given to the levels of endogenous enzyme and the method of blocking this activity should be chosen to have no effect on the antigen under study. Methods of blocking alkaline phosphatase activity have been discussed by Ponder & Wilkinson (1981).

A surprising number of antigenic determinants survive formalin fixation and embedding in paraffin wax, which is of great advantage as routinely processed material can be studied. Other antigens are destroyed by this treatment and frozen sections or alternative methods of processing have to be used. This is frequently the case with lymphocyte surface markers. Further work is needed to find methods of processing which preserve more antigenic determinants as well as preserving good morphology.

OTHER TECHNIQUES

Enzyme histochemical techniques have some valuable applications. The demonstration of acid phosphatase in a metastatic carcinoma for example may be useful in tracing the primary site to the prostate and non-specific esterase is widely employed to identify tumours of histiocytic lineage. However, the widespread distribution of most enzymes means that many of the techniques lack specificity and their susceptibility to formalin fixation and paraffin wax embedding further limits their scope. A notable exception is chloroacetate esterase which can be demonstrated in routine formalin-fixed paraffin wax-embedded sections and is invaluable in identifying neutrophil myeloid cells and tumours derived from them.

Fresh frozen material however, allows a much wider scope for detecting marker enzymes. Application of enzyme histochemistry in tumour pathology are further discussed by Stoward and Lantos in Chapters 2 and 8 respectively.

Touch preparations of fresh tissues and marrow smears are also of use and are suitable for a wide range of techniques (Dearnaley et al 1981).

Mucin histochemistry has a limited but definite value in assessing malignancy in prostate epithelium as well as in the identification of various myxoid soft tissue tumours and in the diagnosis of mesothelioma (see pp. 245, 271).

Fluorescence is usually employed as part of an immunohistological technique but may also be used for demonstrating endogenous fluorescent compounds in tissues. Various amines, for example, become fluorescent after exposure to formaldehyde and other agents; yellow fluorescence is observed with compounds such as 5-hydroxytryptamine which may be present in carcinoid tumours and blue-green fluorescence may be seen in tumours such as neuroblastoma or phaeochromocytoma which produce sympathomimetic amines (see Ch. 8 and 27). These techniques require specialist equipment and also lack specificity unless excitation and emission spectra of the substances in question are plotted accurately; this, of course, renders the technology even more complicated and expensive.

APPLICATIONS TO TUMOUR PATHOLOGY

Classification of tumours

Histochemical stains with tissue specificity are of great value in the classification of tumours especially if they are poorly differentiated morphologically. Specificity may be broad or narrow. The recently described epithelial membrane antigen (EMA), for example, is present in many epithelial tumours but absent in mesenchymal and lymphoid neoplasms (Heyderman et al 1979, Sloane & Ormerod 1981). The monoclonal antibody 2D1 similarly recognizes a surface antigen HLe-1 which is present on all lymphoid cells (Pizzolo et al 1980). These reagents would clearly be useless for classifying carcinomas or lymphomas but have proved very valuable for distinguishing anaplastic carcinomas from malignant lymphomas — a common diagnostic problem. Other stains show narrower specificity and this may have value not only in the classification of tumours but also in predicting the primary site of metastatic carcinomas. A positive immunocytochemical stain for

thyroxine is virtually diagnostic of a thyroid carcinoma. Other stains can give narrow specificity if used in the correct context. Enzyme histochemical stains for acid phosphatase activity may be strongly positive in reactive histiocytes, histiocytosis and certain lymphomas as well as in prostatic carcinomas. However, the presence of large amounts of acid phosphatase in a tumour exhibiting glandular differentiation in a bone biopsy taken from an elderly man points to metastatic prostatic carcinoma. Antisera to prostatic acid phosphatase have recently become available which can be used on routine paraffin wax-embedded sections (see Ch. 24). Such antisera, in our experience, also fail to react with acid phosphatase from other tissues such as histiocytes and T cells.

Alpha-fetoprotein (AFP) can have narrow specificity if used in the correct context. It is usually found in significant quantities in hepatocellular carcinomas and germ cell tumours but the clinical background and the histological appearance of the lesion should usually serve to distinguish these two main possibilities. Further specificity could be obtained using combinations of markers and the presence of β-human chorionic gonadotrophin (β-HCG) as well as AFP in a tumour would easily resolve the differential diagnosis in favour of a germ cell tumour.

Histochemical methods for the classification of tumours of the nervous system are to be found in Chapter 8.

In addition to elucidating the histogenesis of tumours, histochemical methods are increasingly able to reveal heterogeneity within morphologically homogeneous categories. Acute lymphoblastic leukaemia can be classified as B, T or 'common' according to various surface markers and the level of nuclear terminal deoxynucleotidyl transferase. These phenotypic differences relate strongly to response to treatment and prognosis. Other tumours can similarly be sub-divided such as the malignant lymphomas which can be sub-classified with many markers, both enzyme histochemical and immunocytochemical (see Ch. 20). Although this reveals differences of a fundamental biological nature, the clinical significance is not yet firmly established. Many malignant tumours exhibit variable clinical behaviour within morphologically homogeneous categories and it is hoped that histochemical markers which relate specifically to behaviour and prognosis will be identified.

Functional information

The assessment of functional activity of tumours is enhanced by immunocytochemistry because specific tumour products can be identified. Conventional stains may reveal the presence of neurosecretory granules in pancreatic tumours allowing them to be classified as islet cell lesions; immunocytochemical techniques, however, can go a stage further and identify the specific secretory product such as insulin, glucagon etc. The value of immunocytochemistry may also be seen in the classification of pituitary tumours. The cells of the adenohypophysis can be divided into acidophils, basophils and chromophobes using conventional methods. The acidophils produce prolactin and growth hormone but it has been shown that some chromophobe tumours may contain prolactin. It would thus seem that a more useful classification of pituitary tumours could be based on immunohistochemical techniques for the various hormonal products (Halmi & Duello 1976).

Functional characteristics of tumours may be of clinical value in several other ways. They may point to the value of a certain secretory product as a serum marker in monitoring the course of the disease. Many testicular teratomas are associated with elevated serum levels of β-HCG and AFP and the serum level is roughly proportional to the tumour load. The demonstration of these products in a tumour may point to their value as markers in the later stages of the disease even if initial serum levels are not significantly elevated. The demonstration of secretory products may also help in establishing the relationship between tumours and various metabolic abnormalities or clinical syndromes. Some tumours (clear cell carcinoma of the kidney and some neuroendocrinomas) are associated with the secretion of parathormone-like material; the demonstration of such material in a tumour would explain a high serum calcium level and other possible causes, such as bony metastases, would not have to be implicated. It has been shown that some seminomas are associated with elevated levels of β-HCG and prior to the demonstration of this material in the multinucleate

giant cells in these tumours it was generally assumed that a teratomatous component had been missed through sampling error (see Ch. 23).

Additional morphological information

In addition to providing information at the molecular level, histochemical methods may help by revealing structures not normally identifiable with conventional stains. The reticulin stain reveals aspects of the growth pattern of tumours which are less apparent in haematoxylin and eosin stained sections. Endothelial cells can be stained by enzyme histochemical methods for alkaline phosphatase or by immunocytochemical methods for factor VIII; invasion of small blood vessels by tumour can thus be appreciated more readily. We have recently shown that stains for EMA in organs containing metastatic breast carcinoma will reveal a far greater number of tumour cells than can be seen in H & E stained sections. C cells in the thyroid are likewise difficult to identify but their recognition becomes important in assessing C cell hyperplasia in cases of familial medullary carcinoma; this can be achieved readily with conventional silver stains for neurosecretory granules or with immunohistochemical detection of calcitonin.

Assessment of malignancy

Markers of the neoplastic process itself would, of course, be invaluable. One area in which immunohistochemical methods can be successful is in the diagnosis of B cell malignant lymphomas. Although B cells may express more than one class of immunoglobulin on their surfaces or in their cytoplasm this is only ever of one light chain type. Staining for immunoglobulin light chains can thus reveal whether B cell proliferations are monoclonal or polyclonal and thus, by inference, neoplastic or reactive. Presumably, epithelial cell proliferations could be similarly assessed if markers identifying subtypes could be discovered. In the breast, for example, intraductal hyperplasias and dysplasias merge with in situ carcinoma and their distinction is often highly subjective. The distinction is nevertheless an important one from a clinical point of view.

CONCLUSION

In recent years there has been an expansion of histochemical techniques in tumour pathology, much of it due to the improvements in immuno-histochemistry. With the rapid development of monoclonal antibodies this expansion will continue. As outlined in the Introduction, present experience suggests that new markers will be found which define subsets of broad classes of cells — such as epithelial cells or lymphocytes — while other markers may define subsets within more than one class of cell. The immunohistochemical stains can be expected to give new patterns of information which will help to solve certain diagnostic problems although one should not always expect them to produce absolute answers.

REFERENCES

Dearnaley D P, Sloane J P, Ormerod M G, Steele K et al 1981 Increased detection of mammary carcinoma cells in marrow smears using antisera to epithelial membrane antigen. British Journal of Cancer 44: 85–90

Halmi N S, Duello T 1976 'Acidophilic' pituitary tumours. Archives of Pathology and Laboratory Medicine 100: 346–351

Heyderman E 1979 Immunoperoxidase technique in histopathology: applications, methods and controls. Journal of Clinical Pathology 32: 971–978

Heyderman E, Steele K, Ormerod M G 1979 A new antigen on the epithelial membrane: its immunoperoxidase localisation in normal and neoplastic tissue. Journal of Clinical Pathology 32: 35–39

Pizzolo G, Sloane J P, Bevereley P, Thomas J A, Bradstock K E, Mattingly S et al 1980 Differential diagnosis of malignant lymphoma and non-lymphoid tumours using monoclonal anti-leucocyte antibody. Cancer 46: 2640–2647

Ponder B A, Wilkinson M 1981 Inhibition of endogenous tissue alkaline phosphatase with the use of alkaline phosphatase conjugates in immunohistochemistry. Journal of Histochemistry and Cytochemistry 29: 981–985

Sloane J P, Ormerod M G 1981 Distribution of epithelial membrane antigen in normal and neoplastic tissues and its value in diagnostic tumour pathology. Cancer 47: 1786–1795

Taylor C R 1978 Immunoperoxidase techniques: theoretical and practical aspects. Archives of Pathology and Laboratory Medicine 102: 113–121

6

Degenerative diseases of the central and peripheral nervous system

Olga B. Bayliss High

INTRODUCTION

During the last 40 years histochemistry has played an invaluable part in establishing the composition and functional capacity of the different elements in the nervous system. The way in which histochemical techniques have been devised for and applied to the nervous system is excellently reviewed in Adams' *Neurohistochemistry* (1965).

Histochemistry can be diagnostically useful in the pathological nervous system where identification of specific tissue components may assist the pathologist to interpret morphological disturbances observed in conventional histological preparations. Due to the jelly-like consistency of the central nervous system, it has been customary to fix, or ideally to perfuse, whole brain and cord for long periods prior to sampling for celloidin or paraffin wax processing. Such procedures will inactivate or extract certain important tissue constituents — notably enzymes and lipids — which can usually be detected only in frozen sections prepared from unfixed material. Enzyme techniques have been useful to detect early changes in cellular degeneration, atrophy and hypertrophy. Immuno-histochemical methods have also been extensively applied to various neural elements and a considerable amount of research has been devoted to neuroglial enzymes which are more aptly considered in the section on brain tumours (see Ch. 8). It has been suggested that raised levels of certain lysosomal enzymes may predispose to the demyelinating process in multiple sclerosis (Allen & McKeown 1979). It remains to be seen how useful enzyme techniques will prove to be in diagnostic pathology of the degenerative disorders affecting the nervous system, but undoubtedly their value will be somewhat limited

in the routine laboratory where unfixed CNS material is so rarely available.

On the other hand certain selected histochemical techniques do find a valuable practical role in the diagnosis of the following conditions
a. demyelinating diseases
b. degenerative changes involving the neuron
c. metabolic disorders affecting the nervous system.

DEMYELINATING DISEASES

Demyelination is an inevitable consequence of pathological processes such as infarction, haemorrhage and tumours involving the white matter. The myelin sheath undergoes *secondary* degeneration distal to a lesion that separates an axon from its perikaryon, a process that can be demonstrated classically in Wallerian degeneration following peripheral nerve section. In *primary* demyelinating or myelinoclastic disorders on the other hand, the myelin sheath itself is selectively affected in the absence of neuronal injury and the axon is often largely spared. This includes *segmental* demyelination in the peripheral nervous system, a Schwann cell lesion in which internodal lengths of myelin are selectively destroyed.

Primary demyelination can be caused by a variety of toxic, infective and allergic influences, but in spite of the extensive research and speculation, the aetiology of multiple sclerosis — the most common of these conditions — has remained an enigma since Charcot's day. Demyelination and dysmyelination (failure to myelinate properly during development) are features of certain metabolic disorders of the nervous system which will be considered in a later chapter. The more commonly occurring of the demyelinating diseases and their causative agents, where known, are listed in Tables 6.1 and 6.2.

Table 6.1 Demyelinating disorders of the central nervous system

Disorder	Cause and remarks
Multiple sclerosis (including variants: Balo's concentric demyelination, Devic's neuromyelitis optica and Schilder's diffuse cerebral sclerosis)	Precise aetiology not yet established but mechanism probably immunological. Thought to be due to either a 'slow' virus, autoimmune reaction or a metabolic defect
The encephalitides	A variety of viruses usually causing demyelination
Subacute sclerosing panencephalomyelitis	Measles virus
Post-vaccinial encephalomyelitis	Immunoallergic reaction
Progressive multifocal leuco-encephalopathy	Polyoma virus affecting oligodendroglia
Creutzfeldt-Jacob disease	'Slow' virus infection
Deficiency states and intoxications	
Subacute combined degeneration	Avitaminosis B_{12}
Marchiafava-Bignami disease	Alcoholism
Grinker's myelopathy	Carbon monoxide poisoning

Table 6.2 Demyelinating disorders of the peripheral nervous system

Disorder	Cause and remarks
Segmental demyelination (primary)	
Diabetic neuropathy	Schwann cells affected by ischaemia
Diphtheritic neuropathy	Schwann cells affected by diphtheria toxin
Guillain-Barré's syndrome	Autoimmune reaction
Wallerian degeneration (secondary)	Severance or compression of peripheral nerves Organophosphorus compounds, e.g. triorthocresyl PO_4 (TOCP) and diisopropyl fluorophosphate (DFP) Vitamin B_1 deficiency (Beri-beri)
Déjerine-Sotta's disease	Schwann cell hyperplasia

Histochemistry

In spite of considerable variation in the aetiology and pathogenesis of these conditions, a common factor is the loss of myelin. From the histochemist's point of view, the most important feature of demyelination is the transformation of normal myelin, with its distinct *hydrophilic* properties due to the largely polar phospholipid constituents, into *hydrophobic* lipid droplets of cholesterol esters with their affinity for osmium tetroxide and the organotropic Sudan dyes. The newly formed sudanophilic esters which are not present in the normal adult nervous system, provide a convenient hallmark whereby actively degenerating lesions can be distinguished from old, fat-free scars and also from normal myelin itself. Although the myelin sheath physically disintegrates at the outset of demyelination the released fragments do not differ histochemically from the normal myelin sheath until the 10th day of the process when myelin particles are converted by phagocytic cells into sudanophilic cholesterol esters. This catabolic process may last for several weeks and can be demonstrated histochemically — in fresh frozen sections — with combined enzyme and lipid techniques which will pick out macrophages by means of their hyperactive enzymes and the cholesterol esters of degenerating myelin with one of the routine fat stains. The combination that has proved to be most useful in this context is the *β-galactosidase-oil red O method*.

For well over 100 years myelin has been stained by variants of the Weigert Pal method whereby myelin lipids, after chromation, form a lake with haematoxylin. One such variant, the *dichromate-acid haematein (DAH)* method (Baker 1946), has been shown to demonstrate selectively the choline-containing phospholipids of myelin — namely lecithins and sphingomyelins (Roozemond 1971).

Individual myelin lipids can now be identified by selective histochemical procedures, most of which were devised especially to establish the structure and composition of the normal myelin sheath and to monitor the processes of myelination and demyelination in parallel with the biochemical approach to this subject (see Adams 1965). For lecithins and cephalins the gold hydroxamic acid technique is recommended (Adams et al 1963). Sphingomyelins can be demonstrated selectively by the dichromate-acid haematein method when sodium hydroxide hydrolysis is first applied to remove the other reactive myelin lipid — lecithin (Adams & Bayliss 1963). Cerebroside is stained by the regular PAS reaction but staining can be confined to this lipid by means of a series of blockades and extractions which exclude the interference from other PAS-positive compounds (Adams & Bayliss 1963). Details of these and other methods for individual lipids can be found in Bancroft & Stevens (1982). Such techniques find their major application in the diagnosis of the lipid storage diseases (see ch. 7, 16 and 18).

Among the conventional histological techniques, *luxol fast blue* is the method of choice for depicting normal myelin and thereby revealing the contrasting unstained areas that have undergone demyelination (see Fig. 6.1). The copper phthalocyanin (luxol fast blue) reacts with phospholipids and although most of these will be extracted during processing for paraffin wax and celloidin sections, those that remain intimately bound to the proteins in the neurokeratin and proteolipid skeletons of the peripheral and central nervous system myelin sheaths respectively, will be stained. Indeed their reaction is actually enhanced when the other myelin lipids have been removed by the solvents used for processing. The turquoise blue reaction product can be converted to an intense purple complex after treatment with cresyl violet (Klüver & Barrera 1953) or neutral red, both of which incidentally stain other tissue components in conveniently contrasting tones. Although this method may be ideal to delineate areas of demyelination, recognition of actively demyelinating sites containing intracellular cholesterol esters depends upon the histochemical fat stains which can be applied only to sections prepared from frozen material. *Sudan black B* is the most sensitive of these dyes and is an excellent myelin stain per se, because in addition to the fat-staining properties of the Sudan dyes in general, Sudan black B uniquely contains a second dye component which stains certain phospholipids in normal myelin. However, this very feature means that the contrast obtained between normal and degenerating myelin is less marked than with *oil red O* for example, which barely colours the normal myelin sheath so that the cholesterol esters of degenerating myelin can be more precisely localised than with Sudan black B. This dye, however, is considerably more sensitive as a general lipid stain when preceded by bromination as described in the *bromine-Sudan black B (BSB)* method (Bayliss & Adams 1972). Lipids that are soluble in the ethanolic dye solvent — notably lecithins and free fatty acids — are stabilised by bromine; furthermore, free cholesterol which is normally unstained by the Sudan dyes is converted to oily sudanophilic bromo-derivatives, so that the BSB method can be relied upon to detect all the lipids that may be present in a tissue section, thereby enhancing and extending the regular Sudan black B reaction (see Fig. 6.2).

Oil red O has been usefully combined with the *dichromate-acid haematein method* (Bourgeois & Hubbard 1965) to demonstrate phospholipids and triglycerides simultaneously. When applied to the nervous system, this combination method affords a clear distinction between normal myelin (blue) and the cholesterol esters of degenerating myelin which appear red (see Plate 1). Oil red O has also been applied in sequence with the *periodic acid-silver*

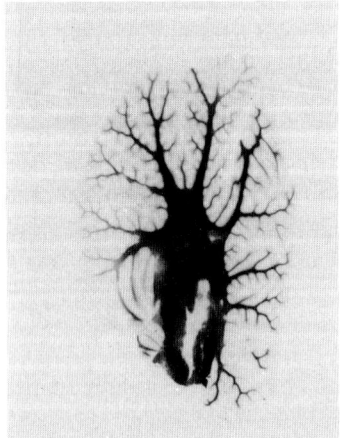 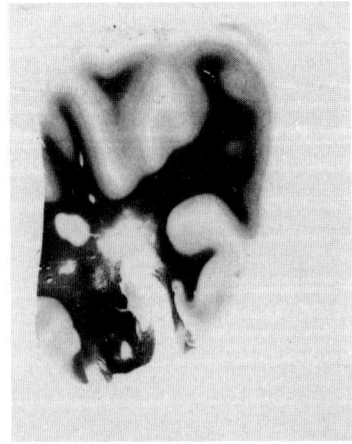

Fig. 6.1 Brain sections from two cases of multiple sclerosis showing areas of demyelination. Luxol fast blue-cresyl violet

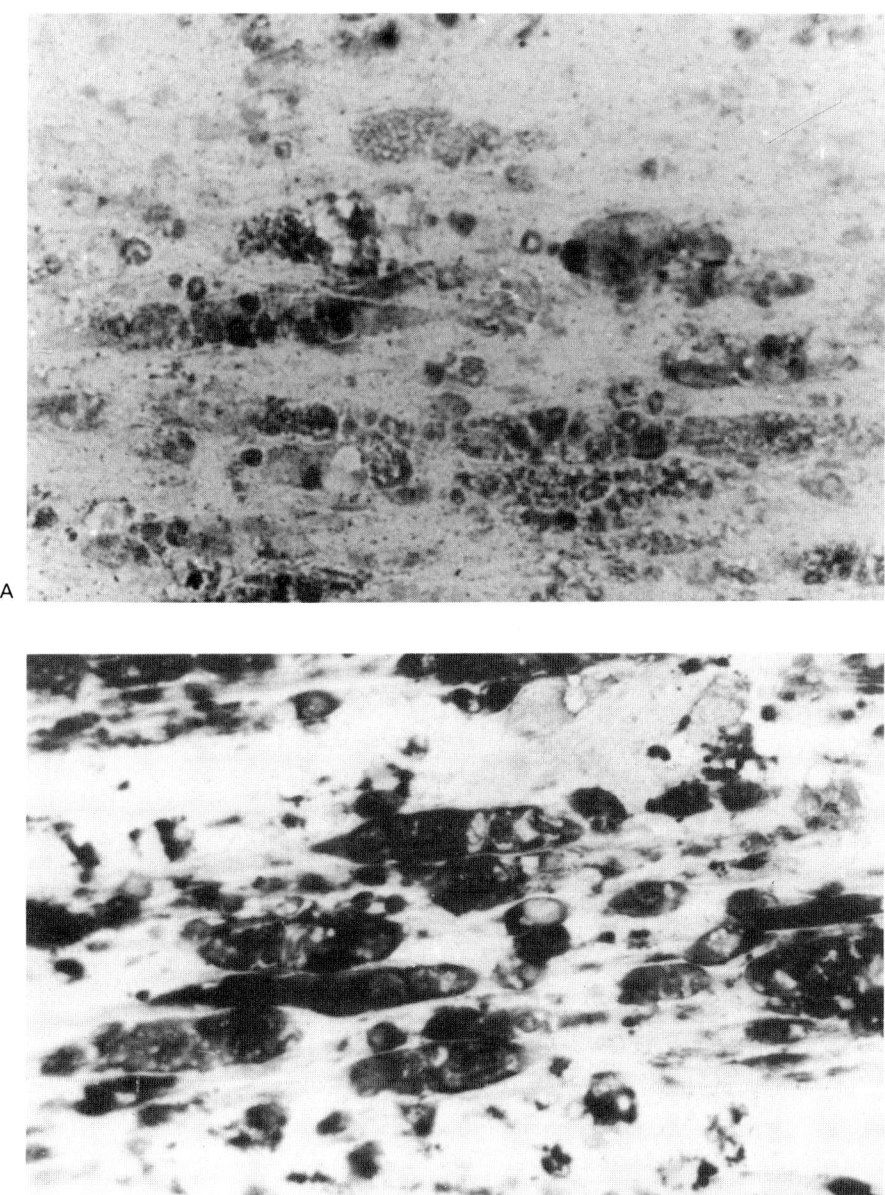

Fig. 6.2 Rat sciatic nerve undergoing Wallerian degeneration 12 days after nerve section. A. Sudan black B ×310.
B. Bromine-Sudan black B ×310

diamine technique (PASDORO) (Bayliss et al 1970).
The silver stain depicts characteristic nuclear patterns
of glia and neurons, whilst fat-laden microglia are
well displayed with oil red O in the actively
demyelinating lesion (see Fig. 6.3). Basement
membrane staining allows the identification of cells
related to capillary endothelium, but although this
may be an advantage in the brain, basement
membrane staining can cause confusion in other
tissues.

Further attempts to combine techniques for
simultaneously depicting normal and degenerating
myelin in contrasting tones suffer practical
limitations. Culling (1974) for example, advocated a

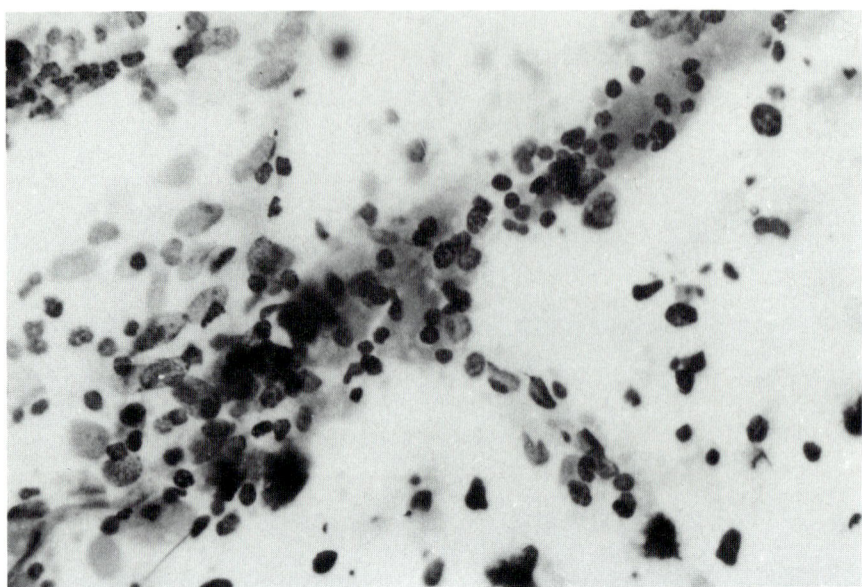

Fig. 6.3 Lipid-laden microglia (grey in picture stained with oil red O) surrounding a capillary within an actively demyelinating lesion in the cerebral cortex of a patient with multiple sclerosis. (Nuclear chromatin appears black with the periodic acid-silver diamine sequence.) Periodic acid-silver diamine-oil red O (PASDORO) ×310

luxol fast blue-oil red O sequence which ought in theory to provide the ideal combination of colours; in fact these two staining procedures are incompatible because the cholesterol esters of degenerating myelin are soluble in the 90% ethanolic luxol fast blue solution necessary for myelin staining. Attempts to stabilise these lipids by bromination or chromation proved to be fruitless.

The classical Marchi reaction depends upon the physicochemical differences between normal and degenerating myelin, but suffers practical limitations because whole tissue slices require osmication for protracted periods. This technique was successfully adapted for tissue sections in the *osmium tetroxide-α-naphthylamine (OTAN)* method by Adams (1959) who elucidated its mechanism as follows: when osmium tetroxide is applied in the presence of a highly polar electrolyte such as potassium chlorate, only the osmium salt will penetrate the *hydrophobic* esters of degenerating myelin whereby it is reduced to black osmium dioxide. Meanwhile the normal *hydrophilic* myelin sheath is accessible to both salts, but the unsaturated polar phospholipids preferentially reduce the chlorate whilst the osmium is retained in its colourless hexavalent form, and is subsequently converted to an orange chelate by α-naphthylamine. In spite of the fact that more

saturated cholesterol esters in other sites have since been shown to react unpredictably, appearing orange rather than black, the OTAN method is highly specific and sensitive when applied to the nervous system. Considering the costly and toxic nature of the reagents employed in the Marchi type of procedure, on balance the method of choice for the simultaneous demonstration of normal and degenerating myelin is therefore the *dichromate-acid haematein-oil red O* sequence. None of the above-mentioned techniques is diagnostic of a specific condition in its own right; nevertheless, the methods do usefully supplement clinical and morphological evidence in the diagnosis of demyelinating diseases.

DEGENERATIVE CHANGES INVOLVING THE NEURON

Loss of neurons may be secondary to neoplastic or cerebrovascular disturbances in the brain, but in this chapter we are concerned with primary neuronal changes. There is a heterogeneous group of disorders arising independently of toxic or metabolic influences, which range from the severely disabling dementias to the insignificant accumulation throughout life of the so-called 'wear and tear' pigment lipofuscin. Lipofuscin granules are not necessarily pathological, but gradually appear in neurons as their lysosomal lipids undergo peroxidation. Melanin in the neurons of the *substantia nigra* can be distinguished from lipofuscin by its persistent staining with Nile blue sulphate after acetone extraction. Chromatolysis, the characteristic manifestation of neuronal degeneration, is readily observed in routine histological preparations and it is in fact in the non-degenerative neurolipidoses that histochemistry finds its major role. This group of storage disorders, in which compounds accumulate as a result of defective neuronal enzymes, will be considered later (Ch. 7).

Meanwhile the strictly degenerative conditions which concern us here include the dementias and Parkinson's disease. The histochemical approach has little to offer for the other neuronal disorders such as the slow-virus infection, Creutzfeldt-Jacob disease; indeed even the dementias are usually sufficiently well-recognised histologically. Nevertheless it may be useful to point out that amyloid in the characteristic plaques (neurofibrillary tangles) of Alzheimer's disease can be demonstrated by Congo red and that the 'Lewey bodies' in Parkinson's disease contain sphingomyelin (Hartog Jager 1969) which is identified by the *NaOH-dichromate acid haematein method* (Adams & Bayliss 1963). Neurons in Pick's presenile dementia have been shown to contain ganglioside which has been demonstrated histochemically with the *modified PAS technique* (Adams & Bayliss 1963), using neuraminidase to exclude interference from another glycolipid, cerebroside (de Groot & Hartog Jager 1980).

The way in which histochemistry can assist in the identification of abnormal intraneuronal materials, which accumulate in certain neuronal disorders, is indicated in Table 6.3.

Table 6.3 Histochemistry of degenerative neuronal changes

Condition	Pathological features	Inclusions	Histochemistry
Alzheimer's disease	cerebrocortical neuronal degeneration, argyrophilic plaques 30–100 nm comprising mass of neurofibrils + central core of amyloid	neurofibrils	PAS
		amyloid	Congo red
Senile dementia	milder and less extensive than above		
Pick's disease	swollen neurons containing argyrophilic bodies	gangliosides	Modified PAS ± neuraminidase
Parkinson's disease	degeneration of pigmented (dopaminergic) neurons of *substantia nigra*	'Lewey bodies' containing sphingomyelin	NaOH-dichromate-acid haematein

REFERENCES

Adams C W M 1959 A histochemical method for the simultaneous demonstration of normal and degenerating myelin. Journal of Pathology and Bacteriology 77: 648–650
Adams C W M (ed) 1965 Neurohistochemistry. Elsevier, Amsterdam
Adams C W M, Bayliss O B 1963 Histochemical observations on the localisation and origin of sphingomyelin, cerebroside and cholesterol in normal and atherosclerotic artery. Journal of Pathology and Bacteriology 85: 113–119
Adams C W M, Bayliss O B, Ibrahim M Z M 1963 Modifications to histochemical methods for phosphoglyceride and cerebroside. Journal of Histochemistry and Cytochemistry 11: 560–561

Allen I V, McKeown S R 1979 A histological, histochemical and biochemical study of the macroscopically normal white matter in multiple sclerosis. Journal of the Neurological Sciences 41: 81–91
Baker J R 1946 The histochemical recognition of lipine. Quarterly Journal of Microscopical Science 87: 441–447
Bancroft J D, Stevens A (eds) 1982 Theory and practice of histological techniques. 2nd ed. Churchill Livingstone, Edinburgh, p 217
Bayliss O B, Adams C W M 1972 Bromine-Sudan black: a general stain for lipids including free cholesterol. Histochemical Journal 4: 505–515
Bayliss O B, Adams C W M, Hallpike J F 1970 The PASDORO method for simultaneously demonstrating DNA and lipids in brain. Histochemical Journal 2: 87–89

Bourgeois C, Hubbard B 1965 A method for the simultaneous demonstration of choline-containing phospholipids and neutral lipids in tissue sections. Journal of Histochemistry and Cytochemistry 13: 571–578

Culling C F A 1974 Handbook of histopathological and histochemical techniques, 3rd edn. Butterworths, London, p 451

deGroot P A, den Hartog Jager W A 1980 A storage product in Pick's presenile dementia. Abstracts of the VIth International Histochemistry and Cytochemistry Congress p 151

den Hartog Jager W A 1969 Sphingomyelin in Lewey inclusion bodies in Parkinson's disease. Archives of Neurology 21: 615–619

Klüver H, Barrera E 1953 A method for the combined staining of cells and fibres of the nervous system. Journal of Neuropathology and Experimental Neurology 12: 400–406

Roozemond R C 1971 The staining and chromium binding of rat brain tissue and of lipids in model systems subjected to Baker's acid haematein technique. Journal of Histochemistry and Cytochemistry 19: 244–251

7

Metabolic disorders of the central and peripheral nervous system

B. D. Lake

A group of diseases that were once thought to be degenerative in origin are now recognised as inherited disorders caused by the deficient activity of certain enzymes (usually lysosomal) involved with lipid and mucopolysaccharide metabolism. Impaired enzyme activity at a particular point in any metabolic chain results in the accumulation of metabolites at the defective stage in the pathway (Ellis 1980). This can variously involve cells of the MP (mononuclear phagocyte) series, perivascular or endothelial cells, as well as those of the nervous system, which are the primary concern of this chapter.

The most important group of metabolic disorders that affect the nervous system are the sphingolipidoses. Sphingolipids include sphingomyelins, cerebrosides, sulphatides and gangliosides, all of which are important constituents of the normal cell

(Fig. 7.1). Structurally they have in common a ceramide moiety derived from the unsaturated aminoalcohol sphingosine, in which one of the amino group hydrogen atoms is substituted by a long-chain fatty acid. Gangliosides, composed of ceramide, hexose molecules, sialic acid and hexosamine, are the lipids responsible for neuronal accumulations in Tay-Sachs disease and the other gangliosidoses. Gangliosides are additionally implicated in neuronal changes secondary to systemic disturbance of mucopolysaccharide metabolism in some types of mucopolysaccharidosis.

Other lipids that accumulate abnormally in the nervous system include cholesterol esters in Wolman's disease and phytanic acid in Refsum's disease. A further important group of disorders of lipid metabolism is Batten's disease in which

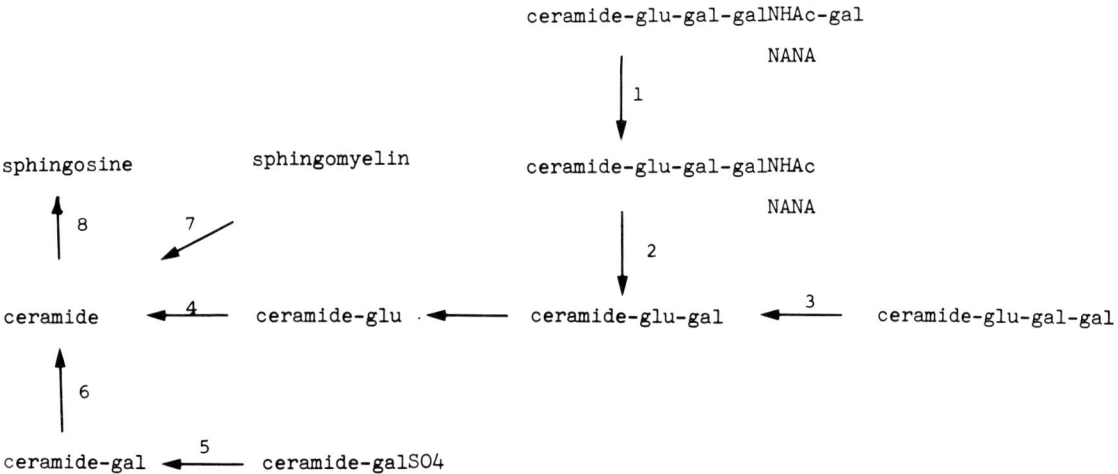

Fig. 7.1 Interrelationships between lipid components and enzyme deficiencies in the lipid storage diseases. 1. β-galactosidase; G_{M1}-gangliosidosis; 2. β-hexosaminidase A; G_{M2}-gangliosidosis (Tay-Sachs); 3. α-galactosidase; Fabry; 4. β-glucocerebrosidase; Gaucher; 5. arylsulphatase A; metachromatic leucodystrophy; 6. galactocerebrosidase; Krabbe leucodystrophy; 7. sphingomyelinase; Niemann-Pick; 8. ceramidase; Farber.

gal = galactose; glu = glucose; galNHAc = N-acetylgalactosamine; NANA = N-acetylneuraminic acid (sialic acid)

lipofuscin-like pigments are deposited in neurons and elsewhere. The major categories of the metabolic disorders involving the nervous system are presented in Table 7.1 to indicate the enzyme defect and metabolites implicated in each case, their sites of deposition, together with selected histochemical methods for their identification in the relevant tissues or body fluids indicated.

The histochemical techniques referred to in Table 7.1 include the highly specific method for demonstrating the activity of the lysosomal enzyme β-galactosidase which has only been feasible in recent years since the appropriate substrate has become commercially available. Among the histochemical methods listed are some relatively non-specific staining techniques such as toluidine blue and the periodic acid-Schiff reaction, which, with judicious use of blockades, extractions and other controls, can become selective for sulphatide and glycolipids respectively. On the other hand the dichromate-acid haematein method (Baker 1946) is specific for choline-containing lipids — lecithins and sphingomyelins — and the reaction can be confined to sphingomyelins if sections are first treated with sodium hydroxide to remove alkali-labile lecithins. The colour of the reaction may vary in relation to the length and degree of unsaturation of the fatty acid side chain, and can range from blue to brown. The perchloric acid-naphthoquinone (PAN) method for cholesterol (Adams 1961) utilises a chemical reaction that is sensitive and highly specific for cholesterol and related sterols. Similarly the calcium-lipase method for triglycerides is selective so long as the lipase preparation is pure and uncontaminated by esterases and provided that control sections are included to distinguish possible cross-reaction of free fatty acids. These are just examples of the methods that are recommended for the demonstration of tissue lipids microscopically. Details of their methodology are given in Appendix 4. For the rationale behind these techniques and their limitations, the reader is referred to Bayliss High (1982).

Before performing the forementioned histochemical procedures it may be useful to confirm that lipid is actually present in the tissue to be examined. The bromine-Sudan black B (BSB) technique is a useful preliminary screening test whereby all the lipid classes can be detected in a single, relatively simple procedure.

Bromine not only reduces the solubility of phospholipids and free fatty acids in the alcoholic staining solution but also converts crystalline-free cholesterol to its liquid bromo-derivatives, which are intensely sudanophilic.

By no means all histochemical methods can be relied upon to identify lipids in the strictly chemical sense but when histochemical observations are taken in conjunction with clinical and morphological evidence, a characteristic pattern will emerge that is diagnostic of a particular disease. In most instances enzyme assays are available and necessary for the precise diagnosis in these cases, since prenatal diagnosis is possible by culture of amniotic fluid cells and suitable enzyme assay. For this reason enzyme assay is mandatory for the diagnosis. Histochemical study can give relatively precise diagnoses but cannot distinguish between the various types of G_{M2}-gangliosidoses. Its use lies in producing a working diagnosis and in saving unnecessary assays. Consequently histochemistry remains a useful diagnostic tool in this field with the one distinct advantage over the biochemical approach — namely the ability to localise metabolites at the cellular level.

Whenever possible snap-frozen tissue, formalin-fixed tissue and wax-embedded tissue should be prepared for examination. Where a choice must be made — due to the small size of the specimen — snap-frozen tissue will provide most information. Thin-layer chromatography of solvent extracts prepared from fresh frozen tissue will effect definite identification of all lipids (Lake & Goodwin 1976) and may be used in some instances on formalin-fixed tissue.

Because the substances involved in this group of conditions are soluble in lipid solvents, routine sections are useful only for morphology. The staining reactions used for identification of the various substances must be performed on sections of snap-frozen or formalin-fixed tissues.

For general background reading, Hers & van Hoof (1973) and Stanbury et al (1978) provide clinical, biochemical and pathological descriptions of most of the disorders mentioned in this chapter.

Gangliosidoses

The ballooned neurons present in the CNS and PNS contain either ganglioside G_{M1} (generalised gan-

Table 7.1 Metabolic disorders of the nervous system

Disorder	Defect	Substances stored	Sites of storage	Tissue for diagnosis by microscopy	Staining methods and other tests
G_{M1}-gangliosidosis (two types)	β-galactosidase	G_{M1}-ganglioside, oligosaccharides, ceramide tetrahexoside	neurons, liver, spleen, kidney	blood, bone marrow, rectal biopsy, liver	PAS, thionin, β-galactosidase, TLC, EM
G_{M2}-gangliosidosis (several types viz. Tay-Sachs, Sandhoff etc)	hexosaminidase	G_{M2}-ganglioside, ceramide trihexoside	neurons	rectal biopsy	PAS, thionin, TLC
Batten's disease (ceroid lipofuscinosis) several types	unknown	lipofuscin-like substances	neurons, muscle, liver, kidney, pancreas, blood cells, sweat glands	rectal biopsy, blood	PAS, Sudan black, luxol fast blue, acid phosphatase, autofluorescence, EM
Gaucher's disease (three types)	glucocerebrosidase (β-glucosidase)	glucocerebroside	spleen, liver, brain stem nuclei (infantile form)	bone marrow, liver	PAS, acid phosphatase, TLC, EM
Krabbe's leucodystrophy	galactocerebrosidase	galactocerebroside	white matter of brain in globoid cells	brain (peripheral nerve)	PAS, acid phosphatase, nerve teasing, TLC, EM
Metachromatic leucodystrophy (sulphatidosis)	aryl sulphatase A	sulphatides	white matter of brain, peripheral nerve, kidney, gall bladder	urine, peripheral nerve (skin biopsy)	Cresyl fast violet, toluidine blue, acriflavine-DMAB, TLC
Niemann-Pick disease type A (infantile)	sphingomyelinase	sphingomyelin, cholesterol	neurons, white matter, spleen, liver, muscle, endothelia	blood, bone marrow, liver, rectal biopsy	Sudan black, acid haematein, acid phosphatase, TLC, EM
Niemann-Pick disease type B (adult)	sphingomyelinase	sphingomyelin, cholesterol	spleen, liver muscle	bone marrow, liver, rectal biopsy (to exclude neuronal involvement)	Sudan black, acid haematein, acid phosphatase, TLC, EM

Disorder	Defect	Substances stored	Sites of storage	Tissue for diagnosis by microscopy	Staining methods and other tests
Niemann-Pick Group C (Ophthalmoplegic lipidosis)	unknown	water-soluble acidic oligosaccharides of unknown composition	neurons, liver, spleen	bone marrow, rectal biopsy, liver	cell. PAS, cell. dig. PAS, acid phosphatase, thionin, TLC, EM
Fabry's disease	α-galactos.dase	ceramide trihexoside	kidney, smooth muscle, spleen	bone marrow, kidney, skin, urine	Sudan black, polarized light, EM, TLC of urine
Pompe's disease (GSD 2)	acid α-glucosidase (acid maltase)	glycogen	in lysosomes everywhere, brain stem nuclei, liver, heart spleen, muscle, endothelia	blood, muscle, heart	cell. PAS, acid phosphatase, EM
Mucopolysaccharidosis (several types)	see Chapter 18	acid mucopolysaccharides (heparan sulphate, chondroitin sulphate, keratin sulphate) several gangliosides	neurons (not mps), liver, spleen, heart	urine, blood, liver, cultured fibroblasts	toluidine blue, colloidal iron, PAS, EM
Farber's disease	ceramidase	ceramides G_{M3}-ganglioside	neurons, liver, lymph nodes	lymph nodes, liver, rectal biopsy,	Sudan black, polarized light, PAS, thionin, TLC, EM
Adrenoleucodystrophy	unknown	very long chain fatty acid esters	adrenal, white matter	adrenal, brain	EM, GLC, oil red O & extractions
Mannosidosis	α-mannosicase	oligosaccharides containing mannose	neurons, liver, spleen, endothelia	blood, bone marrow	cell. PAS, EM, TLC

Abbreviations: TLC = thin layer chromatography; EM = electron microscopy; GLC = gas-liquid chromatography; cell = celloidinized; dig = digested.

gliosidosis) or ganglioside G_{M2} (Tay-Sachs disease and variants). These two gangliosides differ from each other only by a galactose residue and it will be appreciated from Figure 7.2 that the two compounds

with Feyrter's thionin 'enclosure' method. This can be confirmed by a positive Bial reaction (copper, orcinol, HCl) which is not the most reliable of techniques and frequently results in a disintegrated

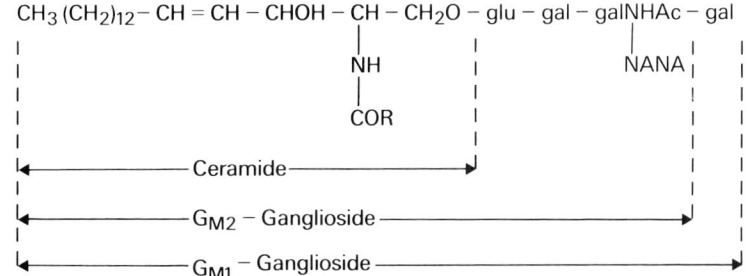

$$CH_3(CH_2)_{12} - CH = CH - CHOH - CH - CH_2O - glu - gal - galNHAc - gal$$

Fig. 7.2   The structure of gangliosides G_{M1} and G_{M2} and ceramide. glu = glucose; gal = galactose; galNHAc = N-acetylgalactosamine; NANA = N-acetylneuraminic acid (sialic acid)

will have very similar staining characteristics and that it will be virtually impossible to differentiate one from the other. Both are readily extracted during processing and formalin-fixed frozen sections, or sections of snap-frozen tissue are necessary to demonstrate neuronal ganglioside accumulation. The ganglioside in glial cells is firmly bound and resists extraction. The high hexose content of gangliosides gives rise to a strong PAS reaction unaffected by diastase digestion (Fig. 7.3). The presence of sialic acid is shown by an immediate rose metachromasia

section. Gangliosides, being particularly polar molecules, are packaged in lysosomes with cholesterol and phospholipid giving rise to the characteristic membranous cytoplasmic body (MCB) seen under the electron microscope. Stains to demonstrate cholesterol are positive and the luxol fast blue-neutral red method also indicates storage of some component (? protein) both in frozen and processed tissue. Neurons in the CNS or elsewhere (rectum) have the same appearance and staining reactions in all types of gangliosidosis. In G_{M1}-gangliosidosis neuronal,

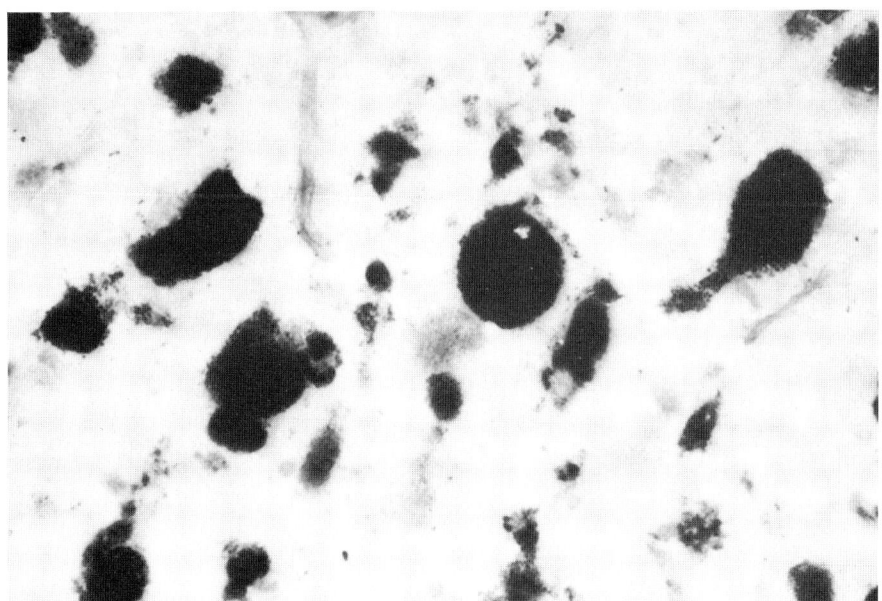

Fig. 7.3   Tay-Sachs disease; cryostat section of brain showing ballooned neurons and glial cells filled with PAS-positive ganglioside. The sections have been lightly counterstained with Carrazzi's haematoxylin. × 350

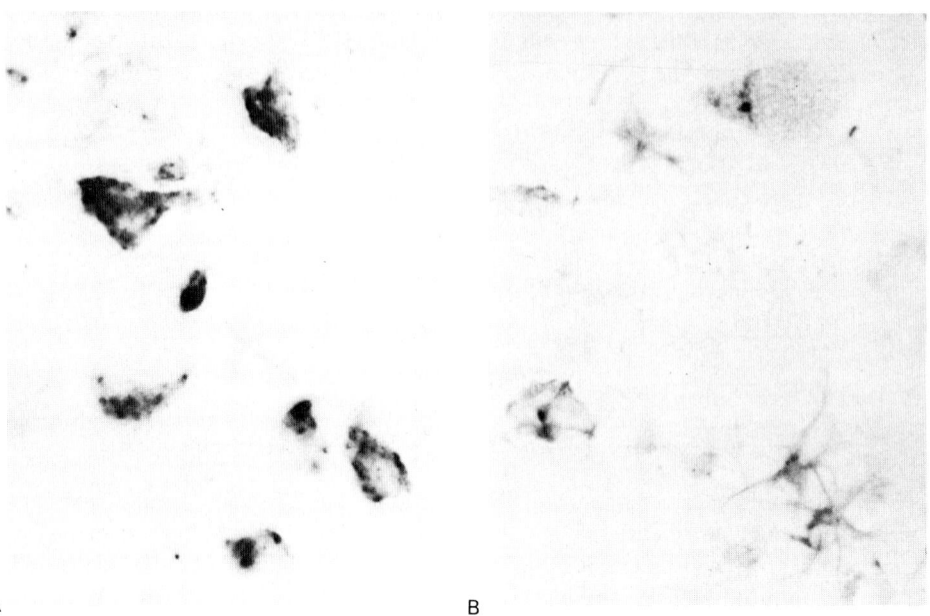

A   B

Fig. 7.4 Cryostat sections of brain stained to show β-galactosidase activity in:
A. normal cortex
B. cortex from a patient with G_{M1}-gangliosidosis.
Normal neurons show strong activity. The ballooned neurons of G_{M1}-gangliosidosis are negative, and only a weak staining is present in reactive astrocytes. The sections have been lightly counterstained with neutral red. ×570

epithelial, endothelial and most mononuclear phagocytic cells show deficient β-galactosidase activity (Fig. 7.4). Reactive astrocytes however retain some β-galactosidase activity which may be of a different isoenzyme. In the lamina propria of the gut many PAS-positive histiocytes are present (frozen or fixed tissue) which show metachromasia (frozen tissue only) with Haust and Landing's toluidine blue method. These histiocytes are not present in G_{M2}-gangliosidosis. Methods for showing hexosaminidase activity do not differentiate between isoenzymes A and B and cannot be used for diagnosis of Tay-Sachs disease. However in Sandhoff's disease no hexosaminidase activity can be detected in brain, rectum, fibroblasts and white blood cells and histochemical methods for hexosaminidase activity reliably show the deficiency. A positive control is of course necessary (Fig. 7.5).

Differentiation of the individual gangliosides can only be achieved by thin-layer chromatography of solvent extracts of fresh (frozen) tissue. Formalin fixation alters the gangliosides and thus fixed tissue is

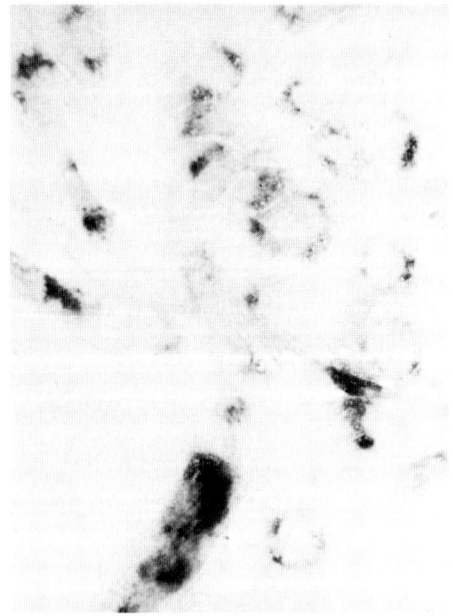

Fig. 7.5 A monolayer of cultured normal fibroblasts stained to show total hexosaminidase activity ×300

less reliable for chromatography although the gross excesses of G_{M1}- and G_{M2}-gangliosides are usually sufficient to mask the effect of slow formalin-induced changes.

Prenatal diagnosis by culture and assay of amniotic fluid cells is possible and fetal tissue from termination of affected pregnancies will show an excess of the ganglioside by TLC and electron microscopy, but routine sections and cryostat sections of snap-frozen tissue show no evidence of neuronal storage.

Metachromatic leucodystrophy

Macrophages throughout the demyelinated white matter of the brain contain deposits of galactocerebroside sulphate (sulphatide) which in frozen sections imparts metachromasia to a wide variety of dyes. Those most commonly used are toluidine blue and cresyl fast violet although thionin may also be used. Wherever sulphatides occur (brain, peripheral nerve, gall bladder, kidney) the metachromasia is yellow-brown with cresyl fast violet (Hirsch Pfeiffer method). However with the more permanent toluidine blue method (Bodian & Lake 1963) the colour varies from red in the brain, yellow red and purple in the macrophages and Schwann cells of peripheral nerves, to yellow and brown in the gall

bladder and kidney tubules (Fig. 7.6 and Plate 2). Neurons of the cortex and gastro-intestinal tract do not show storage of sulphatide, but those of the basal ganglia and dentate nucleus in particular are distended with sulphatide deposition. Sulphatides are also demonstrable with the acriflavine-DMAB method, and have an alcian blue CEC greater than 0.5M magnesium chloride. They are not retained in routinely processed tissue.

The histochemical demonstration of the sulphatase A deficiency is not a reliable procedure because of the presence of sulphatase B which has similar characteristics.

In mucosulphatidosis (multiple sulphatase deficiency), in which there is also urinary mucopolysaccharide excretion, neurons show marked evidence of gangliosidic storage identical to that in the gangliosidoses, in addition to sulphatide accumulation identical to that in metachromatic leucodystrophy.

Krabbe's leucodystrophy

The collections of globoid cells in the demyelinated white matter are PAS-positive in frozen or routine sections (Fig. 7.7). In frozen sections they show no metachromasia, are only weakly sudanophilic, but

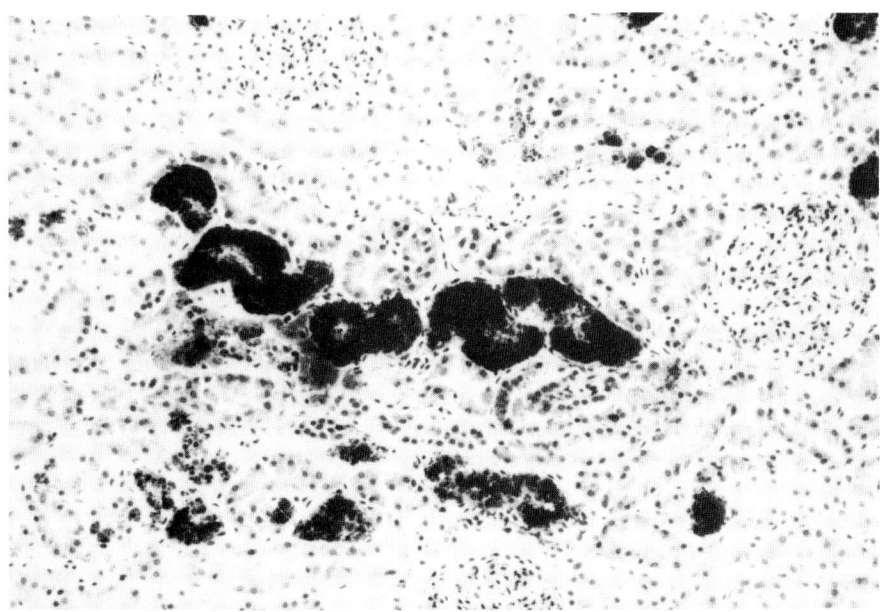

Fig. 7.6 Metachromatic leucodystrophy; cryostat section of kidney stained with toluidine blue to show metachromatic deposits of sulphatide in distal tubules. ×150

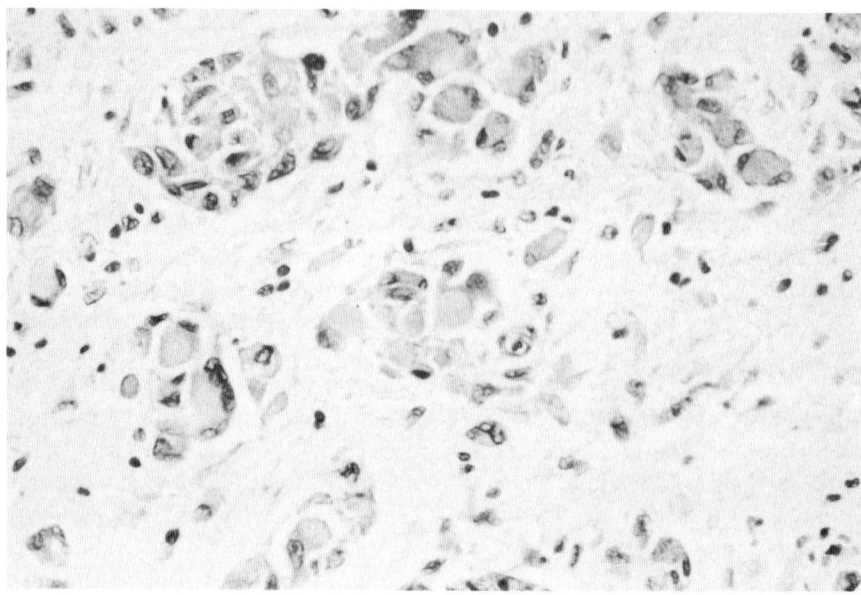

Fig. 7.7 Krabbe's leucodystrophy. Routine section of brain showing weakly PAS-positive perivascular collections of globoid cells in the demyelinated white matter. The section has been lightly counterstained with Carrazzi's haematoxylin. × 370

show strong acid phosphatase activity. No cholesterol or esters are present in the globoid cells. No globoid cells are found in the peripheral nervous system although segmental demyelination is evident. Few macrophages are ever seen in peripheral nerves but the acid phosphatase reaction in Schwann cells is strong, as in all forms of segmental demyelination (Fig. 7.8). This is evident in even the small nerves in skin and muscle and may be used as a diagnostic pointer where the enzyme assay is not available.

Although the deficient enzyme — galactocerebrosidase — is a β-galactosidase, it is

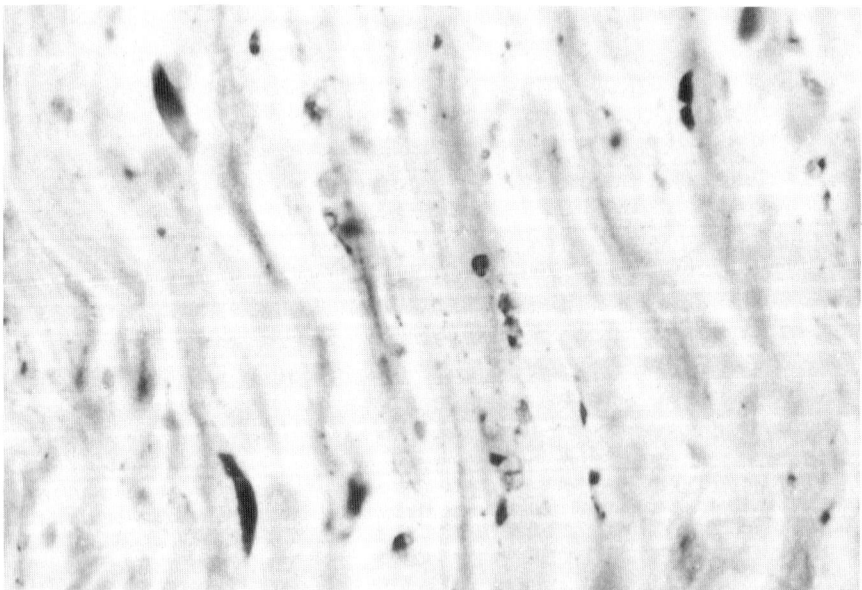

Fig. 7.8 Krabbe's leucodystrophy. Cryostat section of peripheral nerve biopsy showing acid phosphatase activity in a Schwann cell. The section has been lightly counterstained with Carrazzi's haematoxylin. × 880

substrate-specific and the histochemical substrates are not suitable to demonstrate its activity, or lack of it.

Alexander's leucodystrophy

In the younger cases of Alexander's leucodystrophy the strongly eosinophilic Rosenthal fibres, which are present in abundance in the subpial regions, perivascularly in the demyelinated white matter and scattered throughout the cortex, are strongly stained with luxol fast blue (Fig. 7.9). Rosenthal fibres are fibrillary deposits in fibrous astrocytes and are positive with antibodies to glial fibrillary acidic protein (GFAP). The older cases show fewer Rosenthal fibres which stain weakly with luxol fast blue but have the staining characteristics of fibrin. No metachromasia can be shown and neutral fat is only rarely present in macrophages.

Spongy degeneration (Canavan, von Bogaert and Bertrand)

The white matter appears spongy with multiple vacuoles which are also present perivascularly in the cortex. The vacuoles are empty and contain no demonstrable substance. Glycogen deposition may be prominent in perivascular astrocytes, particularly in the cortex.

Batten's disease

This group of disorders has a bewildering variety of names associated with it which include neuronal ceroidlipofuscinosis, Jansky, Bielschowsky, Sjögren, Spielmeyer, etc. (see Brett 1983). They are best classified on the basis of clinical presentation and age, and fall into four main groups, viz. infantile, late infantile, juvenile and adult. They share the common feature of neuronal deposition of a substance (substances) which has the staining characteristics of ceroid or lipofuscin but which is probably neither of these entities.

This material is resistant to processing, and in routine sections of brain stains with Sudan black, PAS, luxol fast blue and shows autofluorescence. The morphological changes in post mortem brain of the infantile form are characteristic with almost total loss of neurons and marked astrocytic storage. However, earlier in the course of the infantile form and in the late-infantile and juvenile forms there is little in the staining reactions of the neuronal storage material to differentiate one form from another.

Electron microscopy on the other hand shows ultrastructural characteristics which serve to differentiate the various forms. Granular osmiophilic deposits (GROD) are present in infantile Batten's

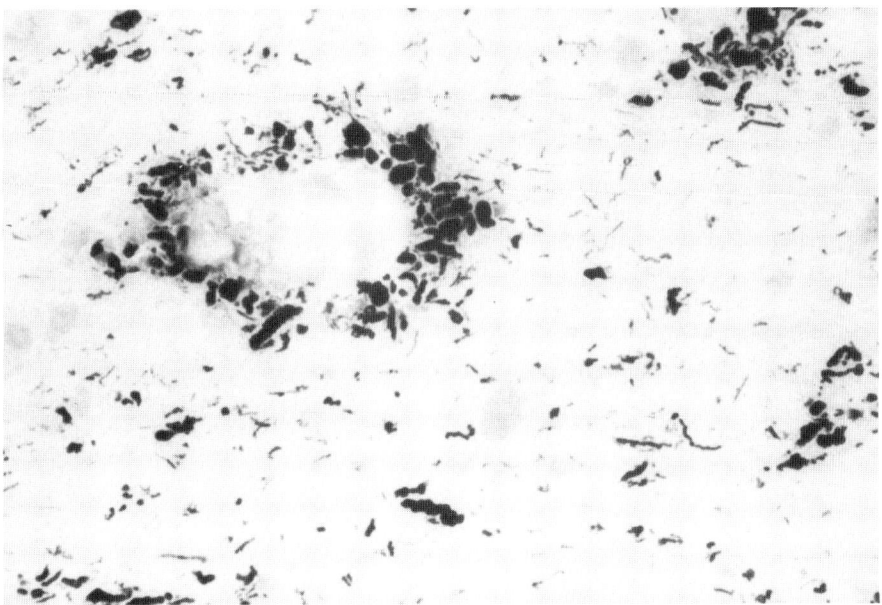

Fig. 7.9 Alexander's leucodystrophy; cryostat section of a brain biopsy stained with luxol fast blue-neutral red showing the numerous perivascular and scattered Rosenthal fibres in the demyelinated white matter. × 350

disease; curvilinear bodies accumulate in late infantile Batten's disease; and predominantly fingerprint bodies are found in the juvenile form (Lake 1981). This is true for brain, skin, lymphocytes and rectum but other sites (skeletal muscle for example), although accumulating storage substance, may show different ultrastructure.

sometimes included under the Niemann-Pick label (as Groups C and D) do not have deficienct sphingomyelinase activity and show no consistent storage of sphingomyelin: these have been referred to as ophthalmoplegic lipidosis.

There is neuronal storage only in type A, type B being entirely visceral. The general appearance of the

Table 7.2 Rectal biopsy: neuronal staining patterns in Batten's disease

	Infantile type	Late infantile type	Juvenile type
PAS	+ +	±	+ +
Sudan black	+ +	+	+ +
Luxol fast blue	–	+	+ +
Ultrastructural	granular osmiophilic deposits	curvilinear bodies	finger-print bodies
Autofluorescence (excitation 360 nm barrier 410 nm)	+ +	+	+ +

Similar depositis are found in smooth muscle cells and endothelia. Histiocytes, staining strongly for acid phosphatase activity, are present among the smooth muscle cells in the juvenile type

The substance is stored in neurons everywhere and diagnosis of the disorder and its type can be made by examination of cryostat sections of snap-frozen biopsies of rectum. In this site the neuronal staining reactions vary from one type to another, and these together with evidence of smooth muscle cell storage and the presence or absence of histiocytes among smooth muscle cells make the differential diagnosis of the different types possible (Figs. 7.10, 7.11 and 7.12). Table 7.2 illustrates the patterns of staining.

The disease is systemic and deposits of the ceroid-lipofuscin like material may be found in liver, pancreatic acinar cells, skeletal muscle, kidney tubules, sweat glands and lymphocytes. The staining reactions vary slightly but the most reliable indicator in all sites, except lymphocytes, is autofluorescence of the stored material.

Niemann-Pick disease

As noted in Chapter 18 there are essentially only two types of Niemann-Pick disease. In types A and B there is a deficiency of sphingomyelinase activity and storage of sphingomyelin. Other conditions

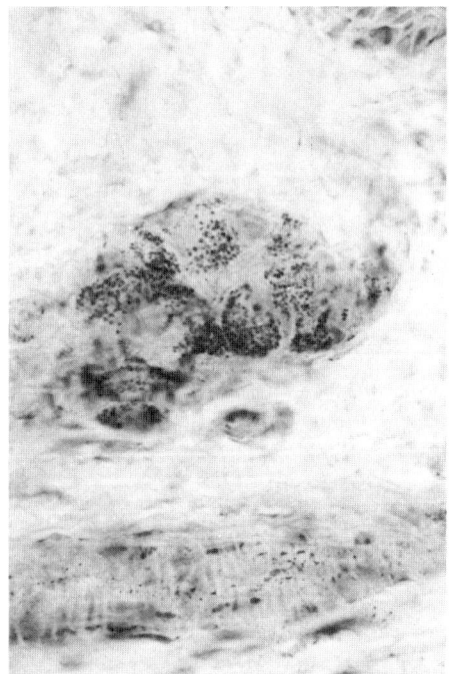

Fig. 7.10 Juvenile Batten's disease; cryostat section of a rectal biopsy stained with Sudan black (and carmalum) showing granular sudanophilic deposits in submucosal neurons and in smooth muscle cells of an arteriole. × 425

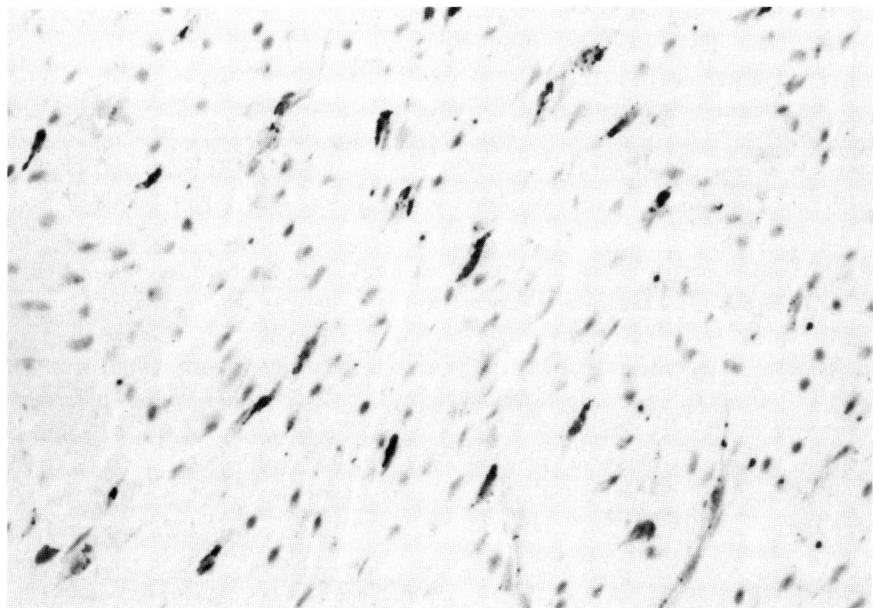

Fig. 7.11 Juvenile Batten's disease; cryostat section of a rectal biopsy stained to show acid phosphatase activity (Gomori). Many small strongly-staining histiocytes are present among the smooth muscle cells of the circular muscle layer. These histiocytes are not present in the late infantile or infantile forms of Batten's disease. The section has been lightly counterstained. × 425

Fig. 7.12 Late infantile Batten's disease; cryostat section of a rectal biopsy stained with Sudan black (and carmalum) showing paranuclear sudanophilic deposits in smooth muscle cells of the circular muscle layer. × 600

neurons is that of the classic ballooned storage cell usually associated with Tay-Sachs disease. Sphingomyelin is stored together with cholesterol and a variable amount of gangliosides, within the neurons and within the numerous histiocytes in the demyelinated white matter. These substances are also found within histiocytes in the spleen, lymph nodes, thymus, lamina propria of the gut and within smooth muscle cells and endothelia.

The presence of gangliosides gives rise to red metachromasia with Feyrter's thionin and a positive PAS reaction. The Schultz and PAN methods for cholesterol are also positive. Sphingomyelin in the CNS is stained blue with the acid haematein reaction, and resists alkaline hydrolysis. Elsewhere the blue colour may vary and in some sites the reaction may be yellow-brown and this may reflect differing degrees of unsaturation of the fatty acid side chains. Sphingomyelin is stained weakly by Sudan black, which imparts a red birefringence in polarized light.

All the affected cells stain for acid phosphatase activity, the reaction product outlining the numerous storage vacuoles in each cell.

Niemann-Pick Groups C & D (Ophthalmoplegic lipidosis)

Although the bone marrow cells in this condition have some similarities to those of Niemann-Pick disease types A & B, and to many other disorders, it represents a separate entity which has so far eluded biochemical identification. Part of the confused classification of this disorder, placing it with Niemann-Pick disease, arose because Crocker and Farber included some cases which are now clearly within this group in their monograph on Niemann-Pick disease (Crocker & Farber 1958).

Within the brain there is morphological evidence of neuronal storage which may be as marked as in classical Tay-Sachs, or may be minimal. Routine sections show no evidence of storage substance, which appears to be a water-soluble oligosaccharide with acidic residues. Formalin-fixed frozen sections have also lost the substance, which can only be demonstrated in protected cryostat sections. In celloidinised sections stained with the PAS reaction the stored material is seen within neurons and in axonal swellings. The cells display strong acid phosphatase activity and exhibit metachromasia with Feyrter's thionin. Luxol fast blue and Sudan black stains are negative and there is no autofluorescence except for the occasional lipofuscin wear-and-tear pigment granules. There is no evidence of demyelination. Neurons throughout the gastro-intestinal tract show storage and a diagnosis can be made on a suction rectal biopsy (Fig. 7.13). Serial

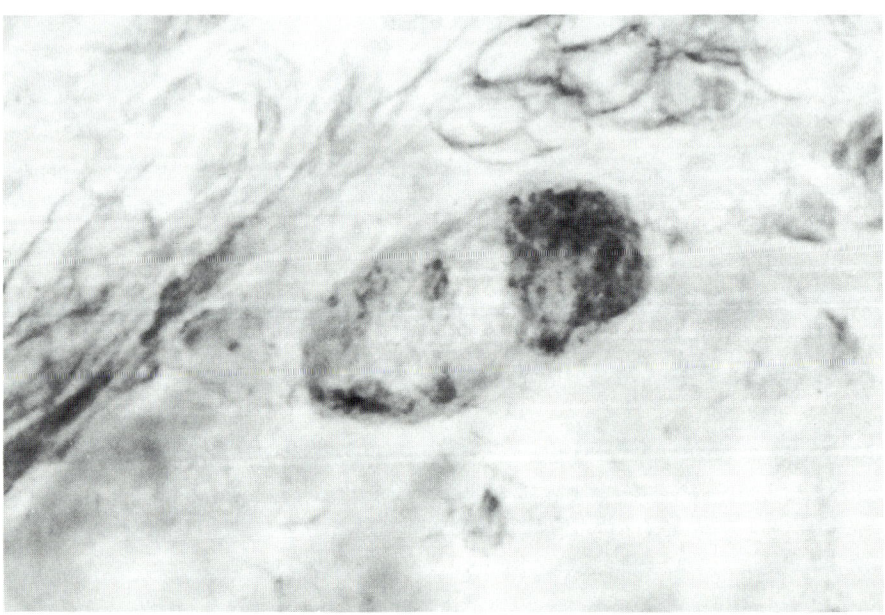

Fig. 7.13 Niemann-Pick Group C (Ophthalmoplegic lipidosis); cryostat section of a rectal biopsy stained with the protected PAS method showing storage in submucosal neurons. This substance is extracted by aqueous media. × 1000

sections are stained by several methods. The neurons show strong acid phosphatase activity, are positive with the protected PAS method (cell PAS) and show metachromasia with Feyrter's thionin method. No staining is observed with Sudan black, luxol fast blue or in sections fixed in formol-calcium prior to staining with PAS. Smooth muscle cells are not involved, and endothelial cells are only positive for acid phosphatase activity in the cases presenting with neurological symptoms in the infantile to late-infantile age group. Foamy histiocytes are present in the submucosa in all groups.

It is vital to establish neuronal involvement because most of the patients present with only visceral symptoms which precede the onset of neurological deterioration by several years. Conjunctival biopsies have been used in diagnosis by study of the ultrastructure of the various cell populations present, but it is my firm belief that in this condition particularly, evidence of neuronal storage — or at least clear signs of neurological deterioration — is essential before the diagnosis can be established.

Gaucher's disease

There is no neuronal storage demonstrable by light microscopy in neurons of cerebral cortex or of the gastro-intestinal tract. Neurons of basal ganglia show some PAS positivity indicative of storage of glucocerebroside only in the infantile form.

Occasional perivascular Gaucher cells may be found, and these have the same staining characteristics as those found in the bone marrow, and are readily detected by an acid phosphatase reaction (Fig. 7.14).

Pompe's disease (GSD 2)

The accumulation of intralysosomal glycogen, best shown in cryostat sections by the celloidin-PAS method, is not in evidence in the neurons of the cerebral cortex, but may be seen in the basal ganglia, brain stem and spinal cord (Figs. 7.15 and 7.16). Neurons of the gastro-intestinal tract show storage throughout. In the cerebral cortex storage of glycogen can be demonstrated in astrocytes. Elsewhere there is gross accumulation of glycogen in the liver, monocyte-phagocyte series of cells of the liver and spleen and in smooth muscle cells everywhere. Endothelial cells also exhibit glycogen storage. Skeletal muscle and heart muscle cells store glycogen in such vast quantities that it may be difficult to recognise the tissue. The glycogen deposition is associated with acid phosphatase activity and there may also be a substance which has the characteristics of an acid mucosubstance demonstrable with alcian blue or toluidine blue. This may be an unusual and so far unrecognised mucopolysaccharide or a glycogen-phosphate compound. The diagnosis is made from

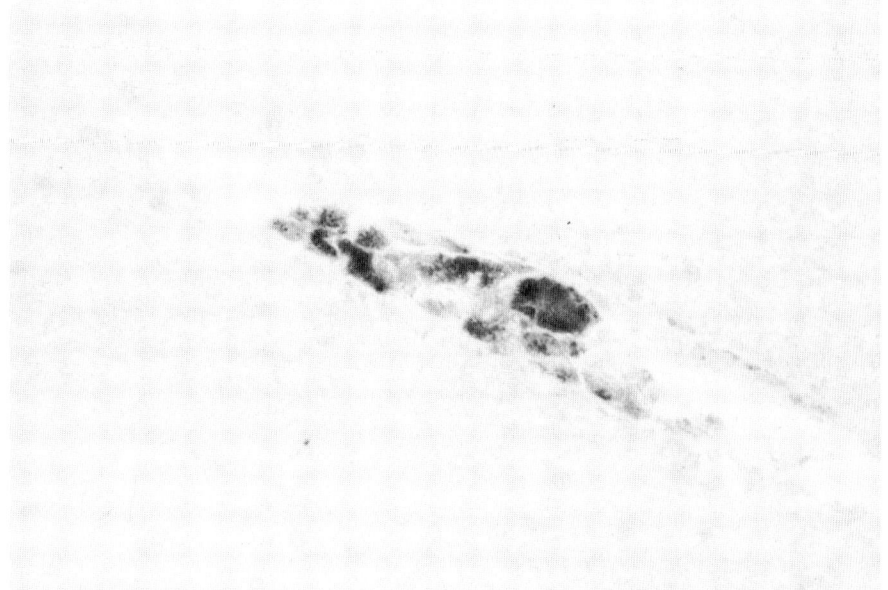

Fig. 7.14 Infantile Gaucher's disease; cryostat section of brain showing a small collection of perivascular Gaucher cells in the white matter. Stained for acid phosphatase activity (Gomori); counterstained with Carrazzi's haematoxylin. × 350

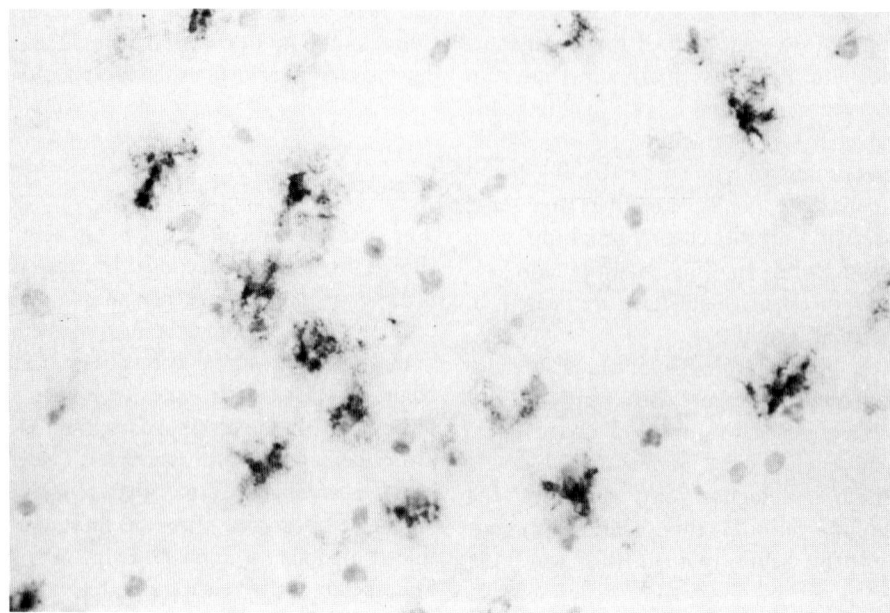

Fig. 7.15 Pompe's disease; cryostat section of brain showing glycogen accumulation in astrocytes in the cortex. No neuronal deposition is found in the cortex. Stained by the protected PAS method with Carrazzi's haematoxylin counterstain. × 450

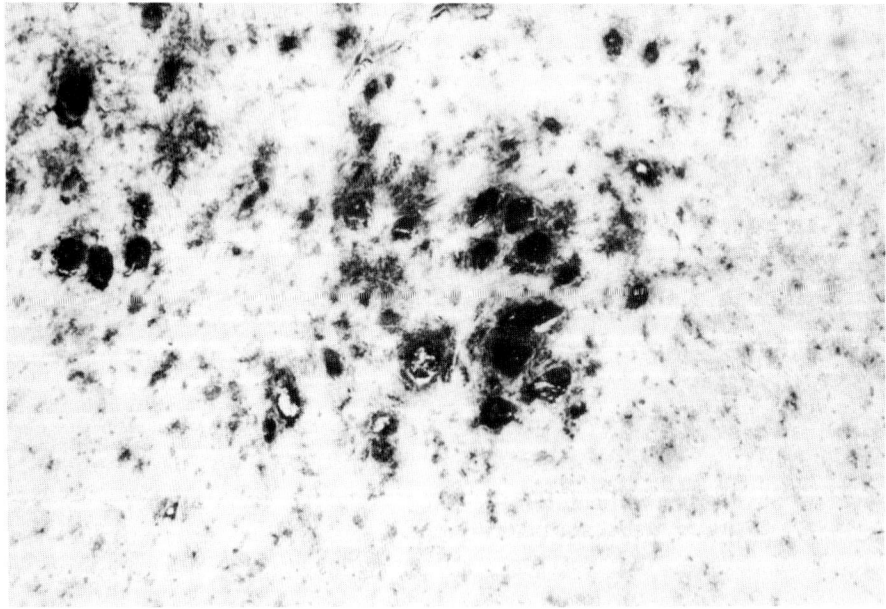

Fig. 7.16 Pompe's disease; cryostat section of spinal cord showing massive glycogen deposition in anterior horn cells. × 90 Stained by the protected PAS method.

blood films in which glycogen is present in vacuolated lymphocytes (Ch. 18).

Aspartylglucosaminuria

Although neuronal storage may be apparent to routine light microscopical examination of the cortex and nuclei of the basal ganglia, only a few granules of a lipofuscin-like material can be demonstrated in the neuronal perikarya. The vacuolar contents do not stain with any of the usual methods.

Fucosidosis

Widespread neuronal ballooning with storage of a soluble oligosaccharide is evident. Fucose-containing compounds are widespread and deposits are found also in histiocytes in spleen, liver, lymph nodes and in the lung where granuloma-like nodules are prominent. All epithelia is affected. With adequate protection to prevent dissolution of the oligo-saccharides the contents of the vacuoles should be PAS-positive. Since the condition is caused by the deficiency of a lysosomal enzyme, α-fucosidase, an

acid phosphatase reaction should indicate those cells affected. The histochemical method for the demonstration of α-fucosidase activity, with 1-naphthyl-α-fucoside as substrate and HPR as coupler, may be helpful but has not been tested in known cases.

Mucolipidoses (ML)

The mucolipidoses show neuronal storage of a ceroid-lipofuscin-like material which can appear similar to that in Batten's disease. The clinical presentations with corneal clouding (ML4), Hurler-like features (ML2, ML3) are however quite different from Batten's disease. In ML2 (I-cell disease) not much evidence of storage is seen in brain or liver but the heart shows deposition of a PAS-positive substance and strong acid phosphatase activity is present in the myofibres (Fig. 7.17). These changes in the heart are less florid than those of Pompe's disease where the myofibres are markedly distended with glycogen deposition, appearing as though the whole heart were a rhabdomyoma.

In ML1 type 1 (sialidosis, cherry-red spot

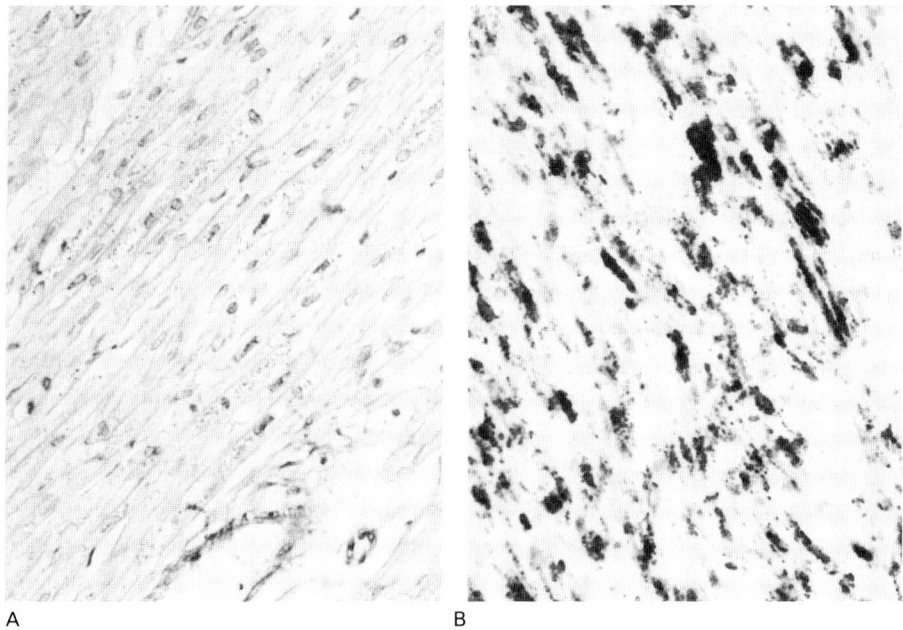

A   B

Fig 7.17 I-cell disease (mucolipidosis 2); heart: A. routine section stained with PAS, showing lipofuscin-like granules in vacuolated muscle cells; B. cryostat section stained for acid phosphatase activity (Gomori) showing strong activity in muscle cells which would normally show no activity. ×290

myoclonus syndrome) there is neuronal storage of a gangliosidic substance giving an appearance and staining reactions identical to those of Tay-Sachs disease. The gross neuronal storage may be found in young patients (by suction rectal biopsy) many years before the onset of dementia.

Adrenoleucodystrophy

This X-linked condition presents in the juvenile age range as a leucodystrophy with adrenal insufficiency (Powers & Schaumberg 1974, Schaumberg et al 1975). The cerebral cortex is normal, but the white matter shows changes of demyelination with sparing of the subcortical fibres. Foci of perivascular inflammatory cells are common. Numerous lipid-laden macrophages may be found throughout the demyelinated white matter. Adrenal atrophy is found in all cases and the cells in the zona reticularis and inner fasciculata are ballooned and contain cytoplasmic striations. The zona glomerulosa is not apparently involved. The striated inclusions, which are also present in the macrophages of the brain, peripheral nerve, testis and many other sites, are birefringent in frozen sections and represent very long chain fatty acid esters of cholesterol (Johnson et al 1976). These esters do not stain with oil red O and are not extracted with acetone, ethanol or methanol. They are soluble in n-hexane, xylene and chloroform. To demonstrate their presence three serial sections are required, stained and extracted as shown in Table 7.3. A neonatal, autosomal recessive form has also been described.

Mucopolysaccharidosis

In those types of mucopolysaccharidosis which have dementia, mental retardation and other signs of central nervous system involvement, the neurons of the CNS and PNS show marked storage resembling Tay-Sachs disease. The neuronal storage is of a mixture of gangliosides and no mucopolysaccharide can be detected. The neurons, in frozen or cyrostat sections, stain positively with PAS, Sudan black and luxol fast blue. Deposition of acid mucopoly-saccharide is found in perivascular regions, particularly in the white matter, and can only be demonstrated in cryostat sections of snap-frozen tissue. The method of choice is the toluidine blue method of Haust & Landing (1961). Macroscopic evidence may be found in brain slices where the perivascular accumulation of mucopolysaccharide has been extracted leaving holes or pits in the white matter (Crome & Stern 1972).

The lamina propria of the gut contains numerous macrophages filled with soluble mucopolysaccharide which is best detected with the Haust & Landing method in cryostat sections. The neurons of the myenteric and submucosal plexuses contain a mixture of gangliosides and stain as the neurons in the brain.

Mannosidosis

Very few reports of the neuropathology of mannosidosis have appeared. Those that have, record the widespread neuronal ballooning, which in routine sections appears very similar to Tay-Sachs disease. In addition there is glial and endothelial storage. The mannosides deposited are very water-soluble and

Table 7.3   Demonstration of esters in adrenoleucodystrophy

Section no	Treatment	Result
1	Stain with oil red O	Neutral fat stained Long-chain esters birefringent
2	Extract with acetone Stain with oil red O	Neutral fat extracted Long-chain esters birefringent
3	Extract with acetone Extract with n-hexane Stain with oil red O	Neutral fat extracted Long-chain esters extracted No birefringence

none remain in routine sections. Protected sections of snap-frozen tissue stained with PAS should show the stored material which will have no other staining properties.

Neuroaxonal dystrophy

This disorder, delineated by Cowen & Olmstead in 1963, is readily diagnosed at post mortem when the axonal spheroids are prominent in the brain stem. In a biopsy of frontal cortex the spheroids (aggregations of microtubules and microfilaments) are not usually visible by light microscopy and the diagnosis has to be made by electron microscopy. The axonal spheroids are present in peripheral nerves and it has been suggested that ultrastructural examination of the nerves in a conjunctival biopsy will make the diagnosis. Although some cases have been diagnosed by this means the consistency of the procedure has not yet been proved. Recently Elleder and Jirasec (personal communication) have shown that the spheroids exhibit strong non-specific esterase activity in a brain biopsy and I have been able to confirm this in two further cases. Short incubation (10 minutes) of cryostat sections shows the spheroids present particularly in the molecular layer of the cortex and also scattered throughout the cortex. Neurons show much weaker activity.

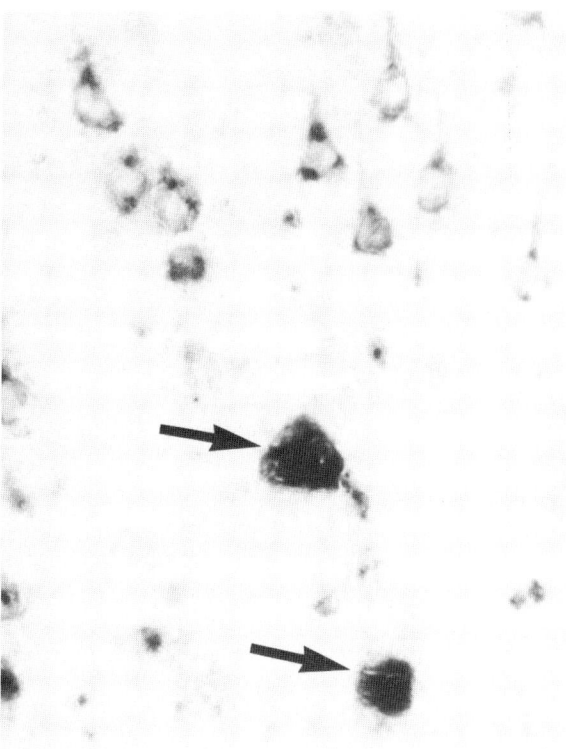

Fig. 7.18 Neuroaxonal dystrophy; cryostat section of a brain biopsy stained for non-specific esterase activity, showing strongly staining axonal spheroids (arrows) with their characteristic paracrystalline clefts. Neurons show much less activity. ×900

REFERENCES

Adams C W M 1961 A perchloric acid naphthoquinone method for the histochemical localization of cholesterol. Nature (London) 192: 331

Baker J R 1946 The histochemical recognition of lipine. Quarterly Journal of Microscopical Science 87: 441–447

Bayliss High O 1982 Lipids. In: Bancroft J D, Stevens A (eds) Theory and practice of histological techniques. Churchill Livingstone, Edinburgh, ch 12

Bodian M, Lake B D 1963 The rectal approach to neuropathology. British Journal of Surgery 50: 702–714

Brett E M 1983 Paediatric neurology. Churchill Livingstone, Edinburgh

Cowen D, Olmstead E V 1963 Infantile neuroaxonal dystrophy. Journal of Neuropathology and Experimental Neurology 22: 175–236

Crocker A C, Farber S 1958 Niemann-Pick disease: a review of 18 patients. Medicine 37: 1–95

Crome L, Stern J 1972 Pathology of mental retardation, 2nd edn. Churchill Livingstone, Edinburgh, p 322

Ellis R (ed) 1980 Inborn errors of metabolism. Croom Helm, London

Haust M D, Landing B H 1961 Histochemical studies in Hurler's disease. A new method for localization of acid mucopolysaccharide and an analysis of lead acetate 'fixation'. Journal of Histochemistry and Cytochemistry 9: 79–86

Hers H G, van Hoof F (eds) 1973 Lysosomes and storage diseases. Academic Press, New York

Johnson A B, Schaumberg H H, Powers J M 1976 Histochemical characteristics of the striated inclusions of adrenoleukodystrophy. Journal of Histochemistry and Cytochemistry 24: 725–730

Lake B D 1981 Metabolic disorders. General considerations. In: Berry C L (ed) Paediatric pathology. Springer-Verlag, Berlin, ch 14

Lake B D, Goodwin H J 1976 Lipids. In: Smith I, Seakins J W T (eds) Chromatographic and electrophoretic techniques, 4th edn. Heinemann, London, vol 1, ch 14

Powers J M, Schaumberg H H 1974 Adrenoleukodystrophy (sex-linked Schilder's disease). American Journal of Pathology 76: 481–500

Schaumberg H H, Powers J M, Raine C S, Suzuki K, Richardson E P 1975 Adrenoleukodystrophy. A clinical and pathological study of 17 cases. Archives of Neurology 33: 577–591

Stanbury J B, Wyngaarden J B, Fredrickson D S (eds) 1978 The metabolic basis of inherited disease, 4th edn. McGraw-Hill, New York

8

Histochemistry of the tumours of the nervous system

P. L. Lantos

Tumours of the nervous system present an extremely varied and somewhat confusing histological spectrum, for they arise from an organ of great cellular complexity and from related structures of considerable diversity. It is for this reason that the new classification by the World Health Organisation (Zülch 1979) is likely to become a useful guide not only for neuropathologists, but also for neurosurgeons, neurologists, neuroradiologists and general pathologists.

The striking variety of neural tumours, posing considerable diagnostic problems, calls for additional investigations by electron microscopy, tissue culture, cytogenetics and histochemistry. The purpose of this chapter is to describe briefly the histology of the commonest tumours, indicating special histological stains relevant to diagnosis, to assess the contribution of histochemistry to the understanding of the biology of neural neoplasms and, at a more practical level, to review histochemical techniques of diagnostic importance.

Histochemistry bridges the gap between morphology and biochemistry. Although neoplasia cannot be recognised by histochemical methods alone, tumours may reveal enzymatic features characteristic of transformed, neoplastic tissues. The deviation from the enzyme pattern of the normal parent tissue may be minimal, particularly in well-differentiated growths; these are the cases in which enzyme histochemistry may help to identify the origin of neoplastic cells. The aim of enzyme histochemistry in tumour diagnosis is therefore threefold: to establish the neoplastic nature of the lesion, to secure the diagnosis and to estimate the degree of malignancy. As a result of the introduction of new, reliable and reproducible techniques in the 1950s and 1960s, various research groups studied the

enzyme patterns of the tumours of the nervous system using both conventional histochemical techniques and quantitative measurements (Wollemann 1974). Although a large body of knowledge has accumulated, enzyme histochemistry has not fulfilled early expectations of becoming a widely and routinely used diagnostic tool.

With the advent of electron histochemistry it was realised that certain enzymes are localised in particular cell organelles: this observation has led to the concept of marker enzymes (Novikoff & Essner 1962). Thus acid phosphatase is a marker for lysosomes, thiamine pyrophosphatase for the Golgi complex and adenosine triphosphatase for the cell membrane (Figs. 8.1–8.3). These enzymes are hydrolases and can be demonstrated by the lead precipitate techniques of Gomori (1939); lead, the reaction product, being an electron-dense heavy metal is an ideal 'stain' for fine-structural studies. The electron microscope offers several advantages over the light microscope in the study of these and many other enzymes. With the low resolution of the light microscope precise localisation of enzyme activity is impossible; consequently the distribution of the reaction product is described in vague terms such as 'intracytoplasmic' or 'perinuclear'. Furthermore, the accuracy of identification of cells with the light microscope is impaired by the reaction product which may partly or completely obscure the cells. With the electron microscope, in contrast, the reaction product can be seen with precision as a well-circumscribed deposit restricted to recognisable cell organelles. Moreover, the relation of the site of enzyme activity to other cell organelles also becomes apparent (Lantos 1974).

The fine-structural localisation of these enzymes has contributed to the understanding of cellular

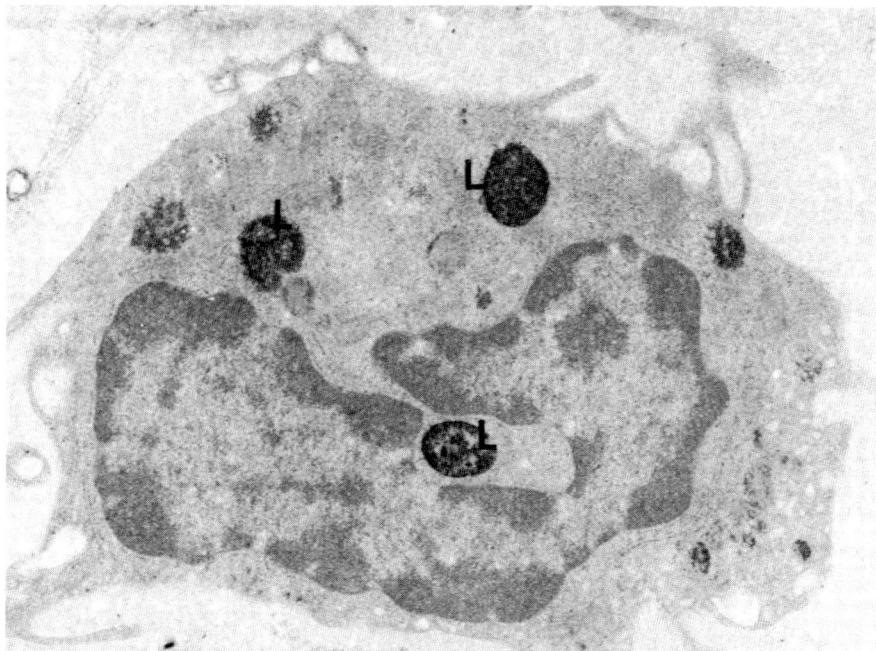

Fig. 8.1 Lysosomes (L) containing acid phosphatase in a macrophage from an experimental glioma. ×15 750.
The tumours were induced transplacentally by 30 mg/kg of ethylnitrosourea injected i.v. into pregnant BD-IX rats on the 15th day
of gestation. Reproduced by courtesy of Springer-Verlag, Heidelberg

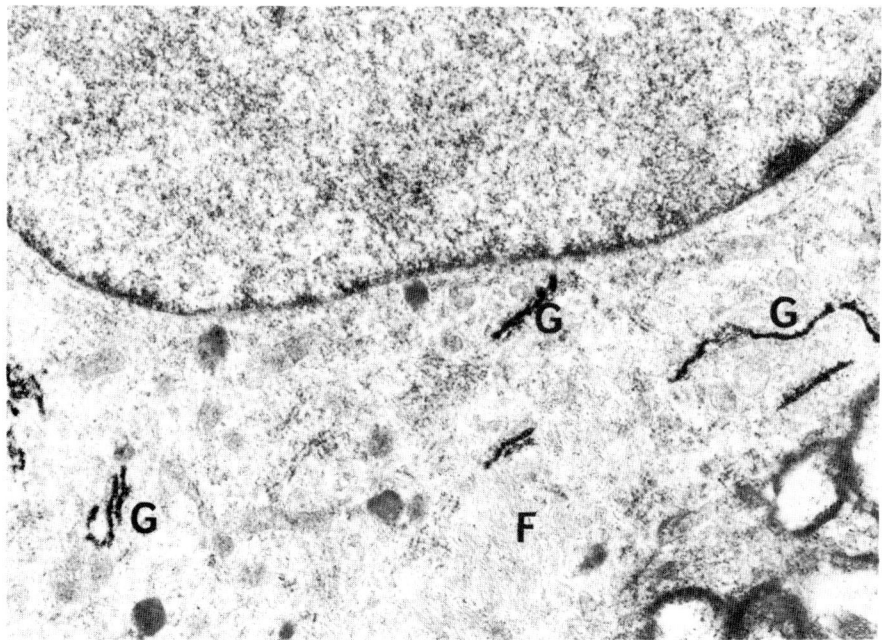

Fig. 8.2 An astrocyte of an experimental glioma reveals thiamine pyrophosphatase activity in the Golgi complexes (G). Many astrocytic
filaments (F) are also present. ×20 000. The tumours were induced transplacentally by 30 mg/kg of ethylnitrosourea injected i.v. into
pregnant BD-IX rats on the 15th day of gestation. Reproduced by courtesy of Springer-Verlag, Heidelberg

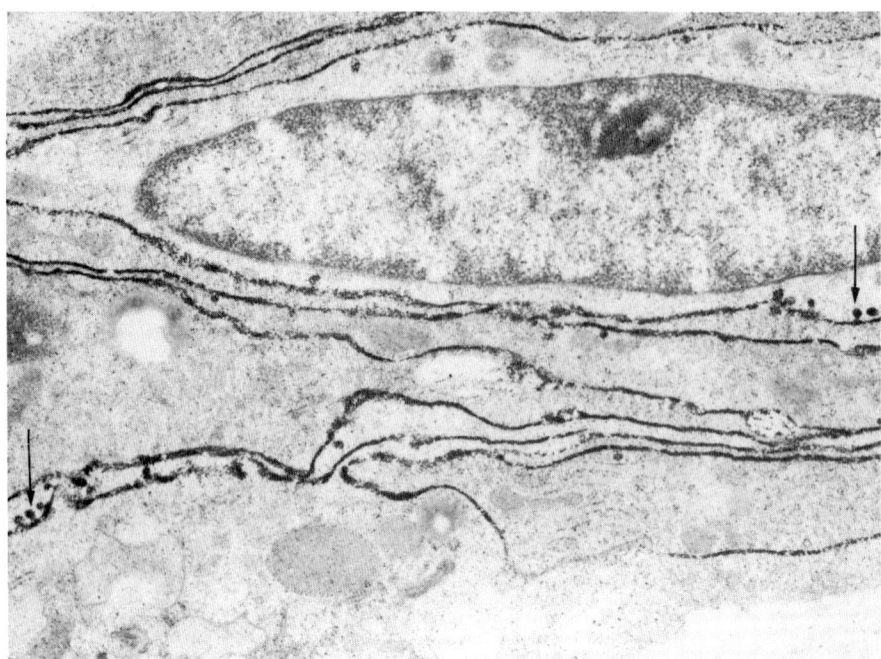

Fig. 8.3 Cell membranes and pinocytotic vesicles (arrows) in the cells of an experimental schwannoma displaying adenosine triphosphatase activity. ×16 500. The tumours were induced transplacentally by 30 mg/kg of ethylnitrosourea injected i.v. into pregnant BD-IX rats on the 15th day of gestation. Reproduced by courtesy of Springer-Verlag, Heidelberg

differentiation and anaplastic changes in tumours of the nervous system, but the technique, requiring special facilities and expertise, has remained beyond the reach of most diagnostic laboratories.

The latest and undoubtedly most important advance in the diagnosis of neural neoplasms is the application of immunohistochemical techniques. Immunohistochemistry affords superior specificity and sensitivity over conventional histological stains and can be adapted to both electron microscopy and tissue culture. It does not require sophisticated equipment or complicated procedures; moreover aldehyde-fixed tissues embedded in paraffin wax and resin as well as cryostat sections and smears can all be examined. However, extreme care must be taken to ensure that the reaction is indeed specific by using purified antigens, absorbed and pre-immunisation sera and a set of controls, both positive and negative. When demonstrating nervous system-specific antigens, control tissues from other organs should also be examined in order to test site specificity (Eng & Bigbee 1978).

Specific tissue antigens can be demonstrated by immunofluorescence and immunoperoxidase; the latter method has advantages over the former in that the reaction product, being electron-dense, can be directly visualised in the electron microscope. Furthermore, it is permanent, allowing convenient, extended and repeated examinations. Of the immunoperoxidase techniques the peroxidase-antiperoxidase and the indirect techniques are used in this laboratory and yield satisfactory results.

The number of antigens specific to the nervous system has been recently reviewed (Eng & Bigbee 1978). It is beyond doubt that in the diagnosis of neoplasms, those antigens which are specific to a particular cell type are the most valuable; with these markers different cell types can be identified and counted even when they form a small proportion of the tumours. Thus neuron-specific enolase and anti-neurofilament antibodies for neurons, glial fibrillary acidic protein (GFAP) and glutamine synthetase (GS) for astrocytes, and galactocerebroside and carbonic anhydrase for oligodendrocytes can best be employed for the identification of tumours originating from the respective cells.

Tissues for examination are obtained either at autopsy or biopsy. Post-mortem material is often of limited value for histochemical studies, for enzyme activity is decreased and autolytic changes prevent precise localisation. In contrast, some antigens, e.g. GFAP, are preserved in sufficient quantities to yield convincingly positive reactions in post-mortem material.

Tumour specimens are removed by needle or open biopsy. The former specimen, because of its small size (which frequently does not exceed 1 cm in length and 0.1 cm in diameter), may not provide adequate material for various examinations. In these cases the biopsy can be divided into two: one portion embedded in paraffin wax for light microscopy to be stained by conventional methods and immuno-histochemical techniques, the other processed in resin for electron microscopy and cytochemistry. With the increasing sophistication of microtomes and adaptation of staining methods, small biopsies can now be entirely embedded in resin, stained with conventional stains for light microscopy and then sectioned for electron microscopy.

Open biopsies usually provide satisfactory material for all the required examinations. Moreover, larger specimens allow topographical orientation: the centre and periphery of tumours, areas of haemorrhage and necrosis as well as invasion and destruction of normal structures can be discerned. Tissue for enzyme histochemistry should be taken from both the centre and periphery: there is different enzyme activity in various parts of the neoplasm reflecting significant regional metabolic differences.

Adequate fixation of tissue and maximal preservation of enzyme activity and antigenicity are conflicting aims which cannot be achieved simultaneously. The fixation time necessary for satisfactory preservation of ultrastructure differs from that required for optimal retention of enzymes and antigens. However, with well-timed fixation and careful selection of fixatives a balance can be established to retain as much activity with as little fixation artefact as possible.

The tissue processed and embedded in paraffin wax for light microscopy is normally fixed in 10% formol-saline. The addition of a phosphate buffer, however, will improve fixation for immunohistochemistry and retain sufficient antigen to yield positive reaction. Immersion fixation, being inferior to perfusion, can result in uneven preservation and leakage of antigen; these, in turn, may cause background or false-positive staining. Tissues for enzyme histochemistry require short fixation depending, of course, upon the enzyme to be demonstrated. A mixture of formaldehyde and glutaraldehyde in cacodylate buffer, one-half strength modified Karnovsky fixative (Karnovsky 1965) has become our fixative of choice for electron microscopy, the length of fixation being 4–5 h. For cytochemistry, the ultrastructural localisation of enzymes, the same fixative can be used, with fixation time reduced to 30 min. A fixative composed of 2% paraformaldehyde and 0.1% glutaraldehyde in cacodylate or phosphate buffer may be the best overall compromise for small biopsies. This solution provides satisfactory preservation of ultrastructure, while retaining demonstrable enzyme and antigen activity.

GLIOMAS

Astrocytomas

These glial neoplasms form a histologically complex group: fibrillary, protoplasmic, gemistocytic, pilocytic and anaplastic astrocytomas can be distinguished.

Fibrillary astrocytomas are composed of stellate or bipolar cells with some pleomorphism, nuclear hyperchromasia and processes filled with astrocytic fibrils. Protoplasmic astrocytomas, in pure form, are rare and the neoplastic cells contain few or no fibrils. The cytoplasm frequently undergoes degeneration and the resulting vacuoles, becoming confluent, form microcysts. The cells of gemistocytic astrocytomas are large and polygonal and the abundant, homogeneously eosinophilic cytoplasm bears a relatively small, eccentric nucleus. Pilocytic astrocytomas are composed of attenuated, bipolar fibrillary cells with fusiform nuclei; the configuration of these cells is influenced by the nerve tracts in which they proliferate. In addition loose-structured tissue, formed mainly by stellate cells and microcystic areas, contributes to the biphasic pattern of these tumours. The anaplastic astrocytomas reveal striking cellular pleomorphism, high mitotic index and changes associated with these malignant tumours: florid endothelial hyperplasia, necrosis and haemorrhage.

Electron microscopy resolves the astrocytic fibres which are composed of filaments of 10 nm; the presence of these cell organelles is the most important diagnostic feature. They can be demonstrated in well-differentiated astrocytomas by Cajal's gold chloride sublimate in frozen sections, but this method is of limited value in anaplastic astrocytomas, as is Holzer's method for fibrillary astrocytes, for it stains fibres of various types rather indiscriminately. The most frequently applied and the most important special stain for neuroglial tumours however is Mallory's phosphotungstic acid haematoxylin (PTAH).

Oligodendrogliomas

Oligodendrogliomas are composed of uniform cells with round or oval nuclei in a clear, empty-looking cytoplasm surrounded by well-defined membranes; these cytological features impart a honeycomb appearance to the tumour. A thin, delicate stroma, in which small blood vessels divide at acute angles, separates groups of tumour cells. Calcification, both in the parenchyma and in the vascular wall, testifies, together with the low mitotic rate, to the slow progression of this neoplasm. Occasionally a considerable astrocytic component warrants the diagnosis of mixed oligo-astrocytoma. Less differentiated, malignant oligodendrogliomas, with cellular pleomorphism, high mitotic rate and endothelial proliferation, also exist. Haematoxylin and eosin (H & E) staining secures the diagnosis in most cases. Mucoid degeneration of neoplastic cells and extracellular mucin can best be demonstrated by the periodic acid-Schiff (PAS) technique or by mucicarmine, while the vascular stroma becomes more prominent in reticulin stains.

Ependymomas

Neoplastic ependymal cells frequently form rosettes around a central lumen or surround blood vessels (pseudorosettes). Even without these diagnostic hallmarks, the neoplastic cells tend to be arranged in groups and cords in the fibrillary stroma. Cellular pleomorphism and mitotic activity are not prominent features. Papillary ependymomas, the myxopapillary tumours in the filum terminale, subependymomas and the occasional truly malignant tumours are now well-recognised variants.

The characteristic rosettes of ependymomas are easily recognisable on H & E stain and should be distinguished from the rosette of Homer Wright found in poorly differentiated neuroectodermal neoplasms (medulloblastomas and neuroblastomas). PTAH will display 'blepharoplasts' under a high-power lens: these tiny, rod-shaped structures are in fact the basal bodies of cilia, the vestiges of the apical, motile organelle of mature ependymal cells. Mucin in neoplastic cells and in the stroma of myxopapillary ependymomas can best be seen in PAS and mucicarmine stains.

Glioblastomas

Glioblastomas, being highly malignant, primitive glial tumours, are composed of undifferentiated fusiform, round or pleomorphic cells, alone or in various combinations resulting in striking cellular pleomorphism, including multinucleate giant cells. While in some tumours there is no evidence of glial differentiation, in others astrocytic or more rarely oligodendrocytic and ependymal areas can be distinguished. Endothelial proliferation, necrosis with pseudopalisading of tumour cells and haemorrhage stamp their mark on the histology. PTAH demonstrates glial fibres when the astrocytic nature of cells is in doubt, although glioblasts may be so undifferentiated as to be devoid of these organelles. Exuberant capillary proliferation, occasionally amounting to a sarcomatous component, is best seen in reticulin stains.

Enzyme histochemistry

The distribution of alkaline and acid phosphatase, aminopeptidase and non-specific esterase in primary tumours of the nervous system, studied by Gluszcz (1963), did not indicate a clear relationship between the grades of malignancy and enzyme activity. Strong activity of alkaline phosphatase and aminopeptidase was found in the mesodermal components of neuroectodermal tumours: proliferating blood vessels in gliomas displayed increased activity. Alkaline phosphatase showed stronger reaction in endothelial cells, while aminopeptidase was more abundant in the other layers of arterial walls, particularly in those with hyperplastic adventitia. In oedematous areas there was a striking decrease in vascular alkaline phosphatase activity suggesting that this enzyme could be used as a marker for the structural integrity of blood vessels. The role of aminopeptidase has remained more controversial: increased activity in viable tumour tissue around necrotic foci was regarded as evidence of proteolytic activity, a function important in the spread of neoplasms. Acid phosphatase and non-specific esterase reactions were strongest at the margin of necrotic or degenerating foci. Their activity did not seem to be related to the degree of malignancy, but reflected catabolic processes which were more frequent and prominent in malignant tumours.

Demonstrating hydrolytic enzymes in human brain tumours, Schiffer et al (1972) suggested that diffuse staining for β-glucuronidase and acid phosphatase reflected 'progressive' changes of protein synthesis, while lysosomal activity represented regressive and catabolic processes.

Of the hydrolytic enzymes, β-glucuronidase is related to malignancy and this positive correlation has been confirmed by quantitative measurements. Moreover, it has emerged from a recent biochemical study that the ratio of N-acetyl-β-glucosaminidase to β-glucuronidase is higher for primary than for secondary neoplasms. This ratio therefore could be of diagnostic importance when histology cannot unequivocally establish whether a brain tumour is primary or secondary (Ramsey et al 1980).

Oxidoreductase enzymes were studied by various research groups (Chason et al 1963, O'Connor & Laws 1963, Viale & Andreussi 1965, Schiffer & Fabiani 1972) and the results were somewhat contradictory. Although the histochemical study of these enzymes has provided useful information on the metabolic activity of tumours of the nervous system, they are of limited importance in diagnostic problems. Viale & Andreussi (1965) observed that the activity of various oxidative enzymes in general was influenced by the degree of cellular differentiation, by disordered structural patterns in areas of rapid growth and by reactive or regressive processes. The undifferentiated cells of medulloblastomas displayed very low enzyme activity, while astrocytic and neuronal differentiation was accompanied by a more intense reaction, activities reaching their highest levels in well-differentiated astrocytomas and nerve cell tumours. β-Hydroxybutyrate dehydrogenase activity particularly closely followed cellular maturation and was therefore considered as an index of differentiation. However, other authors were not

able to obtain positive reaction for this enzyme in any of the gliomas examined (Chason et al 1963). Astrocytomas, on the whole, showed stronger reactions than oligodendrocytomas which, however, contained high levels of glucose-6-phosphate dehydrogenase and 6-phosphogluconate dehydrogenase, two enzymes present in large amounts in normal oligodendrocytes. Neoplastic ependymal cells showed enzyme patterns similar to normal ependymal cells. Staining for various enzymes was weaker in areas of active proliferation within gliomas. Reactive astrocytes displayed strongly positive reactions, while areas of necrosis were devoid of reaction product. Newly formed capillaries, the result of endothelial hyperplasia in malignant gliomas, were well endowed with oxidative enzymes.

O'Connor & Laws (1963) found a consistently low activity of cytochrome oxidase and succinate dehydrogenase in accordance with most histochemical and biochemical investigations. Lactate dehydrogenase activity appeared to be related to malignancy, the strongest reactions being observed in glioblastomas. In a combined biochemical and histopathological analysis the value of lactate dehydrogenase isoenzymes was assessed in the rapid diagnosis of brain tumours. The study of isoenzyme pattern may be helpful in differentiating between neoplastic and gliotic lesions, between primary and secondary neoplasms and in differentiating various types of glioma (McCormick & Allen 1976).

Immunohistochemistry

GFAP, a water-soluble protein, has been isolated from old plaques of multiple sclerosis rich in fibrillary astrocytes. It has become an important marker for normal, reactive and neoplastic astrocytes. Immunohistochemical localisation of GFAP in formalin-fixed, paraffin wax-embedded material is now widely used in the diagnosis of glial neoplasms. By employing the peroxidase-antiperoxidase technique of Sternberger et al (1970), positive reaction was obtained in the neoplastic cells of astrocytomas, astroblastomas, ependymomas and subependymomas; in the astrocytic elements of mixed gliomas, gangliogliomas and gliosarcomas, and in the more differentiated components of glioblastomas and medulloblastomas. Oligodendrogliomas, meningiomas, sarcomas and secondary deposits gave consistently negative results. GFAP can therefore be of considerable value in the following diagnostic problems: the demonstration of the astrocytic nature of components in primitive and highly anaplastic tumours, e.g. glioblastomas and medulloblastomas; the study of mixed tumours in the central nervous system, e.g. mixed gliomas, gangliogliomas, and gliosarcomas; the identification of the glial nature of neoplasms with sparse or absent fibril formations, e.g. ependymomas and astroblastomas; the diagnosis of gliomas invading, or metastasising into, the leptomeninges and extraneural sites, and in the differential diagnosis of nonglial tumours which may resemble astrocytomas (Eng & Rubinstein 1978).

In gliomas, the intensity of immunostaining decreased with increasing anaplasia; this feature may be helpful in assessing malignancy. Neoplastic cells of ependymomas showed variations in staining intensity, being strongest in perivascular and rosette-forming cells and this positive reaction suggested that filaments in ependymal cells could share immunological features with astrocytic filaments

(Velasco et al 1980). GFAP reveals the glial nature of tumours which otherwise could not be diagnosed; being occasionally positive when PTAH does not demonstrate glial fibres. It is likely that GFAP stains subunits or precursors of astrocytic filaments; while PTAH gives positive reaction only with the fibrillary end product. Moreover, PTAH is not a specific stain of astrocytic fibres: it stains myelin sheaths, smooth muscle fibres and extracellular proteins including fibrins (Deck et al 1978).

Glutamine synthetase (GS), a cytoplasmic enzyme which plays an important role in the detoxification of ammonia and in the metabolism of glutamate, a putative excitatory neurotransmitter, has recently been described as being an astrocyte-specific marker in the central nervous system of the rat (Norenberg 1979). GS was therefore studied in a range of human brain tumours and compared with results obtained by staining for GFAP (Figs. 8.4–8.7). All astrocytomas showed positive staining with GS; the intensity of reaction was related to the degree of differentiation and to the amount of cytoplasm of the neoplastic

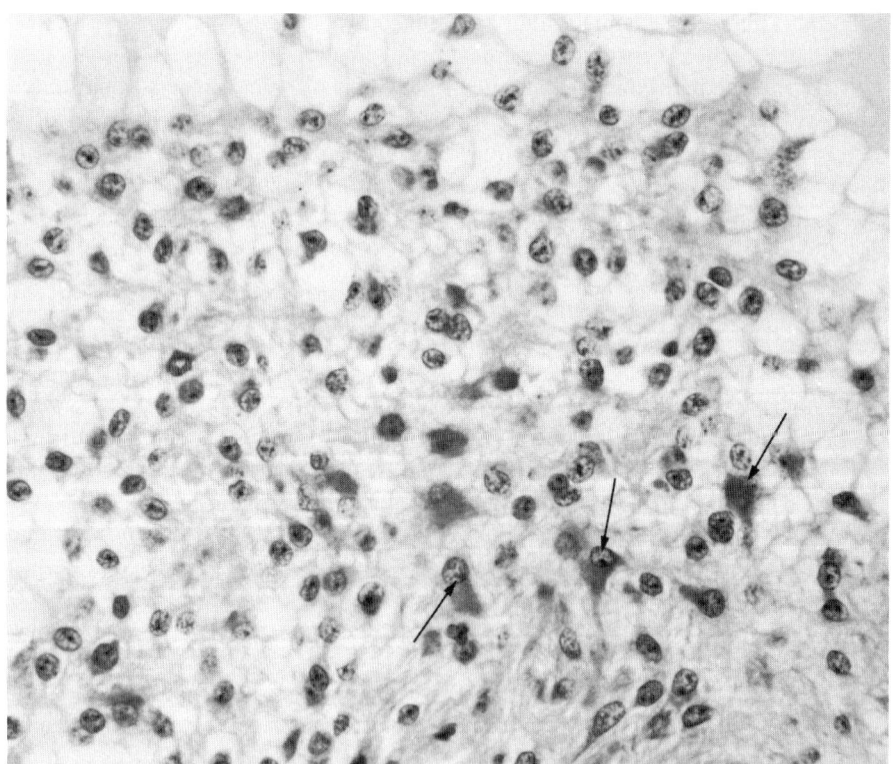

Fig. 8.4 A few cells (arrows) of a cerebellar astrocytoma show glutamine synthetase activity. ×550. (indirect immunoperoxidase method) Reproduced by courtesy of Blackwell Scientific Publications, Oxford.

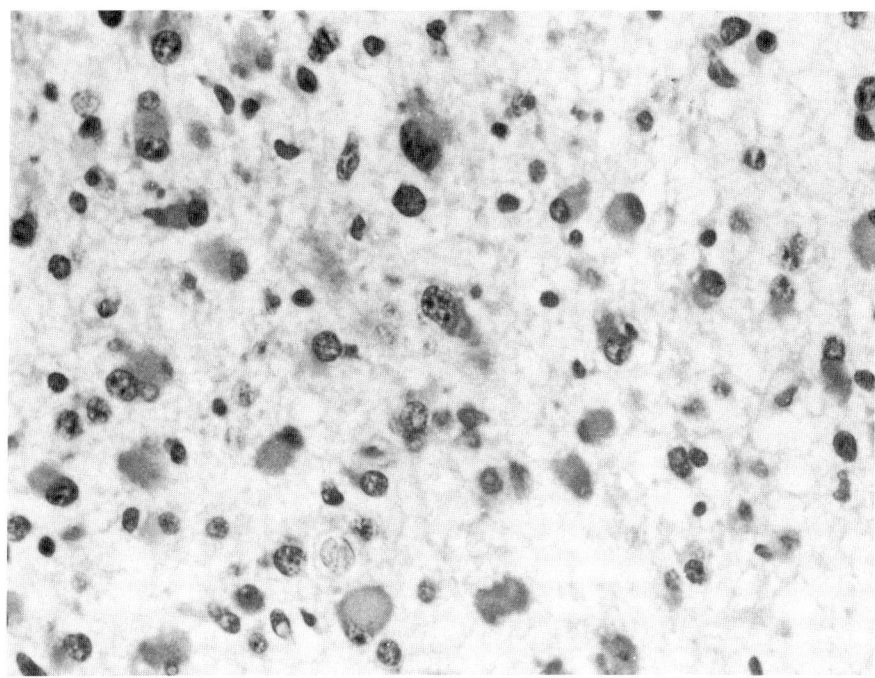

Fig. 8.5   There is strong glutamine synthetase activity in the gemistocytes of an astrocytoma. ×450. (indirect immunoperoxidase method) Reproduced by courtesy of Blackwell Scientific Publications, Oxford.

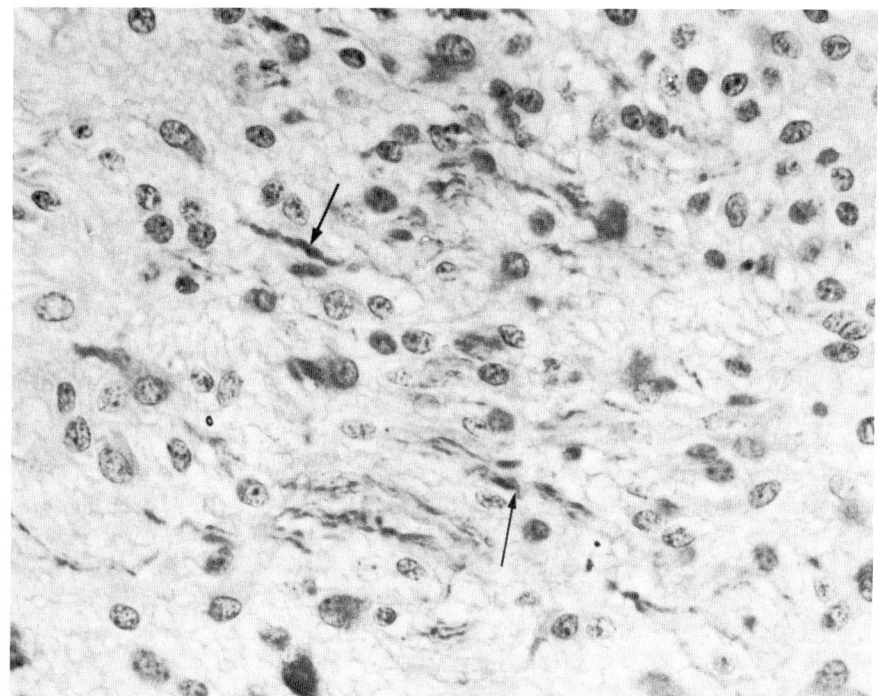

Fig. 8.6   GFAP positive processes (arrows) are seen in the cerebellar astrocytoma demonstrated in Figure 8.4. ×490. (PAP method)

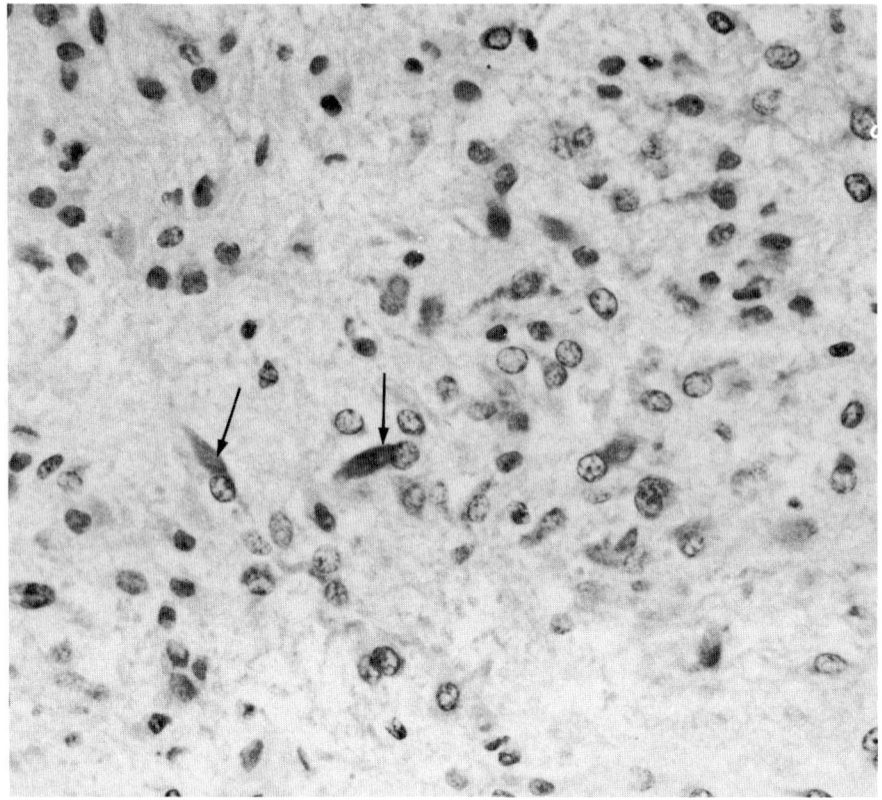

Fig. 8.7   Occasional cells (arrows) are stained for GFAP in a predominantly ependymomatous glioma. (PAP method)

cells. Ependymomas were only weakly positive, although more intense staining was detected around blood vessels. Occasional groups of positively stained cells in medulloblastomas betrayed signs of astrocytic differentiation. Oligodendrocytes and meningiomas were invariably negative. GS and GFAP staining corresponded in virtually all cases; however GS demonstrated poorly-fibrillated cells better than GFAP, but did not mark cell processes so clearly. GS thus offers a practical alternative to GFAP as an astrocyte-specific marker in human cerebral tumours and may be of diagnostic value in cases where poorly differentiated astrocytomas have been found negative with conventional stains for glial fibres or they demonstrate a weak positive reaction to GFAP (Pilkington & Lantos 1981, 1982).

Galactocerebroside and carbonic anhydrase are markers specific for oligodendrocytes. Since oligodendrocytomas are well-differentiated tumours in most cases, the diagnosis can be secured without resorting to histochemistry. In mixed gliomas, chiefly

oligo-astrocytomas, and in malignant, anaplastic oligodendroglial tumours it may be important to demonstrate the oligodendroglial component or to prove the oligodendrocytic origin of anaplastic cells. It is in these cases that the use of oligodendroglial markers may prove to be of value.

MEDULLOBLASTOMAS AND NEURONAL TUMOURS

Medulloblastomas

Although grouped with other undifferentiated and embryonal tumours in the new classification (Zülch 1979), these malignant growths can be more conveniently described with neoplasms of neuronal origin. Medulloblastomas are composed of uniform, small, immature cells with round or oval nuclei surrounded by scanty cytoplasm. The primitive tumour cells, occasionally forming Homer Wright rosettes, are thought to be bipotential: neuronal and glial differentiation has been observed. The desmoplastic variety involves the leptomeninges; the cells are embedded in a fibrous stroma which can best be appreciated in van Gieson and reticulin stains. Neuronal differentiation can be demonstrated by cresyl violet for the Nissl substance in neuronal perikarya and silver impregnation techniques (e.g. Glees and Marsland, Palmgren or Bielschowsky) for nerve fibres.

Neuroblastomas

These malignant tumours rarely occur in the central nervous system, but develop in the sympathetic nerves and adrenals. The peripheral tumours are composed of small undifferentiated cells with round or oval, hyperchromatic nuclei and sparse, ill-defined cytoplasm. The cells have a tendency to form Homer Wright rosettes and to grow neurofibrillary processes. Neuroblasts metabolise, store and secrete catecholamines, and their breakdown products, including vanilmandelic acid and homovanillic acid, are to be found in the urine and serum. The quantitative assessment of these compounds is important in the diagnosis of tumours and in monitoring the efficacy of various therapeutic regimens. Electron microscopy reveals the secretory granules to be dense-cored vesicles of 100 nm diameter.

Gangliocytomas and gangliogliomas

Gangliocytomas or ganglioneuromas are rare, slowly growing, benign tumours. The constituent cells in pure forms are mature, but abnormal neurons, including binucleate and bizarre cells irregularly orientated with anomalous processes also occur, while ganglioneuroblastomas contain a spectrum of nerve cells from undifferentiated precursors to fully mature ganglion cells. The presence of primitive neuroblasts is responsible for the malignancy and poorer prognosis of the latter form. Gangliogliomas are mixed neuronal and glial, predominantly astrocytic, tumours. This latter component has the potential for malignant transformation.

Enzyme histochemistry

Lolova et al (1972) compared the histochemistry of the classical and desmoplastic types of medul-

loblastoma. Demonstrating a wide variety of oxidoreductases and hydrolases, no significant difference was found in the intensity, character and distribution of the reactions. The differentiating neuronal and glial cells, however, showed stronger activity and could therefore be easily discerned among the poorly stained undifferentiated cells.

Acid phosphatase and non-specific esterase gave a slightly positive, finely granular even reaction in medulloblastomas, while in ganglioneuromas the large differentiated cells showed the highest activity, the reaction being weak or even negative in immature cells (Lolova et al 1971a, b).

Immunohistochemistry

Neuron-specific enolase, being localised in neurons, has been proposed as a specific marker for these cells (Schmechel et al 1978) and other cells of the diffuse neuroendocrine system (see Ch. 27). It can therefore be used in the diagnosis of tumours of neuronal origin (neuroblastomas and gangliocytomas), with nerve cell components (gangliogliomas) or those of possible neuronal differentiation (medulloblastomas).

Immunohistochemical localisation of neurofilaments, by using antisera to subunits of these cell organelles, can also contribute to the diagnosis of nerve cell tumours. 'Primitive neuroepithelial' tumours of infancy showed positive reactions as did some cells in medulloblastomas and gangliogliomas (Roessmann et al 1981).

In the diagnosis of neuroblastomas the demonstration of dopamine-β-hydroxylase (DBH) and catecholamines is of considerable importance. DBH can be localised both by immunofluorescence on cryostat sections or by the peroxidase-antiperoxidase technique, this latter allowing ultrastructural studies. Catecholamines can be demonstrated by formaldehyde-condensed fluorescence, glyoxylic acid or by a mixture of formaldehyde and glyoxylic acid (Eng & Bigbee 1978).

SCHWANNOMAS (NEURILEMMOMAS) AND NEUROFIBROMAS

Schwannomas are benign tumours, developing in cranial nerves, spinal roots, and peripheral nerves.

The neoplastic Schwann cells are either fusiform, forming interlacing bundles with palisading in the solid areas of Antoni A pattern, or more pleomorphic, distributed haphazardly in a degenerate, loose matrix of the Antoni B type tissue. Electron microscopy shows the neoplastic cells being surrounded by basement membranes, a feature which is exploited in reticulin stains.

Neurofibromas, occurring as single lesions or part of von Recklinghausen's disease, are benign growths composed of fibroblasts, Schwann cells and abnormal nerve fibres. The presence of these latter structures, demonstrated by silver impregnation, is an important feature in the differential diagnosis from schwannomas in which nerve fibres, although present, are less readily encountered.

Enzyme histochemistry

Acid phosphatase and non-specific esterase are unevenly distributed in schwannomas. Cells in the loose, Antoni B-type tissue possessed more acid phosphatase than cells in the compact bundles of Antoni A type areas (Lolova et al 1971a, b). Some blood vessels exhibited slight alkaline phosphatase and aminopeptidase activity, while the neoplastic cells were devoid of these enzymes. Oxidative enzymes, however, showed a different pattern, the more intensive reaction being localised in the compact areas of Antoni A-type tissue (Schiffer & Fabiani 1972).

Immunohistochemistry

It has been shown that S-100, a protein unique to the nervous system, is present in schwannomas. The immunohistochemical localisation of S-100 may be necessary in areas where unequivocal diagnosis has not been reached by conventional stains. It may be particularly advantageous in the differential diagnosis of peripheral spindle-celled soft tissue tumours: S-100 protein was localised in schwannomas and in the Schwann cell components of neurofibromas but not in fibromas, fibrosarcomas, leiomyomas and leiomyosarcomas (Clark & Hartman 1981).S–100 protein has also been described in melanomas.

MENINGIOMAS

Meningiomas are common, ubiquitous and benign neoplasms. Several types, meningotheliomatous, fibrous, transitional, psammomatous, angiomatous, haemangioblastic, haemangiopericytic and papillary, can be distinguished, but many meningiomas have mixed structures; the histological diversity increases within a single tumour with the number of areas examined. Although enzyme studies have produced much valuable data on the distribution of oxidoreductases and hydrolases, the diagnostic problems of meningiomas can be resolved without histochemistry.

PRIMARY MALIGNANT LYMPHOMAS

This group now includes microglioma and microgliomatosis which have proved to be malignant diffuse lymphomas. The concept of these neoplasms originating from undifferentiated reticulum cells which then mature into metallophilic microglia has been modified by recent immunohistochemical investigations. The neoplastic cells are predominantly B-type lymphocytes at various stages of maturation and produce immunoglobulins. They tend to form perivascular cuffing, but reticulin stain reveals that these cells are also present between layers of the vascular wall. The silver impregnation technique of Weil-Davenport as modified by Scott demonstrates microglial cells intermingled with lymphocytes.

Immunohistochemical investigation is essential in the diagnosis of these tumours. Staining for the kappa and lambda light chains reveals monoclonal immunoglobulin production in most cases, although a few lymphomas synthesise both light chains. This is clear evidence to prove the B cell origin of these lymphomas. That a macrophage component, including microglial cells, is also present can be demonstrated by staining for muramidase, α-1-antitrypsin and α-1-antichymotrypsin. In addition acid phosphatase and non-specific esterase reactions will be positive in phagocytic cells.

In conclusion, enzyme histochemistry has contributed substantially to the understanding of the biology of the tumours of the nervous system. It has provided useful information on the complicated processes of regression, anaplasia and differentiation. However, as a diagnostic tool it is clearly of limited value. Immunohistochemistry, in contrast, has the potential to become part of the histopathologist's armoury. The specificity and sensitivity of the immunostaining, the permanence of the reaction product, the relative simplicity and easy adaptability of the technique ensure that the immunoperoxidase method will gain increasing importance in the diagnosis of neural neoplasms.

REFERENCES

Chason J L, Lander J W, Gonzalez J E, Brueckner G 1963 Respiratory enzyme activity of human gliomas. A slide histochemical study. Journal of Neuropathology and Experimental Neurology 22: 471–478

Clark H B, Hartman B K 1981 S-100 protein as an immunohistochemical marker for neoplasms of glial and Schwann cell origin. Journal of Neuropathology and Experimental Neurology 40: 335

Deck J H N, Eng L F, Bigbee J, Woodcock S M 1978 The role of glial fibrillary acidic protein in the diagnosis of central nervous system tumors. Acta neuropathologica (Berlin) 42: 183–190

Eng L F, Bigbee J W 1978 Immunohistochemistry of nervous system — specific antigens. In: Agranoff B W, Aprison M H (eds) Advances in neurochemistry. Plenum, New York, vol 3, p 43

Eng L F, Rubinstein L J 1978 Contribution of immunohistochemistry to diagnostic problems of human cerebral tumors. Journal of Histochemistry and Cytochemistry 26: 513–522

Gluszcz A 1963 A histochemical study of some hydrolytic enzymes in tumours of the nervous system. Acta neuropathologica (Berlin) 3: 184–201

Gomori G 1939 Microtechnical demonstration of phosphatase in tissue sections. Proceedings of the Society for Experimental Biology and Medicine 42: 23–26

Karnovsky M J 1965 A formaldehyde-glutaraldehyde fixative of high osmolality for use in electron microscopy. Journal of Cell Biology 27: 137A–138A

Lantos P L 1974 Fine structural localisation of acid phosphatase in neural tumours induced by N-ethyl-N-nitrosourea in rats. Acta neuropathologica (Berlin) 29: 45–55

Lolova I, Ivanova A, Bojinov S 1971a Histochemical investigation of non-specific acid phosphatase in cerebral tumors. Journal of Neurosurgery 34: 730–740

Lolova I, Ivanova A, Bojinov S 1971b A histochemical investigation of non-specific esterase in brain tumours. Acta histochemica 39: 239–247

Lolova I, Ivanova A, Bojinov S, Christov V 1972 Some histochemical studies on medulloblastomas with reference to their localization. Neuropatologia Polska 10: 241–247

McCormick D, Allen I V 1976 The value of LDH isoenzymes in the rapid diagnosis of brain tumours. Neuropathology and Applied Neurobiology 2: 269–278

Norenberg M D 1979 The distribution of glutamine synthetase in the rat central nervous system. Journal of Histochemistry and Cytochemistry 27: 756–762

Novikoff A B, Essner E 1962 Pathological changes in cytoplasmic organelles Federation Proceedings 21: 1130–1142

O'Connor J S, Laws Jr E R 1963 Histochemical survey of brain tumor enzymes. Archives of Neurology (Chicago) 9: 641–651

Pilkington G J, Lantos P L 1981 Glutamine synthetase as a specific marker for normal, reacting and neoplastic astrocytes. Neuropathology and Applied Neurobiology 7: 244

Pilkington G J, Lantos P L 1982 The role of glutamine synthetase in the diagnosis of cerebral tumours. Neuropathology & Applied Neurobiology 8: 227–236

, Smith Jr K R, Crafts D C, Chung H D, Fredericks M 1980 Hydrolytic enzyme activities of the nervous system. Archives of Neurology (Chicago) 37: 356–359

Roessmann U, Velasco M E, Autilio-Gambetti L, Gambetti P 1981 Immunohistochemical localization of neurofilament subunit in human neuroepithelial tumors. Journal of Neuropathology and Experimental Neurology 40: 312

Schiffer D, Fabiani A 1972 Contribution of histochemistry to knowledge of brain-tumor metabolism. In: Kirsch W M, Grossi Paoletti E, Paoletti P (eds) The experimental biology of brain tumors. C C Thomas, Springfield, Illinois, p 209

Schiffer D, Fornatto L, Croveri G, Fabiani A 1972 Histoenzymology of human and experimental brain tumours: remarks on the interpretation of hydrolytic enzyme reactions. Neuropatologia Polska 10: 227–233

Schmechel D, Marangos P J, Zis A P, Brightman M, Goodwin F K 1978 Brain enolases as specific markers of neuronal and glial cells. Science (Washington) 199: 313–314

Sternberger L A, Hardy Jr P H, Cuculis J J, Meyer H G 1970 The unlabeled antibody enzyme method of immunohistochemistry. Preparation properties of soluble antigen-antibody complex (horseradish peroxidase-antihorseradish peroxidase) and its use in identification of spirochetes. Journal of Histochemistry and Cytochemistry 18: 315–333

Velasco M E, Dahl D, Roessmann U, Gambetti P 1980 Immunohistochemical localization of glial fibrillary acidic protein in human glial neoplasms. Cancer (Philadelphia) 45: 484–494

Viale G L, Andreussi L G, 1965 Histochemical study of oxidative activity in tumors of the nervous system. Acta neuropathologica (Berlin) 4: 538–558

Wollemann M 1974 Biochemistry of brain tumours. Macmillan, London

Zülch K J 1979 Histological typing of tumours of the central nervous system. World Health Organisation, Geneva

9

The cardiovascular system

Olga B. Bayliss High

INTRODUCTION

Histochemistry has played an essential role in our understanding of the biology and pathology of the cardiovascular system. Since cholesterol deposition and ischaemia are responsible for most cardiovascular accidents it is not surprising that the histochemist has been interested mainly in techniques that demonstrate lipids and enzymes. Such methods have been particularly important to indicate impaired cholesterol metabolism — one of the many factors implicated in the pathogenesis of atherosclerosis.

THE ARTERIAL WALL

Pathology

As cholesterol accumulates in the arterial intima throughout life it contributes to the fibrotic response seen in the typical raised fibro-fatty plaque. Although such lesions will only marginally alter the lumen of the aorta, it is easy to see how stenosis of small-calibre vessels can occur — notably the coronary and cerebral arteries, culminating in myocardial and cerebral infarction respectively. Arteries may also become occluded by thrombus supervening on the surface of a plaque. Alternatively an advanced lesion may become complicated by necrosis, resulting in the formation of an atheromatous ulcer. The ulcer not only provides the ideal nidus for thrombus formation but also may predispose to aneurysmal dilatation or dissection of the arterial wall. Furthermore, the ulcer contents may be discharged into the circulation to embolise elsewhere in the vascular tree. These events are the most frequent consequences of degenerative changes in the vascular system, accounting for about one third of all deaths in man.

Histochemistry

The pathologist will undoubtedly recognise most types of vascular lesions in routine haematoxylin and eosin stained sections; even the distribution of atheroma lipids may be deduced on the assumption that clefts and vacuoles represent sites from which lipids have been extracted during processing for paraffin wax sections. To confirm this impression and to identify the lipids, however, it is necessary to prepare frozen sections from fresh or formalin-fixed material for the so-called 'fat' stains. The most popular of these organotropic dyes are *Sudan black B* and *oil red O*. These Sudan-type dyes have an affinity for lipids that are liquid at staining temperature such as triglycerides and certain cholesterol esters — the so-called 'neutral fats' — but fail to penetrate lipids that are crystalline, notably free cholesterol and its more saturated esters. This deficiency can be overcome by bromination (Bayliss & Adams 1972). Treatment with bromine before Sudan black B staining not only depresses the solubility of free fatty acids and lecithins in the ethanolic dye solvent, but also converts free cholesterol (m.p. 144°C) to bromoderivatives that are liquid at room temperature and, hence, permeable to the Sudan dyes so that both free and esterified cholesterol can be visualised in the atheromatous plaque.

These Sudan dyes are by no means selective for cholesterol or any other individual lipid. The specific histochemical techniques for cholesterol based on the Schultz type of procedure are hampered by transience of the colour reaction and by gaseous distortion of the sections. The *perchloric acid-naphthoquinone (PAN)* method (Adams 1961) utilises an entirely different principle and is preferred for its sensitivity, precision and satisfactory tissue preservation. It is well known that various staining methods for cholesterol require preliminary air-oxidation before they become effective, and this entails considerable delay, but the problem can be overcome if sections are oxidised with ferric chloride for 4 h only, prior to staining with the PAN reagent (Adams & Bayliss High 1980). Furthermore, free cholesterol can be distinguished from its esters by digitonin precipitation followed by acetone extraction to remove selectively cholesterol esters (Adams & Bayliss 1974). The mature atherosclerotic plaque contains both fatty and crystalline lipids which are readily distinguished if a section stained with oil red O is viewed in polarised light with partly crossed polars, so that the birefringence of the unstained cholesterol crystals contrasts with the red-stained sudanophilic esters (see Fig. 9.1). The Sudan dyes can also be used in combination with selected histochemical techniques for marker enzymes. Sudan black B has been used in conjunction with non-specific esterase staining (Bayliss High 1981), but the combination of β-galactosidase (Lake 1974) with oil red O staining is superior in this context. The *esterase-Sudan black B* and *β-galactosidase-oil red O* methods enable the distribution of macrophages and globular lipids to be appreciated simultaneously in foci of inflammation within arterial lesions (see Plate 3). As already

mentioned, partly polarised light permits a further distinction to be made between the crystalline extracellular lipid and the sudanophilic, largely intracellular lipids in these lesions.

A further application of the oil red O method is at autopsy for the gross staining of arteries to emphasise the severity and extent of atheroma in the young subject when the lipid deposits are superficial and less conspicuous to the naked eye than are the raised fibro-fatty plaques of advanced atherosclerosis (see Fig. 9.2). The gross oil red O staining technique may be particularly useful diagnostically in the pulmonary arterial tree to demonstrate the effects of pulmonary hypertension and thereby infer what sort of pulmonary blood pressure the patient had experienced prior to death — a parameter not usually monitored. Lengths of well-rinsed artery, stripped of frank adventitial fat, are treated with oil red O in an extended version of the method used for tissue sections.

For a comprehensive account of the way in which histochemistry has helped to establish the pathogenic mechanisms underlying these and other, rarer, aspects of arterial disease, the reader is referred to Adams (1967). Conditions such as medial degeneration, dissecting aneurysm, 'hyalinisation', thrombosis and diabetic angiopathy have been

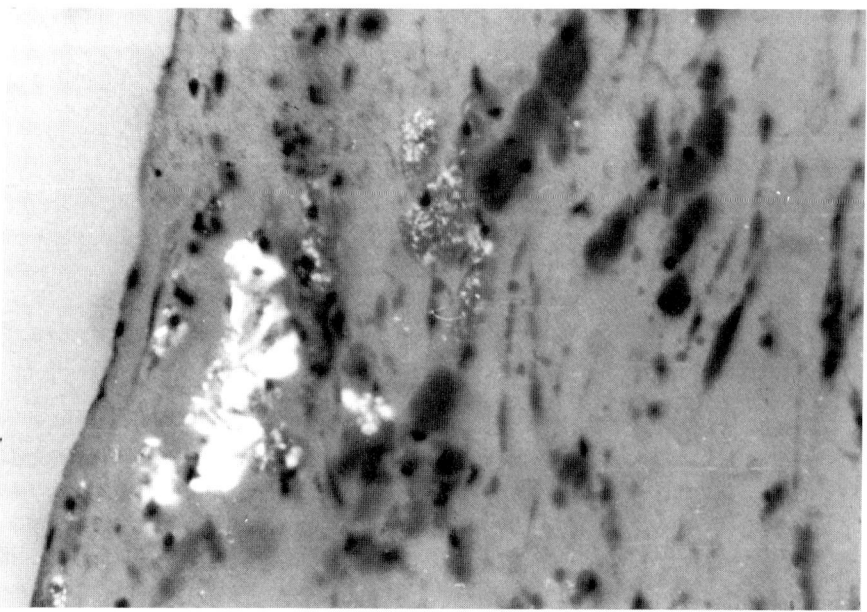

Fig. 9.1 Intracellular sudanophilic lipids and birefringent cholesterol crystals in an atherosclerotic plaque. Oil red O-haemalum in polarised light. ×310

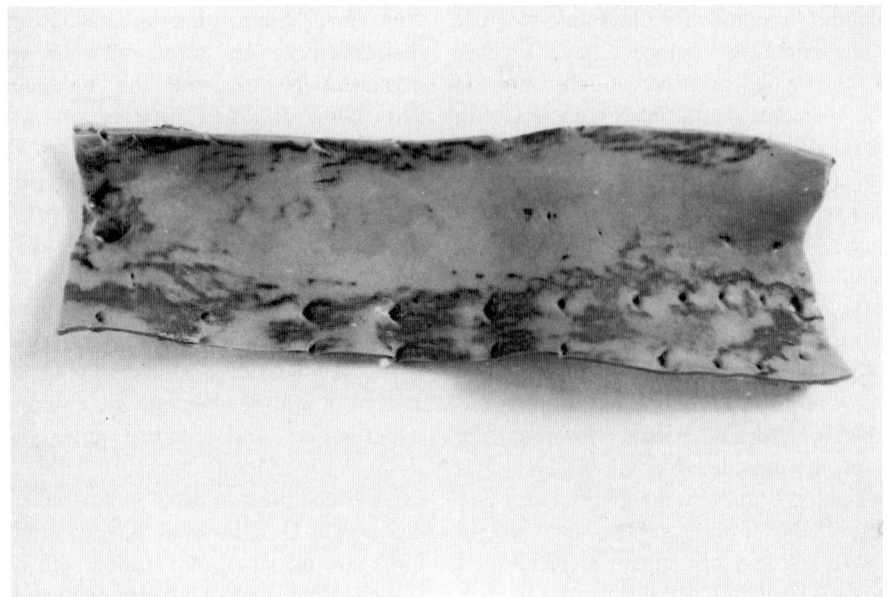

Fig. 9.2   A length of aorta from a 25-year-old man stained in the gross with oil red O.

extensively investigated by a range of histochemical techniques but since the diagnosis of these disorders is usually sufficiently straightforward with conventional histology, their histochemical profiles need not be considered here. Fibrin, derived from plasma fibrinogen, is an important feature of pathological processes involving acute inflammatory exudation and thrombosis. Staining methods for fibrin are well documented elsewhere. Attempts to date thrombi by their variable reaction to fibrin stains have proved too imprecise for general use.

Polyarteritis nodosa, systemic lupus erythematosus and scleroderma comprise a group of vascular disorders which have recently been recognised as autoimmune or hypersensitivity reactions. Characteristic lesions are seen in the renal vasculature and although it is now possible to identify intramural immune complexes by means of immuno-histochemical techniques, such procedures are more appropriately considered elsewhere in this volume.

THE HEART

Myocardial infarction

Pathology

The most common disturbance of cardiac muscle itself is infarction due to stenosis at some point in the coronary tree by atheroma, thrombosis or by a combination of these two factors. The ischaemic myocardium suffers irreversible damage focally, often with fatal consequences. Disintegrating heart muscle cells liberate large amounts of intracellular enzymes into the circulation, a factor used by the biochemist to monitor the severity of an attack.

Histochemistry

Undoubtedly the most important histochemical contribution to diagnostic pathology of the cardiovascular system is the demonstration of early myocardial infarction in the gross at necropsy by means of a dehydrogenase enzyme reaction on slices of unfixed heart (Jestadt & Sandritter 1959). Whereas the biochemist assesses infarction in terms of the transiently raised glutamic-oxalacetic transaminase or β-hydroxybutyrate dehydrogenase levels in serum, the histochemist demonstrates the loss of enzyme activity from the tissue itself and thereby defines the damaged site. A redox indicator is reduced by mitochondrial enzymes within normal heart muscle to produce a staining reaction in all but the affected, infarcted areas. The most reliable and sensitive of a variety of dehydrogenase systems tested is the *nitro blue tetrazolium (NBT) test* (Derias & Adams 1978), using a tetrazolium salt, NAD and endogenous substrate only. The superior results obtained recently can be attributed to adding NAD (Derias & Adams

1982) as this co-enzyme is rapidly lost post mortem. The ischaemic fibres fail to show the dark-blue reaction of normal myocardium and it is thereby easy to demonstrate the site and extent of infarction (see Fig. 9.3). Focally damaged areas appear pale as early as 1 to 5 h after the stated clinical onset of the infarct and this reaction can be achieved even after storage for up to 3 days post mortem. Such early damage would be otherwise undetected when the heart is examined macroscopically at autopsy, but once the lesion has been grossly defined by the NBT test then relevant blocks can be selected for histology.

Primary amyloidosis

This disease affects the tissues of mesodermal origin, including the heart, and is characterised by the deposition of extracellular eosinophilic material with a unique ultrastructure — the β-pleated sheet — now recognised to comprise mainly antibody molecules. As with amyloid in other sites, cardiac deposits have an affinity for the dye Congo red, due to the β-pleated sheet configuration of the amyloid molecule. An alkaline, alcoholic variant of the *Congo red method* (Puchtler et al 1962) is recommended to demonstrate amyloid microscopically; the high chloride ion content of this reagent reduces background staining and intensifies the amyloid-dye binding. Weak staining can be accentuated if the stained section is viewed in polarised light so that the green dichroism of the amyloid-dye complex is visible. Some fluorochrome dyes such as Thioflavin T have been advocated for amyloid staining but, although they may be more sensitive than Congo red, they are in general less selective. Amyloid staining methods are well covered in standard histology texts and need not be considered in depth here.

Glycogen storage diseases

The accumulation of glycogen as a result of congenital enzyme defects is predictably directed towards muscle cells, but of the six known types of glycogen storage, only three actually involve the heart. Of these three conditions, Pompe's disease (GSD type II), due to acid maltase deficiency, is the most frequently encountered. The other two very rare variants are types III and IV, known as Forbes' and Andersen's diseases respectively. The metabolic disorders are considered in detail elsewhere. As usual, glycogen can be demonstrated, in alcohol-fixed tissue, with *Best's carmine method* (Best 1906), or with the *periodic acid-Schiff (PAS) method*, provided that a diastase-digested control section is employed. The PAS method is not of course specific for glycogen, since Schiff's reagent will demonstrate the aldehydes

50 mm

Fig. 9.3 Heart slice stained by the nitroblue tetrazolium (NBT) test, from a 45-year-old man with a recent (clinical history of several hours) extensive anteroseptal infarct. Note also an older scar (arrow)

generated from the 1,2 glycol groups of all hexoses. Therefore comparison with a diastase-treated section is essential to give a measure of the glycogen content. Nor is Best's carmine method entirely specific, yet its high selectivity for glycogen makes it a useful method, particularly when the high concentration of chloride ions and high pH of the dye solution supress interference from basic proteins. Details of glycogen staining can be found in any standard textbook.

Degenerative changes

Fatty degeneration can occur in all oxygen-deprived situations and cardiac muscle cells are particularly vulnerable to the hypoxia induced by anaemia, toxins and of course stenosis of coronary arteries. The speckled appearance of the so-called 'thrush-breast' heart, a characteristic feature of severe anaemia, is due to focal lipid deposits. The accumulation of fat within myofibres is thought to be due to failure to metabolise fat normally transported for energy production. The fatty heart is a feature of myofibrillar degeneration in general and includes some of the cardiomyopathies described below. It is an unexplained feature of Reye's syndrome (Reye et al 1963), in which lipid-laden mitochondria may subsequently become mineralised. Rows of intracellular lipid droplets can be demonstrated

microscopically with the routine fat stains such as *Sudan black B* and *oil red O*.

Brown atrophy. This term derives from the colour of senile heart muscle which has accumulated excessive amounts of lipofuscin. As in other lipid-rich tissues, notably the brain, this pigment is not necessarily pathological — merely an ageing change as lysosomal lipids undergo peroxidation. Lipofuscin can be demonstrated by the *Schmorl, periodic acid-Schiff* and *Ziehl-Neelsen* methods and by its autofluorescence.

The cardiomyopathies

This term covers a heterogeneous group of conditions in which disturbed cardiac function is usually classified clinically as either obstructive, constrictive or congestive. Pathologically the cardiomyopathies may be regarded as disorders of cardiac function due to the following:

Inflammation due to infections (e.g. viral myocarditis).

Obstruction due to a congenital defect of the muscle fibres in the left ventricular outflow tract.

Increased sympathetic innervation of cardiac muscle has been reported in several cases of obstructive cardiomyopathy (Pearse 1964). Such hypertrophic nerve fibres can be demonstrated

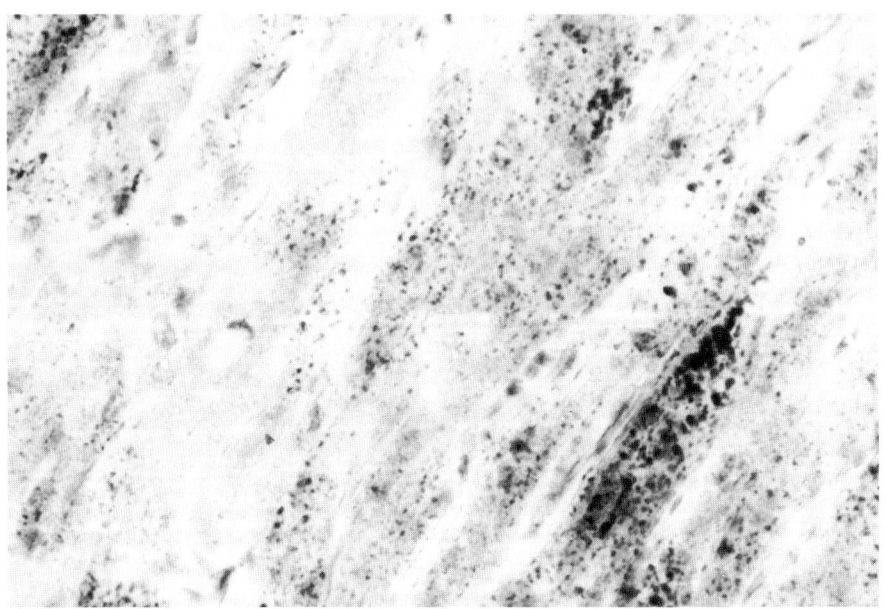

Fig. 9.4 Inter- and intracellular lipid accumulation in the myocardium of a patient with infantile xanthomatous (histiocytoid) cardiomyopathy. Oil red O. ×125

microscopically by the *formaldehyde-induced fluorescence technique for catecholamines* (Falck et al 1962). Neural noradrenaline reacts with formaldehyde to produce tetra-hydroisoquinoline derivatives which further react to give a highly fluorescent, insoluble product.

Disturbed metabolism such as that induced by excessive alcohol consumption, malnutrition and vitamin depletion (e.g. beriberi).

Fatty accumulation is a common consequence of the metabolic cardiomyopathies. In drug- and alcohol-induced disorders, the heart becomes flabby as the degenerate myofibres are replaced by fibrous tissue and triglycerides. Although alcoholic cardiomyopathy is thought to be due to malnutrition and thiamine deficiency, it seems likely that the toxic effect of alcohol itself may impair fatty acid metabolism so that triglycerides accumulate between myofibres. These deposits can be demonstrated with *oil red O* and *Sudan black B*, as with the other cardiolipidoses (see Fig. 9.4). Radford and Chalk

(1980) reviewed 17 cases of a condition designated 'infantile xanthomatous cardiomyopathy' which has a marked female predominance and in which the sudanophilic lipid actually accumulates within the myofibres so that they resemble histiocytes.

Chemotherapy. There is increasing awareness of the important adverse effect of drugs such as adriamycin, cimetidine and methotrexate, which lead to myofibrillar degeneration. Daunorubicin, on the other hand, is an antileukaemic drug which is thought to deposit in cardiac nerve ganglia and thereby reduce cardiac function. An excellent review of the subject of cardiomyopathy can be found in *'Pathology of the cardiomyopathies'* (McKinney 1974).

Although diverse causative agents produce a spectrum of changes in cardiac muscle, few features can be exploited histochemically to assist diagnosis.

Table 9.1 below summarises the way in which histochemical methods may be employed as a diagnostic aid in cardiovascular pathology.

Table 9.1 Application of histochemical techniques in cardiovascular pathology

Condition	Material detected	Technique
Atherosclerosis	free and esterified cholesterol	Sudan black B Bromine-SBB Oil red O + polarised light PAN
Reparative changes in atherosclerosis	macrophages and atheroma lipids	Esterase- Sudan black B β-galactosidase- oil red O
Myocardial infarction	dehydrogenases	Nitroblue tetrazolium test
Primary amyloidosis	amyloid	Congo red
Pompe's disease	glycogen	Best's carmine Diastase-PAS
Fatty degeneration	triglycerides	Sudan black B Oil red O
Brown atrophy	lipofuscin	Schmorl, PAS, Ziehl-Neelsen Autofluorescence
Cardiomyopathies: a. obstructive	adrenergic nerves	Formaldehyde-induced catecholamine fluorescence
b. metabolic (alcohol- and drug-induced) c. infantile xanthomatous (histiocytoid)	triglycerides	Sudan black B Oil red O

REFERENCES

Adams C W M 1961 A perchloric acid naphthoquinone method for the histochemical localisation of cholesterol. Nature (London) 192: 331

Adams C W M 1967 Vascular histochemistry. Lloyd-Luke, London

Adams C W M, Bayliss O B 1974 Lipid histochemistry. In: Glick D, Rosenbaum R M (eds) Techniques of biochemical and biophysical morphology. Wiley Interscience, New York, vol 2

Adams C W M, Bayliss High O B 1980 Preliminary oxidation in histochemical staining methods for cholesterol. Journal of Microscopy 119: 427

Bayliss High O B 1981 The histochemical versatility of Sudan black B. Acta histochemica, suppl. Band xxiv, S 247

Bayliss O B, Adams C W M 1972 Bromine Sudan black (BSB). A general stain for tissue lipids including free cholesterol. Histochemical Journal 2: 87

Best F 1906 Über Karminfärbung des glykogens und der Kerne. Zeitschrift für Wissenschaftliche Mikroscopie und für mikroskopische Technik 23: 319

Derias N W, Adams C W M 1978 Nitro blue tetrazolium test: early gross detection of human myocardial infarcts. British Journal of Experimental Pathology 59: 254

Derias N W, Adams C W M 1982 Macroscopic enzyme histochemistry in myocardial infarction: use of coenzyme, cyanide and phenazine methosulphate. Journal of Clinical Pathology 35: 410

Falck B, Hillarp N-Å, Thieme G, Torp A 1962 Fluorescence of catecholamines and related compounds condensed with formaldehyde. Journal of Histochemistry and Cytochemistry 10: 348

Jestadt R, Sandritter W 1959 Erfahringen mit der TTC (triphenyl tetrazolium chloride) reaktion für die pathologisch-anatomische diagnoses des frichen Herzinfarktes. Zeitschrift für Kreislaufforsch 48: 802

Lake B D 1974 An improved method for the detection of β-galactosidase activity and its application to G_{M1}-gangliosidosis and mucopolysaccharidosis. Histochemical Journal 6: 211

McKinney B 1974 Pathology of the cardiomyopathies. Butterworths, London

Pearse A G E 1964 The histochemistry and electron microscopy of obstructive cardiomyopathy. In: Wolstenholme G E W, O'Connor M (eds) Cardiomyopathies. Churchill, London, p 132

Puchtler H, Sweat F, Levine M 1962 On the binding of Congo red by amyloid. Journal of Histochemistry and Cytochemistry 10: 355

Radford D J, Chalk S M 1980 Infantile xanthomatous cardiomyopathy. Australian Paediatric Journal 16: 123

Reye R D K, Morgan G, Barral J 1963 Encephalopathy and fatty degeneration of the viscera. A disease entity in childhood. Lancet 2: 749

10

Skeletal muscle

Margaret A. Johnson

INTRODUCTION

Muscle pathology has become one of those areas of pathology where histochemical techniques have proved to be not only useful but absolutely indispensable in diagnostic practice. One of the main reasons for the particular value of a histochemical approach to the investigation of muscle disease lies in the nature of skeletal muscle itself. Muscle is not a uniform tissue; its constituent fibres are differentiated into fibre types whose contractile properties and other metabolic characteristics show striking functional differences. Many of these differences can be demonstrated histochemically and this ability to discriminate between the main muscle fibre types in tissue sections can be exploited in the investigation of muscle disorders.

Many neuromuscular diseases selectively affect one or other of the major fibre types and the detection of such selective processes is obviously dependent upon the use of the appropriate histochemical techniques. The normal spatial arrangement of the various fibre types in skeletal muscle is random, but in many neurogenic disorders this random distribution may be superseded by grouping of fibres of uniform type due to remodelling of motor units. Not infrequently this sort of change is the only indication of an underlying neurogenic abnormality and histochemical fibre typing procedures are clearly essential for its detection.

A substantial proportion of congenital myopathies, whose differentiating features are indistinguishable on routine histological examination, owe their initial recognition to the use of histochemical techniques and the number of specific metabolic disorders known to involve skeletal muscle is constantly increasing due to the identification of additional enzyme defects, many of which can be demonstrated histochemically.

The purpose of this chapter is to give a brief account of ways in which histochemical techniques can be used to increase the diagnostic potential of muscle biopsy in the investigation of a wide range of neuromuscular disorders. For detailed descriptions of the pathology of individual muscle diseases, the reader is directed to the various comprehensive monographs on the subject (Dubowitz & Brooke 1973, Bethlem 1980, Walton 1981). Within the framework of the present chapter, only sufficient histopathological detail has been included as will allow the histochemical aspects of the disorders to be set in context.

TISSUE PREPARATION

The preparative techniques used in processing muscle tissue for subsequent histochemical examination are a good deal less time-consuming than those needed for routine histology. Almost all the histochemical techniques used in diagnostic muscle pathology can be carried out on cryostat sections of fresh frozen tissue, though in some instances tissue sections will need to be post-fixed subsequently.

Muscle biopsy techniques

Two types of biopsy procedure are in current use. The first is generally referred to as an 'open' biopsy since the muscle is removed through a comparatively large incision. In the 'needle' biopsy procedure only a very small incision is required and the muscle is removed using a Bergstrom needle or one of its modifications. Correct orientation of the muscle prior to freezing is of paramount importance. If the biopsy has been obtained by means of an 'open' biopsy this generally poses no problem as the longitudinal direction of the muscle fibres should be clearly identifiable. If the specimen has been removed using sutures, any compressed portions from the ends of the biopsy should be removed. The use of muscle biopsy clamps, though invaluable for achieving good fixation in histological preparations, is not necessary for tissue destined to be frozen and merely results in wasted tissue due to compression.

'Open' biopsies should be trimmed to give blocks not exceeding 8 mm³ so that uniformly good freezing can be achieved and the height of the blocks should not exceed their width otherwise the frozen specimens will tend to be unstable. Most diagnostic work entails assessment of fibre sizes and for this, good transversely orientated sections are obligatory; longitudinally orientated blocks of tissue are not often required and need not be provided routinely. The tissue obtained from 'needle' biopsies rarely exceeds 3 mm³ and is frequently fragmented; correct orientation can best be achieved using a dissecting microscope. The fragments are not often large enough to make satisfactory blocks on their own but a convenient way to overcome this problem is to sandwich the muscle, correctly orientated, between two blocks of liver or gelatin.

Tissue freezing and sectioning

Skeletal muscle is not so susceptible to tissue autolysis as some other tissues so immediate freezing is not strictly necessary. However, no more than 30 min should elapse between excision of the biopsy and freezing, and the tissue should be kept cool and not allowed to dehydrate.

Muscle blocks should be placed firmly on filter paper strips; composite blocks of liver (or gelatin) plus sandwiched needle biopsy fragments should be treated in the same way. Freezing is optimally carried out using dichlorodifluoromethane (Arcton 12) cooled to c. −150°C in liquid nitrogen. Liquid nitrogen alone, though adequate for freezing most other tissues, is not recommended for freezing muscle which is particularly prone to ice-crystal artefact. (However, satisfactory freezing in liquid nitrogen can be obtained if the tissue is first liberally dusted with starch powder.) If Arcton 12 is unobtainable, isopentane is an acceptable substitute. Care should be taken to ensure that the temperature of the freezing bath is sufficiently low. This is most easily achieved if

a small container of coolant is first completely frozen in liquid nitrogen and then used as it is thawing. Tissue blocks should be rapidly immersed in the coolant and agitated gently for about 15 sec.

Sectioning of frozen muscle is best carried out at a temperature of −20 to −23°C. A section thickness of 10 μm is suitable for most of the commonly used diagnostic techniques.

HISTOCHEMICAL METHODS

The choice of methods to be used in diagnostic histochemical screening of muscle biopsy tissue is governed by the need to fulfil several objectives. One of the most important of these is to provide a basic 'fibre typing' system which will allow the main metabolic fibre types to be differentiated with reliability in both normal and pathological tissue.

Muscle fibre typing

The method which has proved to be of greatest value in this respect is the myofibrillar ATPase technique. Levels of ATPase activity reflect the basic differences in contractile properties between slow-twitch (Type 1) and fast-twitch (Type 2) fibres. The low ATPase activity in Type 1 fibres is accompanied by high levels of mitochondrial oxidative enzyme activity, a feature which confers some degree of fatigue-resistance on this fibre type. Type 2 fibres, on the other hand, have comparatively low levels of mitochondrial oxidative enzymes, which renders them more prone to fatigue. The low mitochondrial oxidative enzyme content of Type 2 fibres is offset by high myophosphorylase activity so that anaerobic glycolysis represents an important mechanism of energy production in these fibres (Dubowitz & Pearse 1960). The enzyme profiles of Type 1 and Type 2 fibres are summarised in Table 10.1. From this table

Table 10.1 Histochemical fibre types in human muscle

	Type 1	Type 2A	Type 2B	Type 2C*
ATPase pH 9.5				
after MFF fixation	+	+ +	+ + +	+(+)
after pH 4.6 pre-incubation	+ + +	+	+ +	+ +(+)
after pH 4.3 pre-incubation	+ + +	−	−	+(+)
SDH/NADH-TR	+ + +	+ +	+	+ +(+)
Myophosphorylase	+	+ +	+ + +	+(+)

* Normal adult muscle should contain <3% 2C fibres

it can be seen that Type 2 fibres can be classified into two main subtypes on the basis of their ATPase characteristics, glycolytic activity and mitochondrial oxidative enzyme content. Type 2A fibres have lower levels of myophosphorylase than Type 2B, but higher mitochondrial oxidative enzyme activity. The distinction between Types 2A and 2B can be made using the ATPase technique following methanol-free formalin fixation (Hayashi & Freiman 1966) or pre-incubation in pH 4.6 buffer (see Appendix 5 for details of methods). The latter method is generally known as ATPase 'reversal' since after acid pre-incubation the ATPase activity of Type 1 fibres is higher than that of Type 2 fibres (Brooke & Kaiser 1970). The ATPase characteristics of normal muscle fibre types and their correlation with mitochondrial enzyme activity are shown in Figure 10.1.

Marker enzyme techniques

Another aim of a diagnostic protocol is to monitor the state of various muscle components and organelles through the use of 'marker enzyme' techniques. Thus succinate dehydrogenase (SDH) represents a suitable 'marker enzyme' for studying abnormalities in the distribution and activity of mitochondria (Pearse 1972). NADH-tetrazolium reductase (NADH-TR) is also commonly used as an indicator of mitochondrial distribution but it should be borne in mind that since the sarcoplasmic reticulum also contains some NADH-TR activity the enzyme is not exclusively mitochondrial.

Myofibrillar ATPase, whose usefulness in muscle fibre typing has already been discussed, also serves as a marker enzyme for the detection of abnormalities in the myofibrils e.g. areas of focal myofibril disorganisation resulting in loss of ATPase activity.

The distribution of muscle lysosomes can be monitored using techniques for the demonstration of acid phosphatase activity. Simultaneous coupling azo dye methods (Burstone 1958) are to be preferred over lead precipitation methods for use in muscle since it has been demonstrated that non-specific binding of lead salts to myofibrils and other muscle components may occur. In normal human muscle lysosomes are comparatively rare, especially in children's muscle. In young adults a few acid phosphatase-containing granules can be found adjacent to muscle nuclei or immediately under the plasma membrane and the deposition of lipofuscin-like pigments inside lysosomes in later life is accompanied by a gradual increase in the number and size of acid phosphatase-positive granules. The intensity of phagocytic cell reactions in diseased muscle can also be gauged effectively using lysosomal marker techniques. As an alternative to the acid phosphatase technique, lysosomal activity may be demonstrated using methods for the localisation of esterase activity. Moreover, if α-naphthyl acetate is used as substrate it is possible to visualise not only lysosomal esterase but also the sites of endplate cholinesterase.

Detection of metabolic disorders

An important objective in diagnostic muscle pathology is to be able to detect evidence of metabolic disorders, either through identification of specific enzyme defects, or by detection of abnormal tissue components or normal substances stored in abnormal amounts. A classic example of this is the use of the PAS reaction to screen biopsies for abnormal amounts of stored glycogen in cases of glycogenosis. Levels of glycolytic activity can be monitored using techniques for the demonstration of myophosphorylase (Meijer 1968). Similarly screening for lipid storage disorders can be achieved using Sudan black staining; if evidence of lipid accumulation is obtained, the precise nature of the lipid involved can be investigated further.

Diagnostic protocol for muscle biopsies

A routine 'battery' of histochemical techniques should include most, if not all, of those discussed above, viz.

Myofibrillar ATPase — including acid pre-incubation to allow detailed fibre typing

SDH or NADH-TR — for detection of abnormalities of mitochondrial distribution and/or activity

PAS — to screen for abnormal amounts of stored glycogen or other polysaccharides

Myophosphorylase — to indicate glycolytic activity

Sudan black — to screen for excess accumulation of neutral lipid and/or phospholipid

Acid phosphatase and/or esterase — for assessment of lysosomal changes in muscle and extent of phagocytic cell infiltration.

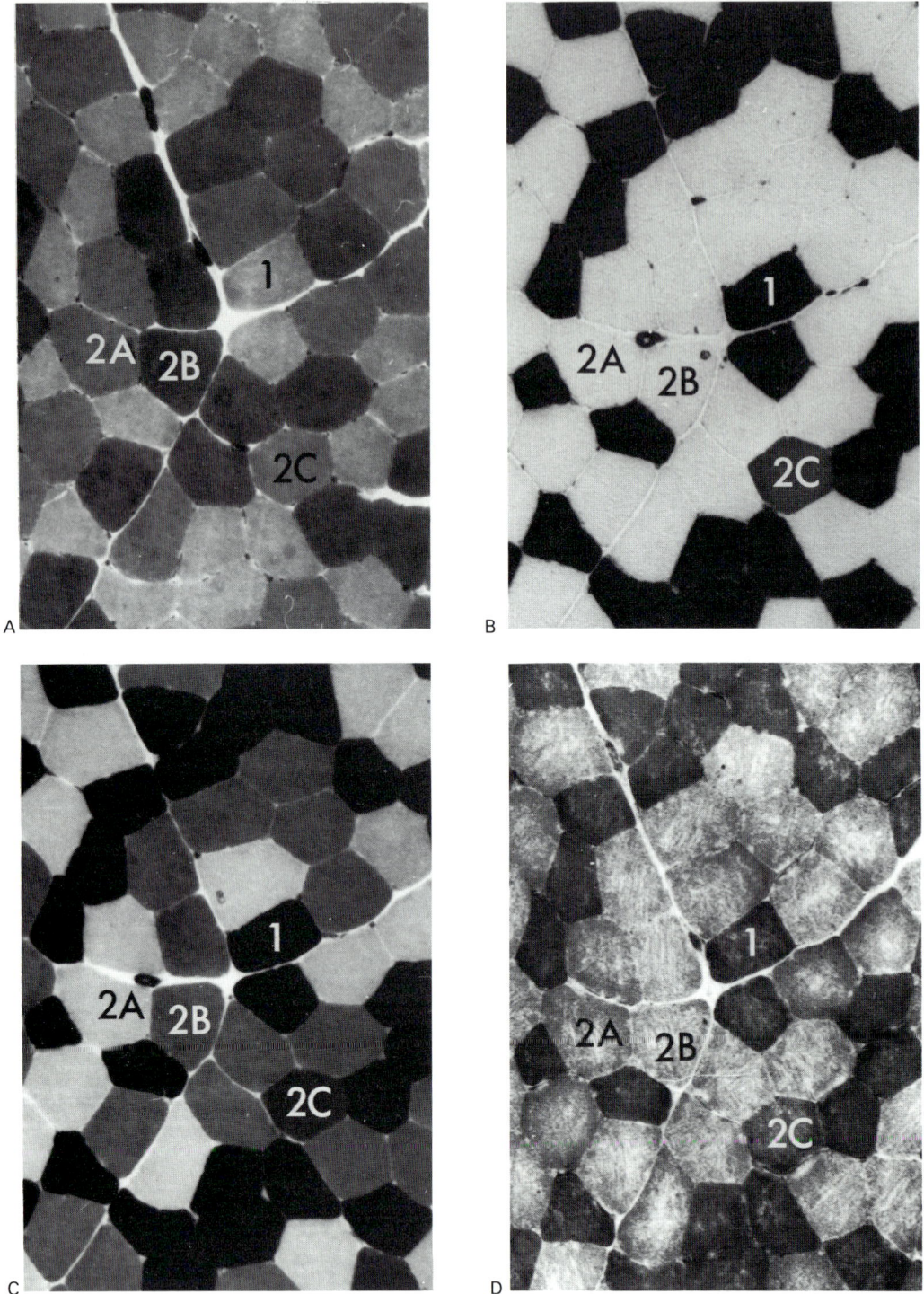

Fig. 10.1 Normal muscle fibre types showing correlation of ATPase characteristics and mitochondrial oxidative enzyme activity. A. ATPase pH 9.5 after MFF fixation; B. ATPase after pH 4.3 pre-incubation; C. ATPase after pH 4.6 pre-incubation; D. NADH-TR. All ×120

These techniques should preferably be carried out on serial sections since it is not uncommon to need to refer from one histochemical preparation to another in order to assess the significance of observed changes, e.g. to determine whether mitochondrial abnormalities (SDH preparations) are confined to Type 1 fibres (ATPase preparations). However, all the above techniques can be completed by one person within 3 h so the demands on laboratory time are not excessive. Finally, although this text specifically excludes histological techniques, sections of the frozen tissue should also be stained using haematoxylin and eosin and/or haematoxylin-Van Gieson and the modified Gomori trichrome method to allow the direct correlation of histochemical and histological features.

DENERVATING DISORDERS

Disorders of skeletal muscle in which denervation is the predominant pathological feature have numerous factors in common even though the site of the lesion may be the anterior horn cells (spinal muscular atrophies) or the peripheral nerve axons or Schwann cells (peripheral neuropathies). It is therefore worth considering some of these common features before examining individual disorders in more detail.

General histochemical features

The main fibre types in normal human muscle (Types 1, 2A and 2B) are randomly arranged, reflecting the random distribution of fibres in normal motor units. The fibres in each motor unit are histochemically uniform but because motor unit territories overlap extensively, the result is a mosaic of the various histochemical fibre types (Fig. 10.2). If only a few motor units are affected by denervation, the distribution of the atrophied fibres will be scattered. There is generally no selective effect of denervating processes on either Type 1 or Type 2 motor units so the atrophied fibres will consist of both fibre types. If a larger proportion of motor units is affected by denervation, groups of atrophied fibres will be encountered; again these will consist of both Type 1 and Type 2 fibres. If, however, these denervated fibres become reinnervated by branches of a single axon the result will be a group of fibres of uniform histochemical fibre type. This phenomenon

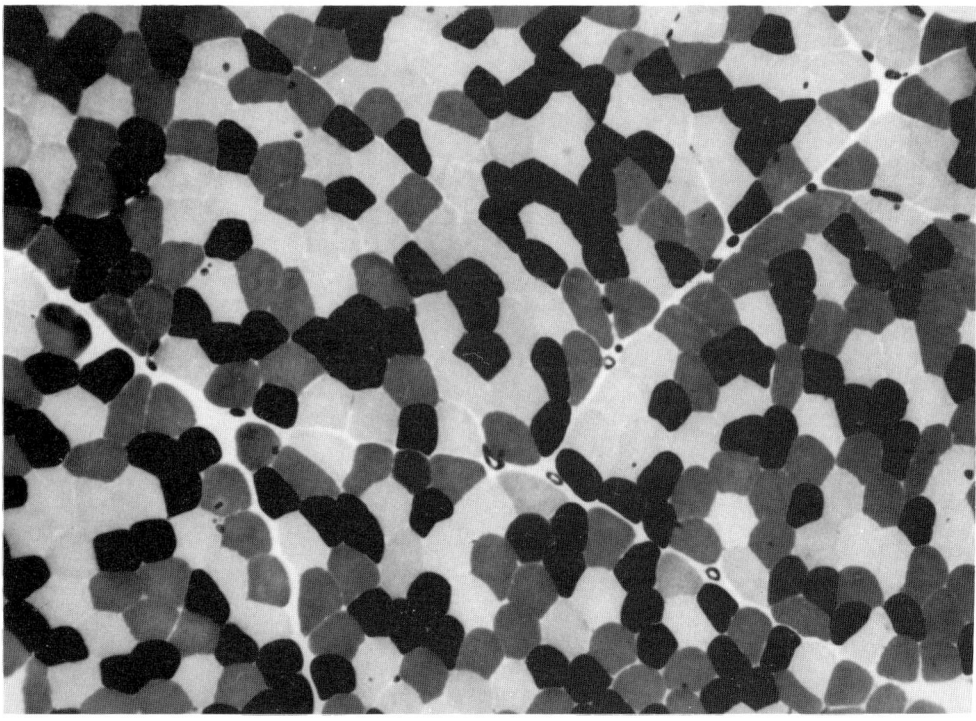

Fig. 10.2   Normal muscle showing random distribution of fibre types. ATPase after pH 4.6 pre-incubation. ×64

of 'uniform fibre type grouping' is an invaluable indication of reinnervation in denervating disorders (Brooke & Engel 1966). Not infrequently muscle biopsies in well-compensated (i.e. reinnervated) denervating disorders may show no signs of atrophied fibres and diagnosis rests on the histochemical detection of fibre type grouping (Fig. 10.3). If subsequent cycles of denervation occur affecting reinnervated fibres, these fibres will be 'grouped' in the physical sense since the territory of a reinnervated motor unit is compact. However, they will also show histochemical fibre type grouping, a feature which serves to distinguish them from groups of atrophied fibres which have resulted from a single cycle of denervation only.

During the process of reinnervation fibres will change from one fibre type to another if the reinnervating axons are of a different motor type from the original innervating axons. Fibres in the process of conversion from Type 1 into Type 2 or vice versa obviously must pass through a transitional stage. This stage is easy to detect in ATPase preparations after pre-incubation at pH 4.3 (see Table 10.1) and such fibres are generally referred to as '2C' fibres. Whereas the ATPase of normal Type 2A or 2B fibres is totally inhibited after pH 4.3 pre-incubation '2C' fibres show activity intermediate between that of Type 1 and Type 2 fibres. A low percentage (< 3%) of '2C' fibres is a common finding in normal human muscle but in the context of denervating disorders, an increased number of '2C' fibres is generally indicative of on-going fibre type conversion as a result of reinnervation. The '2C' ATPase characteristics are also found in various other classes of muscle fibre e.g. in regenerating fibres, since they are indicative of incomplete or transitional states of histochemical fibre type differentiation.

Mitochondrial oxidative enzyme preparations frequently reveal another distinctive abnormality in muscle from many denervating disorders. Affected fibres do not show the usual homogenous distribution of mitochondrial enzyme activity. Instead three concentric zones of low activity (inner zone), high activity (intermediate zone) and normal activity (outer zone) are found; these fibres are known as 'target' fibres (Engel 1961). This type of abnormality is

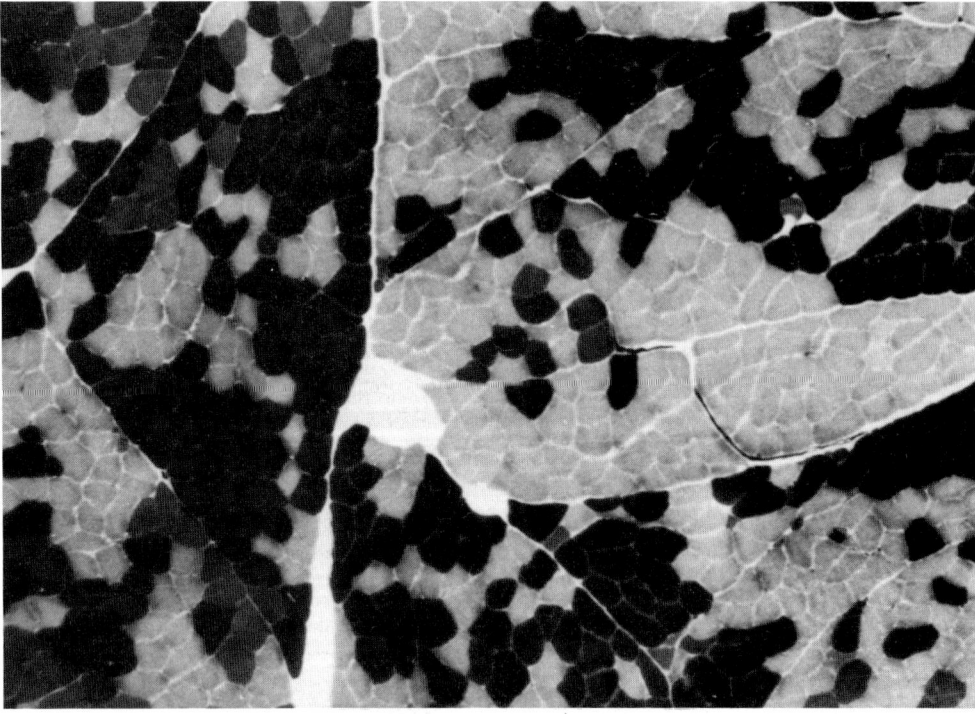

Fig. 10.3   Muscle from patient with peripheral neuropathy showing grouping of fibres of uniform type due to reinnervation. ATPase pH 9.5 after MFF fixation. ×64

almost exclusively confined to denervating disorders (Fig. 10.4) and may be associated with reorganisation of muscle organelles as a consequence of changes in fibre type. Two-zoned 'targetoid' fibres, in which the intermediate high activity zone is absent, are found in a wider range of pathological situations.

Spinal muscular atrophies

The most severe form of spinal muscular atrophy (SMA) has its onset in early infancy (Werdnig-Hoffmann disease). The juvenile form, with onset in later childhood, has a much more benign clinical course (Kugelberg-Welander SMA). There is also an 'intermediate' form of SMA which shares the early onset of Werdnig-Hoffmann disease but is characterised by more gradual progression. In all forms an autosomal recessive mode of inheritance is usual.

Werdnig-Hoffmann disease

This form of SMA involves widespread severe muscle fibre atrophy with no evidence of reinnervation.

Surviving motor units show hypertrophy of fibres which are almost invariably Type 1 (Fig. 10.5). Since there is no evidence of selective sparing of Type 1 motor units, the most likely explanation for this is that surviving units, irrespective of their original fibre type, perform a basically tonic function in the severely weakened muscles and their histochemical profile is adapted accordingly.

Intermediate SMA

Whereas the Werdnig-Hoffmann form of SMA generally proves fatal before the age of 2 years, the more benign 'intermediate' form is associated with survival sometimes into the second decade. The more gradual progression of muscle weakness in the disease is associated with ability of some surviving motor units to enlarge their territories by reinnervating previously denervated fibres. Histochemical evidence of this is provided by the frequent occurrence of uniform fibre type grouping, generally involving Type 1 fibres, though less frequently Type 2 reinnervated motor units may be found. 'Target'

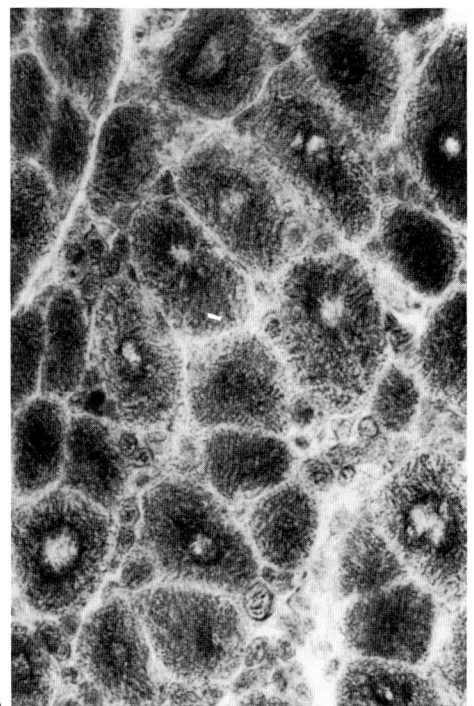
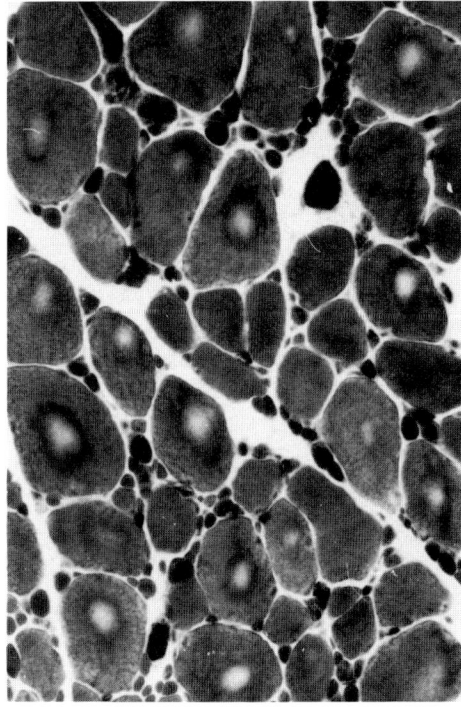

Fig. 10.4 Muscle from patient with the 'intermediate' form of spinal muscular atrophy showing 'target' fibres. ×308
A. NADH-TR; B. ATPase pH 9.5

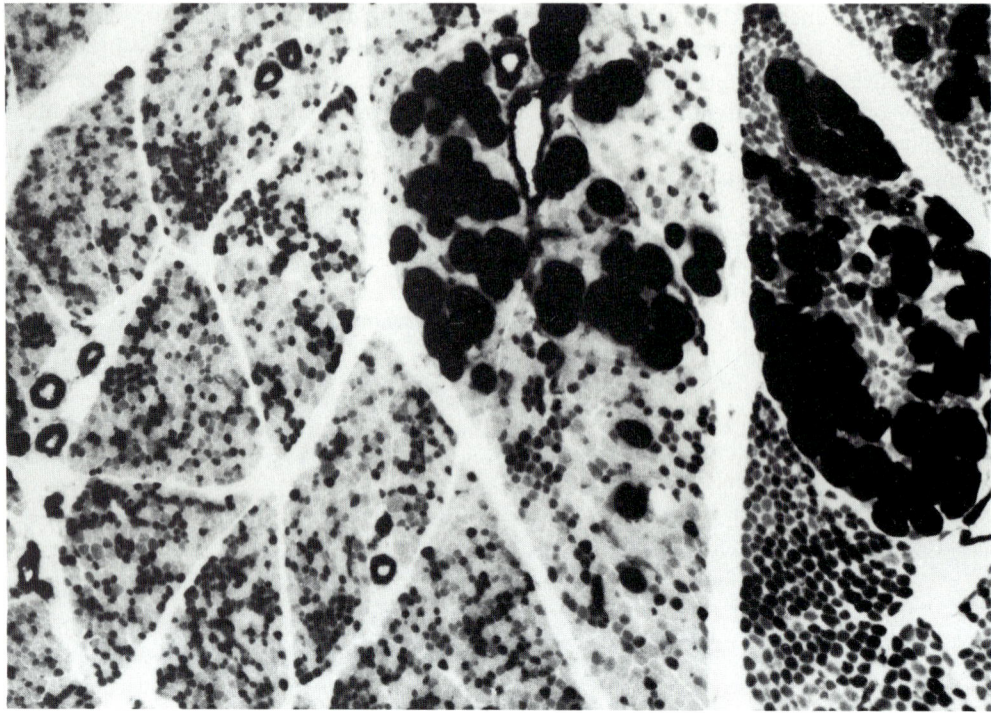

Fig. 10.5   Muscle from patient with the Werdnig-Hoffmann form of spinal muscular atrophy showing widespread atrophy affecting all fibre types. Fibres of surviving motor units are hypertrophied and are all Type 1. ATPase after pH 4.6 pre-incubation. ×120

fibres are not uncommon whereas in Werdnig-Hoffmann disease they are never seen.

Kugelberg-Welander SMA

This form of SMA may have a very gradual clinical course indeed, extending over several decades. Especially in the early stages of the disorder, reinnervation may effectively compensate for the loss of motor units. The more effective the process of reinnervation, the less likely are atrophied fibres to be found in the muscle biopsy sample. Under these circumstances the application of histochemical techniques is essential in order to be able to detect the presence of fibre type grouping in samples of muscle which may appear histologically normal.

Peripheral neuropathies

Although marked denervating changes can be produced in skeletal muscle as a result of peripheral neuropathies of both axonal and demyelinating types, the extent of the denervation produced is generally more restricted in distribution than that associated with central denervating disorders. In denervation of central origin it is not uncommon to see numerous contiguous muscle fascicles consisting totally of atrophied fibres; this finding would not be encountered in any of the peripheral neuropathies. Nevertheless the extent of fibre type grouping may be considerable since multiple cycles of denervation and reinnervation may result from periodic remissions and exacerbations of the disease process. Angulated atrophied fibres frequently show high levels of NADH-TR irrespective of fibre type due to the selective loss of myofibrillar protein and the relative preservation of mitochondrial contents.

MUSCULAR DYSTROPHIES

The muscular dystrophies pose many diagnostic problems for the pathologist, not least because they are diseases of unknown aetiology and the present classification undoubtedly includes disorders within this group which differ widely as regards pathogenesis (Walton 1981). One consequence of this is that whereas the dystrophies have numerous histological features in common, histochemical examination tends rather to reveal the disparate nature of the individual disorders. All the dystrophies are progressive and eventually result in severe loss of muscle bulk. As in chronic denervating disorders, this is commonly associated with increased amounts of interstitial fibrous connective tissue and adipose tissue. However, evidence of frank denervation is lacking; fibre type grouping is not seen and 'target' fibres are consistently absent. Individual forms of muscular dystrophy show different patterns of fibre type involvement which are useful in differentiation and diagnosis.

Duchenne muscular dystrophy (severe X-linked M.D.)

The random variation in fibre size seen in Duchenne M.D. affects both fibre types but in many biopsies the proportion of Type 1 fibres is much greater than normal. Biopsies from pre-clinical cases tend to show more normal fibre type proportions so this feature appears to be progressive and is suggestive either of selective loss of Type 2 fibres or alternatively of functional conversion of fibre types. There is also a progressive loss of the usual clear distinction between Type 1 and Type 2 fibres both in ATPase preparations and in mitochondrial oxidative enzyme preparations. This is partly due to the presence of a high proportion of '2C' fibres which have histochemical characteristics intermediate between those of Type 1 and 2 (Fig. 10.6). Some of the '2C' fibres are clearly identifiable as regenerating fibres whose high cytoplasmic RNA content can be demonstrated using methyl green-pyronin or similar RNA methods. Other '2C' fibres are of more normal diameter and may represent fibres in the process of conversion from Type 2 into Type 1. There is also a severe deficit in numbers of Type 2B fibres in particular; instead of roughly equal proportions of Types 2A and 2B, the type 2B fibres may comprise only about 10% of the total number of Type 2 fibres.

Multiple foci of active necrosis and phagocytosis are features of the early clinical stages of the disorder

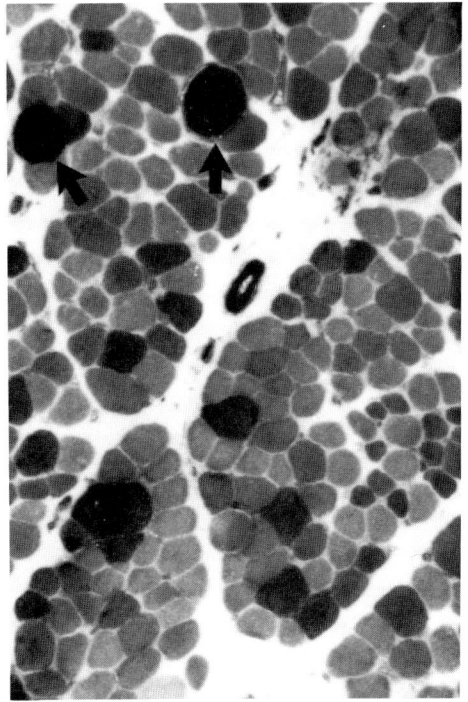 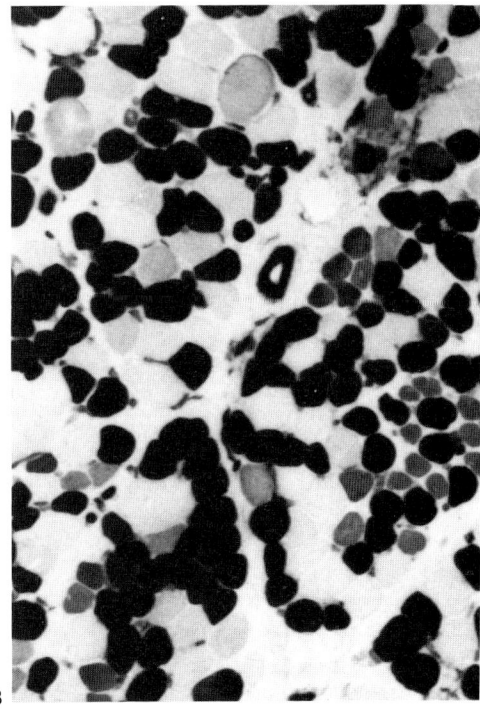

Fig. 10.6 Muscle from patient with Duchenne muscular dystrophy showing rounded, over-contracted fibres (arrows) and numerous '2C' fibres. ×120. A. ATPase pH 9.5; B. ATPase after pH 4.3 pre-incubation

and are strikingly demonstrated using marker techniques for lysosomal activity such as acid phosphatase or esterase.

Numerous rounded intensely-staining fibres can be identified in histological preparations of Duchenne muscle. These fibres give an abnormally high ATPase reaction (Fig. 10.6) and this appears to be due to over-contraction of segments of the affected fibres. A likely cause of this is the influx of calcium due to localised membrane damage; foci of calcium accumulation can sometimes be detected in these fibres by means of the von Kossa, alizarin or fluorescent Morin methods.

Becker muscular dystrophy (benign X-linked M.D.)

Whereas the severe Duchenne form of M.D. is invariably fatal by the late teens or early twenties, the Becker form is characterised by a more gradual progression. One feature associated with the more protracted time-course of the disorder is the greater degree of compensatory hypertrophy which is

observed, affecting mainly Type 2 fibres. Although a Type 1 fibre predominance is seen as in Duchenne M.D., this is not accompanied by a Type 2B fibre deficit. Moreover, the distinction between the histochemical characteristics of the various fibre types is well maintained.

Over-contracted fibres giving abnormally high ATPase reactions are not common and are certainly not present to the same extent as in Duchenne dystrophy. However, two other varieties of abnormal fibre are prevalent. The first is detectable in mitochondrial enzyme reactions (SDH and NADH-TR) and consists of marked irregularities in mitochondrial distribution. This type of abnormality is usually referred to by the whimsical but descriptive term, 'moth-eaten fibres' (Brooke & Engel 1966).

The other type of abnormal fibre is visible on examination of ATPase preparations but is even more striking in methods which demonstrate constituents of aqueous sarcoplasm, e.g. PAS. Displacement of peripheral myofibrils results in the formation of striated rings in affected fibres, hence the term 'ring-fibres', (see Fig. 10.7). It is emphasised that 'ring-

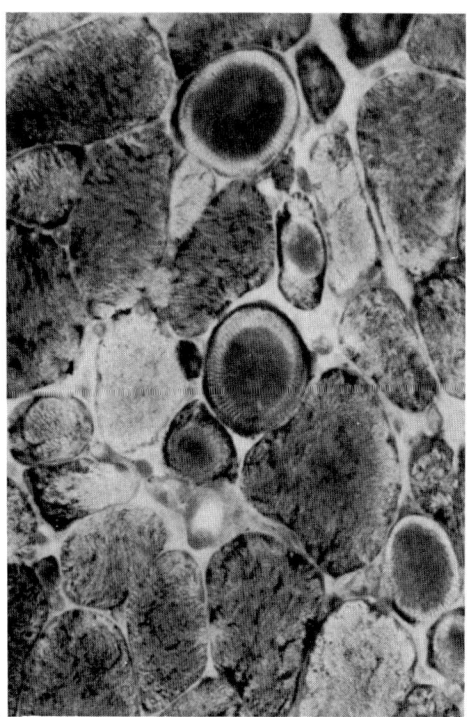
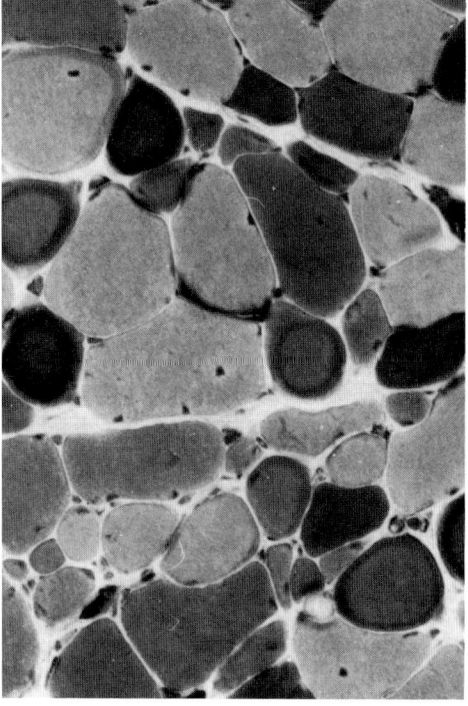

Fig. 10.7 Muscle from patient with Becker muscular dystrophy showing 'ring fibres' containing striated annulets. × 308
A. Periodic acid-Schiff   B. ATPase pH 9.5

fibres' and 'moth-eaten fibres' are not confined to Becker dystrophy but are common in many chronic dystrophic processes.

Facioscapulohumeral dystrophy (and FSH syndrome)

FSH dystrophy is a recognisable clinicopathological entity, with an autosomal dominant pattern of inheritance. However, the clinical manifestations involving weakness of the facial and upper limb-girdle muscles can be seen in other disorders including atypical cases of SMA (Kugelberg-Welander form) and myopathies in which mitochondrial abnormalities are an important feature. These disorders comprise the FSH syndrome and a common problem for the pathologist is their differential histological diagnosis.

FSH dystrophy proper is characterised by increased variation in fibre size, affecting both fibre types; Type 1 fibre predominance is not a feature of this type of M.D. However, Type 1 fibres, especially those affected by atrophy, often show abnormalities of mitochondrial distribution which give the fibres a 'lobulated' appearance in NADH-TR and SDH preparations (see Fig. 10.8). Less frequently, 'moth-eaten' fibres similar to those seen in Becker dystrophy may be encountered. Foci of dense pleomorphic inflammatory cell infiltration are often seen in FSH dystrophy whereas in other dystrophies this is rare.

Cases of FSH syndrome where the pattern of inheritance is autosomal recessive are most often due to chronic SMA. In this context, quantitative assessments of fibre diameters should reveal bimodal size distributions in contrast to the expected findings in FSH dystrophy where variation in fibre size is random and a unimodal size distribution should be preserved.

Limb-girdle syndrome

Weakness of pelvic and femoral musculature is the main feature of a syndrome which has frequently been labelled 'limb-girdle muscular dystrophy'. More recent opinion however, based on assessment of pathological as well as clinical data, tends to regard most cases as variants of either Becker dystrophy or juvenile SMA (Bethlem 1980). It is not surprising

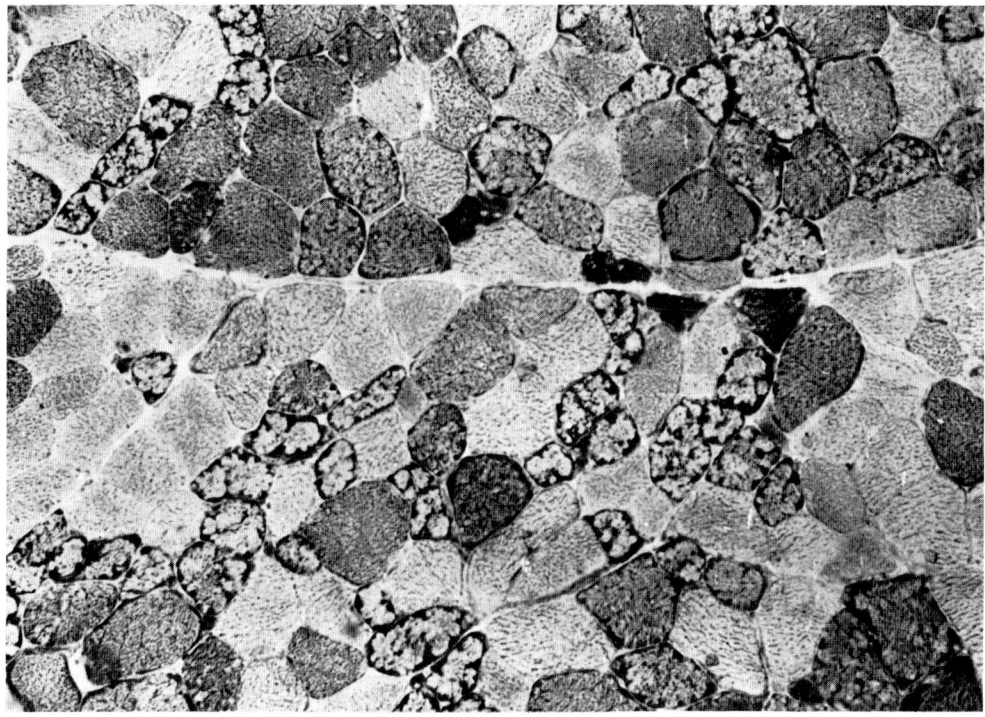

Fig. 10.8 Muscle from patient with facioscapulohumeral dystrophy showing abnormal Type 1 fibres with 'lobulated' appearance due to irregular mitochondrial distribution. NADH-TR. × 160

therefore to find that the histological features in a proportion of cases are indistinguishable from Becker M.D. Similarly, in other cases a bimodal size distribution typical of a neurogenic atrophy is discernible though 'myopathic' features including necrosis and phagocytosis may complicate interpretation of biopsy findings.

MYOTONIC DISORDERS

The main disorders in which myotonia is the predominant clinical feature are myotonic dystrophy (Steinert's disease) and myotonia congenita. These conditions are very dissimilar both in their clinical features as well as histologically and histochemically so that their differentiation presents no real problems (Engel & Brooke 1966).

Myotonic dystrophy

The considerable disparity in fibre size found in this disorder is due to atrophy of Type 1 fibres accompanied by Type 2 fibre hypertrophy affecting both 2A and 2B sub-types. Selective atrophy of Type 1 fibres is much less common in neuromuscular pathology than Type 2 atrophy which is encountered whenever phasic activity is restricted e.g. in simple disuse. Other characteristic features in myotonic dystrophy include the presence of chains of internal muscle nuclei and the presence of very numerous ring fibres. These are frequently associated with 'sarcoplasmic masses' which are zones of afibrillar sarcoplasm outside the rings of misorientated myofibrils. Because these areas contain abnormally dense concentrations of mitochondria and aqueous sarcoplasm they are particularly conspicuous in SDH and PAS preparations; since they contain no myofibrils they appear negative in the ATPase technique. Some authors report on acid phosphatase-positive deposits within muscle fibres, particularly in the infantile form.

Myotonia congenita

In contrast to the findings in myotonic dystrophy, muscle biopsies from patients with myotonia congenita are often histologically normal apart from showing generalised muscle hypertrophy. On histochemical examination however, a total absence of Type 2B fibres is an almost invariable finding, although the total percentage of Type 2 fibres remains within normal limits. These findings are common to both autosomal dominant and recessive forms of myotonia congenita and also occur in several other situations including congenital myotonic dystrophy.

PERIODIC PARALYSIS

Studies of serum potassium levels in cases of periodic paralysis have shown that there are hypokalaemic, hyperkalaemic and normokalaemic forms of this disorder. Hyperkalaemic periodic paralysis is closely allied, though not precisely identical to paramyotonia congenita. The typical pathological findings in periodic paralysis consist of vacuolation of muscle fibres which may be widespread during attacks but at other times muscle biopsy may be diagnostically non-contributory. The vacuoles contain mainly water and electrolytes; glycogen is present only in small amounts due to leakage from the surrounding sarcoplasm. The vacuoles are therefore only weakly PAS-positive and are not likely to be confused with vacuolation due to any of the glycogenoses (see p. 104). The hyperkalaemic and normokalaemic forms of periodic paralysis are more likely to develop a permanent myopathy typified by gross fibre atrophy with some compensatory hypertrophy, Type 1 fibre predominance and a high incidence of internal nucleation.

Both hypo- and hyperkalaemic forms are prone to striking cellular abnormalities referred to as 'tubular aggregates' (Engel et al 1970). These are most conspicuous in NADH-TR preparations, where they appear as areas of extremely dense reaction product (see Fig. 10.9). They are also PAS-positive and may be subsarcolemmal or central in position and may occupy up to one-third of the cross-sectional area of the fibre. The absence of SDH activity from the aggregates allows them to be distinguished with certainty from large clusters of mitochondria. Ultrastructural examination shows them to be composed of double-walled tubules c. 52 nm in diameter packed in hexagonal lattice array. Their high phospholipid content is demonstrable using Sudan black or Baker's acid haematein methods.

Their origin from the sarcoplasmic reticulum has been shown by electron microscopy. The aggregates are confined almost exclusively to Type 2B fibres.

CONGENITAL MYOPATHIES

The group of disorders included under this heading are relatively non-progressive diseases as compared with the muscular dystrophies or central denervating disorders, but are present from birth. The various forms are often difficult or impossible to distinguish clinically, both from each other and from juvenile SMA. Moreover, routine histological examination is generally insufficient to enable their differentiating features to be recognised. Histochemical techniques have proved indispensable for this purpose and have been responsible for the initial recognition of more than one variety of congenital myopathy.

Central core disease

In many cases of this disorder, the muscle biopsy could readily be dismissed as showing no abnormalities in routine histological preparations and the range of fibre size is frequently within normal limits. However, in SDH or NADH-TR preparations, numerous core-like areas devoid of enzyme activity are seen to be present. Sometimes there is only one such area per fibre, or alternatively there may be two, three or even more per fibre (see Fig. 10.10). Cores are usually present in a high proportion of fibres (>30%) and extend along the fibres for distances exceeding 1 mm. Ultrastructural examination shows these areas contain virtually no mitochondria.

It is common for all the fibres in cases of central core disease to be uniformly Type 1 but less frequently cases are found in which there is merely a Type 1 fibre predominance or even where the fibre type proportions are within normal limits. Sometimes older members of a family show only Type 1 fibres on muscle biopsy whereas younger relatives retain some Type 2 fibres. This finding and the fact that raised numbers of '2C' fibres are seen in the younger patients is suggestive of fibre type conversion during the course of the disease. Cores are normally confined to Type 1 fibres or to '2C' fibres. Frequently the

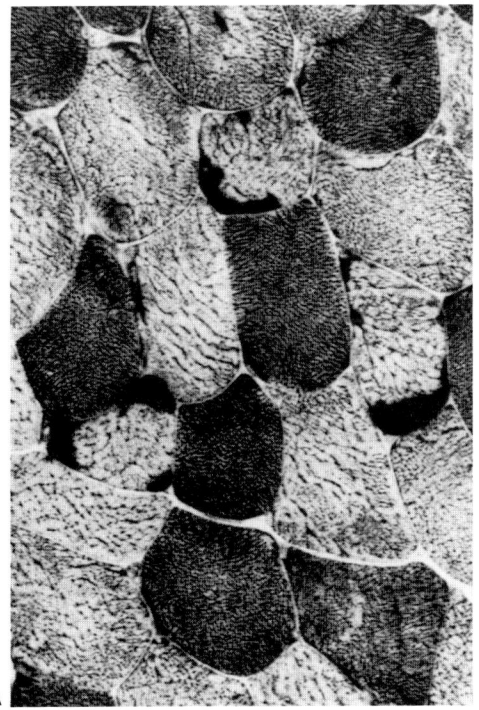
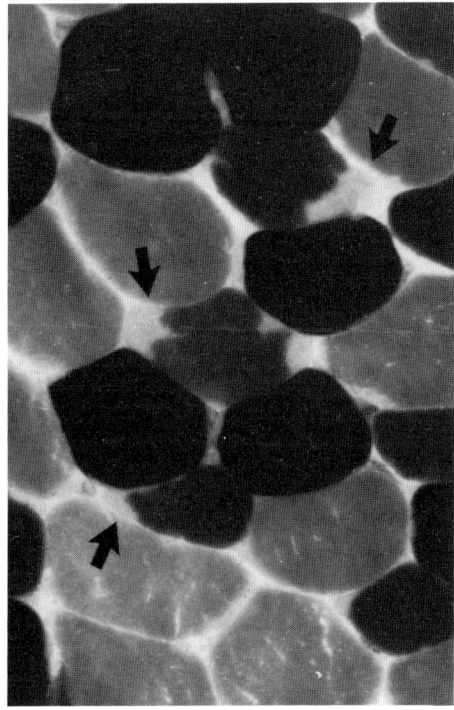

Fig. 10.9 Muscle from patient with hyperkalaemic periodic paralysis showing 'tubular aggregates' in Type 2B fibres (arrows). ×308
A. NADH-TR   B. ATPase after pH 4.6 pre-incubation

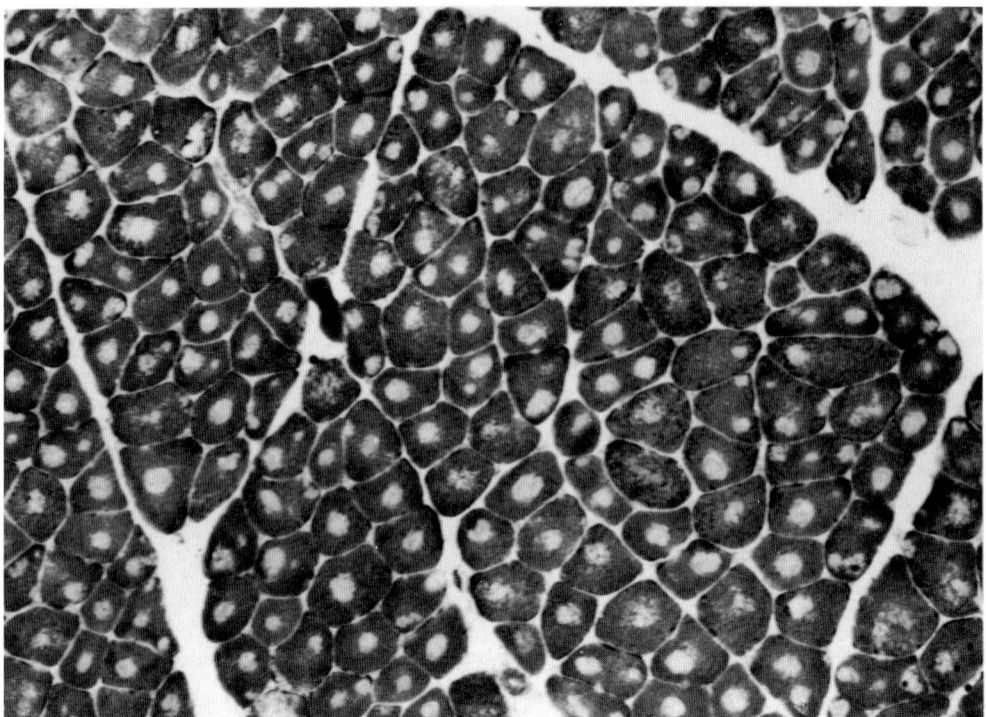

Fig. 10.10 Muscle from patient with 'central core' disease showing absence of mitochondrial enzyme activity from the cores. NADH-TR. ×120

cores are not detectable in ATPase preparations. This type of core is described as 'structured' because the myofibrils in it are either normal or only minimally disrupted. In 'unstructured' cores, the myofibrillar architecture is severely abnormal, resulting in low ATPase activity in the area of the cores (Neville & Brooke 1971).

Multicore disease (Minicore disease)

Multiple small core-like structures are found in another variety of congenital myopathy which has been duly named 'multicore' or 'minicore' disease. Its clinical features differ from those of central core disease in that it tends to be somewhat more severe with a certain degree of muscle wasting. Multicores resemble small unstructured cores extending for less than 100 μm and are thus visible as areas of low activity in ATPase preparations. The disruption of myofibrils is sometimes so severe as to be visible in H & E-stained sections. The margins of the cores are normally much less distinct in mitochondrial enzyme preparations than they are in central core disease.

This correlates with ultrastructural findings that mitochondria, though reduced in number, are not totally absent from the cores. A Type 1 fibre predominance is an almost invariable finding in this disorder but multicores are found in both fibre types.

Centronuclear myopathies

This clinically heterogeneous group of disorders is characterised by the presence of numerous central nuclei in a high proportion of muscle fibres (Bethlem et al 1970). Initially, the name 'myotubular myopathy' was given to one variant of this disorder in which the nuclei were situated in a central zone devoid of myofibrils. These fibres have undoubted morphological similarities with myotubes, but whereas true myotubes are invariably '2C' in histochemical profile since they are incompletely differentiated, the vast majority of these fibres have the histochemical characteristics of mature Type 1 fibres. The term 'centronuclear myopathy' is, therefore, preferable. The central afibrillar zones are prominent in SDH, NADH-TR and myophosphorylase preparations since they contain high concentrations of mitochondria and aqueous sarcoplasm. These zones are devoid of ATPase activity.

Another similar disorder is typified by the presence of very small Type 1 fibres (see Fig. 10.11) in addition to the incidence of central nucleation. It is difficult to know whether these fibres represent the effects of retarded growth (hypotrophy) or whether some loss in diameter has taken place (atrophy), because few cases have been investigated by means of serial biopsy. This type of centronuclear myopathy appears to be more rapidly progressive than the variant without fibre size disparity, but continued careful documentation of these disorders is necessary so that the variants can be more surely identified and prognosis more confidently given.

Congenital fibre type disproportion

This class of muscle disorder is characterised by disparity in fibre size of the main histochemical muscle fibres. Unlike the other categories of congenital myopathy included in this section, CFTD is, however, associated with no specific muscle fibre abnormality other than this disproportion in fibre size (Brooke 1973). Moreover, it is not certain whether the condition represents the effects of aberrant maturation of the motor unit as a whole, rather than a purely myopathic defect. All reported cases share a common clinical picture of congenital hypotonia which is static, slowly ameliorating or in some cases fatal. The first recognised cases were typified by the relative smallness of Type 1 fibres, with Type 2B fibres having the largest diameter. Subsequently, however, other variants with identical clinical histories have been found, in which Type 1 or Type 2A fibres are significantly larger than the other fibre types. A 'significant' degree of difference, in the context of CFTD, is defined as at least 12% between the largest mean diameters of largest and smallest fibre types.

Recent evidence suggests that CFTD may be a histological phenomenon and may not necessarily be related to a specific clinical entity. However, it is particularly important that this type of disorder

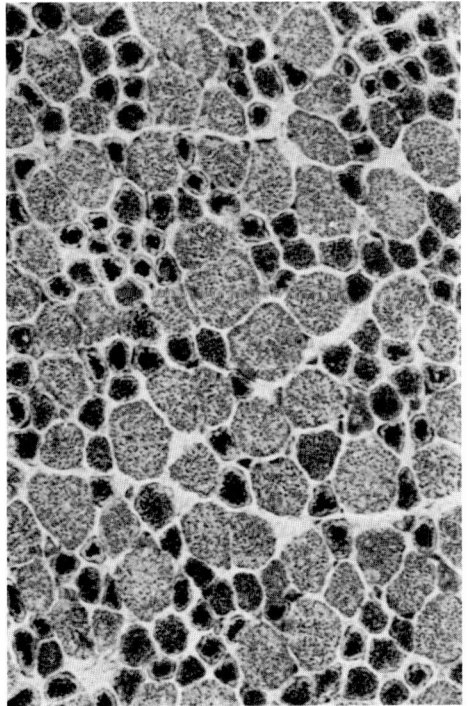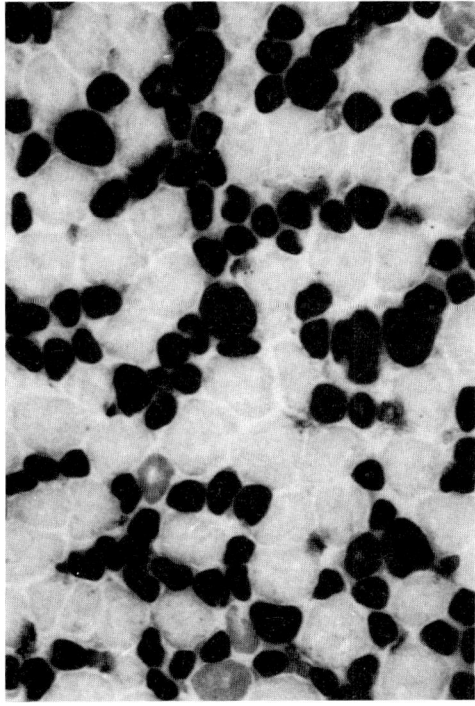

Fig. 10.11 Muscle from patient with centronuclear myopathy showing small myotube-like Type 1 fibres. ×308
A. NADH-TR B. ATPase after pH 4.3 pre-incubation

should be effectively distinguished from atypical cases of infantile SMA. Whereas SMA does not show the selectivity for individual fibre types which is characteristic of CFTD, biopsies from infants less then 3 months old may nevertheless fail to show the pattern of severe atrophy affecting all fibre types which is diagnostic of the Werdnig-Hoffmann form of SMA. The differential histological diagnosis should, however, be straightforward in biopsies taken after the age of 3 months.

Nemaline myopathy

This type of congenital myopathy takes its name from the rod-like protein-containing structures which are found in the disorder often in association with abnormal Z-line material. Although the rods can be seen in trichrome-stained sections, they are not usually apparent in H & E preparations. Severely affected fibres are frequently atrophied as compared with fibres which contain few rods. In the majority of cases Type 1 fibres are selectively affected but this is not an invariable finding and cases have been reported in which both fibre types contain nemaline rods or even where there was selective involvement of Type 2 fibres. Aggregates of nemaline rods show up as negative areas in ATPase sections.

The precise nature of the nemaline rods is still a matter of conjecture, though the fact that methods for tryptophan give negative results and those for arginine and tyrosine are positive has been interpreted as evidence that they resembled Z-line material in composition. Immunocytochemical studies have suggested that a major component may be α-actinin (Sugita et al 1973). Although similar structures may occasionally be found in other neuromuscular disorders, there is little doubt that nemaline myopathy represents a distinct clinico-pathological entity. Its mode of inheritance may be autosomal recessive but is much more often autosomal dominant; in families showing the latter type of genetic transmission the age of onset may vary widely.

METABOLIC MYOPATHIES

Skeletal muscle is involved in a wide range of metabolic disturbances; some of these affect muscle exclusively, while in others additional tissues are involved. Specific enzyme defects in carbohydrate or lipid metabolism and mitochondrial function have been identified; in many cases histochemical methods exist for the demonstration of the enzymes involved and hence the defects can be precisely detected. In many other instances the results of histochemical screening can at least indicate the general area in which further biochemical investigation should be concentrated. An important advantage of histo-chemical screening is that it is remarkably economical as regards use of tissue, since multiple metabolic processes can be investigated on very small tissue samples.

Glycogenoses

Skeletal muscle is affected in a majority of the established forms of glycogenosis. Some of these are characterised by the late onset of neuromuscular symptoms and are compatible with a normal life-span, though the capacity for strenuous exercise may be severely curtailed. In others, however, the onset is in the immediate post-natal period and the prognosis much less favourable (Rowland et al 1971). The main histopathological feature of the glycogenoses is vacuolation of muscle fibres, the severity of which is roughly proportional to the extent of glycogen storage. If individual vacuoles are very large, the stored glycogen cannot be retained in cryostat sections except by protecting them by means of prior celloidinisation.

Type 2 glycogenosis — acid maltase (α-1,4-glucosidase) deficiency

This defect exists in a severe infantile form (Pompe's disease) in which cardiac muscle is also affected, and a more benign late-onset form affecting skeletal muscle only. The infantile form is characterised by severe neonatal hypotonia and on histological examination vacuolation of the muscle is frequently so severe that only a small proportion of the sarcoplasm of each fibre remains, giving the tissue a lace-like appearance. Celloidinised PAS-stained sections show that the vacuolation is due to massive glycogen storage. Both fibre types seem equally affected. There is also some minor storage of a substance with the characteristics of an acid mucosubstance.

In the late-onset form (Fig. 10.12), the extent of glycogen storage and consequent vacuolation is much less severe. In both forms storage is within distended lysosome-like organelles which have high acid phosphatase activity associated with their limiting membranes. In the late-onset form especially, lipofuscin and other dense phospholipid-containing debris may be found in the vacuoles, giving a positive PAS reaction after diastase digestion. Glycogen storage tends to be more prevalent in Type 1 fibres; a possible explanation for this may be that glycogen levels in Type 2 fibres are decreased to a greater extent by the action of myophosphorylase which is unaffected in this disorder. Biochemical confirmation of the suspected enzyme defect is necessary, since there is no histochemical method for its demonstration.

Type 3 glycogenosis — debranching enzyme (amylo-1, 6-glucosidase) deficiency

This disorder presents in infancy with hepatomegaly and hypotonia. The former may subside but residual muscle flaccidity frequently persists. The stored glycogen is highly abnormal in that it contains short external chains and an increased number of branch points (limit dextrin) and is more soluble than normal glycogen. Isolated cases may show Type 1 fibre predominance, and multicore-like structures have also been reported in this disorder (Pellissier et al 1979). Again, the specific enzyme defect is not demonstrable histochemically and biochemical confirmation of the suspected diagnosis is necessary.

Type 4 glycogenosis — branching enzyme (amylo-1, 4 → 1, 6-transglucosylase) deficiency

This disorder is particularly rare; the recorded cases have involved young children and the prognosis is poor. The structure of the storage polysaccharide closely resembles amylopectin, with a vastly reduced number of branch points and long external chains. The presence of such branch points as there are is ascribed to the action of debranching enzyme working in reverse. The glycogen deposited has reduced solubility and is diastase-resistant.

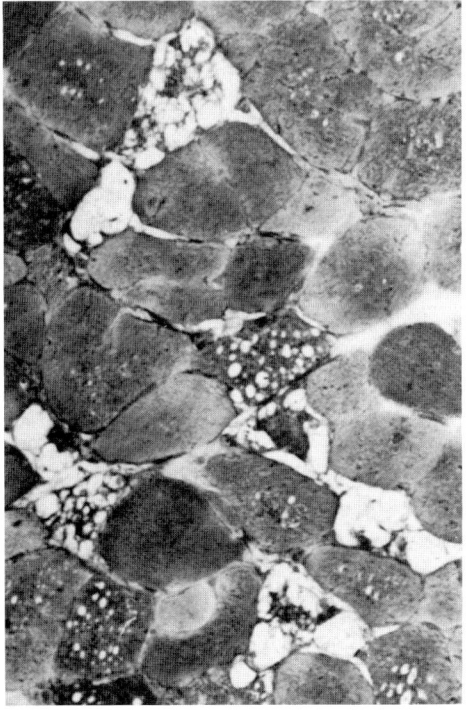 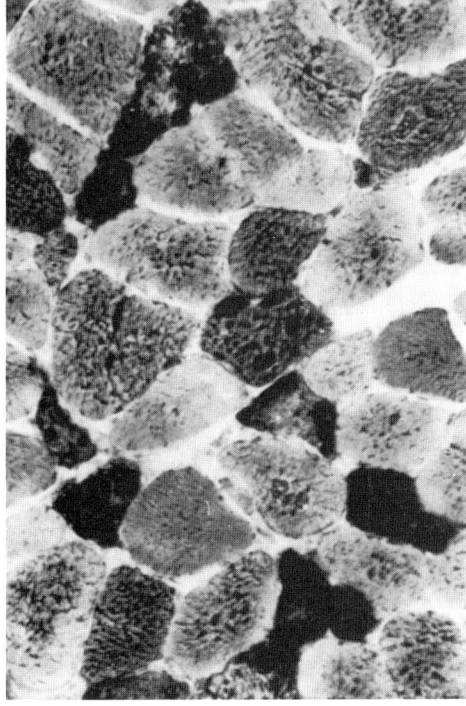

A   B

Fig. 10.12 Muscle from patient with a late onset form of Type 2 glycogenosis (acid maltase deficiency). Vacuolation of the fibres is due to glycogen storage. × 160. A. PAS without celloidinsation; B. PAS after celloidinisation; note retention of glycogen in vacuoles

Type 5 glycogenosis — myophosphorylase deficiency

This type of glycogenosis (McArdle's disease) generally becomes apparent in the teens or twenties because of weakness and pain on exertion. The enzyme deficiency seems to be totally confined to skeletal muscle phosphorylase; on histochemical examination this is completely absent. However, it can be seen that smooth muscle phosphorylase activity in arterioles is unaffected. There are two variants of the disorder (Feit & Brooke 1976). In one the myophosphorylase protein is totally absent, as can be shown by gel electrophoresis. In this form, foci of actual muscle necrosis are common and if regenerating myofibres are present these are found to contain apparently normal phosphorylase activity. The explanation for this is that a 'fetal' isoenzyme is present in myotubes prior to their maturation and that this isoenzyme is unaffected. Because of the occurrence of necrosis and regeneration, '2C' fibres are common in this form of McArdle's disease. The other variant of the disorder involves the production of phosphorylase protein in abnormally large amounts but this protein is enzymatically inert. The muscle is much less 'myopathic' and the only detectable abnormality may be the very modest degree of glycogen storage. Vacuolation is frequently inapparent unless carefully searched for, and is most usually subsarcolemmal.

Type 7 glycogenosis — phosphofructokinase deficiency

The clinical symptoms of this disorder are similar to those of McArdle's disease, with cramps and exercise tolerance being associated with the inability to utilise muscle glycogen fully as an energy source. The glycogen content is variable and may appear to be within normal limits. The enzyme defect can be demonstrated histochemically (Bonilla & Schotland 1970) since PFK activity can be visualised using a multi-step tetrazolium reduction method (see Appendix 5). The enzyme steps involved are PFK itself, aldolase, NAD-linked phosphoglyceraldehyde dehydrogenase and finally NADH-TR. It is obviously necessary to demonstrate that the enzyme pathway from aldolase onwards is functioning normally before concluding that a negative result is due to PFK deficiency. Thus histochemical assays of PFK and aldolase activity should be carried out in tandem, a negative result in the former and positive in the latter indicating a PFK deficiency.

Lipid storage disorders

Although there are numerous conditions in which accumulation of intracellular lipid in skeletal muscle is a predominant feature, the underlying enzyme abnormalities have been identified in comparatively few instances. Utilisation of substrates derived from fatty acids in oxidative phosphorylation is more prevalent in Type 1 fibres than in Type 2. Consequently, it is not surprising that most forms of lipid storage myopathy preferentially affect Type 1 fibres though this is not invariably the case. Screening of muscle biopsies for the presence of excess lipid is most conveniently carried out using Sudan black staining with propylene glycol as solvent since this procedure will allow the demonstration of both neutral and phospholipids. Differentiation of these classes of lipid can then be carried out using oil red O and acid haematein methods (Bourgeois & Hubbard 1965) either separately or in conjunction in the same tissue sections (see Appendix 4). On purely histological examination of cryostat sections, phospholipid deposits are frequently inapparent but neutral lipid droplets are often visible as tiny round vacuoles evenly dispersed throughout the affected fibres.

Carnitine deficiencies

Low levels of carnitine (γ-trimethylamino-β-hydroxybutyrate) in muscle (Engel & Angelini 1973) may be due either to a systemic deficiency caused by impaired hepatic carnitine synthesis or to a more benign condition in which plasma carnitine levels are usually normal. Both conditions are associated with reduced carnitine uptake by skeletal muscle, but the precise enzyme defect responsible for the abnormality has not been identified. Both types of carnitine deficiency give rise to the accumulation of neutral lipid in the intermyofibrillar spaces in muscle. Type 1 fibres are preferentially affected and moderate to severe atrophy of this fibre type may occasionally be encountered in cases of systemic carnitine deficiency. In cases of carnitine deficiency limited to skeletal muscle, lipid storage may be confined to only one sub-population of Type 1 fibres (Fig. 10.13) and fibre diameters are often normal.

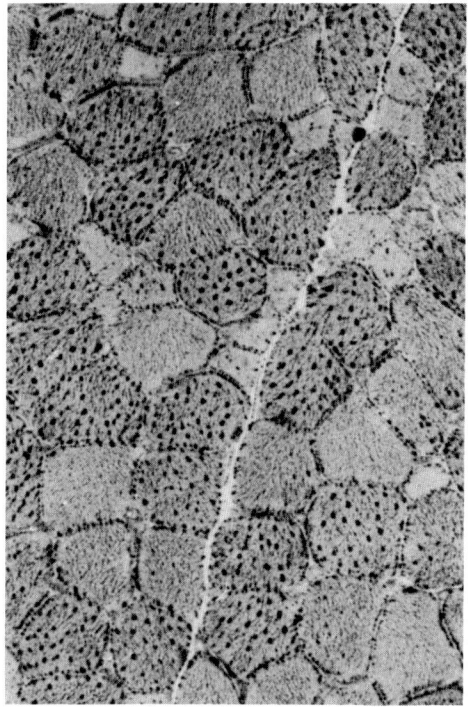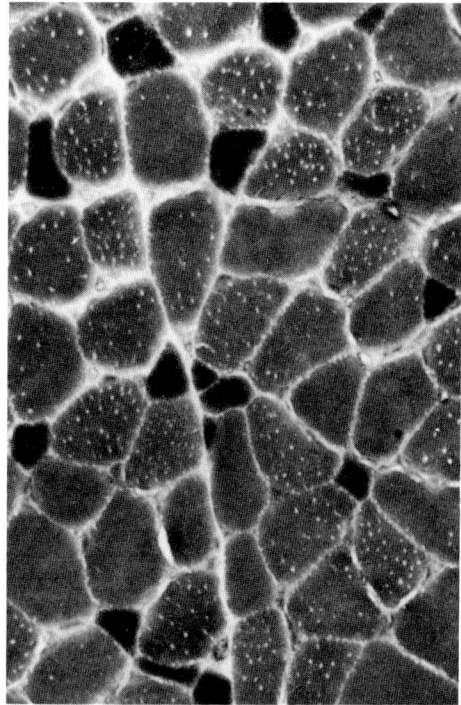

Fig. 10.13 Muscle from patient with carnitine deficiency showing Type 2 fibre atrophy and storage of neutral lipid in a proportion of the Type 1 fibres. A. Sudan black; B. ATPase pH 9.5

Carnitine palmityltransferase (CPT) deficiencies

Two components of CPT have been identified and together participate in the transport of fatty acids across mitochondrial membranes prior to oxidation. CPT 1 is situated on one side of the inner mitochondrial membrane and catalyses the conversion of acyl CoA into acylcarnitine; CTP 2 is situated on the other side of the membrane and catalyses the reverse reaction. Deficiencies in skeletal muscle may involve either or both enzymes (Scholte et al 1979) and fibroblasts and leucocytes are frequently affected. Although the deficiency in muscle causes severe cramps associated with exercise, lipid storage is minimal and may be inapparent except during attacks. The morphology of the muscle fibres is otherwise normal.

Ceroid-lipofuscinoses (Batten's disease)

There are several generalised ceroid-lipofuscinoses (see Ch. 7) in which deposition of ceroid- or lipofuscin-like pigments takes place not only in neuronal tissues but also in skeletal muscle (Carpenter et al 1972). Although deterioration in motor ability is purely secondary to severe diffuse brain damage, muscle biopsy is sometimes undertaken as a diagnostic procedure in preference to biopsy of rectum or appendix. Vast quantities of abnormal pigment-containing material is stored within lysosomes in muscle and is detectable by its autofluorescence, the PAS technique and the Schmörl reaction. Whereas very few lysosome-like organelles are present in the muscle of normal children and the acid phosphatase activity is low, in these disorders the grossly enlarged lysosomes contain high levels of acid phosphatase.

Refsum's syndrome

Moderate storage of neutral lipid may be seen in muscle biopsies from cases of this disorder which is due to a generalised metabolic defect in the handling of phytanic acid (tetramethylhexadecanoic acid), a common constituent of green vegetables and dairy products. Lipid storage in muscle is limited to Type 1 fibres and there is frequent evidence of uniform fibre type grouping since the metabolic defect is also responsible for a peripheral neuropathy due to involvement of Schwann cells and peripheral axons.

Mitochondrial myopathies

Myopathies in which mitochondrial abnormalities are a prominent pathological feature have been recognised largely because of the increasing routine use of histochemical techniques in the investigation of neuromuscular diseases. However, the identification of the precise metabolic defects responsible for the mitochondrial abnormalities has not so far been achieved, except in a small proportion of cases. Morphological and histochemical abnormalities in muscle mitochondria are found in a wide range of disorders including ocular and oculopharyngeal myopathies, some variants of FSH syndrome and myopathies in which lactic acidosis is an important concomitant feature. Until more is known of the underlying causes of these mitochondrial defects, the clinicopathological classification of these disorders is of necessity extremely tentative (Kamieniecka & Schmalbruch 1978, Egger et al 1981).

Ocular myopathies

Two basic types of mitochondrial abnormality are found in the various classes of ocular myopathy, in limb muscles as well as ocular muscles. In the syndrome of ophthalmoplegia in association with retinitis pigmentosa (Kearns-Sayre syndrome), Type 1 fibres are found to contain dense subsarcolemmal aggregates of mitochondria which may extend totally round the periphery of the fibres. These are prominent in NADH-TR and SDH preparations; a high proportion of Type 1 fibres may be affected and the excessive formazan deposition may occasionally be found throughout affected fibres. Fibres containing these broad peripheral zones of clustered mitochondria correspond to the so-called 'ragged-red' fibres seen in trichrome-stained sections (Olson et al 1972). Ultrastructural examination shows that many of these mitochondria contain crystalline inclusions or aggregations of stacked membranes.

A totally different type of abnormality is seen in pure ocular or oculopharyngeal myopathies. In these disorders, NADH-TR and SDH preparations reveal the presence of subsarcolemmal or central vacuoles which are devoid of enzyme reaction product but nevertheless have a prominent rim of formazan deposition round their periphery. These zones are basophilic in H & E-stained sections and red in trichrome preparations and are usually referred to as 'rimmed vacuoles'. Tiny atrophied fibres with extremely high mitochondrial enzyme activity are also common.

FSH syndrome

The mitochondrial changes seen in a small proportion of cases of this syndrome bear a general resemblance to those found in Kearns-Sayre syndrome. However, in contrast to the latter where the range of fibre size is usually within normal limits, in FSH cases the mitochondrial changes are accompanied by widespread variation in fibre size. Detection of mitochondrial abnormalities forms an important feature in the differentiation of this type of disorder from cases of FSH dystrophy and instances of FSH syndrome due to central denervating disorders.

Cytochrome abnormalities

Recent progress has been made in defining the underlying enzyme defects associated with varying degrees of lactic acidosis and marked exercise intolerance with onset in childhood. One such abnormality involves a severe deficit in the level of reducible cytochrome b in muscle, and a specific deficiency of NADH-CoQ reductase has also been reported. Neither of these abnormalities can be identified histochemically, though muscle biopsies from affected patients show mitochondrial changes in SDH preparations resembling those seen in Kearns-Sayre syndrome (Morgan-Hughes et al 1979).

A much more severe defect of cytochrome metabolism is seen in some cases of neonatal hypotonia associated with severe lactic acidosis (DiMauro et al 1980). Muscle biopsy in these infants shows minimal histological changes but randomly distributed fibres showing dense accumulations of reaction product in NADH-TR preparations are seen. The same fibres show histochemical evidence of excess neutral lipid deposition and the PAS reaction is sometimes indicative of gross glycogen depletion throughout the muscle. Histochemical demonstration of cytochrome oxidase activity (Seligman et al 1968, see Appendix 5) shows a complete deficit of this enzyme (Fig. 10.14). The condition proves rapidly fatal, death being due to respiratory and renal failure.

Defects of purine metabolism

Myoadenylate deaminase is important in muscle metabolism as a key enzyme in the regulation of the balance of ATP and ADP in muscle fibres and is also important since muscle has no other anaploteric reaction. Whereas moderate reductions in the level of this enzyme may be seen in various neuromuscular diseases, including muscular dystrophies and denervating disorders, some patients appear to have a total myoadenylate deaminase deficiency (Fishbein et al 1978). This condition gives rise to cramping pains and weakness on exertion but muscle biopsy reveals no histological abnormalities. The enzyme defect may be demonstrated histochemically using a method which involves reduction of a tetrazolium salt by ammonia in the presence of dithiothreitol (see Appendix 5).

Muscle cramps and pain on exercise are also encountered in another defect of purine metabolism, namely xanthine oxidase deficiency. However, levels of xanthine oxidase are very low even in normal muscle and the defect is more reliably demonstrated in biopsies of jejunum. Occasionally crystalline inclusions which probably contain xanthine and hypoxanthine can be found in the intermyofibrillar spaces in muscle fibres.

ENDOCRINE MYOPATHIES

This group of neuromuscular disorders is the source of numerous diagnostic problems because the effects of hormonal abnormalities on muscle may be exerted directly, via the central or peripheral nervous system or via the muscle vasculature. On occasions the effects of a combination of more than one of these mechanisms may be encountered in the same patient. Histochemical techniques have proved invaluable in this context, in differentiating between selective fibre type atrophies and those due to neurogenic

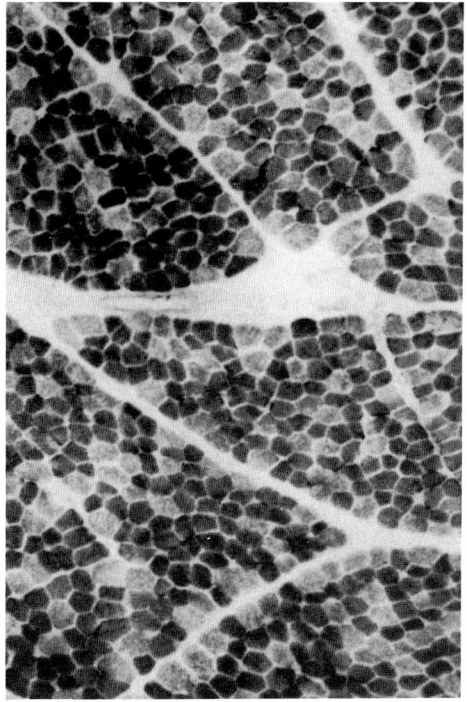
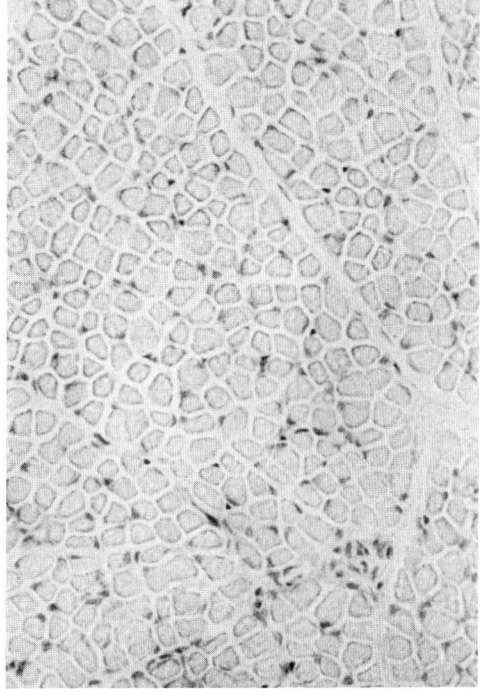

Fig. 10.14 Muscle from patient with a mitochondrial myopathy due to cytochrome oxidase deficiency. ×308
A. Control muscle showing normal cytochrome oxidase activity; B. Complete absence of activity in muscle of affected patient

mechanisms, and in detecting hormonal effects on particular enzymes.

Steroid myopathies

Muscle weakness and wasting are well-documented side-effects of steroid therapy and are also frequently associated with the endogenous glucocorticoid excess found in Cushing's syndrome. Selective Type 2 fibre atrophy is the most common manifestation of both conditions, though in severe Cushing's syndrome atrophy of Type 1 fibres may also occur (Pleasure et al 1970). The atrophied Type 2 fibres generally show reduced levels of myophosphorylase, a feature which helps to distinguish the Type 2 atrophy in this situation from that due to simple disuse. In both Cushing's syndrome and iatrogenic steroid myopathies moderate neutral lipid accumulation may also occur, affecting Type 1 fibres almost exclusively.

Neuromuscular abnormalities in diabetes

The direct effects of insulin deficiency on muscle tend, not surprisingly, to parallel those of glucocorticoid excess. Again Type 2 atrophy is the predominant pathological feature and the myophosphorylase activity of these fibres may be severely reduced. The distribution of mitochondria is often irregular and some degree of lipid storage is also a common feature. These 'direct' effects of diabetes on the muscle fibres may, however, be complicated by the presence of a neuropathy, the underlying cause of which is damage to the vasculature of the peripheral nerves. Under these circumstances small group atrophy involving both Type 1 and Type 2 fibres is seen but uniform fibre type grouping or large group atrophy are not normally encountered.

Dysthyroid myopathies

Abnormal thyroid hormone levels are responsible for a variety of pathological changes in skeletal muscle. Neuropathies of the compression type may be encountered in severe hypothyroidism (myxoedema) and mixed motor and sensory neuropathies have also been documented in hyperthyroidism. In these cases the pathological findings are indistinguishable from other peripheral neuropathies. However, in the absence of any electrophysiological evidence of neuropathy, it is often found that fibre type proportions are abnormal in dysthyroid patients. The number of Type 1 fibres is generally raised in hypothyroid subjects, whereas hyperthyroid patients often show increased numbers of Type 2 fibres. The mechanism of fibre type conversion appears to be a neurally-mediated effect of iodothyronine levels on ATPase activity in muscle.

INFLAMMATORY MYOPATHIES

It might be thought that the inflammatory myopathies would represent one area of muscle pathology in which histological techniques alone should suffice for the purposes of diagnosis. However, the pathogenesis of the various forms of myositis varies greatly and histochemical techniques are particularly useful in analysing the patterns of muscle damage seen in each. The dermatomyositis of childhood and adolescence is a fairly well-defined clinico-pathological entity, in the pathogenesis of which vascular damage plays an important role (Carpenter et al 1976). In adult dermatomyositis a more variable pattern of muscle destruction is apparent with only some cases showing evidence of vascular damage. This group has a high incidence of associated malignant disease and considerable overlap with the clinical and immunological features of connective tissue disease. In adult polymyositis, uncomplicated by skin changes or other evidence of multi-system disorders, the muscle pathology does not seem to be associated with vascular lesions but the underlying mechanism is at present not known.

Juvenile dermatomyositis

This disorder is characterised by recurrent capillary damage affecting skin, skeletal muscle and the alimentary tract. If muscle biopsies are taken during an early phase of capillary necrosis the vascular impairment can be seen to cause loss of ATPase and mitochondrial oxidative activity from the centre of affected fibres. The distribution of these fibres is often peri-fascicular and the detection of this type of change may enable the diagnosis to be made even in the absence of inflammatory cell infiltrates which may prove elusive. A subsequent stage of muscle damage often involves total necrosis and regeneration of these fibres. The regenerating fibres are readily distinguishable from normal fibres on account of their small diameter and high sarcoplasmic RNA content. This feature is clearly demonstrated in methyl green-pyronin preparations; the prominent nucleoli of regenerating fibre nuclei also show a very high RNA content. During maturation these fibres appear as '2C' fibres in ATPase preparations; differentiation into mature Type 1 or Type 2 fibres occurs after reinnervation but their central nucleation and small diameter tend to persist. If restitution of capillary damage is sufficiently rapid it is probable that total muscle fibre necrosis does not occur, but such fibres may show moderate to severe atrophy. Perifascicular atrophy in a quiescent stage of the disorder may be the only indication of previous muscle damage. The precise mechanism of vascular damage in dermatomyositis is not known, but a high incidence of positive immunostaining for IgG, IgM and C_3 in the vessels of the juvenile form and some adult cases is suggestive of complex-mediated cell damage (Whitaker & Engel 1972).

Adult dermatomyositis

In dermatomyositis with onset in adult life, the incidence of capillary necrosis varies greatly from case to case. In certain circumstances it is widespread enough to cause muscle infarction affecting whole fascicles. Loss of myophosphorylase and glycogen are the earliest indicators of irreversible muscle necrosis; subsequently ATPase and mitochondrial enzyme activity is lost and the fibres degenerate totally. The post-acute stage of this type of change is marked by the prevalence of regenerating myogenic elements, demonstrating high RNA content. Regenerating capillaries are conspicuous because of their high ATPase activity.

Large foci of inflammatory cells may be found in dermatomyositis complicating connective tissue disorders such as rheumatoid arthritis or systemic lupus erythematosus. Almost invariably there are signs of muscle damage adjacent to these foci which, if not detectable histologically, are seen clearly enough in histochemical preparations because of the enzyme abnormalities involved (Fig. 10.15). This is

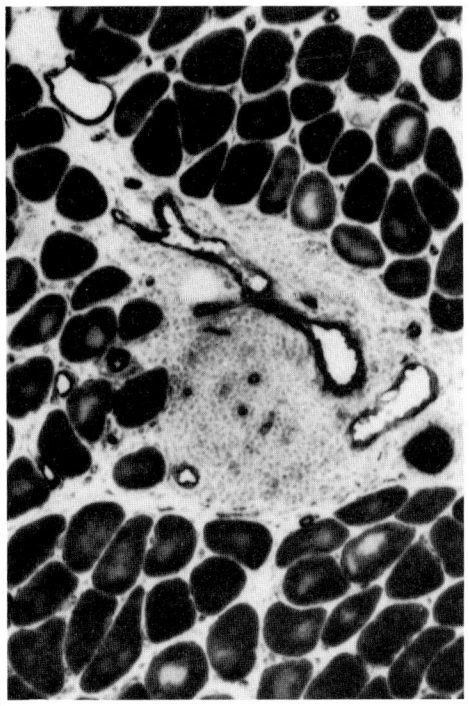
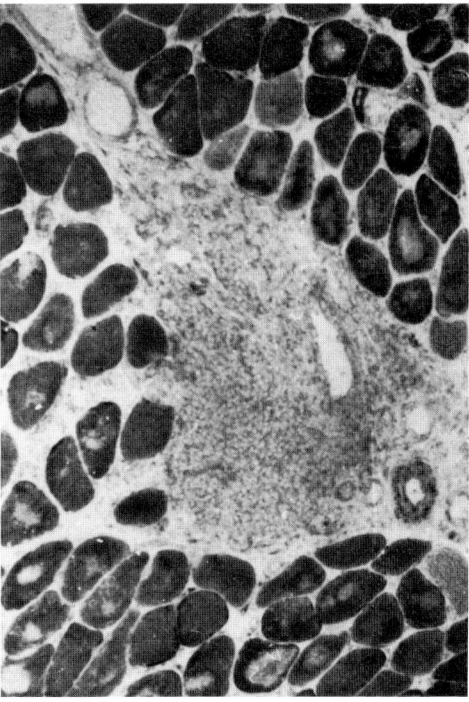

Fig. 10.15 Muscle from patient with dermatomyositis showing a large perivascular inflammatory focus and loss of enzyme activity from the centres of adjacent fibres. ×120 A. ATPase pH 9.5; B. NADH-TR

an important factor in the differentiation of myositis occurring in association with connective tissue disorders from polymyopathies where inflammatory foci, e.g. rheumatoid nodules, are not associated with any muscle destruction.

Polymyositis

In polymyositis uncomplicated by features of connective tissue disease, isolated fibre necrosis is prevalent rather than affected fibres being grouped in distribution. In contrast to almost all cases of juvenile dermatomyositis and many cases of adult dermatomyositis, in which inflammatory foci are confined to connective tissue septa, similar foci are also found within muscle fascicles. The type of necrosis is also substantially different from that seen in association with vascular damage. Vacuolation of fibres is a prominent feature but central loss of ATPase activity is not seen. Necrosis of individual fibres seems more protracted and occasional intracellular calcification, detectable in von Kossa preparations, may be apparent. As in all inflammatory muscle disorders, methods for the demonstration of ribonucleoprotein are useful in the estimation of numbers of active plasma cells and similarly the number of actively phagocytic cells can be gauged using lysosomal enzyme techniques. There is little evidence of immune complex deposition in the vessels of 'pure' polymyositis but a postulated mechanism of muscle damage involves cell-mediated cytotoxic effects on some component of the muscle fibres themselves. A more complete investigation of this type of process using immunocytochemical techniques may result in clarification of the pathogenesis of this condition and greater precision in diagnosis.

REFERENCES

Bethlem J 1980 Myopathies, 2nd edn. Elsevier/North Holland, Amsterdam
Bethlem J, Van Wijngaarden G K, Mumenthaler M, Meijer A E F H 1970 Centronuclear myopathy with type 1 fiber atrophy and 'myotubes'. Archives of Neurology 23: 70–73
Bonilla E, Schotland D L 1970 Histochemical diagnosis of muscle phosphofructokinase deficiency. Archives of Neurology 22: 8–12
Bourgeois C, Hubbard B 1965 A method for the simultaneous demonstration of choline-containing phospholipids and neutral lipids in tissue sections. Journal of Histochemistry and Cytochemistry 13: 571–578
Brooke M H 1973 Congenital fiber type disproportion. In: Kakulas B A (ed) Clinical studies in myology. Excerpta Medica, Amsterdam, p 147–159
Brooke M H, Engel W K 1966 The histologic diagnosis of neuromuscular diseases: a review of 79 biopsies. Archives of Physical Medicine and Rehabilitation 47: 99–121
Brooke M H, Kaiser K K 1970 Three 'myosin adenosine triphosphatase' systems. The nature of their pH lability and sulfhydryl dependence. Journal of Histochemistry and Cytochemistry 18: 670–672
Burstone M S 1958 Histochemical demonstration of acid phosphatases with naphthol AS phosphates. Journal of the National Cancer Institute 21: 523–531
Carpenter S, Karpati G, Andermann F 1972 Specific involvement of muscle, nerve and skin in late infantile and juvenile amaurotic idiocy. Neurology (Minneapolis) 22: 170–186
Carpenter S, Karpati G, Rothman S, Walters G 1976 The childhood type of dermatomyositis. Neurology (Minneapolis) 26: 952–962
DiMauro S et al 1980 Fatal infantile mitochondrial myopathy and renal dysfunction due to cytochrome c oxidase deficiency. Neurology (Minneapolis) 30: 795–804
Dubowitz V, Brooke M H 1973 Muscle biopsy, a modern approach. Saunders, Philadelphia
Dubowitz V, Pearse A G E 1960 Reciprocal relationship of phosphorylase and oxidative enzymes in skeletal muscle. Nature (London) 185: 701–702
Egger J, Lake B D, Wilson J 1981 Mitochondrial cytopathy: a multi-system disorder with ragged red fibres. Archives of Disease in Childhood 56: 741–752
Engel A G, Angelini C 1973 Carnitine deficiency of human skeletal muscle with associated lipid storage myopathy. Science 173: 899–902
Engel W K 1961 Muscle target fibres, a newly recognised sign of denervation. Nature (London) 191: 389–390
Engel W K, Brooke M H 1966 Histochemistry of the myotonic disorders. In: Kuhn E (ed) Progressive Muskeldystrophie, Myotonie, Myasthenie. Springer-Verlag, Stuttgart, p 203–222
Engel W K, Bishop D W, Cunningham G C 1970 Tubular aggregates in type II muscle fibers: ultrastructural and histochemical correlation. Journal of Ultrastructural Research 31: 507–525
Feit H, Brooke M H 1976 Myophosphorylase deficiency: two different molecular etiologies. Neurology (Minneapolis) 26: 963–967
Fishbein W N, Armbrustmacher V W, Griffin J L 1978 Myoadenylate deaminase deficiency: a new disease of muscle. Science 200: 545–548
Hayashi M, Freiman D G 1966 An improved method of fixation for formalin-sensitive enzymes with special reference to myosin adenosine triphosphatase. Journal of Histochemistry and Cytochemistry 14: 577–581
Kamieniecka Z, Schmalbruch H 1978 Myopathies with abnormal mitochondria: a clinicopathologic classification. Muscle and Nerve 1: 413–415
Meijer A E F H 1968 Improved histochemical method for the demonstration of the activity of α-glucan phosphorylase. Histochemie 12: 244–252
Morgan-Hughes J A, Darveniza P, Landon D N, Land J M, Clark J B 1979 A mitochondrial myopathy with a deficiency of respiratory chain NADH-CoQ reductase activity. Journal of the Neurological Sciences 43: 27–46

Neville H E, Brooke M H 1971 Central core fibers: structured and unstructured. In: Kakulas B A (ed) Basic research in myology. Excerpta Medica, Amsterdam, p 497–511

Olson W, Engel W K, Walsh G O, Einaugler R 1972 Oculocraniosomatic neuromuscular disease with 'ragged-red' fibers. Archives of Neurology 26: 193–211

Pearse A G E 1972 Histochemistry, theoretical and applied, 3rd edn. Churchill Livingstone, Edinburgh

Pellisier J F, DeBarsy T, Faugere M C, Rebuffel P 1979 Type III glycogenosis with multicore structures. Muscle and Nerve 2: 124–132

Pleasure D E, Walsh G O, Engel W K 1970 Atrophy of skeletal muscle in patients with Cushing's syndrome. Archives of Neurology 22: 118–125

Rowland L P, DiMauro S, Bank W J 1971 Glycogen storage disease of muscle. Problems in biochemical genetics. Birth Defects Original Articles Series 7: 43–51

Scholte H R, Jennekens F G I, Boury J J B 1979 Carnitine palmityltransferase II deficiency with normal carnitine palmityltransferase I in skeletal muscle and leucocytes. Journal of the Neurological Sciences 40: 39–51

Seligman A M, Karnovsky M J, Wasserkrug H L, Hanker J S 1968 Non-droplet ultrastructural demonstration of cytochrome oxidase activity with a polymerising osmiophilic reagent, diaminobenzidine (DAB). Journal of Cell Biology 38: 1–14

Sugita H, Masaki T, Ebashi S, Pearson C M 1973 Protein composition of rods in nemaline myopathy. In: Kakulas B A (ed) Basic research in myology. Excerpta Medica, Amsterdam, p 298–302

Walton J N (ed) 1981 Disorders of voluntary muscle, 4th edn. Churchill Livingstone, Edinburgh

Whitaker J N, Engel W K 1972 Vascular deposits of immunoglobin and complement in idopathic inflammatory myopathy. New England Journal of Medicine 286: 333–338

11

Skin

A. McQueen & E. Heyderman

With the establishment of dermatopathology as an important branch of general pathology has come the development of sophisticated techniques in the investigation of skin diseases. Histochemical methods are among such techniques many of which are still experimental but may well in the future prove valuable in specific diagnostic situations. The section on the immunopathology of the skin will deal in detail with immunofluorescence methods using fresh frozen tissue and reference will also be made to immunoperoxidase methods on paraffin wax-embedded material.

The prerequisite for successful interpretation of skin histology is a good biopsy, and there should be no hesitation in reporting as a 'failed biopsy' any skin specimen which has been severely traumatised by forceps, or which lacks orientation so that tangential cutting obscures or distorts the total picture. In skin

disease many clinically dissimilar conditions may display at some time in their natural history very similar histological appearances, so that close collaboration between dermatologist and pathologist is essential. In the case of non-neoplastic disease of the skin the biopsy site should be selected bearing in mind that very early or late lesions may well lack diagnostic criteria, and that the clinically established representative lesion usually provides the best information at the histological level. The use of adrenalin-free local anaesthetic is important on account of the artefacts which may be produced in collagen by adrenalin, and it is useful for the biopsy to be taken through the edge of a lesion so as to include a portion of the adjacent clinically normal skin (Fig. 11.1). If tissue is required for cryostat sections as well as for routine histology, the specimen may be divided before removal. For both purposes it is essential that

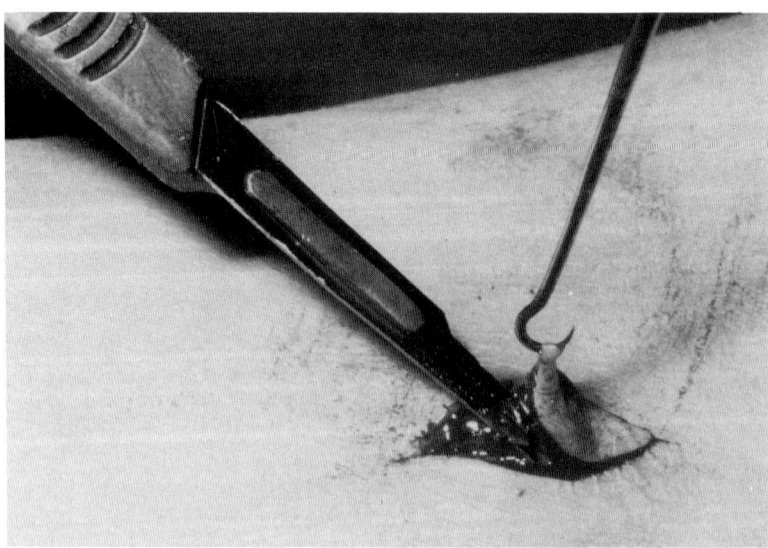

Fig. 11.1 Biopsy taken through the edge of a lesion to include adjacent clinically normal skin

the specimen be orientated properly by placing it dermis down on to a small piece of blotting paper, before snap-freezing in liquid nitrogen, or any other suitable quenching material, or placing it in 10% neutral formalin complete with blotting paper.

If difficulty is encountered in sectioning skin biopsies, an 8% phenol in methylated spirits step may be included in the processing schedule. However, it does sometimes happen that this softening has been inadequate, and in this case treatment of the paraffin block — for 1 h immediately before cutting — in Mollifex© (a BDH product) is advisable.

STAINS

The great majority of dermatopathological problems may be solved by routine histological methods. Dermatopathologists tend to develop their own personal system of diagnostic staining profiles, and the methods suggested below are by no means comprehensive but have been found over many years to provide useful information whereby often specific diagnosis may be suggested.

Periodic acid-Schiff

Probably the most commonly used stain after H & E, the PAS stain, has a number of applications in dermatopathology apart from its most obvious application in distinguishing between glycogen and mucin. It is a useful routine practice to have a PAS preparation to detect fungal elements which may be present in keratin, especially when dealing with biopsies of vulvar skin. The basement membrane at the epidermo-dermal interface stains well with PAS, and the thickened basement membrane of lupus erythematosus is strikingly demonstrated. When a healing lesion is suspected, the increase in the glycogen content of the keratinocytes over dermal granulation tissue is well seen, and it is of benefit to become familiar with the expected amount of glycogen seen at various sites and under various conditions. In porphyria, both the erythropoietic variety and in porphyria cutanea tarda, PAS-positive material is present around the blood vessels in the upper dermis, and is diastase-resistant (see also p. 157). In diabetic microangiopathy, fine granular deposits of PAS-positive material may be seen within the walls of small blood vessels and may suggest an unexpected diagnosis of diabetes. The granules of granular cell myoblastoma, not always clearly seen on H & E, are also PAS-positive and diastase-resistant, and the presence of large atypical PAS-positive cells in the epidermis should raise suspicion of areolar or extramammary Paget's disease. Finally, the use of topical steroids for lengthy periods produces changes in the dermis among which is a perivascular mantling of diffuse pale PAS-positive material.

Alcian blue

This is but one method whereby glycosaminoglycans may be demonstrated, and there are several conditions in which the use of the stain will either be diagnostic or will narrow the differential diagnosis. The following are such examples:

Follicular mucinosis or alopecia mucinosa occurs in two forms; as a benign primary disorder of the pilosebaceous unit, or as a manifestation of lymphoma which may not yet be overt. Histologically it is characterised by the presence of mucin in the outer root sheath and the sebaceous gland, first as beads or strings, later as more definite small pools.

Myxoedema and pretibial myxoedema

Reticular erythematous mucinosis has a non-specific perivascular infiltrate of lymphocytes and histiocytes in the upper and mid-dermis but shows a diffuse accumulation of mucin throughout affected areas.

Other conditions where minute to moderate amounts of mucin in the form of beads and strings are seen in the dermis include:

Lupus erythematosus and *dermatomyositis,* the latter in association with an increase in the number of small vascular clusters seen in the papillary dermis.

The so-called *lupus profundus (lupus erythematosus panniculitis),* where subcutaneous nodules are found which consist of inflammatory infiltrates of lymphocytes, plasma cells and histiocytes obliterating the fat and bundles of hyalinised sclerotic collagen in which streaks of mucin may be demonstrated.

Papular mucinosis or lichen myxoedematosus, a condition characterised by asymptomatic papules usually on the face and arms and not associated with thyroid disease, shows stellate fibroblasts and thickened collagen bundles between which substantial amount of mucin can be identified.

Scleromyxoedema with marked thickening of the

dermal collagen and numbers of fibroblasts irregularly scattered throughout may be difficult to differentiate from a fibroma unless mucin is looked for in the dermis.

Lymphocytic infiltration of the skin (*Jessner*) also shows coarse beads of mucin in the dermis.

Urticaria pigmentosa, where the infiltration of mast cells is shown by the alcian blue (pH 1) method.

IMMUNOPATHOLOGY OF THE SKIN

Within the last 20 years immunofluorescence techniques (IF) have become established as major contributions to the diagnosis of certain bullous skin diseases as well as some 'collagen' diseases, notably lupus erythematosus (Harrist & Mihm 1979). These will be discussed in detail here. In addition there are a number of conditions in which immunofluorescence findings have been reported but have no diagnostic value. It is essential that the clinician performing the biopsy has a clear understanding of what is required. The skin biopsy must be obtained using an atraumatic technique, particularly in the case of bullous diseases. Further details of handling the specimen and transport to the laboratory are given in Appendix 1.

Serum specimens are obtained by separation from 10 ml of clotted venous blood within 1 h of collection and put into a sterile container. Serum need not be frozen, and may be sent to the laboratory if necessary by mail. It should be emphasised that whole blood specimens should only be delivered to the laboratory if it is possible to do so within 1 h of collection since lysed blood contaminates the serum and may invalidate the test. Freezing of the whole blood and keeping it overnight is not an acceptable method of preserving a specimen.

For bullous diseases, a divided skin specimen, half for IF and half for routine histology, is recommended. The half for IF should be immediately snap-frozen. An early representative lesion is chosen, preferably no more than 12 h old, and the biopsy should include both lesion and adjacent clinically normal skin which together with a serum sample will usually provide the diagnosis.

If chronic discoid lupus erythematosus (CDLE) is suspected, a biopsy should be taken from a lesion which is at least 1 month old and upon which there

has been no topical steroid treatment for at least 2–4 weeks before biopsy; very early lesions may give negative results, as may those lesions treated by topical steroids.

When systemic lupus erythematosus (SLE) is suspected, a biopsy obtained from clinically normal exposed skin such as the wrist may be useful in diagnosis, but the results must be considered along with other parameters such as the presence of anti-nuclear antibody, DNA binding capacity of serum, and complement levels. A diagnosis of LE should not be made solely on the basis of positive IF findings.

For direct IF tests for general diagnostic purposes the profile of conjugates should comprise anti-human IgA, IgG, IgM and complement (C3), with any other conjugates added when specifically required e.g. fibrinogen. A typical conjugate labelling for a specimen would read as follows: IgA, IgG, IgM, IgC3: IgA, IgG, IgM, IgC3, thus permitting four superficial and four deeper sections from the block to be tested. Details of methods are given in Appendix 6.

The bullous diseases

Until the advent of IF methods the classification and differential diagnosis of bullous diseases remained arbitrary at best: now, with the application of agreed criteria, specific diagnosis can be reached in the majority of these conditions (Table 11.1).

Table 11.1 Diagnostic immunofluorescence tests in bullous skin disease

	Direct	Indirect
Pemphigus group	+ve IC	+ve IC
Bullous pemphigoid	+ve BMZ	+ve BMZ
Cicatricial pemphigoid	+ve BMZ	−ve
Herpes gestationis	+ve BMZ	−ve
Dermatitis herpetiformis	+ve BMZ or DP	−ve
Chronic bullous disease of childhood	+ve BMZ	+ve BMZ

IC = Intercellular; BMZ = Basement membrane zone; DP = Dermal papillae

The pemphigus group (Fig. 11.2)

In all forms of pemphigus, circulating antibodies directed against the intercellular substances of squamous epithelium are demonstrable during active

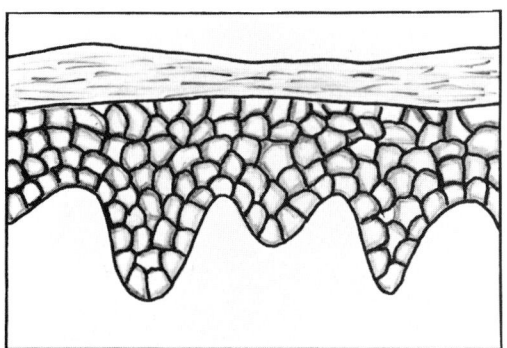

Fig. 11.2 Pemphigus group: diagram to show the intercellular deposition of IgG in the squamous epithelium. (Direct IF test)

phases of the disease by the indirect IF test. (Figs 11.3 and 11.4). The antibodies are of the IgG class, and while in general the titre reflects the severity of the disease, this is not absolute.

Pemphigus-like antibodies may be found in patients with severe burns, practolol eruptions, penicillin eruptions, trichophyton rubrum infections, pemphigoid and with high titre anti-A and anti-B blood group sera. Such findings are rare, and usually are seen weakly at titres of up to 1/40. Certain drugs have been incriminated as factors in the production of

pemphigus, notably penicillamine, and others such as phenylbutazone, rifampicin, and captopril.

Direct IF tests will demonstrate intercellular IgG in the epidermis of perilesional skin as well as complement in virtually all patients in whom the disease is active. In the rare variety known as pemphigus erythematosus (Senear-Usher syndrome), both intercellular and basement membrane deposition of immunoglobulin and complement may be demonstrated by direct immunofluorescence, especially if the biopsy is obtained from light-exposed skin. In mixed bullous disease, also known as pemphigus herpetiformis, the clinical features of dermatitis herpetiformis are present, but the immunofluorescence findings are those found in pemphigus.

Bullous pemphigoid (Fig. 11.5)

Circulating basement membrane antibodies of the IgG class can be demonstrated by indirect IF tests in about 70% of cases (Fig. 11.6), but there is no consistent correlation between severity of the disease and the IF antibody titre; indeed positive findings may be obtained after clinical recovery. Perilesional

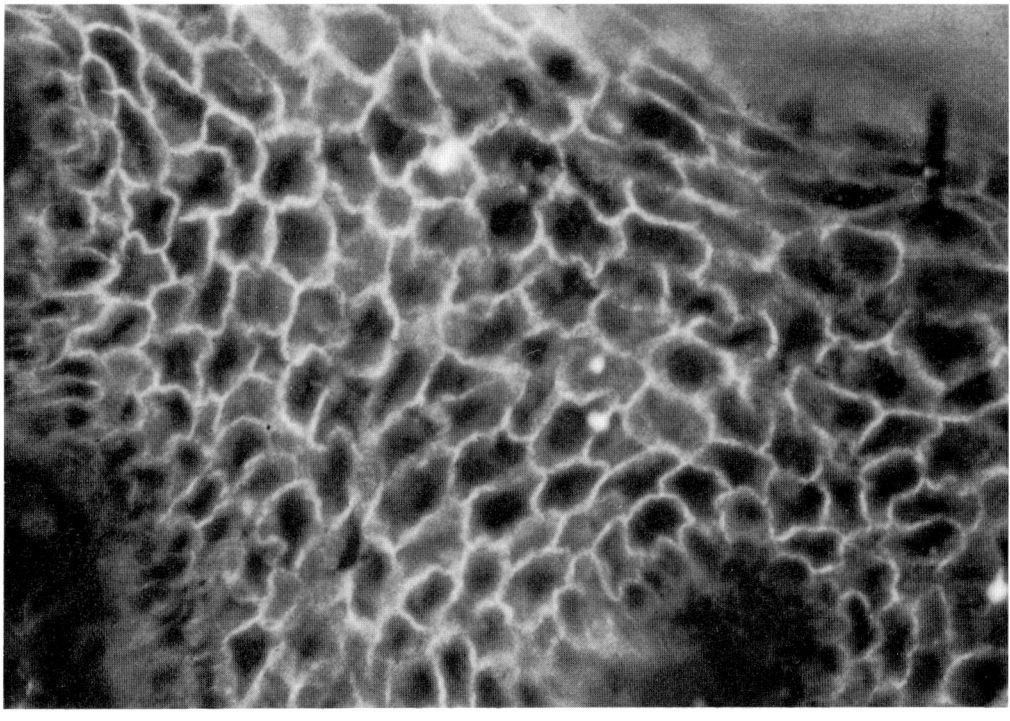

Fig. 11.3 Pemphigus: Indirect IF test to show the intercellular deposition of IgG in the epidermis

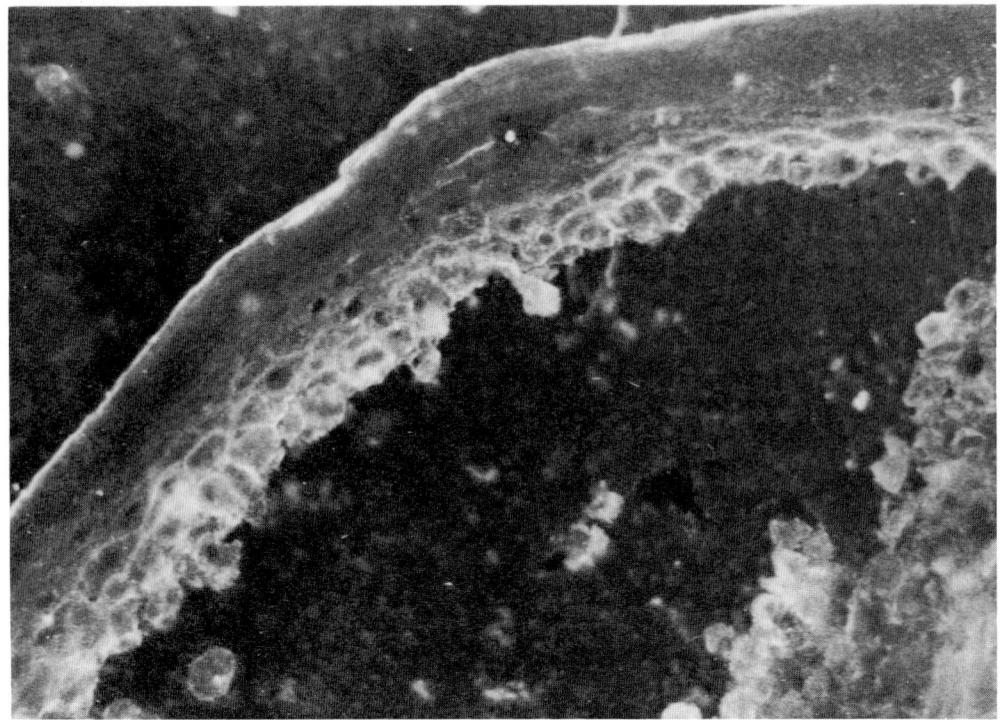

Fig. 11.4 Pemphigus (bulla): Direct IF to show IgG deposition in the epidermis

skin biopsy shows linear deposition of immuno-globulin, usually IgG, and complement at the basement membrane zone on direct IF.

In *cicatricial pemphigoid*, indirect IF tests are usually negative, while on direct IF testing, the majority of patients with the disease will show immunoglobulin and complement in a linear fashion at the basement membrane zone of affected skin or mucosa. IgG is most frequently found, but IgA and rarely IgM have been reported.

Herpes gestationis is a rare blistering disease seen

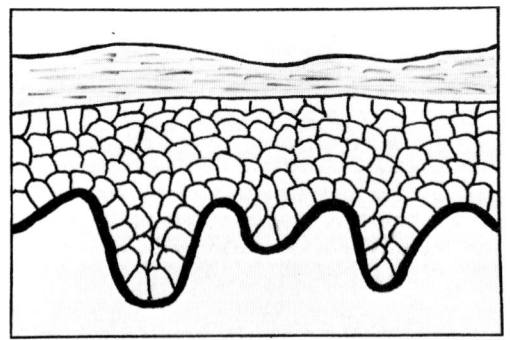

Fig. 11.5 Bullous pemphigoid: diagram to show basement membrane zone deposition of IgG in linear fashion on direct IF

during pregnancy or the puerperium, has a tendency to recur during subsequent pregnancies, and may also recur if the patient uses an oral contraceptive. Characteristically, direct IF tests on perilesional skin show linear deposition of C3 at the basement membrane zone, and very occasionally IgG. Indirect tests are negative in this disorder.

Dermatitis herpetiformis (Fig. 11.7)

This is an intensely itchy disorder in which symmetrical crops of urticated papules and vesicles occur on the back, extensor aspects of the limbs, the external genitalia and the scalp. The great majority of patients also suffer from a gluten-sensitive enteropathy with jejunal villous atrophy. Histo-logically the vesicles are seen to be subepidermal in location.

In a perivesicular lesion or in perilesional skin, direct IF tests show a characteristic deposition of IgA in a granular fashion in the dermal papillae in over 80% of cases (Fig. 11.8), but a linear band of IgA deposition at the basement membrane zone may be seen in a few cases. A high percentage of patients will

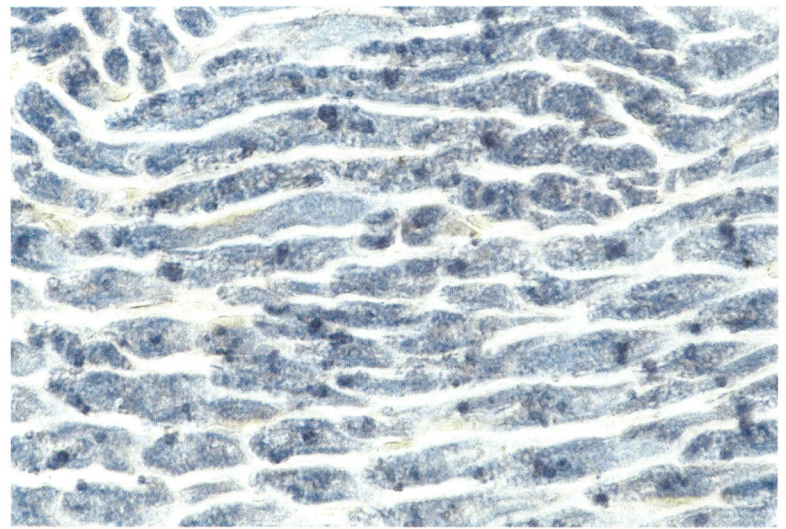

A

Plate 1 Secondary demyelination
(Wallerian degeneration) in rat sciatic
nerve A. Normal nerve B. 10 days after
nerve section C. 14 days after nerve
section. Dichromate–acid haematin–Oil
Red O–Methyl Green ×846. Note the
progressive loss of normal myelin (blue
staining) and a corresponding increase
in the cholesterol esters (red).

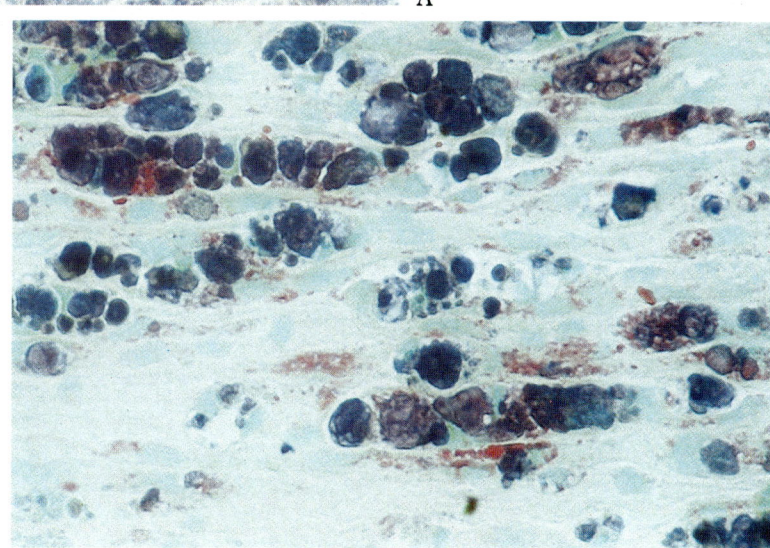

B

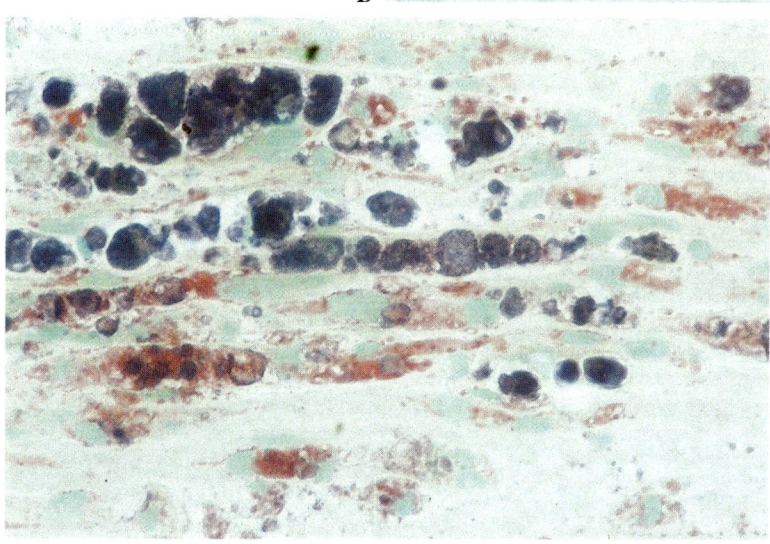

C

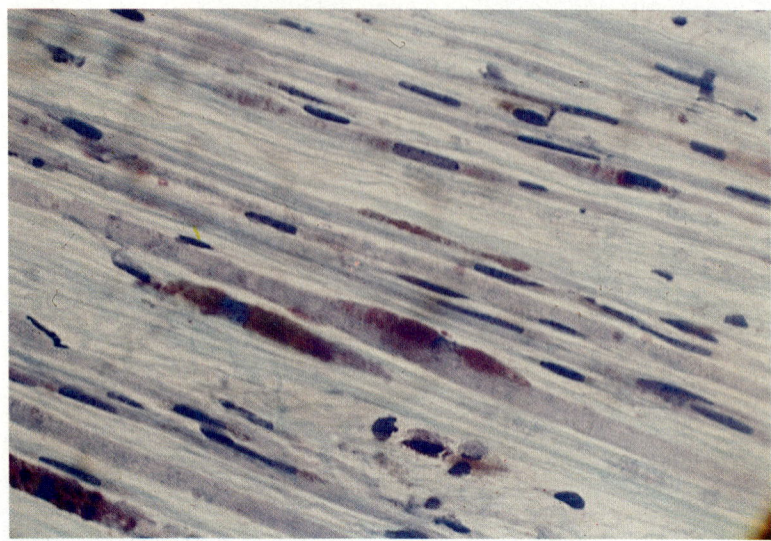

Plate 2 Metachromatic leucodystrophy; cryostat section of a sural nerve biopsy stained with toluidine blue (Bodian & Lake) to show yellow, brown and red metachromatic deposits of sulphatide in Schwann cells. ×1404

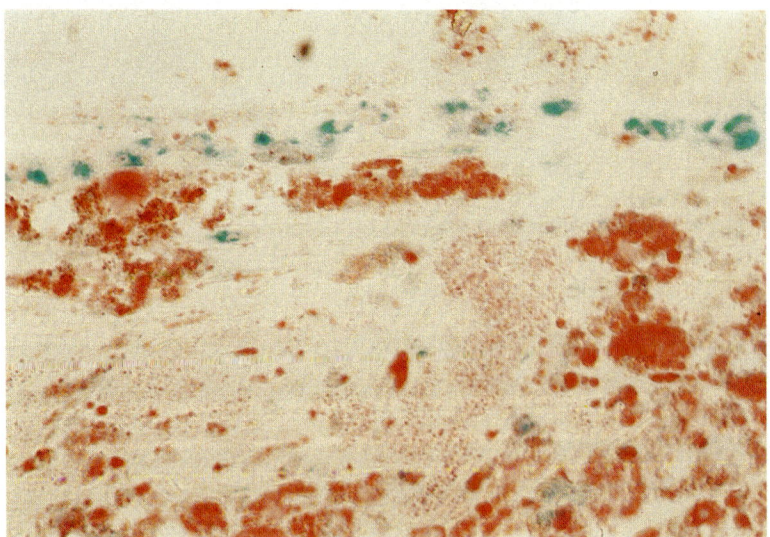

Plate 3 Lysosomal β-galactosidase activity in macrophages (blue) and sudanophilic lipid (red) in an advanced complicated atherosclerotic lesion.

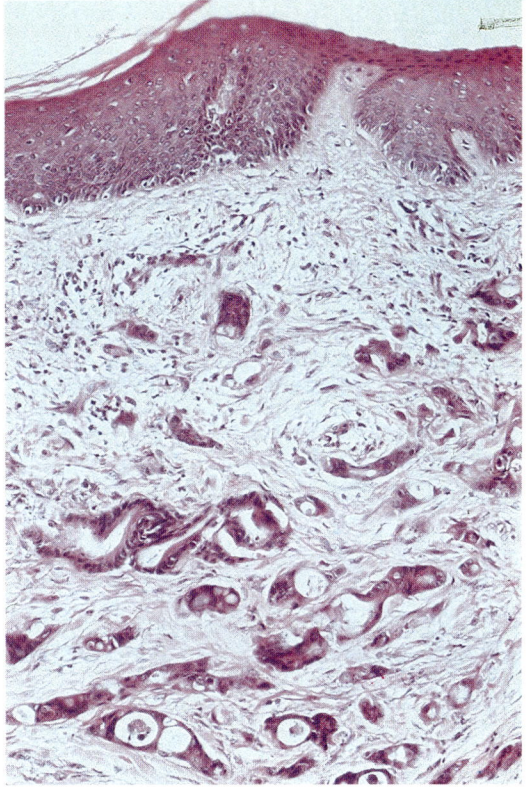

4

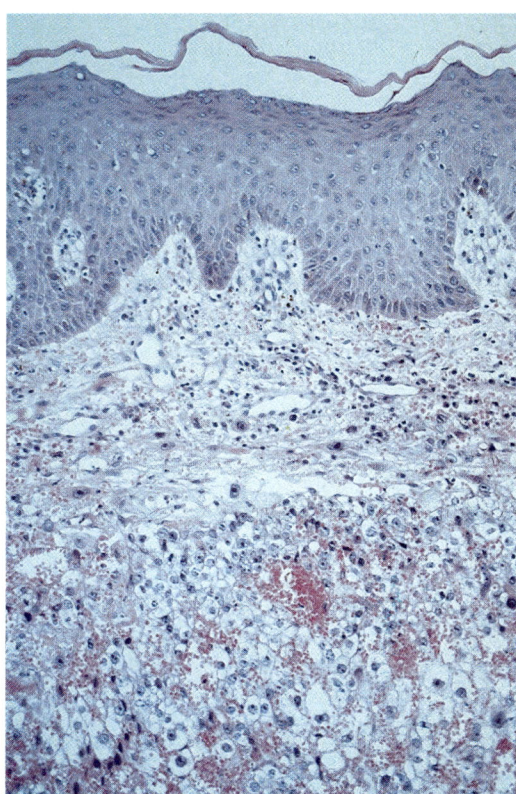

Plate 6 Secondary deposit of colon carcinoma in the umbilicus — 'Sister Joseph's nodule'. ×252 (H & E)

5

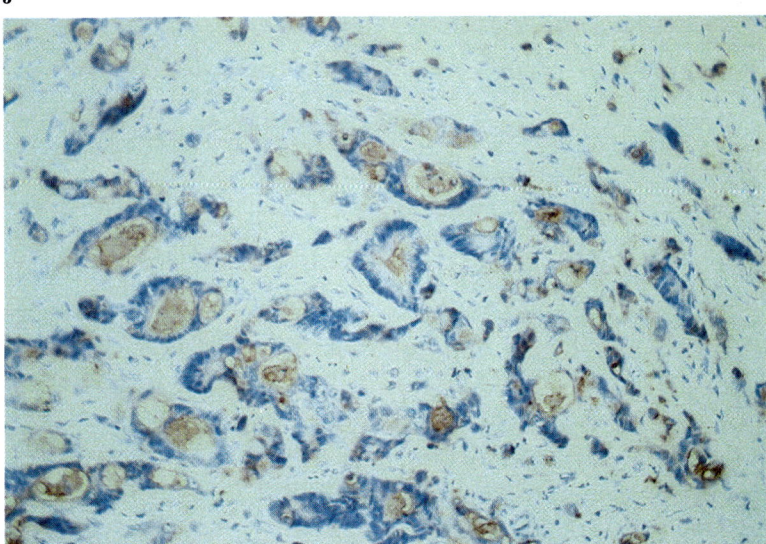

Plate 4 and 5 Skin deposit of metastatic breast carcinoma. There were multiple lytic lesions in the skull but the immunoperoxidase stain for epithelial membrane antigen (EMA) in Plate 5 confirms the epithelial origin of the tumour and excludes multiple myeloma. This single cell pattern of invasion is typical of lobular carcinoma of the breast, which in this patient was only found clinically some months after presentation with skin metastases and bone involvement.
Plate 4 H & E ×252, Plate 5 EMA immunoperoxidase ×1008

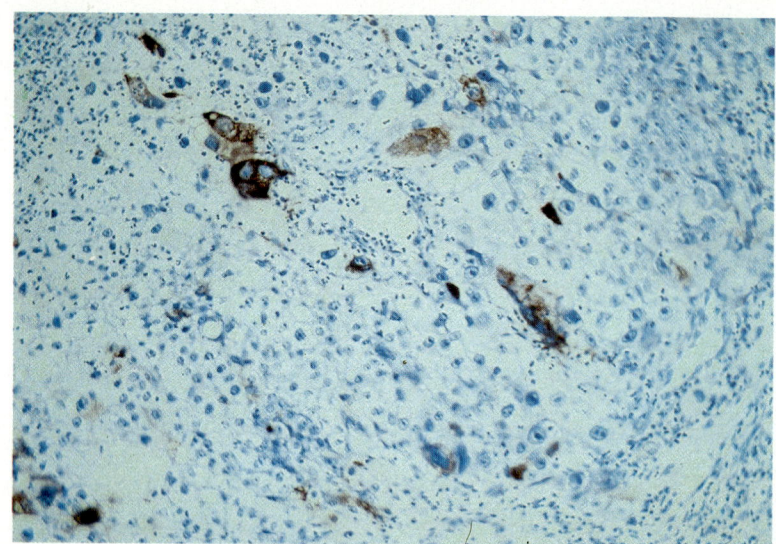

Plate 7 The colon carcinoma deposit stained with an indirect immunoperoxidase technique for carcinoembryonic antigen (CEA). The pattern of staining of the luminal membrane and necrotic debris in the acini is typical of gut carcinomas but may be seen in other adenocarcinomas. ×630

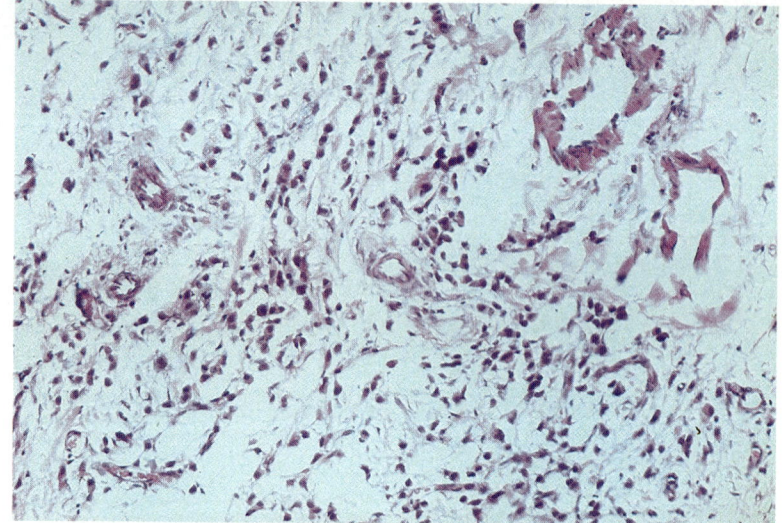

Plate 8 Vulval deposit of metastatic gestational choriocarcinoma. The appearances were somewhat atypical and suggested possible metastatic anaplastic carcinoma. ×630 (H & E)

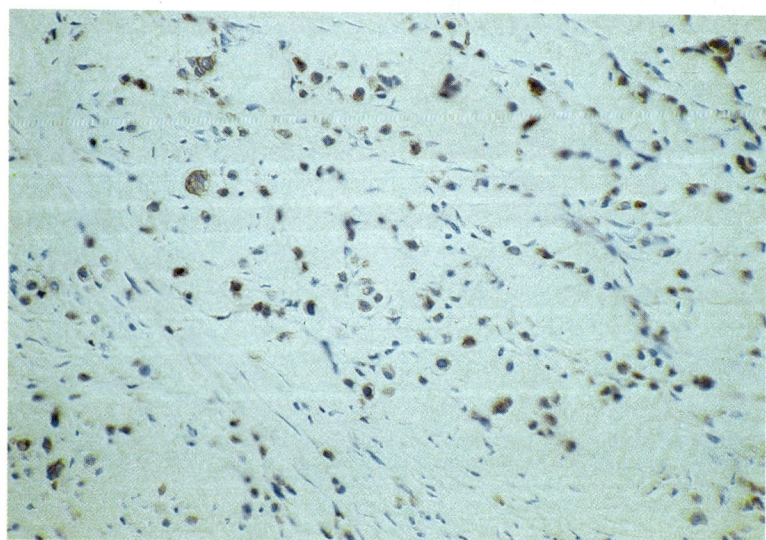

Plate 9 The metastatic choriocarcinoma stained with antisera to the specific beta subunit of HCG reveals abundant HCG-positive syncytial giant cells in close association with nests of clear cells, the malignant cytotrophoblast. ×1008

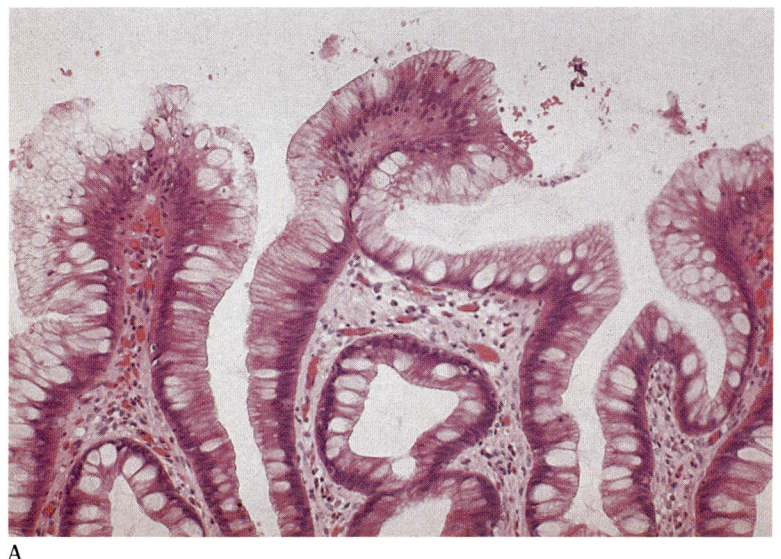

A

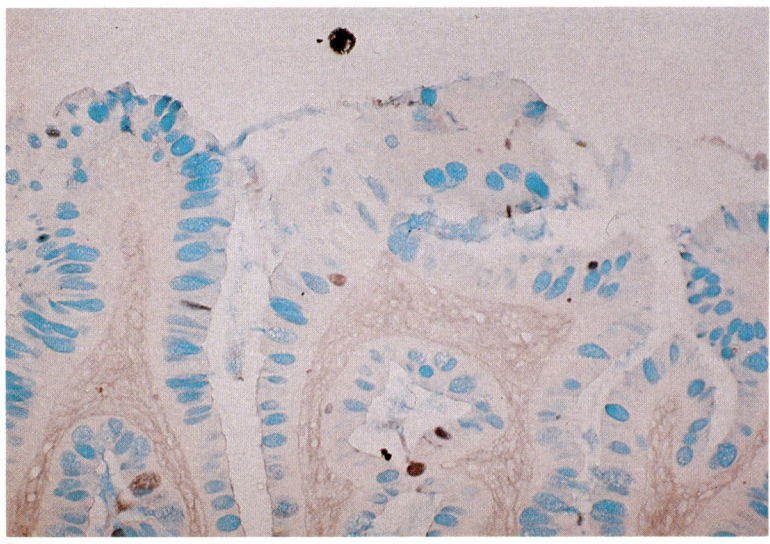

B

Plate 10 Gastric mucosa showing Type I or complete intestinal metaplasia: it resembles small intestine with goblet cells secreting sialomucins (blue). The columnar cells do not contain mucin and remain unstained. ×360 A. H. &. E; B. HID-AB method

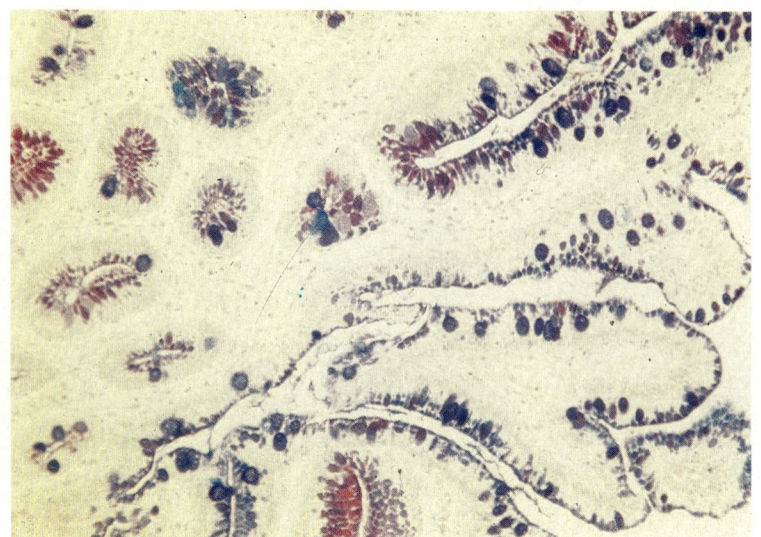

A

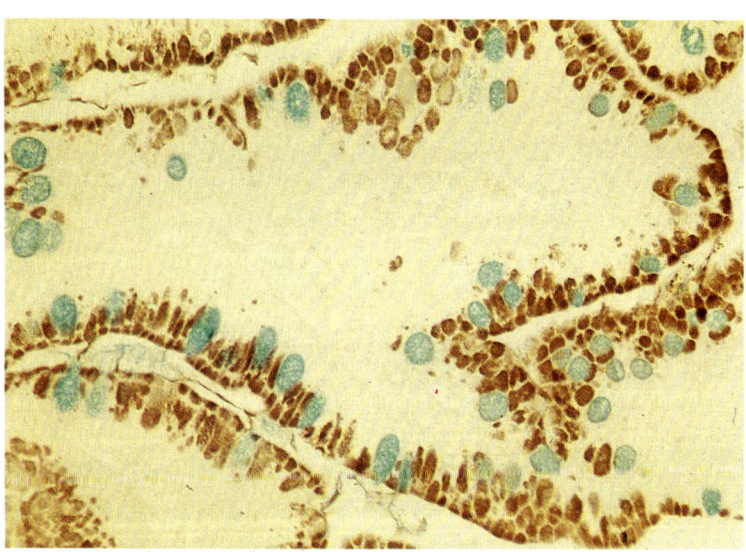

B

Plate 11 Gastric mucosa showing Type III or incomplete intestinal metaplasia. It contains elements of both gastric and intestinal epithelia in various degrees of maturation as illustrated in the section (A) stained with AB pH 2.5 + PAS. The range of staining characteristics from magenta through shades of purple to blue reflect the mature and hybrid forms of gastric and intestinal epithelia present. In the parallel section (B) stained with HID-AB goblet cells secrete both sialo- (blue) and sulphomucins (brown) and the intervening columnar cells secrete acid mucins with a high content in sulphated material (brown).

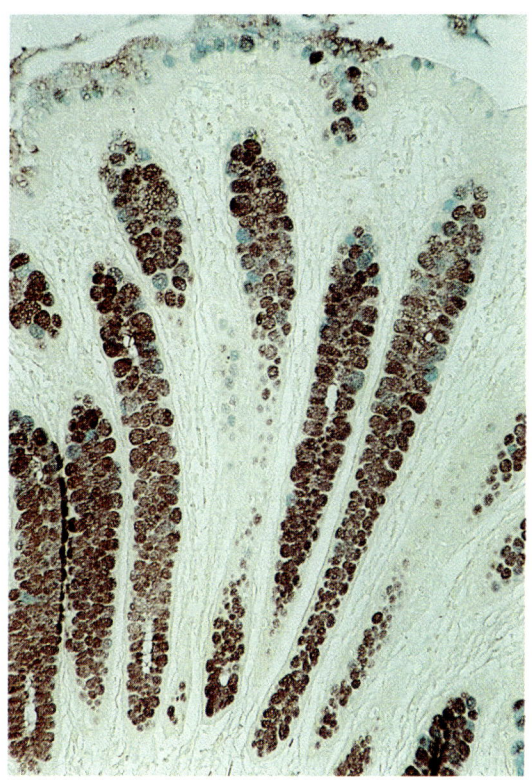

Plate 12 Normal colonic mucosa showing a predominance of sulphomucins (brown). (HIB-AB)

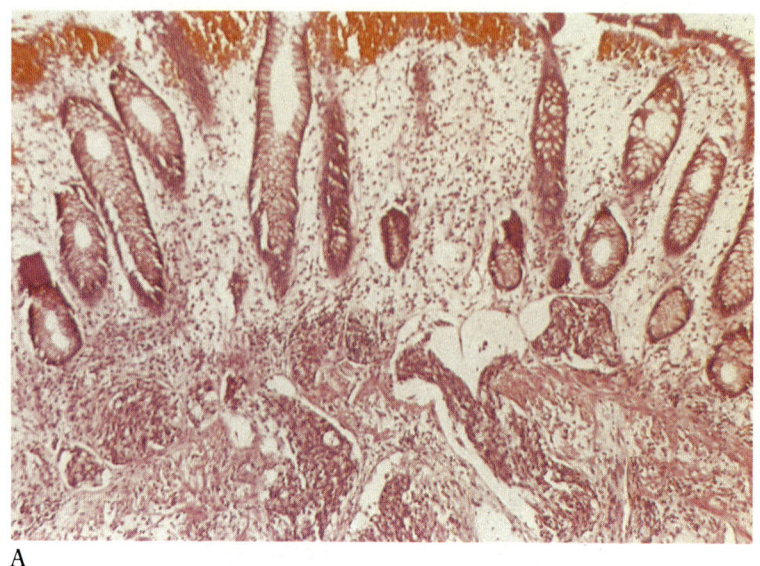

A

B

Plate 13 Colonic adenocarcinoma (in the submucosa) and the overlying mucosa show an abnormal mucus pattern with predominance of sialomucin (blue). Compare with Plate 12. A. H & E; B. HID-AB.

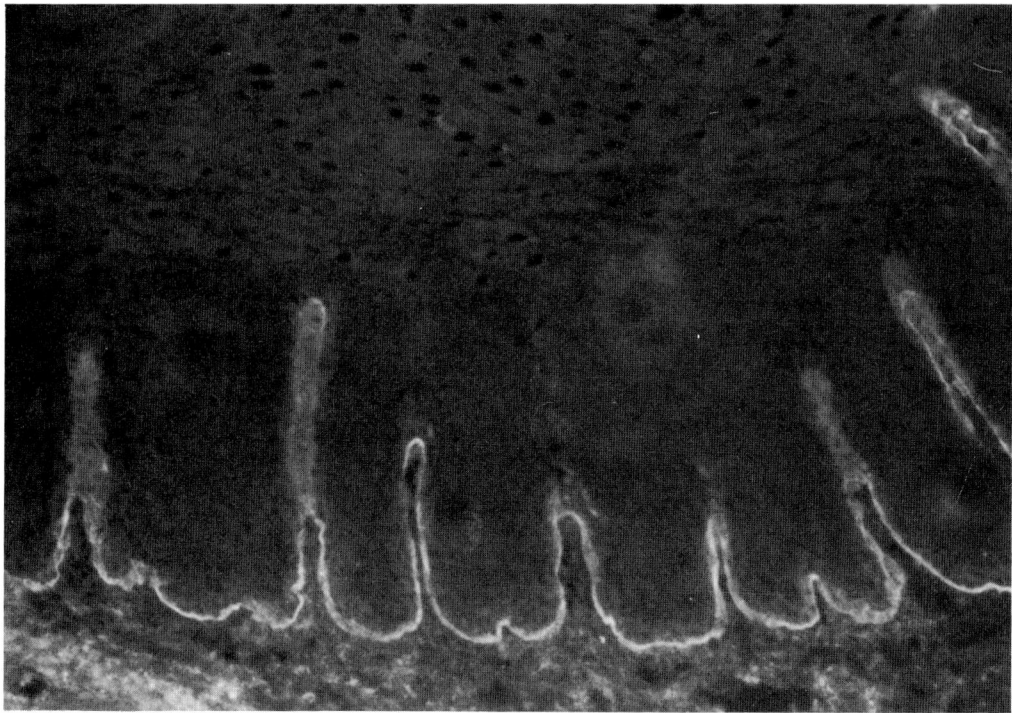

Fig. 11.6 Pemphigoid: IgG deposition in linear fashion in the basement membrane zone. (Indirect IF)

have C3 demonstrable either in a granular or linear fashion in the dermal papillae or along the basement membrane zone. Only a few, especially where there is linear banding, will contain circulating IgA in their sera, demonstrable at low titre at the basement membrane zone on indirect IF testing. In chronic bullous dermatosis of childhood (CBDC) bullae are found usually in the pelvic area, and the condition may recur over 2 or 3 years, but is self-limiting.

Direct IF tests show linear deposits of IgA along the basement membrane zone in the majority of cases (see Fig. 11.5) as well as circulating IgA type antibody directed against the basement membrane zone in about 70% of cases. Patients with CBDC do not suffer from an enteropathy such as that seen with dermatitis herpetiformis.

Lupus erythematosus (Table 11.2)

Two forms of this disease are recognised. In chronic discoid lupus erythematosus (CDLE) lesions are found only in the skin and the course of the disease is benign. Systemic lupus erythematosus (SLE), however, is a multi-system disease and may have a

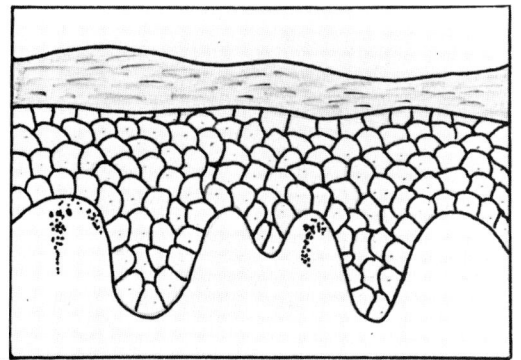

Fig. 11.7 Dermatitis herpetiformis: diagram to show granular deposits of IgA in the dermal papillae. (Direct IF)

Table 11.2 Direct immunofluorescence tests in lupus erythematosus

	Lesion skin	Clinically normal exposed skin
CDLE	+ve BMZ	−ve
SLE	+ve BMZ	+ve BMZ

CDLE = Chronic discoid lupus erythematosus; SLE = Systemic lupus erythematosus; BMZ = Basement membrane zone

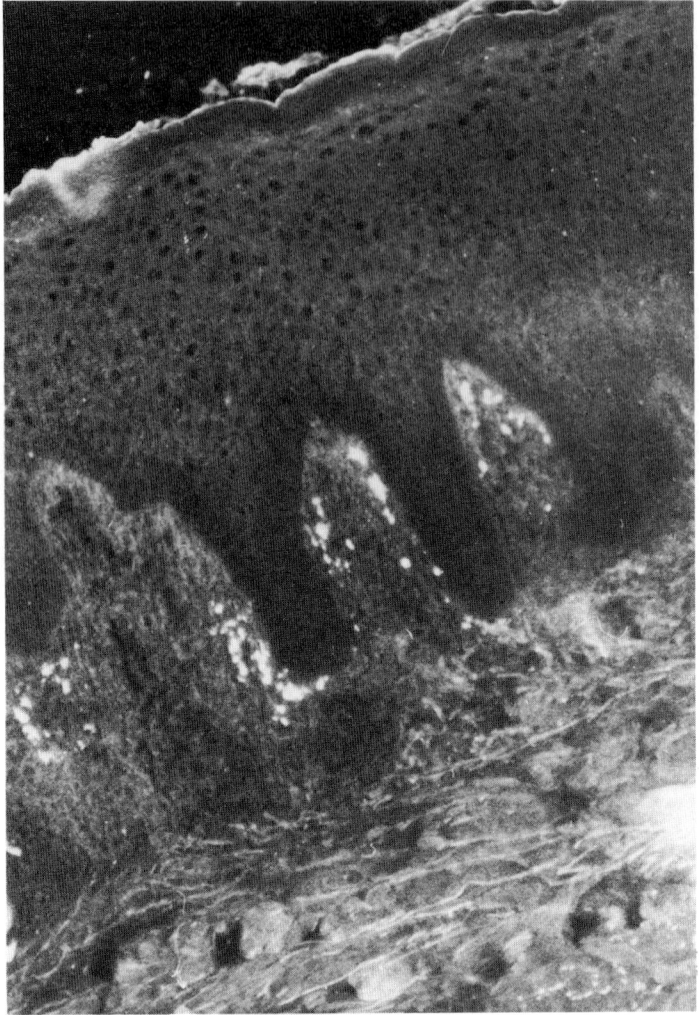

Fig. 11.8 Dermatitis herpetiformis: characteristic deposition of IgA in a granular fashion in the dermal papillae. (Direct IF)

fatal outcome. In each type similar histological appearances are found in the skin lesions, and direct IF tests are useful not only in confirming a diagnosis of LE, but in distinguishing between the two types. In CDLE positive results are obtained only from a biopsy of lesional skin, whereas in active SLE both lesional and clinically normal exposed skin biopsies will yield positive results. These consist of the deposition of IgG or IgM at the basement membrane zone as a coarsely granular or microgranular linear or homogenous band which has a very characteristic appearance (Figs 11.9 and 11.10). Various components of complement including C3 are also found and often non-specific colloid bodies (also seen in lichen planus) are present in the upper dermis.

Direct IF tests are positive in over 90% of cases of CDLE and SLE, and about 80% of cases of SLE give positive results in light-exposed clinically normal skin. Biopsies obtained from specimens of non-exposed clinically normal skin yield positive IF results in about 40% of cases of SLE and are considered to indicate a worse prognosis.

Other diseases which may demonstrate basement membrane zone positive IF findings include porphyria, polymorphous light eruption, rosacea, rheumatoid arthritis and senile keratosis, as well as some drug eruptions, but usually not in drug-induced LE.

These tests are invaluable in the differential diagnosis of LE because of the characteristic

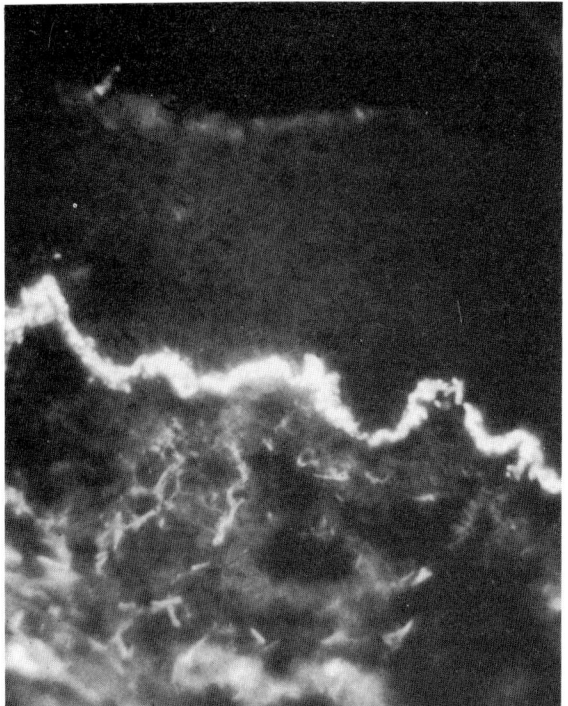

Fig. 11.9 Systemic lupus erythematosus (SLE): deposition of IgG as a coarsely granular band along the basement membrane zone. (Direct IF)

appearance of the IF findings, and in those rare cases of transition from CDLE to SLE, but it must be emphasised that they should be interpreted in the full knowledge of the clinical status of the patient involved.

Much information has accrued concerning IF studies in a large number of other skin diseases, but since they present no consistent pathognomonic features they will not be considered here.

Lymphomas (see also Ch. 20)

Reference has already been made to the various methods for distinguishing between certain cell types in these disorders, and the same methods may be applied to skin manifestations of lymphoma. In particular, the use of the non-specific esterase, acid phosphatase, and chloroacetate esterase techniques for identification of T cells (mycosis fungoides), B cells and monocytes as applied to both frozen sections and paraffin wax-embedded material, is recommended in cases where the diagnosis is in doubt. In addition the chloroacetate method provides a very effective method for the recognition of mast cells in tissue sections by the appearance of bright

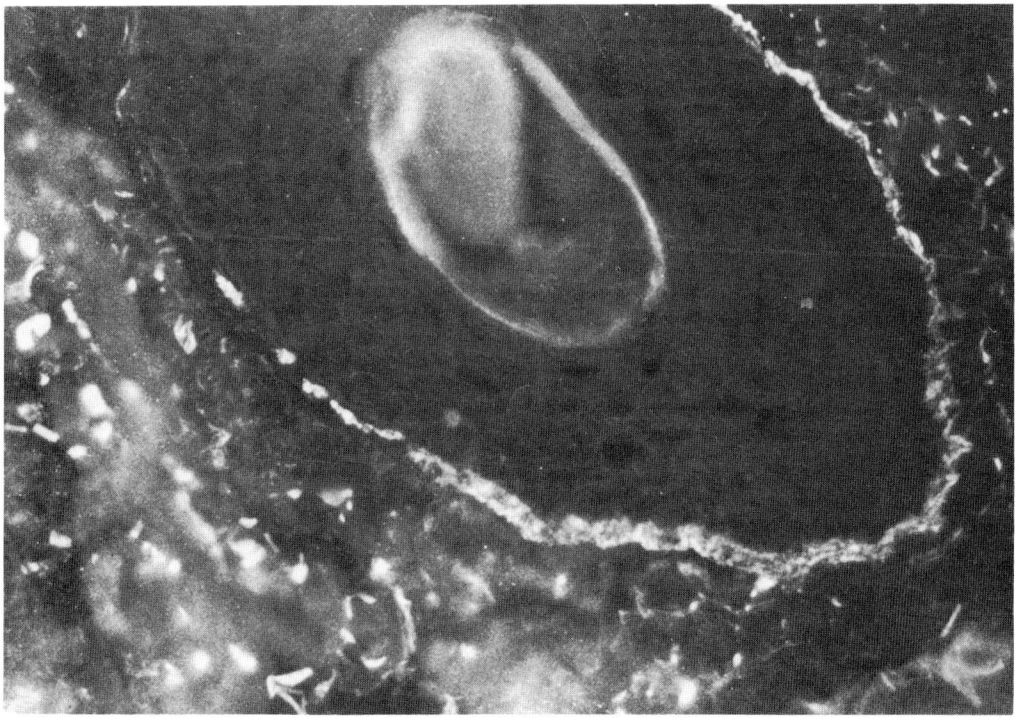

Fig. 11.10 Systemic lupus erythematosus (SLE): deposition of IgG around a hair follicle. (Direct IF)

orange-red granules in their cytoplasm. The muramidase technique is useful for the recognition of histiocytes, not commonly a problem in skin lymphomas, and myeloid cells, which may well be difficult to distinguish from cells of the monocyte series in some circumstances.

The ATPase method (on frozen material) shows the Langerhans cells, the high-level dendritic cells of the epidermis which are dopa-negative, and which recent work has shown to be increased in number in mycosis fungoides. Recent work suggests that monoclonal antibody OK-T6 may be a useful marker for Langerhans cells and of importance in the diagnosis of histocytosis X (Murphy et al 1981) (see also Chs 18 & 20).

In *mycosis fungoides* the epidermal keratinocytes show surface HLA-DR (Ia-like) antigen and an increase in numbers of Langerhans cells in the epidermis. Some Langerhans cells may also be found in the dermis.

The use of immunohistochemical techniques is now widely established, and T cells, B cells, and monocytes may be identified by their surface markers. Moreover, the identification of T cell sub-populations is now possible by the use of the monoclonal antibody technique, but at the present time can hardly be advocated for routine use on account of the expense of the antisera.

Cutaneous metastasis

Common sites of metastasis from malignant tumours include lymph nodes, lung, liver, brain and bone. Cutaneous metastases are rare. In men the commonest source of a cutaneous metastasis is bronchus (24%), the large bowel (19%), melanoma (13%) and squamous carcinoma of the oral cavity (12%). In women the commonest primary site is breast (69%), then large bowel (9%), melanoma (5%) and ovary (4%) (Brownstein & Helwig 1972).

Although cutaneous metastases are rare, they constitute an interesting source of tumours which present as secondary deposits of a 'silent' primary. If there is a known primary tumour the pathologist can compare the histology of the skin biopsy with that of the original tumour. Where the primary is unknown the pathologist is faced with the problem of suggesting likely sites of origin.

Cawley and his colleagues (1964) suggested

a haematoxylin–phloxine B–alcian blue–orange G (HPAO) method for the identification of the primary site of tumours presenting with skin metastases. Although the method may show adenocarcinomatous or squamous differentiation in what appears to be an undifferentiated tumour, like many other histological stains it lacks specificity.

Immunoperoxidase approach

An alternative approach is to use a variety of specific immunocytochemical antisera to attempt to determine the primary site of origin. The indirect immunoperoxidase method (Heyderman 1979) uses an unlabelled specific first antibody, and a second, anti-species antibody conjugated to the very stable enzyme, horseradish peroxidase (Fig. 11.11). Affinity-purified first and second antibodies increase both the specificity and readability of the preparations (Heyderman et al 1981). Details of the method are given in Appendix 6.

Antisera. It is apparent that since tumours may secrete several tumour markers a variety of antisera need to be employed for their identification. Immunoperoxidase profiles can be constructed for different types of tumour, and possible primary sites included or excluded on this basis. Taken together with the morphology and clinical picture, a much more precise identification may be possible.

Epithelial membrane antigen (EMA) is an antigen or antigens, defined by an antiserum raised against mild fat globule membrane and present in many normal and neoplastic secretory epithelia (Heyderman et al 1979). It has been demonstrated in all breast carcinomas so far studied, and in some squamous cell tumours, especially those with pseudoglandular differentiation (personal data). If a skin metastasis is negative for EMA it is unlikely to be of breast origin. We have seen two lobular carcinomas of the breast present with widespread bone involvement, with the small primary lesion detected only after some months. In the second case, there were widespread skin metastases and multiple lytic lesions in the skull. The differential diagnosis in a skin biopsy was between a poorly differentiated multiple myeloma, or lymphoma, or metastasis from an occult breast carcinoma (Plate 4). Some of the tumour cells contained PAS-positive diastase-resistant material, but this could have been compatible with

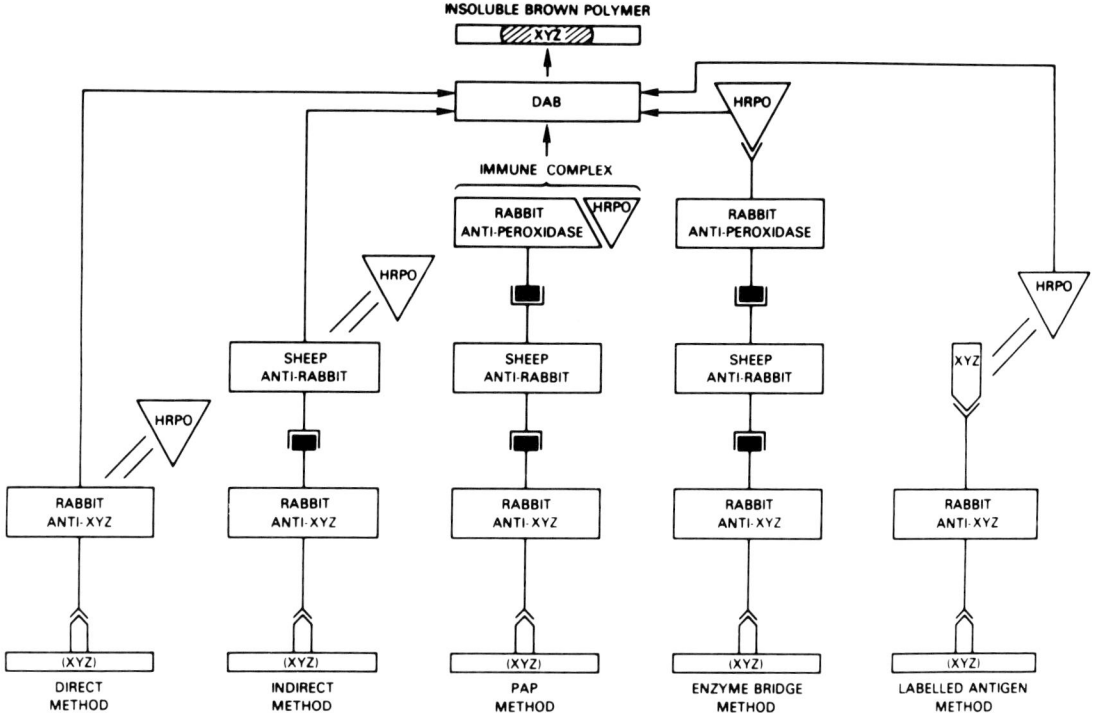

Fig. 11.11 Immunoperoxidase methods: reprinted with kind permission of the editors of the Journal of Clinical Pathology

any of the differential diagnoses. The indirect immunoperoxidase technique was used to investigate the presence of a variety of products. It was positive for EMA, thus consistent with a metastatic adenocarcinoma (Plates 4 and 5). In a woman the likely primary source was the breast, and this was confirmed some months later.

Similarly, in our experience all the gut adenocarcinomas are positive for carcinoembryonic antigen (CEA) (Heyderman & Neville 1977). Although CEA has been demonstrated in a wide variety of other tumours (Goldenberg et al 1976, Heyderman & Neville 1977), its absence in an adenocarcinoma virtually rules out a gut origin (Plates 6 and 7).

Other tumours which present initially with cutaneous metastases may similarly be identified more precisely using these techniques. It is important to remember that tumours may secrete inappropriate products, so the presence, for example, of the specific beta subunit of human chorionic gonadotrophin (HCG) (Vaitukaitis et al 1972) is not necessarily pathognomonic of metastasis from a trophoblastic tumour (Braunstein et al 1973). However, taken in

combination with the morphology of malignant syncytiotrophoblast and cytotrophoblast in close apposition, the diagnosis of atypical metastatic gestational choriocarcinoma could be made (Plates 8 and 9).

Markers other than CEA, EMA and placental proteins which may be of value in the investigation of cutaneous metastases include thyroglobulin (LoGerfo et al 1978, Burt & Goudie 1979); immunoglobulins (Taylor 1976), though staining for these gives rise to problems with collagen staining in skin biopsies (personal data), and prostatic acid phosphatase (Jöbsis et al 1978). In time many other markers will become available for this study. The problem discussed here is that of suggesting a possible origin of a cutaneous metastasis; that of differentiating between a primary skin tumour and metastatic disease may be addressed in the same way.

Miscellaneous

The diagnosis of a variety of other skin conditions and/or the identification of cell structures and substances in the skin may be helped by the use of 'routine' histochemical methods (Bancroft & Stevens 1982). These methods include stains for *amyloid* and in the skin, the thioflavin T method with an acid pH to increase selectivity has proved both simple and reliable. They are however not specific, and positive results may be found in patients with porphyria, colloid milium or lipoid proteinosis.

The DOPA technique gives good permanent preparations in sections of lightly fixed tissue. Alternatively cryostat sections may be used. Both methods demonstrate *melanocyte* activity well and may be of great help in detecting tumours derived from melanin-producing cells. Melanomas are also rich in α-mannosidase activity (Elleder, personal communication).

The alkaline phosphatase method is most useful in identification of capillary endothelium in difficult cases where *Kaposi's idiopathic haemorrhagic sarcoma* is suspected. However, a more specific indicator of the disorder is anti-factor VIII related antigen (Nadji et al 1981).

Enzyme histochemistry may also be useful in the differential diagnosis of *basal cell carcinoma*, particularly the absence of lactate dehydrogenase, acid and alkaline phosphatase activities in contrast with the significant levels found in breast carcinomas and in those of adeno-adnexal origin (Elias et al 1980).

Haemosiderin deposition in the skin may occur in many circumstances, but its presence in variable amounts may alert the pathologist to a diagnosis of *pigmented purpuric dermatosis* (Schamberg's disease), the *lichenoid purpura* of Gougernot and Blum, or even some cases of *histiocytic dermatofibroma*.

Mast cells in *urticaria pigmentosa* are best shown by the toluidine blue metachromatic method and by chloroacetate esterase or alcian blue at pH 1.0 in paraffin wax sections (Fig. 11.12).

The histochemical detection of microsomal arylsulfatase C (MAS) activity in the skin may prove of value in the study of keratotic conditions. Recent studies have shown MAS deficiency in the epidermis of patients with X-linked ichthyosis as well as in carriers in contrast with a positive MAS reaction found in other keratotic lesions such as psoriasis, actinic keratosis, and seborrheic wart (Jöbsis et al 1980).

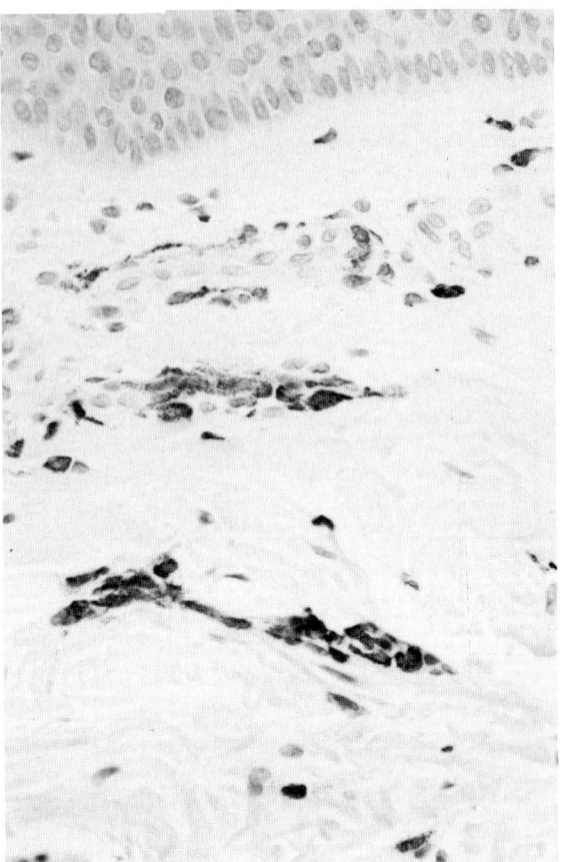

Fig. 11.12   Skin — mastocytosis: Chloroacetate esterase method applied to paraffin wax sections to visualise mast cells in the dermis

Graft versus host disease (GVH)

There are a few helpful findings in GVH disease in addition to those seen in routinely processed tissue. In the epidermis there is a loss of the ATPase positive cells (Langerhans cells which can also be identified by their α-mannosidase activity); the loss may be focal and confined to a very small area of a biopsy. The basal keratinocytes acquire the antigen HLA-DR (Ia like antigen) and there is an increase of HLA-DR positive (also ATPase positive) cells in the dermis. The dermal infiltrate of lymphocytes is almost exclusively of the suppressor/killer type which bear the antigen OKT8 in addition to OKT3 antigen common to all lymphocytes.

REFERENCES

Bancroft J D, Stevens A (eds) 1982 Theory and practice of histochemical techniques, 2nd edn. Churchill Livingstone, Edinburgh

Braunstein G, Vaitukaitis J L, Carbone E P, Ross G T 1973 Ectopic production of human chorionic gonadotropin by neoplasms. Annals of Internal Medicine 78: 39–45

Brownstein M H, Helwig E B 1972 Patterns of cutaneous metastasis. Archives of Dermatology 105: 862–868

Burt A, Goudie R B 1979 Diagnosis of primary thyroid carcinoma by immunohistological demonstration of thyroglobulin. Histopathology 3: 279–286

Crawley E P, Hsu U T, Weary P E 1964 The evaluation of neoplastic metastases to the skin. Archives of Dermatology 90: 262–265

Elias E A, Elias R A, Bijlsma P J, Tazelaar D J 1980 The enzyme histochemistry of metastasizing basal cell carcinoma of the skin. Journal of Pathology 131: 235–241

Goldenberg D M, Sharkey R M, Primus F J 1976 Carcinoembryonic antigen in histopathology: immuno-peroxidase staining of conventional tissue sections. Journal of the National Cancer Institute 57: 11–22

Harrist T J, Mihm M C 1979 Cutaneous immunopathology: the diagnostic use of direct and indirect immunofluorescence techniques in dermatologic disease. Human Pathology 10: 625–653

Heyderman E 1979 Immunoperoxidase technique in histopathology: applications, methods and controls. Journal of Clinical Pathology 32: 971–978

Heyderman E 1980 Immunocytochemistry in cancer diagnosis. In: Symington T, Williams A E, McVie J G The Tenth Pfizer International Symposium: Cancer assessment and monitoring. Chucchill Livingstone, Edinburgh, p 147–171

Heyderman E, Monaghan P 1979 Immunoperoxidase reactions in resin embedded sections. Journal of Investigative Cell Pathology and Biology 2: 119–122

Heyderman E, Neville A M 1977 A shorter immunoperoxidase technique for the demonstration of carcinoembryonic antigen and other cell products. Journal of Clinical Pathology 30: 138–140

Heyderman E, Gibbons A R, Bulman A S 1981 Immuno-cytochemical identification of cells. New approaches to laboratory medicine. VIIIth Merz & Dade International Symposium Rosalky S B, Darmstadt, p 135–148

Heyderman E, Steele K, Ormerod M G 1979 A new antigen on the epithelial membrane: its immunoperoxidase localisation in normal and neoplastic tissues. Journal of Clinical Pathology 32: 35–39

Jöbsis A C, De Vries G P, Anholt R R H, Sanders G T B 1978 Demonstration on the prostatic origin of metastases. An immunohistochemical method for formalin-fixed embedded tissue. Cancer 41: 1788–1793

Jöbsis A C, DeGroot W P, Tigges A J, De Bruijn H W A, Rijken Y, Meijer A E F H, Marinkovic-Ilsen P 1980 X-linked ichthyosis and X-linked placental sulfatase deficiency: a disease entity. Histochemical observations. American Journal of Pathology 99: 279–290

LoGerfo P, Li Volsi V, Colaccio D, Feind C 1978 Thyroglobulin production by thyroid cancers. Journal of Surgical Research 24: 1–6

Murphy G F, Bhan A K, Sato S, Mihm M C Jnr, Harrist T J 1981 A new immunologic marker for human Langerhans cells. New England Journal of Medicine 304: 791–792

Nadji M, Morales A R, Ziegles-Weissman J, Penneys N S 1981 Kaposi's sarcoma. Immunohistologic evidence for an endothelial origin. Archives of Pathology and Laboratory Medicine 105: 274–275

Taylor C R 1976 An immunohistological study of follicular lymphoma, reticulum cell sarcoma and Hodgkin's disease. European Journal of Cancer 12: 61–75

Vaitukaitis J L, Braunstein G D, Ross G T 1972 A radioimmunoassay which specifically measures human gonadotrophin in the presence of human luteinising hormone. American Journal of Obstetrics & Gynecology 113: 751–758

12

Malignant and inflammatory diseases of the gastro-intestinal tract

M. Isabel Filipe

MALIGNANCY

The problems associated with the early diagnosis and control of gastro-intestinal (GI) cancer are of prime importance as colorectal cancer constitutes one of the most common causes of death in Western Europe and United States. In Japan gastric cancer is common. For successful cancer control it is essential:

1. to detect early malignant change
2. to recognise precancerous lesions
3. to identify groups of patients with higher risk of developing malignancy.

The role of the pathologist is to identify epithelial dysplasia, defined as an atypical cell growth with potentiality to evolve to malignancy. The general features of dysplasia are agreed but nonetheless the uncertainty of prognosis, and the lack of precise criteria to distinguish neoplastic dysplasia from reactive epithelial hyperplasia, raise problems of interpretation. Attempts have been made to find new methods to help solve some of these difficulties. Alterations in carbohydrate metabolism (Filipe & Branfoot 1974, Isselbacher 1974, Kim & Isaacs 1975, Gorman & LaMont 1978, Rogers et al 1978, Filipe 1979) and the demonstration of antigens (Burtin et al 1972, Goldenberg 1976, Isacson & Le Vann 1976, Neville & Laurence 1980) associated with GI neoplasia may prove useful markers of malignant change and are discussed in this chapter.

Good preparation of the biopsy material is important if the maximum information is to be obtained. The samples should be correctly orientated by placing the tissue, mucosal face upwards, on pieces of card or frosted glass, fixed in formal-saline or other suitable fixative and embedded in paraffin wax in such a way as to allow sections to be cut with longitudinally open full-length glandular tubules.

Mucins

Mucins in the human GI tract are glycoproteins formed by a protein backbone with attached side chains of sugar residues D-mannose, D-galactose, L-fucose, N-acetyl derivatives of hexosamines and often a terminal sialic acid group. In some, ester sulphates are also present.

Histochemically, the mucins are divided into neutral and acid, the latter being either rich in sialic acids (sialomucins) or sulphate groups (sulphomucins). Sialomucins can be further subdivided according to the proportion of N- or O-acetyl derivatives of sialic acid in the molecule and whether or not they are liable to neuraminidase digestion. A battery of histochemical techniques is available to separate these different groups (Pearse 1968). However, in our experience the main types of GI mucins can be easily identified by a combination of the methods shown in Table 12.1. All these methods offer the advantage of being carried out in formalin-fixed tissues routinely embedded in paraffin wax.

Lectins

At present the use of lectins is limited to research. They are, however, briefly mentioned here, as I believe they will play an increasing role in understanding malignant transformation and distinguishing normal from malignant cells. Lectins are proteins found primarily in plants (and in some invertebrates and vertebrates) which bind more or less specifically to particular sugar molecules or groups of molecules on the cell surface. Their main importance in medical research stems from their ability preferentially to agglutinate malignant cells.

The use of immunoperoxidase techniques to

demonstrate a battery of lectins already available, which differ in their specificity, may reveal a profile of changes indicative of malignant transformation (Sachs et al 1974, Shoham & Sachs 1974, Lotan & Nicholson 1978, Boland & Kim 1983).

Tumour-associated antigens

An increasing number of antigens associated with GI tract neoplasia have been described and of particular interest are the two oncofetal antigens — alpha-fetoprotein (AFP) and carcinoembryonic antigen (CEA). The former will be dealt with in relation to hepatic tumours (see Ch. 15), whilst CEA, because of its relevance to colonic cancer, is discussed in this chapter. The chemical and physical properties and the clinical applicability of immunodiagnostic tests for CEA have been the subject of many studies and were recently reviewed by Goldenberg (1976). I will restrict myself to describing the localisation of CEA in the tissues by immunoperoxidase techniques and its value in the routine diagnosis of malignant change.

CEA is a glycoprotein and contains sialic acid in its molecule. It is probably the best-known and most investigated marker for colonic malignancy. Unfortunately its levels in sera do not detect early malignant change and although there is a general relationship between recurrence and rising levels of antigen, this is insufficiently precise in the individual patient to form the basis of clinical action (Goldenberg 1976, Neville & Lawrence 1980). However, it may be of value as a prognostic indicator (Wanebo et al 1978).

The use of immunoperoxidase techniques for the identification of CEA in the tissues has shown conflicting results, with some claiming it to be a reliable indicator of malignant change in colonic mucosa (Isaacson 1976, Isaacson & Le Vann 1976, Filipe et al 1981, O'Brien et al 1981).

The differences are probably due to the different techniques used by the various authors and to

Table 12.1 Histochemical methods to visualize epithelial mucins in the gastrointestinal tract*

Method	Interpretation
Diastase-Periodic Acid-Schiff (D-PAS) (Pearse 1968)	Magenta: A. All MS-containing hexoses and deoxyhexoses with vicinal glycol groups B. Neutral MS C. Some non-sulphated acid MS (sialomucins)
Periodate-borohydride/ saponification/PAS (PB/KOH/PAS) (Reid et al 1973, Culling et al 1974)	PB: Reduces the periodate-engendered aldehydes in the tissues, thus abolishing their Schiff reactivity A. O-acyl esters from a potential vicinal diol, or B. a sialic residue linked glycosidically to a potential vicinal diol PAS: Activity after PB/KOH may indicate presence of O-acylated sialic acids
Alcian blue pH 2.5 (AB 2.5) (Pearse 1968)	Blue: A. Weakly sulphated MS B. Carboxyl groups of sialomucins
Alcian blue pH 2.5-Periodic acid Schiff (AB 2.5-PAS) (Pearse 1968)	Magenta: neutral (MS) Blue: acid (MS) Purple-blue: neutral and periodate reactive acid MS
High iron-diamine (HID) (Spicer 1965)	Brown-black: sulphated MS Sialomucins unstained
High iron-diamine-alcian blue pH 2.5 (HID-AB) (Spicer 1965)	Brown-black: sulphated MS Blue: non-sulphated acid MS (sialomucins)

MS = Mucosubstances
* This is not an exhaustive list of the techniques available but it includes the most common and those used routinely in our laboratory

antibody specificities. In our experience (Filipe et al 1981) and others (Isaacson 1976, Isaacson & Le Vann 1976, O'Brien et al 1981) in routine formalin-fixed and paraffin wax-embedded tissues, CEA content seems to be a reliable indicator of malignant change in colonic mucosa. The fact that it has also been reported in other GI and extra-GI tract neoplasias reduces its value in detecting the primary origin of a tumour but it still helps in the differential diagnosis (Goldenberg 1976).

STOMACH

The mucous secretory cells of the *normal* gastric epithelium secrete neutral mucins (D-PAS-positive) almost exclusively. Traces of acid mucins (sialo- and possibly sulphomucins) can be seen in the proliferative epithelial zone of the base of the pits and mucous neck cells in the body, occasionally in the lower pit in the antrum and in the glandular elements in the oesophageal-gastric junction. Predominance of sulphated groups is found in the submucosal oesophageal glands.

A. Detection of malignant cells. These are easily recognised in routine histological preparations. However, in cases of undifferentiated or signet-ring cell types of carcinoma, particularly if the biopsy is poorly orientated and superficial, groups of isolated malignant cells may not be detectable on H & E stained sections. In this instance mucin stains are helpful. I find the combined AB pH 2.5-PAS technique for both neutral and acid mucins preferable to the D-PAS alone to avoid false-positives with normal gastric cells (D-PAS-positive only) or false-negatives in cases of carcinoma secreting predominantly acid sulphated mucins (D-PAS-negative). These techniques may also be useful in smears.

B. The premalignant lesions in the stomach are the adenomas and intestinal metaplasia. Adenomas are rare and will not be discussed here.

Intestinal metaplasia (IM) has been shown to be associated with gastric cancer but it is also common in benign conditions such as chronic atrophic gastritis and peptic ulcer and therefore has little practical value in gastric cancer control. However, the majority of gastric carcinomas seem to arise from areas of IM

and the identification of a subgroup more consistently related to cancer would have important implications in the selection of patients at higher risk of developing malignancy and thus in need of careful follow-up.

The use of mucin stains, AB pH 2.5-PAS and HID-AB, reveal three variants of IM according to histological cell types and mucus-secreting patterns (Jass & Filipe 1979, Jass 1980) (Plates 10 and 11):

1. Type I, or complete, resembles small intestine and secretes non-sulphated acid mucins (sialomucins) (Plate 10A & B).

2. & 3. Types II and III, or incomplete, contain elements of both gastric and intestinal epithelia in various degrees of maturation. Goblet cells secrete either sialomucins (Type II) or sulphomucins or a mixture of the two (Type III). The intervening columnar cells reveal the presence of neutral and acid mucins in varying proportions, with sulphomucins being present in III IM, but absent in II IM, (Plate 11A and B).

The presence of N- and o-acetyl sialic acid derivatives also varies as revealed by the PB/KOH/PAS technique. Goblet cells in Type I IM contain both N- and o-acetyl sialic acid variants, whilst the latter is absent in Types II and III IM. Our studies (Jass & Filipe 1979, 1981, Jass 1980) and of Taglbjaerg & Nielsen (1978) and Sipponen et al (1980) suggest that Type I IM is more commonly found in benign conditions, while the other two types, particularly those with marked secretion of sulphomucins, show a significant association with carcinoma, and may represent a variant with a higher malignant potential.

Also of relevance is the description of *Tumour antigens* in 'normal' mucosa bordering gastric cancer within IM. These include alpha fetoprotein (AFP) (Kitaoka et al 1973), carcinoembryonic antigen (CEA) (Burtin et al 1973), goblet cell antigen (GMC) (Rapp et al 1979) and, of particular interest, fetal sulphoglycoprotein antigen (FSA) (Bara et al 1978, Hakkinen et al 1980). It is not clear whether FSA also occurs in metaplastic epithelium. However, it has been suggested that this could be related to the sulphomucins of type III IM (type IIB) (Jass 1980). Recent reports by Hakkinen et al (1980) of FSA content in gastric juice from over 53 000 unselected individuals of 'cancer-age' in Finland seem to demonstrate its value (a) as a mass-screening test for gastric cancer in a high-risk population, (b) in the differential diagnosis between benign and malignant gastric ulcer and (c) in detecting early cancer, probably even in a pre-malignant phase.

In spite of some unresolved questions such as the presence of FSA in the gastric juice of 7% of the rural population and in 14% of gastric ulcer patients in Finland, its potential value deserves attention.

LARGE BOWEL

Normal colonic mucosa

Goblet cells in the lower two-thirds of crypts secrete predominantly sulphomucins (Plate 12). Sialomucins are present in the upper crypt and surface epithelium and contain N- and o-acylated sialic acid derivatives (Rogers et al 1978, Filipe 1979). The high content of the latter distinguishes colonic from other GI tract mucins (Culling et al 1975). Deviations from the 'normal mucous pattern' have been described in colonic mucosa in a variety of pathological conditions.

This abnormal secretion is characterised histochemically by increased sialomucins, decreased or absent sulphated material and in some lesions lower levels of o-acylated sialic acids are also observed (Culling et al 1977, Dawson et al 1978, Filipe 1979). It has been suggested that these qualitative mucin changes could be of value (a) as indicators of malignant transformation and (b) in the differential diagnosis of inflammatory bowel disease (Filipe 1979).

In our experience, using an immunoperoxidase technique (Filipe et al 1981) in normal colonic mucosa, carcinoembryonic antigen is usually absent or found only in traces. When present it is seen in the luminal aspect of the epithelial cells at the bottom of the crypt. The presence of increasing amounts of CEA in the colonic mucosa in association with dysplasia and malignancy seems to be a useful complement to the histological criteria in assessing malignant transformation and possibly even predicting dysplasia (Isaacson 1976, Isaacson & Le Vann 1976, Filipe et al 1981).

Precancerous lesions

Increased sialomucin secretion has been consistently observed in the so-called 'Transitional' (TR) mucosa adjacent to colorectal carcinoma (Plate 13A & B). We first introduced the term 'transitional' mucosa (Filipe 1969) to describe non-neoplastic mucosa adjacent to colonic cancer where abnormal mucous secretion was found in the absence of morphological atypia. We now believe it may represent early pre-polypoid adenomatous change. Histochemical and biochemical studies in normal, 'transitional' and carcinoma tissues demonstrated the presence of five variants of sialic acid in all three regions which differ from each other in the relative proportions of N- and o-acetyl derivatives of sialic acid, with the values for 'transitional' mucosa graded between those from normal tissue and tumours. Moreover, a higher proportion of neuraminidase-sensitive sialic acids is found in both 'transitional' and carcinoma tissues, the latter revealing a decrease in the more heavily substituted sialic acids (Dawson et al 1978, Rogers et al 1978). Estimation of sialyltransferases in the same tissues also reveal a gradual increase from low levels in the normal to a higher content in the tumour, with 'TR' values in between (Filipe & Rogers 1979).

The fact that these changes, though more marked in the mucosa around carcinomas, are also seen in areas distant from them, suggests a primary rather than a local secondary effect of tumour growth. Furthermore, the extent of mucin changes, aggressiveness of the tumour and prognosis seem to be related (Filipe & Branfoot 1974, Greaves et al 1980). It has been suggested that tumour cells are coated by neuraminidase-sensitive sialic acids, which may not only hide their antigens from recognition by the host, but also shield them from his immunocompetent cells (Currie & Bagshawe 1968). Thus, sialic acids may play an important role in immunological reactions to tumour-specific antigens and the control of tumour growth.

Further supporting evidence that 'transitional' mucosa may represent a pre-dysplastic or pre-adenomatous change, is the increased CEA content, in contrast with its absence in the normal epithelium (Filipe et al 1981, O'Brien et al 1981).

In 23 colectomy specimens from cases of *familial polyposis coli* a similar abnormal mucin pattern is observed in the non-neoplastic mucosa between polyps. Excess sialomucin is more marked in the left than in the right colon and involves more extensive areas of mucosa in patients who had developed carcinoma compared with those free of malignancy. The predominance of sialomucins is apparently not related to the degree of dysplasia of the adjacent polyp (Filipe et al 1980).

In rats with *chemically induced colorectal cancer*, variations in mucin composition with increased sialomucins are also described in zones of dysplasia (Filipe 1975).

Adenomas

It is generally accepted that adenomas are important precursors of colorectal cancer. It is equally known that only a small percentage will progress to invasive carcinoma, and it is not yet possible to predict which polyp will become malignant. Size, degree of epithelial dysplasia and pattern of growth (tubular or villous) are useful criteria for assessing malignant potential but no precise marker exists. The presence of increased values of tumours antigens such as CEA (Isaacson & Le Vann 1976, Sharkey et al 1977, Filipe et al 1981, O'Brien et al 1981, Skinner et al 1981), particularly if assessed in combination with variations in mucin composition (Culling et al 1977, Filipe et al 1981), may serve as useful indicators of early malignant change.

We have investigated in parallel the value of these two parameters, mucin secretion and CEA content, in 74 colonic adenomas, and compared with 'transitional' mucosa, carcinoma and non-neoplastic polyps (Filipe et al 1981). The results are encouraging. We found, as in the normal colon, a high content of o-acylated sialomucins in 'TR' mucosa in 91% of the cases examined. In adenomas a gradual loss of o-acylated sialic acids parallels the degree of dysplasia, with the lowest values being seen in adenomas showing malignant transformation and in invasive carcinomas (Figs, 12.1A & B, 12.2).

Although a predominance of sialomucins shown on HID-AB staining is consistently found in 'TR' mucosa, this technique is apparently of no value in predicting malignant transformation in adenomas, as no relationship could be established between the pattern of mucin secretion and the degree of dysplasia or malignancy of the adenoma.

In the majority of the cases examined the CEA content of the 'TR' mucosa was low or absent as in the normal, but in 21% of the cases a moderate amount of CEA could be found in the luminal aspect

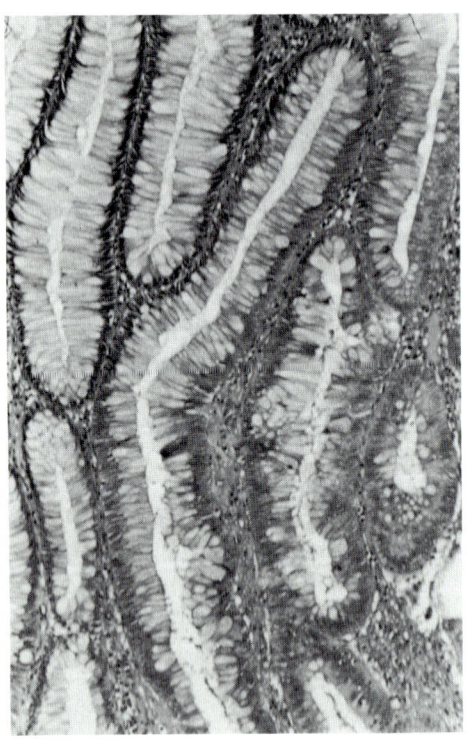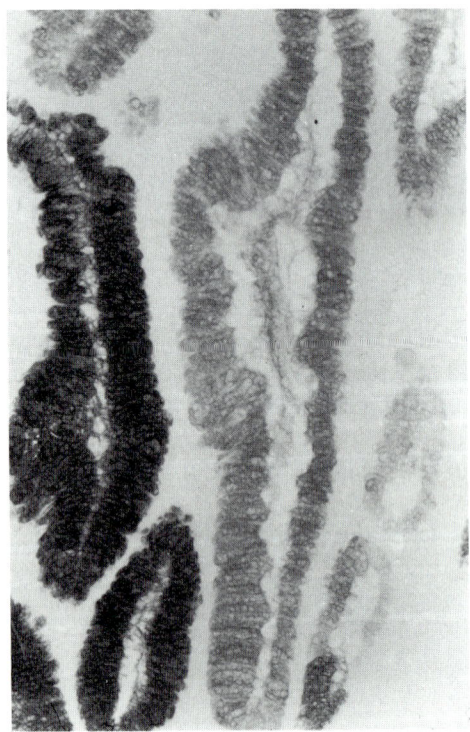

Fig. 12.1 Adenomatous polyp: areas of mild dysplasia showing strong PAS staining after PB/KOH in contrast to the weak PAS reaction seen in severely dysplastic glandular epithelium. A. H & E; B. PB/KOH/PAS

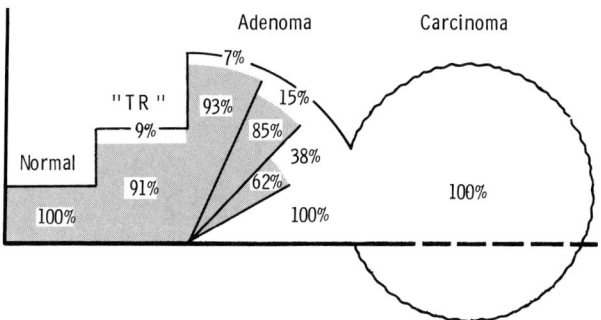

VARIATIONS IN O-ACYLATED SIALIC ACID CONTENT
IN THE SEQUENCE OF TRANSITIONAL - ADENOMA - CARCINOMA

Fig. 12.2 Variations in O-acylated sialic acid content in the sequence of transitional-adenoma-carcinoma. A decrease in the O-acylated sialic acids correlates well with the severity of dysplasia and malignancy.
Key — Semi-quantitative visual assessment on paraffin wax sections stained with PB/KOH/PAS method. (High content: shaded area; low content or absent: blank.)
  % refers to the number of cases showing either a high or low content. In adenoma the segments represent increasing degrees of dysplasia. Horizontal axis = muscularis mucosae. Vertical axis = mucosal height

of the crypt epithelium. In adenomas the CEA content gradually increases in direct relationship with the severity of dysplasia, being highest in malignancy (Figs 12.3, 12.4). However, caution should be taken in the interpretation of malignant change, as the assessment is subjective and exceptions may occur. In adenomas, the intracellular distribution of the CEA-positive material is either diffuse throughout the cytoplasm and on the surface glycocalyx or concentrated in the apical cytoplasm. There are

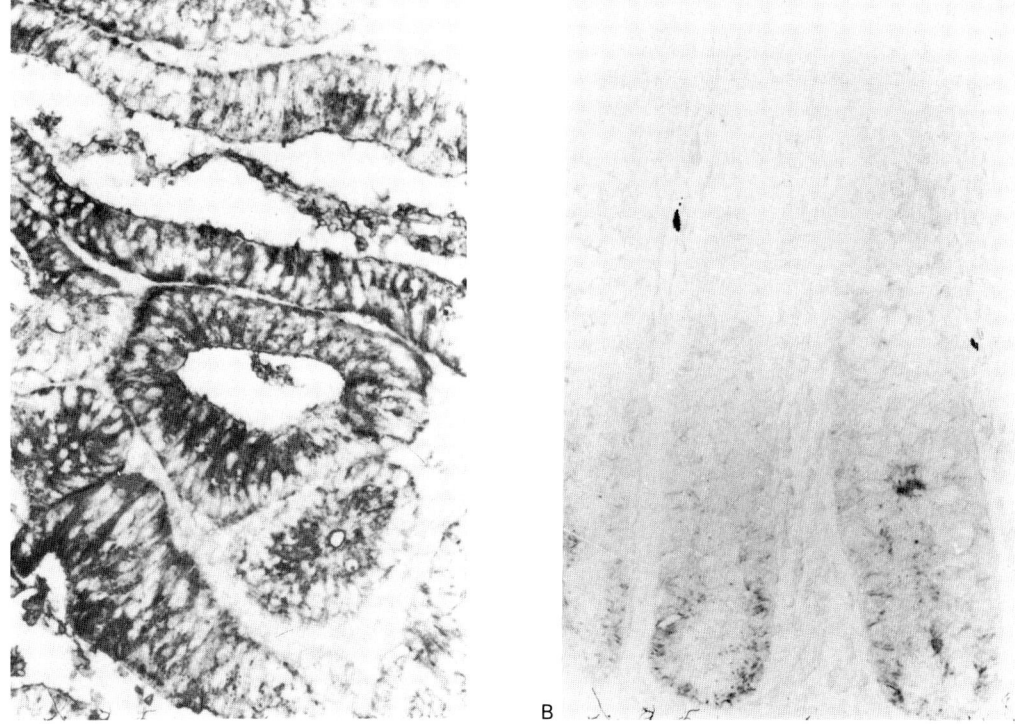

Fig. 12.3 Colonic adenocarcinoma (A) showing high content in CEA (immunohistochemical indirect alkaline phosphatase method) in contrast with normal colon (B).

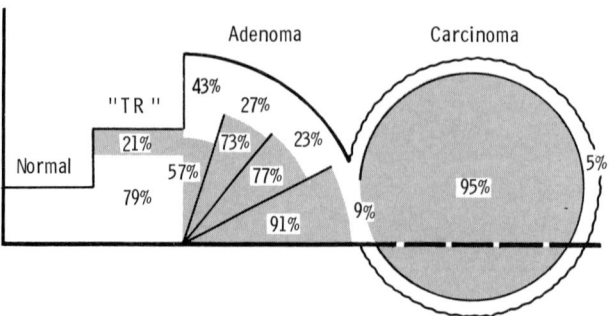

Fig. 12.4 Variations in CEA content in the sequence of transitional-adenoma-carcinoma. An increase in CEA correlates well with the severity of dysplasia and malignancy.
Key — Semi-quantitative visual assessment on paraffin wax sections stained by the indirect alkaline phosphatase method for CEA. (High content: shaded area; low content: blank.)
  % refers to the number of cases showing either a high or a low content. In adenoma the segments represent increasing degrees of dysplasia. Horizontal axis = muscularis mucosae. Vertical axis = mucosal height

similarities and differences between our findings in normal, transitional and adenomatous colonic epithelia and those described by other authors (Burtin et al 1972, Isaacson & Le Vann 1976, Sharkey et al 1977, O'Brien et al 1981, Skinner et al 1981). There is evidence that the differences in CEA content between normal and malignant mucosa are quantitative rather than qualitative (Burtin et al 1972, Sharkey et al 1977), therefore differences in technique could explain the disparity of results reported by various authors.

Mucous secretion and CEA content in inflammatory polyps is similar to normal.

In the metaplastic polyps, the composition of mucus in the goblet cells is similar to that found in the normal colonic epithelium but, unlike the normal, some of the interspersed absorptive-type cells in the upper crypt show secretory activity predominantly of sialomucins with a low content in o-acylated types. Various amounts of CEA are also seen in the majority of the metaplastic polyps, particularly in the upper half of the crypt. Skinner et al (1981) reported similar findings.

From the observations mentioned above, a relationship seems to exist between alterations in the glycoproteins and CEA content of the colonic epithelial cells and malignant transformation.

The use of both mucin stains and immunoperoxidase techniques for CEA may prove complementary to the histological criteria for detecting early malignant transformation and possibly predicting dysplasia (Figs. 12.2 and 12.4).

Identification of origin of metastatic carcinoma

The composition of mucin secreted by tumour cells along the GI tract reveals no histochemical characteristics which could be related to the organ of origin or histological type and attempts to identify the origin of adenocarcinomas have not been successful. The demonstration by Culling et al (1975) of o-acetyl derivatives of sialic acid in the lower gastro-intestinal tract carcinomas and its absence from all others could be a useful means of differential diagnosis. However, this is not a reliable criterion as the degree of acylation is related to the degree of differentiation of the tumour, being absent in the poorly differentiated carcinomas (Montero & Segura 1980). A positive result however is in favour of a colorectal origin.

Although CEA is found in various malignant tissues and even in some benign and non-neoplastic conditions (Goldenberg 1976), its presence and concentration in formalin-fixed paraffin wax sections using the immunoperoxidase technique can still be of value in (a) detecting malignant cells in needle biopsy or aspirates and (b) deciding the site of origin of metastatic carcinoma.

The finding of CEA-negative carcinoma cells makes the diagnosis of colorectal carcinoma unlikely. Conversely, a CEA-positive reaction strongly suggests colorectal, gastric or lung carcinoma. Other malignant tumours from pancreas, ovary and endometrium may show weak CEA reaction. However, in those cases other markers help to discriminate between the various tumours: CSpA (colon-specific antigen) and TA (colonic neo-antigens) for colorectal cancer, FSA for gastric cancer, POA (pancreatic oncofetal antigen) for pancreatic cancer, to cite just a few examples (Goldenberg 1976, Heyderman 1980, Neville & Laurence 1980, Heyderman et al 1981).

It is clear from the literature that no single marker specific for a certain type of tumour exists. Reading through the various chapters in this book, it is apparent that accuracy of diagnosis can be better achieved by a combination of 'tumour markers', antigens, enzymes, hormones and others.

INFLAMMATORY BOWEL DISEASE

In biopsy material the differentiation of various inflammatory bowel diseases is sometimes difficult. The two main entities, *ulcerative colitis (UC)* and *Crohn's disease (CD)* present many common features and in the absence of granulomas, the distinction between the two in a single rectal biopsy is not always possible. The preservation of mucin secretion in CD even in the presence of marked inflammation and its depletion in UC, though helpful, is not a reliable criterion as a similar feature is found in the resolving phase of UC. There is histochemical and biochemical evidence of qualitative changes in mucin composition in both CD and UC but at present they are of no diagnostic value (Filipe 1979, Clamp 1980, Ehsanullah et al 1982b).

Alteration of mucin secretion with predominance of sialomucins were also found in *solitary ulcer syndrome* (SUS) (Ehsanullah et al 1982a). In contrast a normal mucin pattern was consistently seen in *non-specific proctitis* in spite of inflammation. Although histology remains the most important investigation in the diagnosis of SUS, mucin changes provide valuable additional evidence.

Dysplasia in ulcerative colitis

One of the problems in the histological interpretation of biopsy material in ulcerative colitis is the assessment of cancer risk in patients with epithelial dysplasia. As was mentioned at the beginning of this chapter, its value is limited by the focal nature of the

lesion, its presence in the absence of carcinoma and vice versa and the lack of precise histological criteria to discriminate between neoplastic dysplasia and reactive hyperplasia.

It has been suggested that a study of qualitative differences in mucin secretion might help in predicting the development of cancer in patients with UC (Filipe 1979). In a recent study of 120 rectal biopsies from 65 patients with UC (Ehsanullah et al 1982b), we observed a predominance of sialomucins apparently related to the histological severity, the extent and duration of the disease. All biopsies with dysplasia, except one, show abnormal mucin staining (Table 12.2).

Table 12.2 Ulcerative colitis: correlation between disease activity (inflammation), dysplasia, and mucin secretion patterns

	Normal mucus pattern Sulphomucins	Sialomucins
Remission (20)	20	0
Remission + dysplasia (5)	0	5
Active disease (31)	17	14
Active disease + dysplasia (33)	1	32

( ) = Total number of biopsies

The important question here is whether sialomucins are related to dysplasia or inflammation? In support of the latter is the finding of similar alteration in mucin composition in the absence of dysplasia in Crohn's disease and in solitary ulcer syndrome. Supporting evidence for a relationship between dysplasia and abnormal mucins is the presence of normal mucins in non-specific proctitis and in other miscellaneous conditions of large bowel even when inflammation is severe. Further evidence of this relationship is the co-existence of dysplasia and sialomucins in rectal biopsies taken from the patients with UC during the remission phase in the absence of inflammation (Ehsanullah et al 1982b).

Some patients with active disease but no dysplasia showed sialomucin staining. Whether such cases indicate changes preceding the morphological atypia, and features of dysplasia will develop in subsequent biopsies or, alternatively, the mucin changes represent a non-specific reaction which regresses when the inflammation ceases, we do not know

(Listinsky & Riddell 1981). However, preliminary follow-up studies indicate that in some cases altered mucin secretion precedes dysplasia and malignancy.

However, I feel that the use of mucin stains (HID-AB) patterns serves to indicate an underlying pathology which must be further investigated by more sophisticated methods (Ehsanullah et al, in preparation).

The demonstration of CEA by the immuno-peroxidase technique may be of further help in the interpretation of true dysplasia and premalignant changes in ulcerative colitis and deserves more attention (Isaacson 1976).

SECRETORY IMMUNOGLOBULINS

In the large intestine, IgA is the main immunoglobulin present in both epithelium and plasma cells and constitutes the first barrier of immunological defence in the gastro-intestinal mucosal surface. The secretory IgA molecule consists of an immunoglobulin synthesised and secreted by the plasma cells in the lamina propria and a non-immunoglobulin part, the secretory component (SC), a glycoprotein synthesised by the epithelial cells (Poger & Lamm 1974, Brandtzaeg & Baklien 1977).

In the mucosal colonic epithelium, IgA and SC are present in the columnar cells but not in the goblet cells (Brandtzaeg 1974, Poger & Lamm 1974). A decrease in their synthesis and transport has been described in some colonic adenocarcinomas inversely correlated with the degree of differentiation and in areas of severe dysplasia in adenomatous polyps, indicative perhaps of early malignant change (Poger et al 1976, Weisz-Carrington et al 1976).

REFERENCES

Bara J, Gardais A P, Loisillier F, Burtin P 1978 Isolation of a sulfated glycopeptidic antigen from human gastric tumours. Its localization in normal and cancerous gastro-intestinal tissues. International Journal of Cancer 21: 133

Boland C R, Kim Y S 1983 Lectin markers of differentiation and malignancy. National large bowel cancer project: Workshop. Raven Press (in press)

Brandtzaeg P 1974 Mucosal and glandular distribution of immunoglobulin components: differential localisation of free and bound SC in secretory epithelial cells. Journal of Immunology 112: 1553–1559

Brandtzaeg P, Baklien K 1977 Intestinal secretion of IgA and IgM: a hypothetical model. In: Immunology of the gut. Ciba Foundation Symposium 46. Elsevier/North Holland, Amsterdam

Burtin P, Martin E, Sabine M C, Von Kleist S 1972 Immunological study of polyps of the colon. Journal of the National Cancer Institute 48: 25–29

Burtin P, Von Kleist S, Sabine M C, King M 1973 Immunological localization of carcinoembryonic antigen and non-specific cross-reacting antigen in gastro-intestinal normal and tumoral tissue. Cancer Research 33: 3299–3305

Clamp J R 1980 Gastro-intestinal mucus. In: Wright R (ed) Recent advances in gastro-intestinal pathology. Saunders, London, p 47–58

Culling C F A, Reid P E, Burton J D, Dunn W L 1975 A histochemical method of differentiating lower gastro-intestinal tract mucin from other mucins in primary or metastatic tumours. Journal of Clinical Pathology 28: 656–658

Culling C F A, Reid P E, Writh A J, Dunn W L 1977 A new histochemical technique of use in the interpretation and diagnosis of adenocarcinoma and villous lesions in the large intestine. Journal of Clinical Pathology 1056–1062

Currie G A, Bagshawe K D 1968 The role of sialic acid in antigen expression: further studies of the Landschutz ascites tumour. British Journal of Cancer 22: 843–853

Dawson P A, Patel J, Filipe M I 1978 A quantimet and statistical analysis of variations in sialomucins in the mucosa of the large intestine in malignancy. Histochemical Journal 10: 559–572

Ehsanullah M, Filipe M I, Gazzard B 1982a Morphological and mucous secretion criteria for differential diagnosis of solitary ulcer syndrome and non-specific proctitis. Journal of Clinical Pathology 35: 26–30

Ehsanullah M, Filipe M I, Gazzard B 1982b Mucin secretion in inflammatory bowel disease: correlation with disease activity and dysplasia. Gut 23: 485–489

Filipe M I 1969 Value of histochemical reactions for mucosubstances in the diagnosis of certain pathological conditions in the colon and rectum. Gut 10: 577–586

Filipe M I 1975 Mucous secretion in rat colonic mucosa during carcinogenesis induced by dimethylhydrazine. A morphological and histochemical study. British Journal of Cancer 32: 60–77

Filipe M I 1979 Mucins in the human gastro-intestinal epithelium. A review. Investigative and Cell Pathology 2: 195–211

Filipe M I, Branfoot A C 1974 Abnormal patterns of mucous secretion in apparently normal mucosa of large intestine with carcinoma. Cancer 34: 282–290

Filipe M I, Rogers C 1979 Mucous-secretory patterns associated with colonic cancer. In: Schauer R (ed) Proceedings of the 5th International Symposium on Glycoconjugates. Georg Thieme, Stuttgart, p 651 (abstract)

Filipe M I, Abbas S, Greaves P 1981 Mucin composition and CEA content in assessing malignant potential of colonic adenomas. VIIIth European Congress of Pathology, Helsinki, p 22 (abstract)

Filipe M I, Mughal S, Bussey H J R 1980 Patterns of mucous secretion in the colonic epithelium in familial polyposis. Investigative and Cell Pathology 3: 329–343

Goldenberg D M 1976 Oncofetal and other tumour-associated antigens of the human digestive system. In: Morson B C (ed) Current topics of pathology. Springer-Verlag, Heidelberg, 63: 289–342

Gorman T A, LaMont J T 1978 Glycoprotein synthesis and secretion in human colonic cancers and normal colonic mucosa. Cancer Research 38: 2784–2789

Greaves P, Filipe M I, Branfoot A C 1980 Transitional mucosa and survival in human colorectal cancer. Cancer 46: 764–770

Hakkinen I P, Heinonen R, Inberg M V et al 1980 Clinicopathological study of gastric cancers and precancerous states detected by fetal sulphoglycoprotein antigen screening. Cancer Research 40: 4308–4312

Heyderman E 1980 Immunocytochemistry in cancer diagnosis. In: Symington T, William A E, McVie J G (eds) The 10th Pfizer International Symposium. Cancer assessment and monitoring. Churchill Livingstone, Edinburgh, p 147–171

Heyderman E, Gibbons A R, Bulman A S 1981 Immunocytochemical identification of cells. New approaches to laboratory medicine. VIIIth Merz & Dade International Symposium Rosalky S B, Darmstadt 135–148

Issacson P 1976 Tissue demonstration of carcinoembryonic antigen (CEA) in ulcerative colitis. Gut 17: 561–567

Isaacson P, Le Vann H P 1976 The demonstration of carcinoembryonic antigen in colorectal carcinoma and colonic polyps using an immunoperoxidase technique. Cancer 38: 1348–1356

Isselbacher K J 1974 The intestinal cell surface: some properties of normal, undifferentiated and malignant cells. Annals of Internal Medicine 81: 681–686

Jass J R 1980 Role of intestinal metaplasia in the histogenesis of gastric carcinoma. Journal of Clinical Pathology 33: 801–810

Jass J R, Filipe M I 1979 A variant of intestinal metaplasia associated with gastric carcinoma: a histochemical study. Histopathology 3: 191–199

Jass J R, Filipe M I 1981 The mucin profiles of normal gastric epithelium, intestinal metaplasia and gastric carcinoma. Histochemical Journal 13: 931–938

Kim Y S, Isaacs R 1975 Glycoprotein metabolism in inflammatory and neoplastic diseases of the human colon. Cancer Research 35: 2092–2097

Kitaoka H, Hattori N, Mucojima J 1973 Alpha-fetoprotein content in tissues from patients with gastric cancer. Tumour Research 8: 171–177

Listinsky C M, Riddell R H 1981 Patterns of mucin secretion in neoplastic and non-neoplastic diseases of the colon. Human Pathology 12: 923–929

Lotan R, Nicholson G L 1978 Membrane glycoproteins — dynamics and affinity isolation. In: Glycoproteins and glycolipids in disease processes. American Chemical Society Symposium Series 80, p 256–271

Montero C, Segura D I 1980 Retrospective histochemical study of mucosubstances in adenocarcinomas of the gastro-intestinal tract. Histopathology 4: 281–291

Neville A M, Laurence D J R 1980 Tumour markers and the gastro-intestinal tract. In: Wright R (ed) Recent advances in gastro-intestinal pathology. Saunders, London, p 255–266

O'Brien M J, Zamckeck N, Burke B 1981 Immunocytochemical localization of carcinoembryonic antigen in benign and malignant colorectal tissues. Assessment of diagnostic value. American Journal of Clinical Pathology 75: 283–290

Pearse A G E 1968 Histochemistry — theoretical and applied, 3rd edn. Churchill Livingstone, Edinburgh, vol 1

Poger M E, Lamm M E 1974 Localization of free and bound secreting component in human intestinal epithelial cells. A model for the assembly of secreting IgA. Journal of Experimental Medicine 139: 629–642

Poger M E, Hirsch B R, Lamm M E 1976 Synthesis of secreting component by colonic neoplasms. American Journal of Pathology 82: 327–338

Rapp W, Windisch M, Peschke P, Wurster K 1979 Purification of human intestinal goblet cell antigen. Its immunological demonstration in the intestine and in mucus-producing adenocarcinomas. Virchows Archiv für pathologische Anatomie und Physiologie 382: 163–177

Rogers C M, Cooke K B, Filipe M I 1978 The sialic acids of human large bowel mucosa: o-acylated variants in normal and malignant states. Gut 19: 587–592

Sachs L, Inbar M, Shinitzky M 1974 Mobility of lectin sites on the surface membrane and the control of cell growth and differentiation. In: Clarkson B, Baserga R (eds) Control of proliferation in animal cells. Cold Spring Harbor, New York, p 283–296

Sharkey R M, Hagihara P E, Goldenberg D M 1977 Localization by immunoperoxidase and estimation by radioimmunoassay of carcinoembryonic antigen in colonic polyps. British Journal of Cancer 35: 179–189

Shoham J, Sachs L 1974 Differences in lectin agglutinability of normal and transformed cells in interphase and mitosis. In: Clarkson B, Baserga R (eds) Control of proliferation in animal cells. Cold Spring Harbor, New York, p 297–304

Sipponen P, Seppala K, Varis K 1980 Intestinal metaplasia with colonic-type sulphomucins in the gastric mucosa, its association with gastric carcinoma. Acta pathologica et microbiologica scandinavica 88: 217–224

Skinner J M, Whitehead R 1981 Tumour-associated antigens in polyps and carcinoma of the human large bowel. Cancer 47: 1241–1245

Taglbjaerg P S, Nielsen H O 1978 'Small intestinal type' and 'colonic type' metaplasia of the human stomach. Acta pathologica et microbiologica scandinavica 86: 351–355

Wanebo H J, Rao B, Pinsky C M 1978 Preoperative carcinoembryonic antigen level as a prognostic indicator in colorectal cancer. New England Journal of Medicine 299: 448–451

Weisz-Carrington P, Poger M E, Lamm M E 1976 Secreting immunoglobulins in colonic neoplasm. American Journal of Pathology 85: 303–316

13

Malabsorption

Z. Lojda

INTRODUCTION

Appropriate nutrition depends on the adequate intake of food, its effective splitting on digestion and proper absorption and utilisation of breakdown products. A deficiency in one or more of these processes leads to manifestations of malnutrition. In clinical terminology the set of symptoms which occur either regularly or occasionally in diseases with disorders of digestion, absorption, secretion, and motility of the small intestine is designated as *malabsorption syndrome* (MS). Its classification is not easy (Jeffries et al 1969, Lojda et al 1971, Morson & Dawson 1979): the enterocytes of the small intestine are responsible not only for absorption but also for the terminal digestion of nutrients, and the pathophysiological mechanisms leading to manifestations of MS appear in many cases in various combinations. In every individual case the diagnosis is achieved on recognition of the nosological unit which causes the malabsorption symptomatology. The information on whether the malabsorption is caused by a defect in the enterocytes (primary MS) or outside them (secondary MS) is of basic importance and can often be resolved by intestinal biopsy. A close collaboration between clinician and pathologist is essential to yield the most useful information.

Diagnostic assessments based on purely morphological analysis of biopsies are limited and have been recently summarised by Morson & Dawson (1979). Histochemistry has broadened them substantially (Lojda et al 1971, Lojda 1974, 1976a, 1981) and an outline of this approach is given in this chapter.

BIOPSY OF THE SMALL INTESTINAL MUCOSA

Specimens of the jejunal mucosa are usually taken with the aid of a suction biopsy capsule. The capsule is positioned just beyond the duodenojejunal flexure and its position checked by X-ray. Several types of biopsy capsules are available. The Crosby-Kugler capsule, which is the most widely used, enables the biopsy of a larger specimen from one site (about 4×7 mm, 20–25 mg) while with other instruments several smaller biopsies from different sites can be obtained. Biopsies are taken from fasting patients in the morning (to preclude the influence of the circadian rhythm on enzyme activities). In adult patients and older children, biopsies can be performed in the outpatient department. In babies and smaller children a short stay in hospital is necessary. The value of the information obtained by biopsy far surpasses the difficulties and potential hazards of the intervention. Of decisive significance are biopsies in diffuse conditions. In focal lesions a negative result does not exclude a suspected disorder. Duodenal biopsies are not suitable for the diagnosis of MS, because of a proximodistal gradient of enzyme activities (particularly of disaccharidases) in the enterocytes from the duodenum to the jejunum (Lojda 1976a, Lojda et al 1979b). For example, in alactasia the biopsy must be from the jejunum since there is sometimes no activity in the normal duodenum (Lojda 1974, 1976a). On the other hand when a deficiency of enteropeptidase (enterokinase) is suspected, the duodenal mucosa should be examined, because in man the activity of this enzyme is normally low in the proximal jejunum (Lojda 1976a).

PROCESSING AND EVALUATION OF THE BIOPSY

The biopsy is carefully spread mucosal surface upwards on filter paper, a gelatine foil, or plastic micromesh. The dissecting microscope appearances are not as important as formerly contended and observation of unfixed specimens ought to be as short as possible to prevent autolytic changes. Small biopsy specimens are quenched directly in light petroleum chilled in an acetone- or ethanol-dry ice mixture. If the specimen is larger it can be divided into two or more pieces some of which are fixed (details in Appendix 1).

Good cryostat sections enable morphological evaluation so that the making of paraffin wax sections is not necessary in routine practice.

Cryostat sections are cut perpendicularly to the mucosal surface and processed as shown in Table 13.1.

In *haematoxylin and eosin*-stained sections, the regularity of the surface, size and shape of villi as well as depth of crypts are assessed. Other features easily recognised in routine preparations are the enhanced basophilia of the enterocytes, increased mitotic activity in the crypts and increased numbers of intra-epithelial lymphocytes, features which are associated with hyperregeneration states (e.g. coeliac disease). The latter can be easily identified by the reaction for ATPase (Lojda 1976b; Fig. 13.1A, B). Enterocytes with enlarged nuclei are seen in vitamin B_{12} deficiency and greater numbers of goblet cells are found in patients with cystic fibrosis and coeliac disease. Paneth cells can likewise be observed in H & E-stained preparations and their granules demonstrated by the reaction for tryptophan. At present they have

Table 13.1 Methods in diagnosis of intestinal biopsies. Abbreviations: C, cryostat section from snap-frozen specimen; ClA, postfixed in chloroform-acetone (1:1) at 4°C 5 min; F, postfixed in 4% formaldehyde at room temperature 5 min; O, unfixed; P, paraffin wax section from fixed specimen; R, if required

Method	Specimen treatment
Obligatory reactions	
Haematoxylin-eosin	CF or P
PAS-reaction	CClA or P
Lactase (indigogenic or azo-dye)	CO or CClA
Trehalase (GO-PO-DAB)	CO
Sucrase (GO-PO-DAB or azo-dye)	CO or CClA (azo-dye)
Enteropeptidase (indirect method) R	CO
Fettrot 7B	CF
Immunoglobulins R	CO
Amyloid (Congo red) R	CF or P
Useful reactions	
Acid phosphatase (azo-dye)	CF (fixation at 4°C)
ATPase (Wachstein and Meisel)	CF (fixation at 4°C)
DAP IV (azo-dye with Gly-Pro-4-methoxy-2-naphthylamine)	CClA
Methyl green-pyronin	P

no great significance in the diagnosis of the MS. The same holds true for endocrine cells which are therefore not considered here.

In the lamina propria hyperaemia and oedema may be artefacts induced by the suction biopsy. *Assessment of the cellularity* is an important parameter. Plasma cells are revealed by the methyl green-pyronin reaction and if required immunoglobulins can be identified by immunohistochemical methods (see Ch. 20) (Fig. 13.2 A–C). Macrophage activity can be demonstrated by the reaction for acid phosphatase

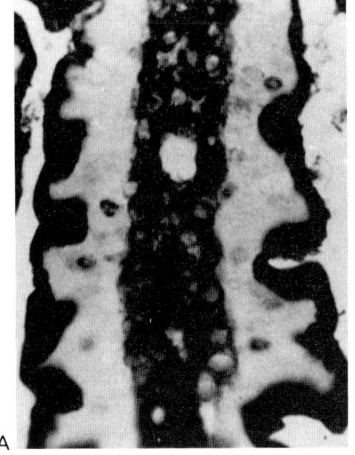
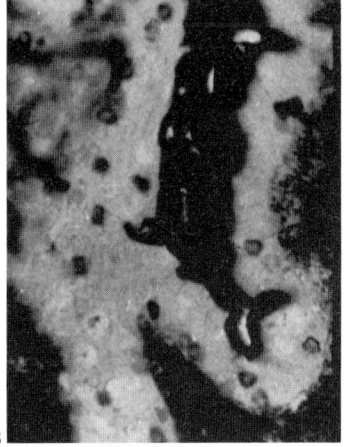

Fig. 13.1 ATPase revealed with the Wachstein-Meisel method. CF. A. Normal jejunal mucosa. B. Jejunal mucosa of a patient with coeliac disease. In addition to a strong positive reaction in the brush border and in the lamina propria there is a remarkable positivity in the cell membranes of epithelial lymphocytes, the number of which is almost doubled in the patient with coeliac disease. ×410

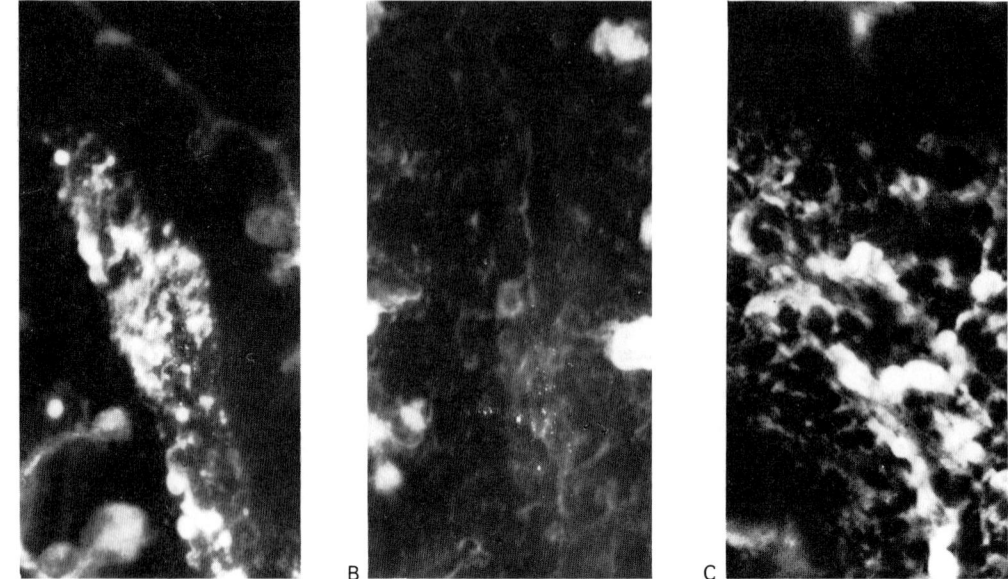

Fig. 13.2   IgA demonstrated with an indirect immunofluorescence method. CO. A. Normal mucosa. Strong fluorescence in plasmocytes of the lamina propria, a weaker fluorescence of the brush border. ×610. B. Mucosa of a patient with IgA deficiency. Fluorescence in the lamina propria is substantially reduced. ×610. C. Flat mucosa of a patient with coeliac disease. Numerous IgA-positive plasmocytes are present in the lamina propria. ×610

(azo-coupling reaction using naphthol AS-BI phosphate and hexazonium-*p*-rosaniline (Lojda et al 1979a). Neutrophils and eosinophils are selectively shown by the reaction for myeloperoxidase (using N-phenyl-*p*-phenylene-diamine and α-naphthol (Lojda et al 1979a) or by the reaction for elastase-like activity (neutrophils only) using naphthol AS-D chloroacetate and hexazonium-*p*-rosaniline (Lojda et al 1979a). Resting T-lymphocytes show a positive reaction for dipeptidyl (amino) peptidase (DAP) IV (Lojda 1981). DAP IV is also present in the capillary endothelium in the normal jejunal mucosa. A decrease of DAP IV activity in the lamina propria points to inflammatory changes (Fig. 13.3A, B).

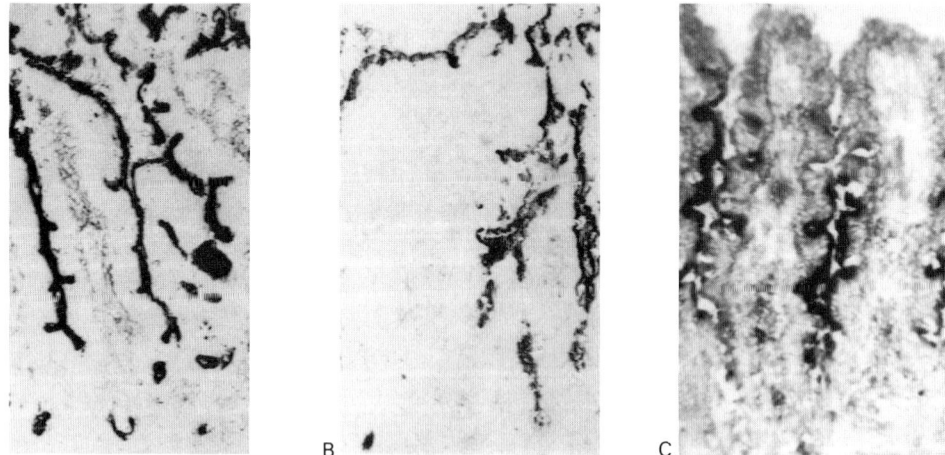

Fig. 13.3   DAP IV (A, B) and γ-glutamyl transferase (C). CC1A. DAP IV was demonstrated according to Lojda with Gly-Pro-MNA and fast blue B, γ-glutamyl transferase also according to Lojda with γ-Glu-MNA, Gly-Gly and fast blue B. CClA. A. Normal jejunal mucosa. Strong activity in the brush border. In the lamina propria a very distinct reaction can be found in capillary endothelium and in T-lymphocytes. ×120. B. Jejunal mucosa of a patient with coeliac disease. Decreased activity in the brush border and also in the lamina propria in which only some lymphocytes (helper cells) react. ×120. C. Jejunal mucosa of a patient with intractable diarrhoea. No reaction for γ-glutamyl transferase is present in the brush border of enterocytes covering upper halves of villi. ×120

In H & E-stained preparations dilated lymphatic vessels can be recognised. When the dilation is caused by an obstruction of the lymph flow the detection of lipids using Sudan dyes (preferably Fettrot 7B) can be useful. In these states lipids (both intra- and extracellular) are found in the lamina propria of fasting patients. The finding of *lipid* droplets in enterocytes covering normal villi points to a-β-lipoproteinaemia.

The demonstration of *disaccharidases* is of fundamental importance. In routine practice the histochemical demonstration of these enzymes is more convenient than the biochemical determination of disaccharidase activities in homogenates of the jejunal mucosa. It requires much less material (the demonstration of one disaccharidase can be performed using 1–3 sections 10 μm thick) and enables, at the same time, the assessment of the morphological pattern. A semiquantitative evaluation is sufficient. For routine practice the assessments of lactase, sucrase and trehalase suffice. The method of choice for the demonstration of *lactase* is the indigogenic procedure with 4-Cl-5-Br-3-indolyl-β-D-fucoside (Lojda & Kraml 1971, cf. Lojda et al 1979a) (see Appendix 5) (Figs. 13.4A–C and 13.7B).

These reactions enable the demonstration of lactase deficiency. Lactase is the most sensitive indicator of the degree of differentiation and injury of enterocytes. In pathological processes its activity is the first to be affected and the last to be restored. The second in respect to sensitivity is *trehalase*. Its activity is demonstrated with its natural substrate (trehalose) in the coupled glucose oxidase-peroxidase-diaminobenzidine (GO-PO-DAB) method (Lojda 1972, Lojda et al 1979a) (Appendix 5) (Fig. 13.5B). *Sucrase* appears the most resistant of disaccharidases to

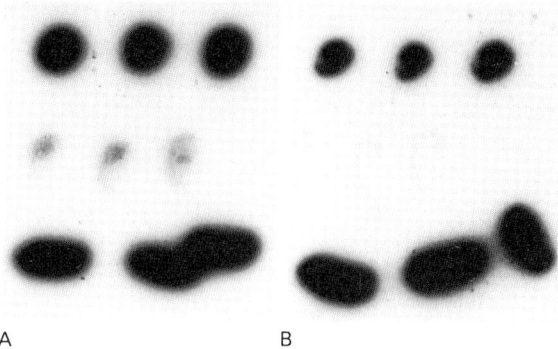

A   B

Fig. 13.5 Disaccharidases. CO. GO-PO-DAB method. Macrophotographs: 3 consecutive sections of normal jejunal mucosa (bottom row), of jejunal mucosa from a patient with coeliac disease (central row) and from a patient after 1 year on a gluten-free diet (upper row). The degree of the coloration corresponds to the enzyme activity. A. sucrase; B. trehalase. ×2

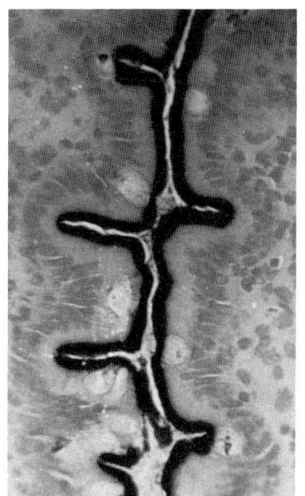
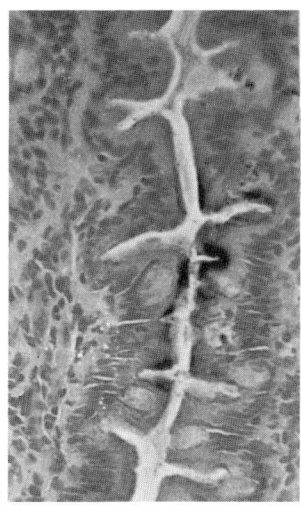
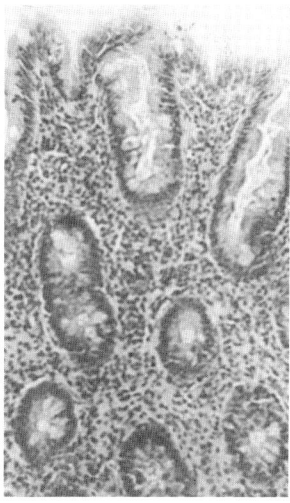

A   B   C

Fig. 13.4 Lactase. CO. Indigogenic method with 4-Cl-5-Br-3-indolyl-β-D-fucoside according to Lojda and Kraml. Counterstained with Kernechtrot. A. Normal mucosa. Strong reaction in the brush border. ×410. B. Jejunal mucosa of a patient with isolated lactase deficiency. Two enterocyte populations can be seen. In one no reaction is apparent, in the other some reaction is present. ×410. C. Jejunal mucosa with subtotal villous atrophy in a patient with coeliac disease. The reaction is completely negative. ×120

injury. It can be demonstrated either with its natural substrate (sucrose) using the GO-PO-DAB method (Lojda 1972, Lojda et al 1979a) or with a synthetic substrate 6Br-2-naphthyl-α-D-glucoside employed in the simultaneous azo-coupling method with hexazonium-p-rosaniline (Lojda 1965, Lojda et al 1979a; Fig. 13.6A–C) (Appendix 5). Although this substrate is also cleaved by glucoamylase (Lojda et al 1973) deficiency of sucrase (and isomaltase) can well be detected by it (Fig. 13.6B).

The demonstration of *enteropeptidase* activity requires a two-step reaction according to Lojda & Mališ (1972) and Lojda et al (1979a) in which the activity of trypsin originating from trypsinogen is demonstrated, with benzyloxycarbonyl-Gly-Gly-Arg-4-methoxy-2-naphthylamide as substrate and fast blue B (Lojda 1981) (Appendix 5). This method is more sensitive than the direct demonstration of enteropeptidase with Gly-Asp-Asp-Asp-Asp-Lys-2-naphthylamine (Antonowicz 1979). For a better characterisation of changes in enterocytes, particularly in the long-term follow-up of patients with coeliac disease, it is advisable to perform other reactions, for the demonstration of brush border proteases (Lojda 1981) or lysosomal enzymes (Lojda et al 1971, Lojda 1974). This, however, goes beyond routine diagnosis and therefore will not be considered here further.

DIAGNOSTIC OUTLINE OF INDIVIDUAL FORMS OF MS IN ENTEROBIOPSIES

Isolated enzymopathies

Isolated disaccharidase deficiencies

These are either congenital or acquired and occur in childhood and in adults. Their incidence is race-dependent, being more frequent in coloured individuals. The morphological appearance of the mucosa is usually normal, or with only minor irregularities of villi, a somewhat larger number of goblet cells and a slightly increased cellularity in the lamina propria.

Lactase deficiency (hypolactasia, alactasia) is the most common. Histochemically three types are recognised. Most frequently lactase is not detectable. In others there is a general diffuse reduction of lactase activity. In the third type there are two populations of enterocytes: one, which prevails numerically, is lactase-negative, the other is lactase-positive (Fig. 13.4B). It is not possible to decide by biopsy alone whether the deficiency is congenital or acquired, although the last type is usually acquired.

Isolated sucrase-isomaltase deficiency occurs much less frequently. It is usually congenital. Sucrase is present in traces or not detectable at all (Fig. 13.6B).

Isolated trehalase deficiency is the least frequent. It is inborn and manifests itself after eating mushrooms.

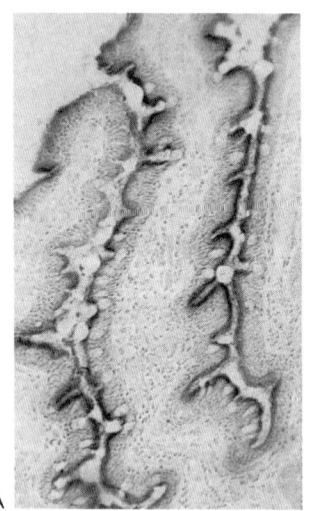
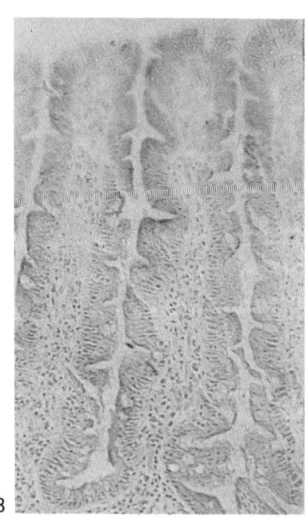
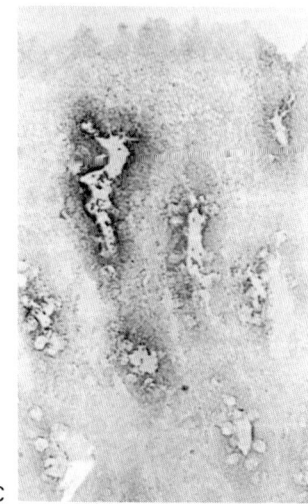

Fig. 13.6 α-glucosidases (mainly sucrase). CClA. 6-Br-2-naphthyl-α-D-glucoside and hexazonium-p-rosaniline. A. Normal jejunal mucosa. Strong reaction in the brush border. ×120. B. Jejunal mucosa from a patient with isolated sucrase-isomaltase deficiency. The reaction in the brush border is missing. ×120. C. Jejunal mucosa from a patient with coeliac disease, with total villous atrophy. Reaction present in the brush border of enterocytes in the middle part of the crypts. ×120

Trehalase activity is not detectable. A diminished activity of trehalase (isolated or combined with hypolactasia) can be found in some patients with cystic fibrosis whose jejunal mucosa is otherwise normal.

Combined disaccharidase deficiencies

These deficiencies occur particularly in coeliac disease, tropical sprue, cow's milk protein intolerance, Whipple's disease, radiation enteritis and after administration of some drugs. In these cases, however, changes are already seen in H & E-stained preparations.

Enteropeptidase deficiency

Congenital enteropeptidase deficiency in children is rare. Enzyme activity is completely missing. Secondary deficiencies (decreased activity) occur in some patients with partial and total atrophy of duodenal villi and in intractable diarrhoea of infancy. It is to be stressed that no judgement about enteropeptidase activity can be made on the morphological appearance of the duodenal mucosa. In many patients with coeliac disease displaying flat mucosa the enzyme activity is within normal limits. Conversely, in some patients with only minor changes in the mucosa the activity can be low (Lojda & Jodl 1974). It is suggested that hypoproteinaemia and oedema in some patients with protein malabsorption might be related to low entero-peptidase activity.

A-β-lipoproteinaemia

This is a rare disorder caused by defective synthesis of the apoprotein(s) in enterocytes necessary for the formation of chylomicrons. Jejunal biopsy usually shows no gross abnormality of the villi. Enterocytes of fasting patients display a foamy appearance in sections stained with H & E. This is due to lipid droplets which can be demonstrated by staining with Fettrot 7B.

Coeliac disease (non-tropical sprue, coeliac sprue, gluten-induced enteropathy, gluten-sensitive enteropathy, primary malabsorption syndrome sensu stricto)

Changes in the jejunal mucosa depend on the stage of the disease and its treatment. The characteristic morphological features are the flattening of the mucosal surface caused by changed villi which are short, broad and sometimes absent (partial, subtotal or total atrophy). The enterocytes on the surface facing the lumen are smaller and crowded together so that the surface appears more than one layer thick. There is deepening of crypts, increased mitotic activity of enterocytes which are basophilic, increased numbers of intra-epithelial lymphocytes, and thickening of the basement membrane. In the lamina propria oedema and increased cellularity consisting of plasmocytes, lymphocytes, macrophages and eosinophils is found. These morphological changes have their cytochemical counterpart, which is described in detail in previous communications (Lojda et al 1971, Lojda 1974).

For the diagnosis of gluten-sensitive enteropathy it is required that the pattern should improve after a gluten-free diet and deteriorate after a gluten challenge.

Although there is no diagnostic difficulty in sections stained with H & E, histochemistry can aid in a more subtle assessment of the changes and of the improvement of the pattern after treatment. One cannot judge the extent of biochemical changes and their improvement solely on the basis of morphological evaluation. The demonstration of some enzymes is very helpful. Of the brush border enzymes lactase is the most severely affected and in the acute phase activity is virtually absent (Fig. 13.4C). Trehalase is similar in this respect and its activity is present in traces or may be lacking (Fig. 13.5B). The activity of sucrase is low (Figs. 13.5A, 13.6C) and in some patients cannot be demonstrated. DAP IV enables a very rapid assessment of the involvement of the lamina propria. More serious pathology is indicated by a greater decrease of activity (Fig. 13.3B). An expression of coeliac disease to varying extents occurs also in patients with *dermatitis herpetiformis Duhring*.

Protein intolerance

A pattern closely resembling coeliac disease, which cannot be differentiated from it solely on the basis of the evaluation of a single jejunal biopsy, occurs in patients with an *intolerance to cow's milk protein and soy bean protein*. In these patients changes are usually of a somewhat milder degree.

Tropical sprue

Similar changes in the jejunal mucosa are found also in patients with *tropical sprue (endemic sprue)* which is endemic in south-western Asia, Indonesia, some parts of India, Caribbean area, Nigeria, and the Middle-East. It affects natives as well as immigrants and visitors. The disease can manifest itself in countries where it does not occur normally, in persons who return from endemic regions. Tropical sprue does not respond to gluten-free diet and responds well to antibiotics, folic acid and vitamin B_{12} treatment. No clear-cut distinction of tropical sprue from coeliac disease is possible in biopsies (Lojda et al 1971, Lojda 1974). Lipids present in the basement membrane of the enterocytes, and the presence of brush border enzymes claimed by some to be characteristic for tropical sprue, are also found in coeliac disease.

Immunodeficiency disorders

MS can occur also in *immunodeficiency disorders*. In some of them the morphological pattern of the jejunal mucosa is normal with only a deficiency in a specific immunoglobulin-(usually IgA)-producing plasma cell population (Fig. 13.2B). In these cases isolated alactasia occurs more frequently than in patients with a normal composition of plasma cells in the lamina propria. In some patients there are conspicuous morphological villous changes and completely flattened mucosa may be found. About 1.5% of patients with coeliac disease responding to a gluten-free diet have IgA deficiency in the lamina propria which is sometimes referred to as hypogamma-globulinaemic sprue. This deficiency can be demonstrated immunohistochemically. In these patients a compensatory increase in number of IgM plasmocytes is found.

Radiation enteritis and cytotoxic agents

The lesions with flat mucosa described above belong to the hyperregeneratory type. There are cases with flattened mucosa, however, that belong to the hyporegeneratory type. These lesions occur *after* X-ray irradiation and after treatment with folic acid antagonists such as methotrexate. The cytoplasm of enterocytes does not display enhanced basophilia. Disaccharidase deficiencies are present. In the lamina propria a decrease of DAP IV positivity is remarkable.

Drug-induced changes

Other drugs such as *chloramphenicol, aureomycin, terramycin, and pheninedione* produce slight morphological changes. Hypolactasia or alactasia is a very common histochemical finding. After *neomycin* treatment these changes are more pronounced and a hypotrehalasia can be demonstrated in addition to alactasia. A very impressive finding is the activation of macrophages in the apical region of villi well demonstrated with the acid phosphatase reaction. The overall cellularity of the lamina propria is not increased. On the contrary it seems decreased. Severe changes of subtotal villous atrophy has been described with nefenamic acid, which return to normal after withdrawal of the drug.

Whipple's disease

This is a rare disorder usually classified under secondary malabsorption syndrome in which the cause of malabsorption lies beyond the absorptive cells of the intestine. It is diagnosed very easily by the presence of sickle-form particle-containing cells (SPC-cells) in the lamina propria which are strongly PAS-positive (Fig. 13.7A). In the brush border of enterocytes no lactase is present. A significant decrease of trehalase and sucrase also occurs pointing to the participation of enterocytes in the malabsorption. After treatment with broad-spectrum antibiotics the number of SPC-cells diminishes substantially. However, dilated lymph vessels and alactasia persist (Fig. 13.7B).

Cystic fibrosis

In some patients with *cystic fibrosis* the pattern in jejunal biopsies is completely normal. However, in more than 50% a decreased activity of trehalase, sometimes combined with hypolactasia, can be found even in cases in which there is no mucus hypersecretion. In about one half of patients the number of goblet cells is increased and PAS-positive mucus fills crypts and adheres to the surface of villi. In some patients the pattern of coeliac disease is found.

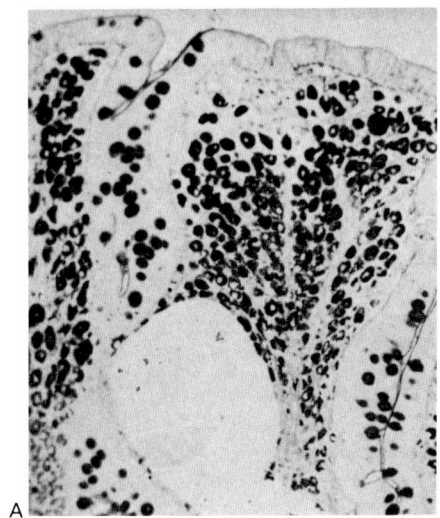 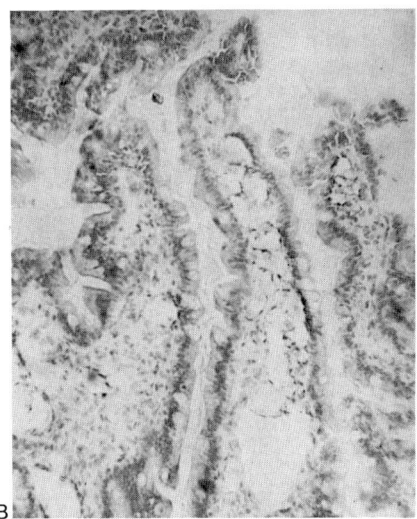

Fig. 13.7 Whipple's disease. CClA. A. PAS reaction before the treatment with antibiotics. Strong positivity in SPC cells in the lamina propria. The goblet cells and brush border of enterocytes also show a positive reaction. Note a dilated lymphatic. ×230. B. Same case after the treatment with antibiotics. Although there is a remarkable improvement of the morphological pattern lactase activity is lacking and dilated lymph vessels still persist. Indigogenic method of Lojda and Kraml. Counterstained with Kernechtrot. ×120

Efferent loop

In evaluating changes in the *mucosa in the efferent loop* it should be borne in mind that all changes within 20 cm of the efferent loop have no diagnostic value. Alactasia is found sometimes in the mucosa beyond 20 cm.

Other disorders

No constant changes are found in the mucosa of patients with *glucose-galactose malabsorption*, in *Hartnup disease, cystinuria* and other rare disturbances of the amino acid metabolism. Patchy, almost zero activities of some brush border enzymes found in enterocytes in the apical region of villi, particularly in cystinuria, are not constant. These changes cannot be considered diagnostic for these diseases because they occur also in some patients with intractable diarrhoea in which no disturbances of amino acid metabolism are found (Fig. 13.3C).

Amyloid

Demonstration of *amyloid* (best using Congo red with dichroism in polarized light) is performed when secondary amyloidosis is suspected, although a positive finding in jejunal biopsy is very rare.

REFERENCES

Antonowicz I 1979 The role of enteropeptidase in the digestion of proteins and its development in human fetal small intestine. In: Ciba Foundation Symposium 70. Excerpta Medica, Amsterdam, p 169–183
Jeffries G H, Weser E, Sleisenger M H 1969 Malabsorption. Gastroenterology 56: 777–785
Lojda Z 1965 Some remarks concerning the histochemical detection of disaccharidases and glucosidases. Histochemie 5: 339–360
Lojda Z 1972 An improved method for the demonstration of disaccharidases with natural substrates. Histochemie 30: 277–280
Lojda Z 1974 Cytochemistry of enterocytes and of other cells in the mucous membrane of the small intestine. In: Smith D H (ed) Biomembranes. Intestinal absorption. Plenum Press, London, vol 4A, p 43–122

Lojda Z 1975 Suitability of the azocoupling reaction with 1-naphthyl-β-D-glucoside for the histochemical demonstration of lactase (lactase-β-glucosidase complex) in human enterobiopsies. Histochemistry 43: 349–353
Lojda Z 1976a Der Wert der Darmbiopsie für die Diagnostik und Beurteilung der Therapie des Malabsorptionssyndroms. Ergebnisse der experimentellen Medizin 27: 189–196
Lojda Z 1976b Cytochemical study on epithelial lymphocytes of the human jejunum. Sborník lékařský 78: 263–268
Lojda Z 1981 Proteinases in pathology. Usefulness of histochemical methods. Journal of Histochemistry and Cytochemistry 29: 481–493
Lojda Z, Jodl J 1974 Histochemical investigation of enterokinase in duodenal and jejunal biopsies of children with malabsorption syndrome. Československá Gastroenterologie a Výživa 28: 532–539

Lojda Z, Kraml J 1971 Indigogenic methods for glycosidases. III. An improved method with 4-Cl-5-Br-3-indolyl-β-D-fucoside and its application in studies of enzymes in the intestine, kidney and other tissues. Histochemie 25: 195–207

Lojda Z, Mališ F 1972 Histochemical demonstration of enterokinase. Histochemie 32: 23–29

Lojda Z, Gossrau R, Schiebler T H 1979a Enzyme histochemistry. A laboratory manual. Springer-Verlag, Heidelberg

Lojda Z, Kociánová J, Mařatka Z 1979b Histochemistry of the human duodenal mucosa with special reference to the gradient of activities of the brush border enzymes.

Scandinavian Journal of Gastroenterology 14, supplement 54: 7–13

Lojda Z, Frič P, Jodl J, Chmelík V 1971 Cytochemistry of the human jejunal mucosa in the normal and in malabsorption syndrome. Current Topics in Pathology 52: 1–63

Lojda Z, Slabý J, Kraml J, Kolínská J 1973 Synthetic substrates in the histochemical demonstration of intestinal disaccharidases. Histochemie 34: 361–369

Morson B C, Dawson I M P 1979 Gastrointestinal pathology, 2nd edn. Blackwell, London

14

Acetylcholinesterase in the diagnosis of Hirschsprung's disease and other gastro-intestinal disorders

B. D. Lake

HIRSCHSPRUNG'S DISEASE

The diagnosis of Hirschsprung's disease by demonstration of the absence of ganglion cells in the distal portions of the large intestine is the standard method employed in most routine histopathology laboratories.

Full-thickness biopsies of rectum, performed under general anaesthetic, present few problems to the pathologist but have the disadvantage that such large biopsies may cause scarring and surgical complications in the subsequent pull-through procedure. Also, many patients present in the neonatal period, are often severely ill and a general anaesthetic is an unnecessary additional hazard. For these reasons suction rectal biopsies are now generally preferred and these may be taken without anaesthetic on the ward or as an out-patient procedure. The density of ganglion cells varies in the last few cm of the rectum and studies by Aldridge & Campbell (1968) emphasise the importance of knowing as precisely as possible the site of the biopsy in relation to the anal verge. The diagnosis of 'not Hirschsprung's disease' is straightforward since the presence of ganglion cells excludes Hirschsprung's disease. However, as there is a normal hypoganglionic region extending 1–2 cm from the anal verge, a minimum of 50 serial sections of an adequate biopsy must be examined.

Such problems have led to other approaches to diagnosis. Early reports by Meier-Ruge and his colleagues (1972, 1974) suggested that the diagnosis could be made by assessment of the numbers of nerve fibres showing acetylcholinesterase activity in the lamina propria. These claims were not upheld by all investigators but our experience at the Hospital for Sick Children in London over a period of 9 years

(1975–1983) has confirmed Meier-Ruge's observations. Provided the guidelines outlined below are followed, and the biopsies are taken at known, carefully defined distances from the anal verge, there should be no false-positives and no false-negatives.

The method for demonstrating the acetyl-cholinesterase-positive nerve fibres should be permanent and show the fibres readily with good contrast from the background. Koelle's modification of Gomori's method tends to fade, particularly if mounted in synthetic media, and Karnovsky and Roots' direct colouring method sometimes has poor contrast. In 1972 Hanker, Anderson and Bloom described a method for ultrastructural studies in which the reaction product (copper ferrocyanide) of the Karnovsky and Roots' medium was made more electron-dense by formation of osmium black.

Lake et al (1978) used this method for the light microscopic demonstration of acetylcholinesterase-positive nerve fibres in a series of suction rectal biopsies for the diagnosis of Hirschsprung's disease. Cryostat sections of the suction rectal biopsies snap-frozen (supported on animal liver, thin wedges of 12% gelatin or in OCT compound) are cut at 10 μm and air-dried. Fixation in formol calcium for 30 sec is followed by a brief wash and incubation for 1 h in the medium as described in Appendix 5. The intensifying agent is *p*-phenylene diamine rather than diaminobenzidine, which is carcinogenic.

Although an inhibitor (iso OMPA) is added to reduce pseudocholinesterase staining, the inhibition is not complete and a diffuse reaction is found in the smooth muscle. The inhibitor appears to have no effect on the esterases found in cells of lymphoreticular aggregates and these give a strongly positive reaction. In the normal rectum and descending colon only a few fine nerve fibres are found in the lamina propria and muscularis mucosae. Occasional nerve trunks and fibres are found in the submucosa and may be prominent close to the anal verge in the area where there is paucity of ganglion cells. On routine histological examination this can give rise to a false impression of Hirschsprung's disease. Fine acetylcholinesterase-positive nerve fibres may be found around blood vessels. Ganglion cells are well stained, having a stippled appearance with nerve fibres among the groups of cells, sometimes with fibres entering — or emerging. The acetylcholinesterase method is particularly helpful in the easy identification of the clusters of 'immature' ganglion cells sometimes found in the neonate (Yunis et al 1976).

In both short-segment (up to the rectosigmoid junction) and long-segment (beyond the rectosigmoid junction) Hirschsprung's disease there is an increase in the numbers and usually in size of the nerve fibres in the lamina propria and muscularis mucosae. This increase may be dramatic. Nerve trunks are also found in increased numbers in the submucosa. No indication of the severity of the disease can be gained from the appearance of the biopsy, numerous nerve fibres being found in the older, milder and in the severe neonatal cases. It is important to note that there should be an increase both in the muscularis mucosae and lamina propria for a positive diagnosis of Hirschsprung's disease.

In total colonic aganglionosis the changes may appear particularly mild with only a few submucosal nerve trunks. The acetylcholinesterase-positive nerve fibres may show no increase in the lamina propria and only a mild-to-moderate increase in the submucosal aspect of the muscularis mucosae. However, this apparently mild change is not found in all cases of total colonic aganglionosis, and they can show the changes of the more common long-segment Hirschsprung's disease.

OTHER CONDITIONS

Intestinal neuronal dysplasia

Scharli & Meier-Ruge (1981) have drawn attention to a condition they call neuronal intestinal dysplasia which has in the past been labelled neuronal colonic dysplasia, or more properly colonic neuronal dysplasia. These patients present with enterocolitis, as Hirschsprung's disease or as residual symptoms of obstruction following adequate resection of an aganglionic segment (Puri et al 1977). Suction rectal biopsies from these patients show an increase in fine acetylcholinesterase-positive nerve fibres in the lamina propria with no increase in the muscularis mucosae.

Scattered ganglion cells may be found in the lamina propria, and an apparent excess of submucosal ganglion cells and ganglioneuromata should be seen before a diagnosis of intestinal neuronal dysplasia can be made. The symptoms are alleviated by resection of the affected segment, though this may be difficult to

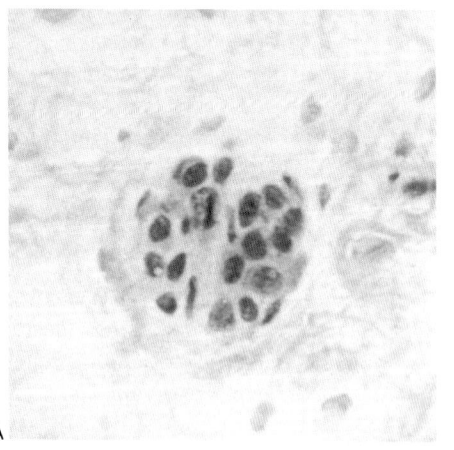 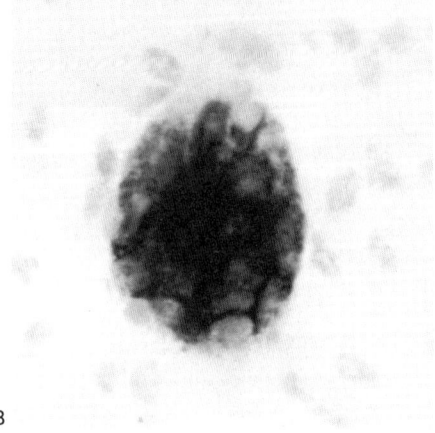

A   B

Fig. 14.1 'Immature' ganglion cells in the submucosa of a suction rectal biopsy from a neonate. Serial cryostat sections stained with H & E (A) and for acetylcholinesterase activity (B). The ganglion cells are more readily identified in acetylcholinesterase preparations. ×450

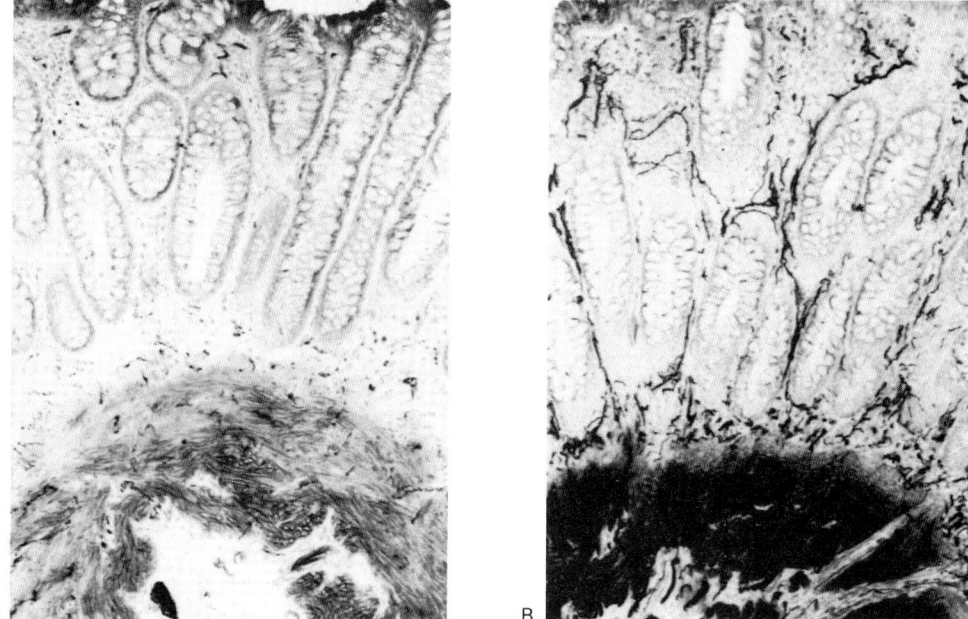

Fig. 14.2 Cryostat sections of suction rectal biopsies stained to demonstrate acetylcholinesterase activity. The normal (A) shows only a few fine nerve fibres in the lamina propria and muscularis mucosae. Ganglion cells are present in the submucosa. In Hirschsprung's disease (B) the increase in nerve fibres is readily apparent. × 100

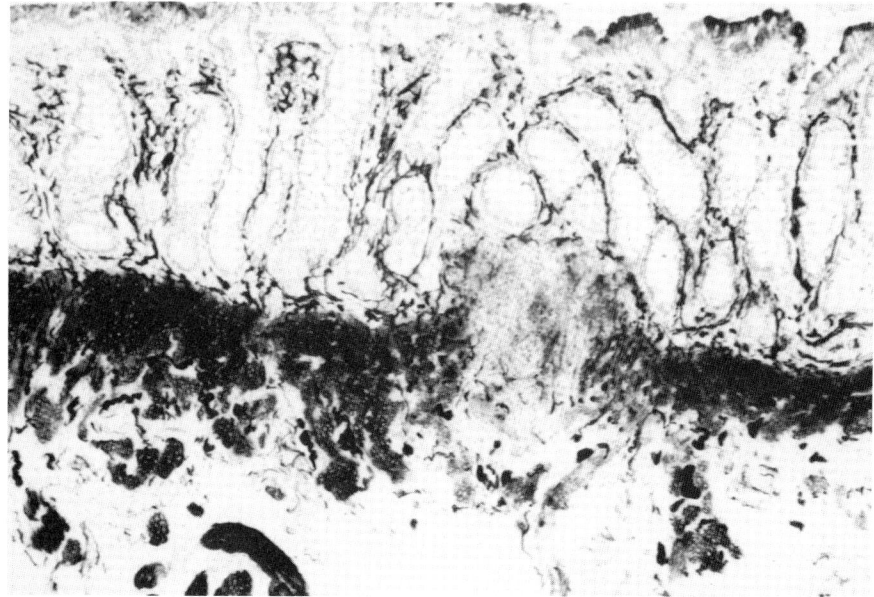

Fig. 14.3 Cryostat section of a suction rectal biopsy from a patient with Hirschsprung's disease. Acetylcholinesterase-positive nerve fibres can be seen in the lamina propria and muscularis mucosae, and nerve trunks are present in the submucosa. × 90

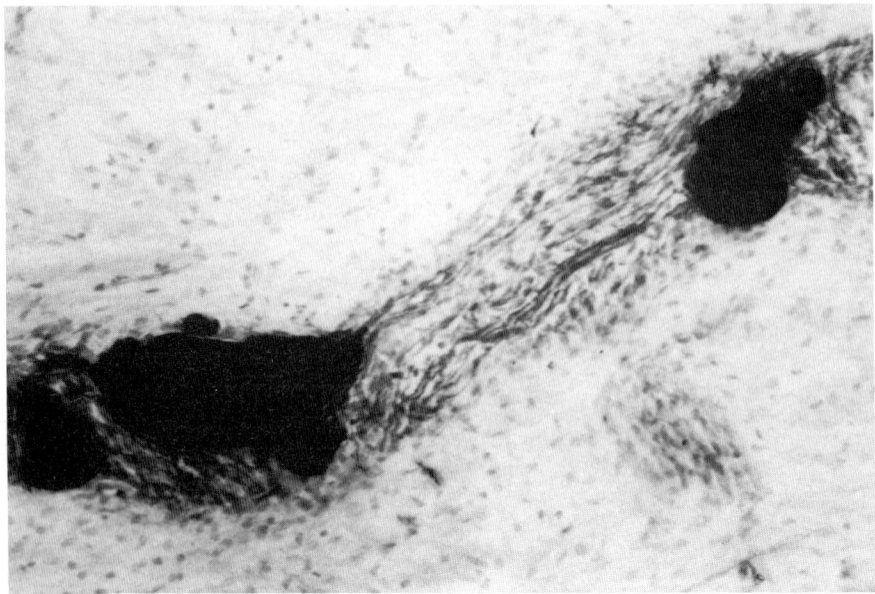

Fig. 14.4 Cryostat section of resected colon from a patient with intestinal neuronal dysplasia. Numerous ganglion cells present within a submucosal nerve trunk (ganglioneuroma) are shown by the acetylcholinesterase method. ×150

define since the degree of abnormality may not correlate with the severity of symptoms. Submucosal ganglioneuromata are also associated with phaeochromocytoma and medullary thyroid carcinoma (as part of the multiple endocrine neoplasia syndrome). If ganglioneuromata are found, the serum calcitonin and/or urinary VMA levels should be measured.

Inflammatory bowel disease

A marked increase in fine nerve fibres in the lamina

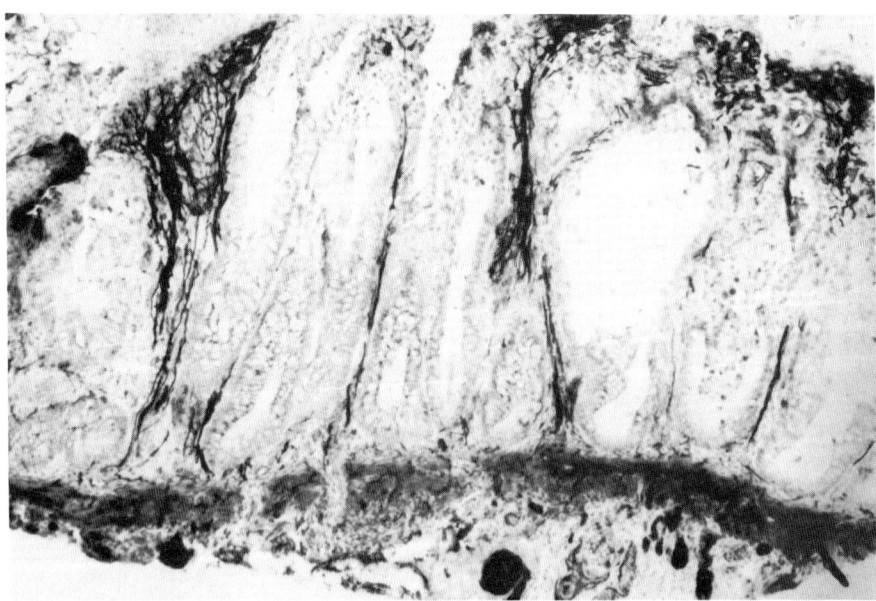

Fig. 14.5 Colonoscopy biopsy from an infant with allergic colitis. The histological picture is of ulcerative colitis, but there is a marked increase of thickened ribbon-like nerve fibres in the lamina propria shown by the acetylcholinesterase method. The muscularis mucosae shows no increase in nerve fibres and ganglion cells are present in the submucosa. A crypt abscess is present. ×90

propria with normal muscularis mucosae may be found in some cases of necrotising enterocolitis. We have also seen several infants with allergic colitis showing a marked increase in large broad AChE-positive fibres in the lamina propria with normal muscularis mucosae and normal submucosal ganglion cells. Similar milder changes may be seen in ulcerative colitis. Occasional increases in acetyl-cholinesterase-positive fibres are found in Crohn's disease but there is a more dramatic increase in fibres rich in vasoactive intestinal polypeptide (VIP) in the lamina propria (Bishop et al 1980).

Adynamic bowel syndrome

Some patients with this syndrome show absence of the argyrophilic myenteric plexus. There may also be a gross depletion of acetylcholinesterase-positive nerve fibres in the myenteric plexus and an absence of fibres in the circular and longitudinal muscle coats (Puri et al 1977). Other patients show ultrastructural change in the circular muscle with degeneration of smooth muscle cells and an increase in connective tissue. These latter patients are said to have a visceral myopathy. Ganglia are normal.

REFERENCES

Aldridge R T, Campbell P E 1968 Ganglion cell distribution in the normal rectum and anal canal. Journal of Pediatric Surgery 3: 475–489

Bishop A, Polak J M, Bryant M G, Bloom S R, Hamilton S 1980 Abnormalities of vasoactive intestinal polypeptide-containing nerves in Crohn's disease. Gastroenterology 79: 853–860

Hanker J S, Anderson W A, Bloom F E 1972 Osmiophilic polymer generation: catalysis by transition metal compounds in ultrastructural cytochemistry. Science 175: 991–993

Lake B D, Puri P, Nixon H H, Claireaux A E 1978 Hirschsprung's disease. An appraisal of histochemically demonstrated acetylcholinesterase activity in suction rectal biopsies as an aid to diagnosis. Archives of Pathology and Laboratory Medicine 102: 244–247

Meier-Ruge W 1974 Hirschsprung's disease. Its aetiology, pathogenesis and differential diagnosis. In: Current topics in pathology. Springer, Berlin, vol 59, p 131–179

Meier-Ruge W, Lutterbeck P M, Herzog B, Morger R, Moser R, Scharli A 1972 Acetylcholinesterase activity in rectum suction biopsies as diagnostic in Hirschsprung's disease. Journal of Pediatric Surgery 7: 11–17

Puri P, Lake B D, Nixon H H 1977 Adynamic bowel syndrome. Report of a case with disturbance of the cholinergic innervation. Gut 18: 754–759

Puri P, Lake B D, Nixon H H, Mishalany H, Claireaux A E 1977 Neuronal colonic dysplasia: an unusual association of Hirschsprung's disease. Journal of Pediatric Surgery 12: 681–685

Scharli A F, Meier-Ruge W 1981 Localized and disseminated forms of neuronal intestinal dysplasia mimicking Hirschsprung's disease. Journal of Pediatric Surgery 16: 164–169

Yunis E J, Dibbins A W, Sherman F E 1976 Rectal suction biopsy in the diagnosis of Hirschsprung's disease. Archives of Pathology and Laboratory Medicine 100: 329–333

15

Liver: malignant tumours and chronic liver disease

M. Isabel Filipe & B. D. Lake

In the majority of liver biopsies, the diagnosis can be achieved by H & E and routine histological methods, i.e. collagen and reticulin stains, PAS with or without diastase digestion and Perls' reaction (MacSween et al 1979). However, in certain conditions, described below, the interpretation of liver biopsies presents difficulties even for a skilled pathologist and histochemical and immunohistochemical techniques are a useful additional tool.

MALIGNANT TUMOURS

Metastatic tumours in the liver are more common than primary cancers. In adults the commonest primary malignant tumour is the hepatocellular carcinoma (HCC). Its incidence varies in different parts of the world, being highest in Africa, particularly in Mozambique, and relatively uncommon in the western world.

Most HCC arise in cirrhotic liver, more frequently of the macronodular type, while in others an association with hepatitis B in the absence of cirrhosis has been reported and it may also complicate haemochromatosis (Anthony 1978, Scheuer 1980, Sherlock 1981).

Morphologically, the tumour cells in most cases of HCC mimic normal hepatocytes, often arranged in a trabecular pattern similar to normal liver, and are readily diagnosed in routine H & E sections. The tumour cells may contain fat and glycogen, and bile secretion is not uncommon. Mallory's hyaline may be found and PAS diastase-resistant globular inclusions are seen in about 15% of cases (Scheuer 1980). However, difficulties may arise in differentiating HCC, cholangiocarcinoma and metastatic carcinoma.

Diastase-PAS and alcian blue pH 2.5 stains may reveal mucins present in cholangiocarcinomas and metastatic adenocarcinomas but absent in HCC. In HCC reticulin is scanty or absent in contrast with the increased connective tissue stroma in cholangio-carcinomas. In addition, the reticulin pattern is most useful to assess malignant change versus adenoma or dysplasia.

The search for markers of HCC has included alpha-fetoprotein (AFP), carcino-embryonic antigen (CEA), α_1-antitrypsin (AAT) and various enzymes. None of them is specific in itself but a multiparameter approach can in certain cases be a useful complement to the morphological and clinical data.

Alpha-fetoprotein

One of the major fetal serum glycoproteins, this is detected in 4-week-old fetus, reaches its peak at 13–15 weeks, then falls to zero 2–5 weeks after birth.

In the fetus, AFP is synthesised in the liver, yolk sac and in lesser amounts in the thymus and gastro-intestinal tract. Small amounts of AFP are also produced in fetal lung. In the adult AFP is largely produced in HCC and yolk sac tumours of the testis and ovaries, less frequently in GI tract tumours and in a small number of bronchogenic carcinomas (Okuda & Nakashima 1979).

With the less sensitive immunodiffusion assays, the serum levels of AFP showed a great degree of specificity for HCC, yolk sac tumours of testis and ovaries and liver metastasis particularly from GI tract cancers. The development of more sensitive radioimmunoassays increased the number of false-positives (though not reaching the high levels associated with HCC), including other malignant tumours and various common non-malignant diseases, in particular viral hepatitis, chronic active hepatitis and cirrhosis. However, from recent surveys of many tumours it seems that high serum levels of AFP in excess of 200 ng/mg are a useful diagnostic tool in HCC and gonadal yolk sac tumours. Serial AFP estimations have an important role in the early detection of metastasis, monitoring the effects of therapy and possibly as a prognostic sign in non-malignant diseases such as hepatitis (Goldenberg 1976, Neville & Laurence 1980).

In the tissues AFP appears in the tumour cells of HCC as PAS-positive diastase-resistant globules similar to α_1-antitrypsin. Immunohistochemical techniques in formalin-fixed paraffin wax-embedded tissues reveal its presence in most cases of HCC, diffusely or as coarse granules in the cytoplasm of the tumour cells, in the cell membrane and in the perinuclear zone. AFP is almost always present in hepatoblastomas (Fig. 15.1). It is not detected in normal liver cells or liver adenomas, while there are conflicting reports regarding its finding in areas of dysplasia, non-neoplastic cells adjacent to HCC and in the so-called transitional cells in liver cell regeneration (Peyrol et al 1978, Thung et al 1979). No AFP is demonstrated in cholangiocarcinomas.

Carcinoembryonic antigen (CEA)

This has also been demonstrated in HCC tumour cells, in about 30% of the cases according to Thung et al (1979). It has not been detected in normal hepatocytes but is found in the apical cytoplasm and luminal surface of epithelial cells in bile ducts and ductules. CEA is not a specific marker for any particular neoplasm and therefore its diagnostic value is limited. However, in liver metastases its absence virtually excludes a primary tumour in the GI tract (Gerber & Thung 1978).

α_1-Antitrypsin (AAT)

This is a strong protease inhibitor produced by the liver. Similar to AFP, it appears in the hepatocytes as PAS-positive diastase-resistant cytoplasmic inclusions. By immunohistochemical techniques AAT is a common finding in HCC as fine granules in the cytoplasm of the tumour cells. It is also present in the hepatocytes in non-neoplastic areas adjacent to HCC, in liver adenomas and in focal nodular hyperplasia associated with the use of oral contraceptives (Palmer & Wolfe 1976).

Hepatitis B virus antigens

HBsAg but not HBcAg have been detected in HCC tumour cells (Thung et al 1979). Using double immunohistochemical methods, it is possible to demonstrate the presence of two or more antigens

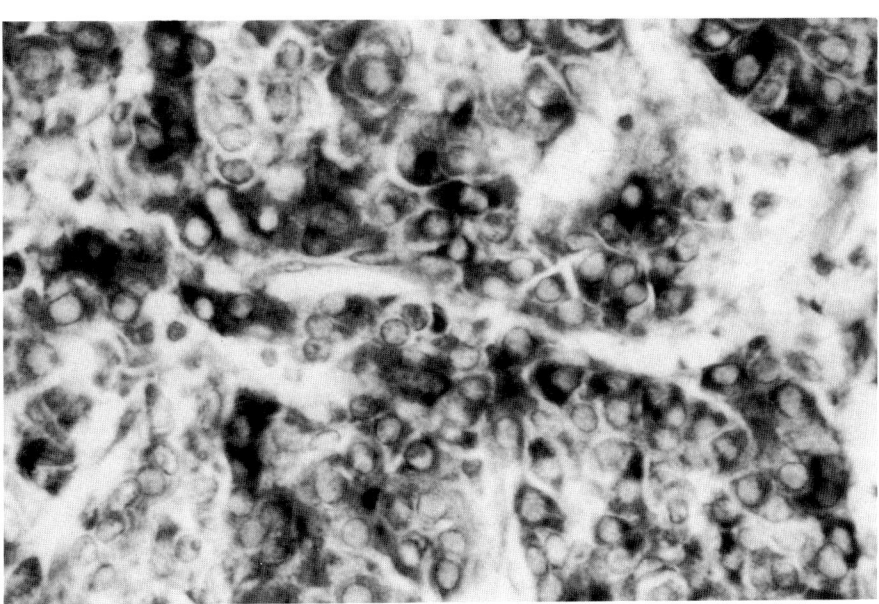

Fig. 15.1 Hepatoblastoma; routine section stained to show alphafetoprotein (AFP) by the indirect PAP method. Deposits can be seen in several tumour cells. The appearance varies from one area to another.

simultaneously in HCC in the same or in different tumour cells.

Enzymes

Histochemical methods for a variety of enzymes such as γ-glutamyl transpeptidase (GGTPase), canalicular adenosine triphosphatase (ATPase), glucose-6-phosphatase (G6Pase) and alkaline phosphatase applied to fresh frozen cryostat sections for HCC and other neoplastic and non-neoplastic liver tissues reveal changes in the amount and distribution of these enzymes (Gerber & Thung 1980). In the normal liver tissue G6Pase is demonstrated in the cytoplasm of hepatocytes throughout the lobules, ATPase in the bile canaliculi, and GGTPase in the bile canaliculi and in the epithelium of bile ducts and ductules. Similar activity and distribution are seen in liver cell adenoma and in the non-neoplastic areas adjacent to HCC. In contrast HCC shows abnormal enzyme patterns in various combinations, the most common finding being decreased or absent G6Pase and ATPase and increased GGTPase activities. The fibrolamellar variant of HCC with a better prognosis has been recently described (Paradinas et al 1982). This variant is characterized by numerous mitochondria in the tumour cells, and reactions for succinate dehydrogenase and NADH tetrazolium reductase reflect the increased numbers of mitochondria and may be useful pointers of this important variant of HCC. In cholangiocarcinomas there is weak or absent G6Pase and ATPase activities and a strong reaction for GGTPase. The activity of G6Pase is usually absent from hepatoblastoma.

In normal liver tissue alkaline phosphatase activity is seen in the endothelial cells of sinusoids and large vessels and sometimes focally in bile canaliculi but not in bile duct epithelium. This sinusoidal enzyme activity increases in various liver diseases. Of diagnostic significance perhaps is the presence of a marked canalicular alkaline phosphatase activity (not ATPase) in association with primary or secondary malignant tumours in the liver and less marked and patchy in collagen diseases, long-standing extrahepatic biliary obstruction and in α_1-antitrypsin deficiency (Hägerstrand 1975, 1976).

From the data accumulated it is apparent that there is no single antigen and enzyme specific for malignant tumours in the liver and furthermore tumour cells in HCC seem to be markedly heterogenous in their antigenic and enzymatic expression, which may be a reflection of subpopulations of tumour cells or tumour cells in various degrees of differentiation.

CHRONIC HEPATITIS

A simplified classification of chronic hepatitis describes two types — chronic persistent hepatitis (CPH) and chronic active hepatitis (CAH). The morphological features of either CPH and CAH are not specific and may be a reflection of various conditions such as virus hepatitis (type B, non-A, non-B), primary biliary cirrhosis, metabolic disorders (e.g. Wilson's disease) etc. (Scheuer 1980, Sherlock 1981).

The accuracy of the diagnosis assessing the severity and determining the cause is important because of its prognostic and therapeutic implications. Table 15.1 shows some of the features commonly found.

Table 15.1 Distribution of substances commonly found in hepatocytes in liver disease

	Fat	PAS + globules	Mallory's hyaline	HBAg + orcein	CAP	Copper	Iron
Viral hepatitis (HBAg)	−	−	−	+	−	−	−
PBC	−	−	±	−	+	+	−
Wilson's disease	±	−	+	−	+	+	+
α-1-antitrypsin deficiency	±	+	±	−	+	±	−
Alcoholism	+	−	+	−	−	−	±
Reye's syndrome	+						

Type B viral hepatitis

The presence of HB core and surface antigens can be demonstrated in liver cells by the orcein method (Deodhar et al 1975) and by both direct or indirect immunofluorescent and immunoperoxidase techniques in either unfixed frozen tissues or in formalin-fixed paraffin wax-embedded material (Burns 1975, Huang 1975, Nayak & Sachdeva 1975). Formalin fixation does not seem to inactivate the antigens*. Parallel studies with the three methods have confirmed the specificity of the orcein method for HBsAg. The staining mechanism is probably related to the presence of disulphide bonds (Fig. 15.2).

outlined inclusions of various shapes and sizes, as vesicles and tubules or in a fibrillar pattern, which correlates well with the electron microscopical observations.

The parenchymal distribution of the HBsAg may be diffuse throughout or spotty or both and therefore sampling errors may occur. There is a good correlation between HBsAg + cells and 'ground-glass' hepatocytes and an inverse relationship with areas of liver cell necrosis.

In hepatitis patients various patterns of intracellular and parenchymal distribution of HBsAg have been reported in association with each type of chronic

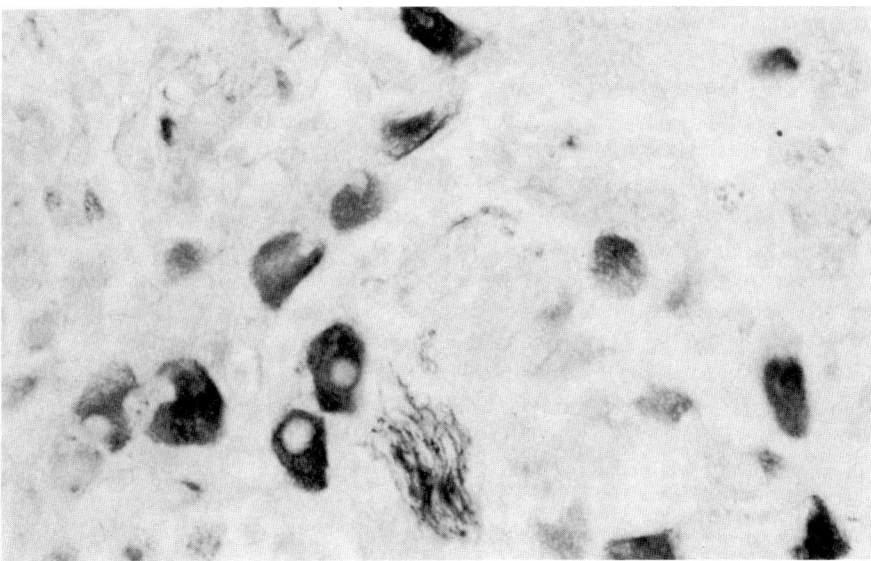

Fig. 15.2 Liver biopsy. Routine section stained with orcein (Doedhar modification of Shikata's method) to show HBs Ag in a patient with viral hepatitis. The 'ground glass' cells in H & E preparations stain darkly with orcein. Elastic fibres are also seen (bottom centre).

The HBcAg (core antigen) and HBsAg (surface antigen) of the Dane particle (which is probably the virus itself), are recognised as distinct components with proper morphological and antigenic expressions.

The patterns of distribution of HBsAg and HBcAg are different and their relative proportions in the liver seem to be related to the host immune response and the histological types of hepatitis. HBcAg is present in the nuclei and rarely in the cytoplasm. HBsAg is localised in the cytoplasm and membrane of liver cells and occasionally in Kupffer cells but never in the nuclei. The deposits of antigen may present as clearly

hepatitis and in acute hepatitis, hence useful parameters for differential diagnosis. HBsAg filling the whole cytoplasm with a diffuse lobular distribution is found in CPH and cirrhosis with little activity whilst its localisation in the cell membrane is a feature seen in CAH, active cirrhosis and in acute hepatitis with possible transition to chronicity. The presence of focal accumulations of surface antigen in the cytoplasm is described in acute hepatitis and in biopsies with chronic hepatitis in HBsAg serum-negative patients (Ray et al 1976a, b).

HBcAg, similarly to membrane-bound HBsAg, is

* Although immunoperoxidase methods in formalin-fixed paraffin wax-embedded tissue seem to produce a weaker stain reaction than in frozen material.

more prevalent in CAH and active cirrhosis. Both are virtually absent in acute hepatitis. A significant relationship seems to exist between patterns of distribution of both HBc and HBs antigens in the liver tissue and the immunological state of the host and this may be reflected in the various histological types of hepatitis with important prognostic implications (Gudat et al 1975, Ray et al 1976a).

Primary biliary cirrhosis (PBC)

Copper-associated protein

In the fetus and newborn the liver copper level is higher than the normal adult level which is attained by 1 year of age. Copper is bound to protein (copper-associated protein, CAP) to protect the cell from toxic damage (Evans et al 1980).

CAP is not seen in routinely stained sections but appears as small (less than 1 μm) grey granules in Perls' iron stain, and is PAS-positive after removal of glycogen with amylase. Oxidation of the sulphydryl and disulphide bonds of CAP with acidified permanganate solutions produces sulphuric acid residues making CAP basophilic and demonstrable with alcian blue. CAP is selectively stained with Shikata's orcein method (Deodhar et al 1975), and appears as coarse brown or black-brown granules in the cytoplasm of periportal hepatocytes (see Fig. 15.5).

The presence of CAP is an important diagnostic feature of PBC particularly in the differential diagnosis from CAH (Sipponen et al 1976). CAP is increased in primary biliary cirrhosis, conditions with long-standing cholestasis and α_1-antitrypsin deficiency but negative or only weakly positive in most specimens from patients with CAH or other types of cholestatic and non-cholestatic liver disease (e.g. Wilson's disease, biliary atresia, extra-hepatic obstruction) (Ludwig et al 1979). In primary biliary cirrhosis the amount of CAP correlates with the copper concentration measured by neutron activation analysis. However, in Wilson's disease (total copper >4 μmol/g dry weight, normal levels 0.4 μmol/g) no such correlation exists either with CAP or copper staining (Jain et al 1978). Furthermore, the morphology of the orcein positive material in PBC, differs from the inclusion bodies of Australia antigen (HBsAg) (Salaspuro & Sipponen 1976).

OTHER LIVER CONDITIONS

Wilson's disease

Wilson's disease is characterised by a low serum caeruloplasmin level and accumulation of copper in the liver and other organs. It is transmitted in an autosomal recessive fashion.

In the well-established cases the liver is cirrhotic (macronodular) and resembles the macronodular cirrhosis of other causes. Mallory bodies are common. Iron lipofuscin and bile pigment may be present in addition to variable amounts of copper and copper-associated protein (CAP) in hepatocytes and macrophages. CAP may be shown by the orcein method, and copper should be shown by at least two methods. The colorimetric methods for copper are not entirely specific and hence it is important to use several methods giving different colours. Iron pigment should also be shown to be separate from the copper deposition. In routine sections of liver the rubeanic acid method (green/black) and the p-dimethylaminobenzylidine rhodanine (red/orange) are reliable (Fig. 15.3). In cryostat sections sodium diethyldithiocarbamate solutions give an immediate yellow colour which is visible macroscopically, but which gives poor localisation.

In the precirrhotic stage many glycogenic nuclei are present in the periportal areas but this feature is common in neonates, infants, diabetic patients and in some of the glycogen storage disorders. Iron is often present in Kupffer cells (as a result of the haemolytic anaemia) and copper, though present and increased by quantitative methods, is not always increased by cytochemical staining procedures. This may reflect differences in distribution between cytosolic copper (more soluble) and lysosomal copper (less soluble). With progression of the disorder more copper becomes bound within lysosomes. The liver is moderately to markedly fatty.

α_1-antitrypsin deficiency

A deficiency of serum α_1-antitrypsin leads either to lung pathology (panlobular emphysema) or to liver pathology (fibrosis, cirrhosis, neonatal hepatitis) or both. α_1-antitrypsin (AAT) is the major protease inhibitor (Pi) in serum and several allelic variants are known; deficiency is regarded as less than 20% of normal activity and is usually associated with the S and Z alleles.

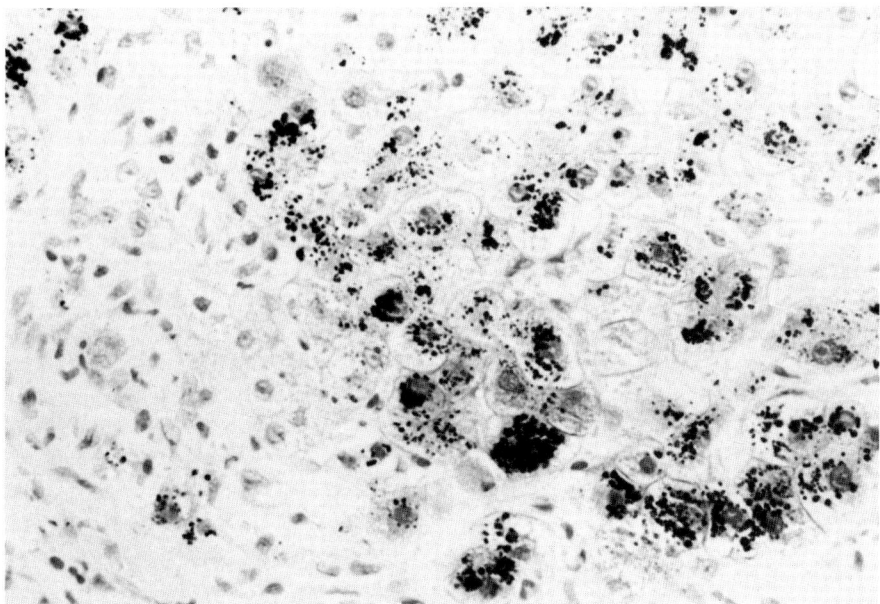

Fig. 15.3 Wilson's disease. Routine section of a liver biopsy stained with p-dimethylaminobenzylidine rhodanine for copper. Many copper-containing granules are present in liver cells in a cirrhotic liver. A very similar appearance is seen in Stage 3 primary biliary cirrhosis.

In the liver there is fibrosis, often cirrhosis, and hepatocytes in the periportal regions contain cytoplasmic droplets which are eosinophilic and PAS-positive after diastase digestion. These deposits can be patchily distributed. Fatty change and canalicular bile stasis can also be seen. In childhood the presentation may be of 'neonatal hepatitis'.

The size of the droplets is an important consideration (Clausen et al 1980) since PAS-positive granules may occur in a variety of situations. Droplets over 3 μm in diameter are pathognomonic for the PiZ allele but these droplets occur only in about one third of cases (Fig. 15.4). Droplets of 1–3 μm in diameter occur more frequently, and if the small dust-like droplets are included then up to 85% of cases will be detected. However, as the size of the droplets acceptable for diagnosis decreases, there is an increased risk that other liver disorders may be included.

The PAS-positive droplets react with antisera to AAT and may be identified as AAT by an immunoperoxidase technique. Although the substance in the liver cells has immunoreactivity it is not physiologically active in serum because of the genetic defect which results in the synthesis of a defective AAT glycoprotein which cannot be transported out of the cell. The defective AAT accumulates in dilated endoplasmic reticulum.

It is uncommon to detect the PAS-positive droplets in affected infants under 2–3 months of age and it is possible that the more sensitive immuno-histochemical method may be of more value in the diagnosis in this age range.

Lipofuscin granules and copper associated protein deposits (Fig. 15.5) are often present in addition to the α_1-antitrypsin droplets in periportal hepatocytes, and each of these substances is PAS positive after diastase or saliva digestion to remove glycogen. In these circumstances by using the staining sequence 'digestion — Schmorl reaction — PAS', the lipofuscin and copper associated protein deposits are stained blue leaving the α_1-antitrypsin droplets bright red.

Reye's syndrome

The changes in the liver from patients with Reye's syndrome consist of an excess of fat (often described as being in both micro- and macrovesicular form), mild portal inflammatory infiltrate and retention of bile pigment in hepatocytes. Occasional periportal necrosis may be present. Glycogen is usually

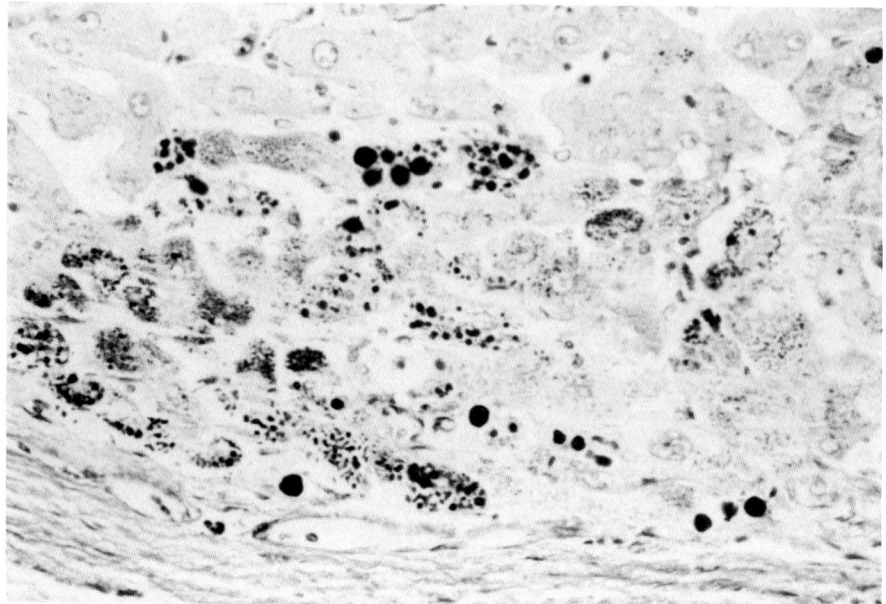

Fig. 15.4 α_1-antitrypsin deficiency: routine section of liver stained with PAS after amylase digestion. Numerous droplets of varying sizes are present adjacent to a fibrotic portal area.

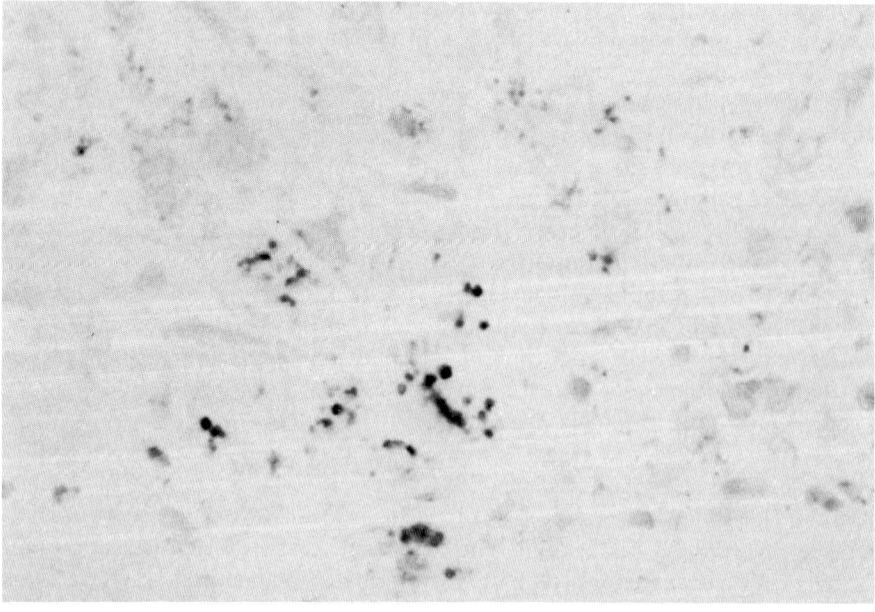

Fig. 15.5 α_1-antitrypsin deficiency: routine section of liver showing copper-associated protein in hepatocytes stained with orcein. Copper is also present.

depleted; however in a patient who has been vigorously treated with IV dextrose etc., the glycogen content may be normal. Fat deposits may also occur in the heart and in the renal tubular epithelium. Reye's syndrome has been described in infants — where aflatoxin B has been isolated from liver samples — and in older children where a variety of viruses have been implicated.

In Reye's syndrome there are marked mitochondrial disturbances leading to a defective urea cycle and consequent high ammonia levels. The disturbance to mitochondrial function, probably of a toxic nature, can be readily shown by staining for succinate dehydrogenase activity in sections of snap-frozen tissue. No activity can be demonstrated in liver biopsies from patients with Reye's syndrome (Fig. 15.6), although it is possible that in the occasional patients who recover, a biopsy in the recovery phase might show some evidence of return of mitochondrial function with weak SDH activity. Other conditions in which there is a toxin affecting mitochondria (e.g. mushroom poisoning) may also show loss of succinate dehydrogenase activity.

In the genetic defects of the urea cycle resulting in high blood ammonia levels (deficiencies of carbamoyl phosphate synthetase, ornithine carbamoyl transferase, arginino succinate synthetase, or arginino-succinate lyase), no fat is usually seen and the activity of succinate dehydrogenase is normal. The glycogen content is normal and only minimal amounts of bile pigment are seen.

Porphyrias

These are disorders associated with abnormal biosynthesis of haem, in which an excess of porphyrin precursors, porphobilinogen and delta amino-laevulinic acid (ALA) is excreted in the urine.

They may be primary and inherited or secondary. In the former various types are described according to the enzymatic defect and the clinical presentation (Table 15.2). Those with liver involvement (chronic hepatic porphyrias, CHP) are the porphyria cutanea tarda (PCT) and the erythropoietic protoporphyria (EPP) (Scheuer 1980, Sherlock 1981). In PCT there is usually evidence of liver disfunction with lesions ranging from chronic persistent hepatitis to chronic active hepatitis and cirrhosis. Siderosis and fatty changes may also be present. Deposits of uroprophyrin in the liver cells can be shown by red fluorescence in UV light or as needle-shaped birefringent crystals (Fig. 15.7). The amount, the

Table 15.2

Porphyrias	Defect
Acute intermittent porphyria	Uroporphyrinogen I synthetase
Hereditary coproporphyria	Coproporphyrinogen oxidase (A1A-S activity > in liver)
Porphyria variegata	Protoporphyrinogen oxidase (A1A-S > in liver)
Porphyria cutanea tarda	Uroporphyrinogen decarboxylase (Deposits of uroporphyrin in liver)
Erythropoietic protoporphyria	Haem synthetase (Deposits of protoporphyrin in liver)

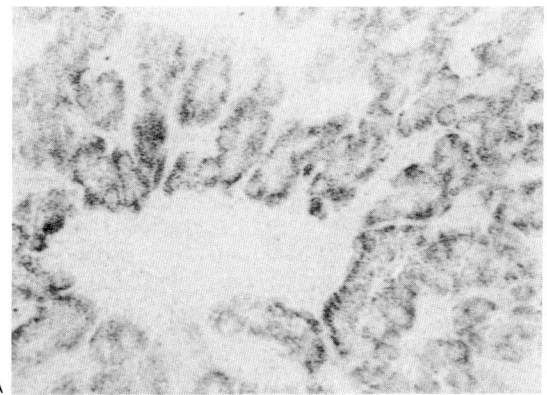

Fig. 15.6 Cryostat sections of liver stained for succinate dehydrogenase activity. A. Normal liver showing fine granular activity in hepatocytes. A portal tract in the centre is unstained. B. Section from a patient with Reye's disease. No SDH activity is present. A few clumps of bile pigment are seen.

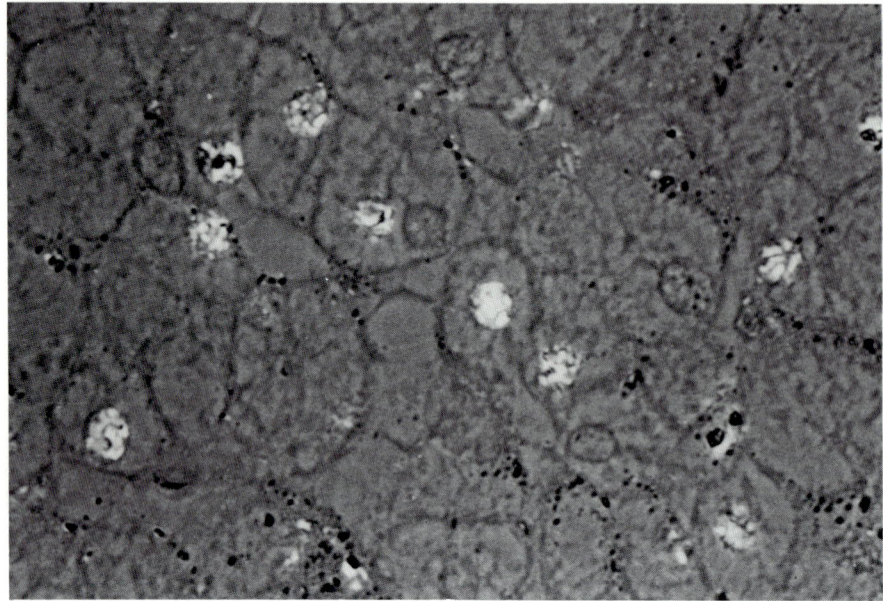

Fig. 15.7 Liver in porphyria cutanea tarda showing birefringent crystals in the hepatocytes. (Unstained paraffin wax section. Polarizing microscopy.)

type and the distribution of porphyrin deposits in the liver varies with the phase of CHP. Red fluorescent spots are typical of subclinical forms of CHP type A and B whereas the fine-reticular to the uniform fluorescence is present in fully manifest PCT. There is evidence that the proportion of uro- and hepta-isomers vary with the phase of the disease (Look & Doss 1976).

In EPP the liver lesions may be serious and cirrhosis may develop. Focal dark-brown deposits of pigment containing protoporphyrin are seen in the liver cells and in some cases in Kupffer cells, portal macrophages and bile canaliculi. The porphyrin

nature is revealed by fluorescent microscopy (UV light) or as birefringent crystals on polarising microscopy (Bruguera et al 1976).

Secondary coproporphyrias may be caused by heavy metal intoxication, particularly lead, in sideroblastic anaemias, various liver diseases, the Dubin-Johnson syndrome and as a complication of drug therapy.

For fluorescence microscopy (UV light) use unstained sections from paraffin wax-embedded material or air-dried unfixed smears. For polarising microscopy use either unstained or H & E-stained paraffin wax sections.

REFERENCES

Anthony P P 1978 Tumours of the liver. In: Anthony P P, Woolf N (eds) Recent advances in histopathology. Churchill Livingstone, Edinburgh, vol 10, p 213–233

Bove K E 1975 The character and specificity of the hepatic lesion in Reye's syndrome. In: Pollack J D (ed) Reye's syndrome. Grune & Stratton, New York, p 93–116

Bruguera M, Esquerda J E, Mascaró J M, Piñol J 1976 Erythropoietic protoporphyria. A light, electron and polarization microscopical study of the liver in three patients. Archives of Pathology and Laboratory Medicine 100: 587–589

Burns J 1975 Immunoperoxidase localisation of hepatitis B antigen (HB) in formalin-paraffin processed liver tissue. Histochemistry 44: 133–135

Callea F, Ray M B, Desmet J 1981 Alpha-1-antitrypsin and copper in the liver. Histopathology 5: 415–424

Clausen P P, Gad I, Lindskov J, Orholm M, Ring-Larsen H 1980 The specificity of alpha-1-antitrypsin globules as a marker of the Pi-Z gene. In: Lake B D, High O B, Holt S J, Stoward P J (eds) Abstracts of the VIth International Histochemistry and Cytochemistry Congress. Royal Microscopical Society, Oxford, p 69

Deodhar K P, Tapp E, Scheuer P J 1975 Orcein staining of hepatitis B antigen in paraffin sections of liver biopsies. Journal of Clinical Pathology 28: 66–70

Dvorackova I, Proks C 1979 Reye's syndrome in the newborn. In: Crocker J F S (ed) Reve's syndrome, 2nd edn. Grune & Stratton, New York, p 319–338

Evans J, Newman S P, Sherlock S 1980 Observations on copper-associated protein in childhood liver disease. Gut 21: 970–976

Gerber M A, Thung S N 1978 Carcino-embryonic antigen in normal and diseased liver tissue. American Journal of Pathology 92: 671–680

Gerber M A, Thung S N 1980 Enzyme patterns in human hepatocellular carcinoma. American Journal of Pathology 98: 395–400

Goldenberg D M 1976 Oncofetal and other tumour-associated antigens of the human digestive system. In: Morson B C (ed) Current topics in pathology. Springer-Verlag, Heidelberg, vol 63, p 289–342

Goldfischer S, Sternlieb I 1968 Changes in the distribution of hepatic copper in relation to the progression of Wilson's disease (hepatolenticular degeneration). American Journal of Pathology 53: 883–901

Gudat F, Bianchi L, Sonnabend W, Thiel G, Aenishaenslin W, Stalder G 1975 Pattern of core and surface expression in liver tissue reflects state of specific immune response in hepatitis B. Laboratory Investigation 32: 1–9

Hägerstrand I 1975 Distribution of alkaline phosphatase activity in healthy and diseased human liver tissue. Acta Pathologica et Microbiologica Scandinavica section A 83: 519–526

Hägerstrand I 1976 Bile canalicular alkaline phosphatase and disease. Acta Pathologica et Microbiologica Scandinavica section A 84: 271–277

Huang Shao-Nan 1975 Immunohistochemical demonstration of hepatitis B core and surface antigens in paraffin sections. Laboratory Investigation 33: 88–95

Irons R D, Schenk E A, Lee J C K 1977 Cytochemical methods for copper. Archives of Pathology and Laboratory Medicine 101: 298–301

Ishak K G, Sharp H L 1979 Metabolic errors and liver disease. In: MacSween R N M, Anthony P P, Scheuer P J (eds) Pathology of the liver. Churchill Livingstone, Edinburgh, p 104–106

Jain S, Scheuer P J, Archer B, Newman S P, Sherlock S 1978 Histological demonstration of copper and copper-associated protein in chronic liver diseases. Journal of Clinical Pathology 31: 784–790

Look D, Doss M 1976 Fluorescence and biochemical findings in liver biopsies in chronic hepatic porphyrias. In: Porphyrias in human diseases. 1st International Porphyrin Meeting Freiburg/Br. Karger, Basel, p 325–327

Ludwig J, McDonald G S A, Dickson E R, Elveback L, McCall J T 1979 Copper stains and the syndrome of primary biliary cirrhosis. Evaluation of staining methods and their usefulness for diagnosis and trials of penicillamine treatment. Archives of Pathology and Laboratory Medicine 103: 467–470

MacSween R N M, Anthony P P, Scheuer P J 1979 Pathology of the liver. Churchill Livingstone, Edinburgh

Nayak N C, Sachdeva R 1975 Localization of hepatitis B surface antigen in conventional paraffin sections of the liver. American Journal of Pathology 81: 479–492

Neville A M, Laurence D J R 1980 Tumour markers and the gastro-intestinal tract. In: Wright R (ed) Recent advances in gastro-intestinal pathology. Saunders, London, p 255–266

Okuda K, Nakashima T 1979 Hepatocellular carcinoma: a review of the recent studies and developments. Progress in Liver Disease 6: 639–650

Palmer P E, Wolfe H Y 1976 Alpha-1-antitrypsin deposition in primary hepatic carcinomas. Archives of Pathology and Laboratory Medicine 100: 232–236

Paradinas F J, Melia W M, Wilkinson M L et al 1982. High serum vitamin B12 binding capacity as a marker of the fibrolamellar variant of hepatocellular carcinoma. British Medical Journal 285: 840-842

Peyrol S, Grimand J A, Chayvialle J A, Veyre B, Paliard P, Lambert R 1979 Tissular immunoenzymatic detection of hepatic alphafetoprotein in human hepatomas. Digestion 18: 351–370

Ray M B, Desmet V J, Bradburne A F, Desmyter J, Fevery J, De Groote J 1976a Differential distribution of hepatitis B surface antigen and hepatitis B core antigen in the liver of hepatitis B patients. Gastroenterology 74: 462–467

Ray M B, Desmet V J, Fevery J, De Groote J, Bradburne A F, Desmyter J 1976b Distribution patterns of hepatitis B surface antigen (HBsAg) in the liver of hepatitis patients. Journal of Clinical Pathology 29: 94–100

Salaspuro M, Sipponen P 1976 Demonstration of an intracellular copper-binding protein by orcein staining in long-standing cholestatic liver diseases. Gut 17: 787–790

Scheuer P J 1980 Liver biopsy interpretation, 3rd edn. Baillière Tindall, London

Sherlock S 1981 Diseases of the liver and biliary system, 6th edn. Blackwell Scientific, London

Sipponen P, Salaspuro M P, Makkonen H M 1975 Orcein-positive hepatocellular material in histological diagnosis of primary biliary cirrhosis. Annals of Clinical Research 7: 273–277

Sipponen P, Salaspuro M P, Makkonen H 1976 Histological characteristics of chronic hepatides and primary biliary cirrhosis with special reference to orcein-positive hepatocellular accumulations. Annals of Clinical Research 8: 200–205

Thung S T, Gerber M A, Sarno E, Popper H 1979 Distribution of five antigens in hepatocellular carcinoma. Laboratory Investigation 41: 101–105

16

Metabolic disorders of the liver

B. D. Lake

STORAGE DISORDERS AFFECTING THE LIVER

Storage disorders affect the liver in one of several ways. The enzyme defect may be manifest in the hepatic parenchymal cells alone, producing enlarged hepatocytes containing a stored product which is usually detectable in appropriately prepared sections. In extreme examples — phosphorylase kinase deficiency for instance — routine sections may appear to be composed of plant cells. In the disorders affecting the cells of the mononuclear phagocyte (reticuloendothelial) system the hepatic parenchymal cells may not be involved and the picture is then of infiltrating storage cells usually in the portal areas but also throughout the sinusoids. Gaucher's disease (all types) is an example of this category. In other situations hepatocytes are additionally involved and are enlarged and foamy with storage cells present in sinusoids and portal areas. In a few conditions — for example Farber's disease — routine sections show no abnormality even though marked changes are detectable, particularly in sections of snap-frozen tissue.

In all of these circumstances histochemical methods are important to discover whether a storage disorder is present and to define the nature of the disorder. As mentioned above, routine sections may show a storage disorder is present but to define the type of disorder snap-frozen tissue is essential. In the absence of snap-frozen tissue, formalin-fixed frozen sections can provide much useful information, but with good co-operation between the clinician, surgeon and pathologist snap-frozen tissue should always be available.

GLYCOGEN STORAGE DISEASES

The glycogen storage diseases are a group of disorders (Stanbury et al 1978) which result in the deposition of variable amounts of glycogen of differing structures. All require biochemical assay of the enzymes involved for final proof, but histochemical methods can go a long way towards the diagnosis and can eliminate those cases which do not fall into the glycogen storage group, thus saving the biochemists unnecessary assays. Routine histological examination will give an indication only in severe examples. Where the diagnosis is suspected before the biopsy is undertaken only a small portion of tissue should be put into formalin, the majority being frozen for histochemistry and biochemical analysis.

The three commonly encountered types of glycogen storage disease affecting the liver are types 1, 3 and 6 with two or three minor types also being found on occasions. Table 16.1 lists the various types and their main clinical, enzymatic and pathological features. Figure 16.1 shows the pathways of glycogen metabolism.

Demonstration of glycogen

The glycogen found in the liver in the several types of glycogen storage disease varies in structure and consequently in its solubility in aqueous solvents also varies. Normal glycogen structure is found in types 1 and 6, a more soluble limit dextrin in type 3, and an insoluble amylopectin in type 4. Much of the glycogen in type 2 is intralysosomal and in the β-particle form (monoparticulate) and is thus more soluble than the usual α-particles (rosettes) of glycogen. The glycogen in the muscle in types 5 and

Table 16.1 Glycogen storage diseases

Type	Enzyme defect	Major clinical features	Major pathological features
1A Von Gierke	glucose-6-phosphatase	hepatomegaly, hypoglucosaemia, lactic acidosis	fatty liver, renal proximal tubules contain much glycogen
1B	? glucose-6-phosphate translocase	as type 1A	as type 1A
2 infantile, Pompe	acid α-1:4-glucosidase	cardiomegaly, hypotonia, hepatosplenomegaly	marked excess of glycogen in MP system, hepatocytes and cardiac and skeletal muscle
juvenile	acid α-1:4-glucosidase	muscular weakness	vacuolar myopathy with excess glycogen
adult	acid α-1:4-glucosidase	muscular weakness, respiratory difficulty	vacuolar myopathy with excess glycogen
3	debranching enzyme (amylo-1:6-glucosidase)	hepatomegaly, mild fasting hypoglycaemia	marked excess glycogen in liver, acinar fibrosis
4	branching enzyme	hepatosplenomegaly, jaundice	cirrhosis, indigestible glycogen, bile stasis
5 McArdle	myophosphorylase	muscle weakness and cramps on exercise	vacuolar myopathy with excess glycogen
6A	liver phosphorylase kinase unclassifiable	hepatomegaly, very mild fasting hypoglycaemia variable	marked excess glycogen in liver, slight fat accumulation variable
7	phosphofructokinase	muscle weakness and cramps on exercise	vacuolar myopathy with excess glycogen

7 has a normal structure. Routine fixation will extract variable amounts of glycogen and routine sections cannot be relied on to provide an accurate reflection of the glycogen content. Special fixatives, for example Bouin, Carnoy or alcohol, may provide a better guide to glycogen content but this will waste valuable biopsy material which would be better used for cryostat sectioning and enzyme assay.

The protected PAS method (PAS celloidinised) on cryostat sections of snap-frozen tissue is the most accurate way of assessing the glycogen content. Normal glycogen levels range up to 6% (wet weight basis) and in this amount the hepatocyte appears full but not distorted. Very high contents of glycogen (up to 15%) are found in types 3 and 6, and in some patients with type 1, although fat is usually the predominant feature with a normal amount of glycogen in the latter (Fig. 16.2). High glycogen contents (> 7% wet wt) may also be found in the liver from diabetic patients and in infants who have been given intravenous dextrose feeds prior to surgery. The glycogen in all types (except type 4 and in an odd single-case report) is readily digestible through the protective celloidin coating by salivary amylase. The

amylopectin in type 4 is not hydrolysed by amylase or diastase but is inexplicably attacked by pectinase preparations.

In types 5 and 7 the muscle glycogen may not be visibly in excess (see Ch. 10).

Pompe's disease (GSD2) shows glycogen within the Kupffer cells as well as the hepatocytes. Smooth muscle cells of the vasculature and endothelial cells also show increased amounts of glycogen. The heart is grossly affected, the whole of the cardiac muscle showing vacuolar change as does skeletal muscle. Deposits of an acid mucopolysaccharide may also be shown in skeletal muscle and heart by the toluidine blue method of Haust & Landing (1961).

Glucose-6-phosphatase activity

Normal glucose-6-phosphatase activity is distributed throughout the hepatocyte with accentuated activity in the nuclear envelope. The reaction shows a gradient of activity from the terminal hepatic vein increasing towards the portal space and Zone 1 of the acinus. An incubation time of 20 min is adequate, and is short enough not to show interference from non-

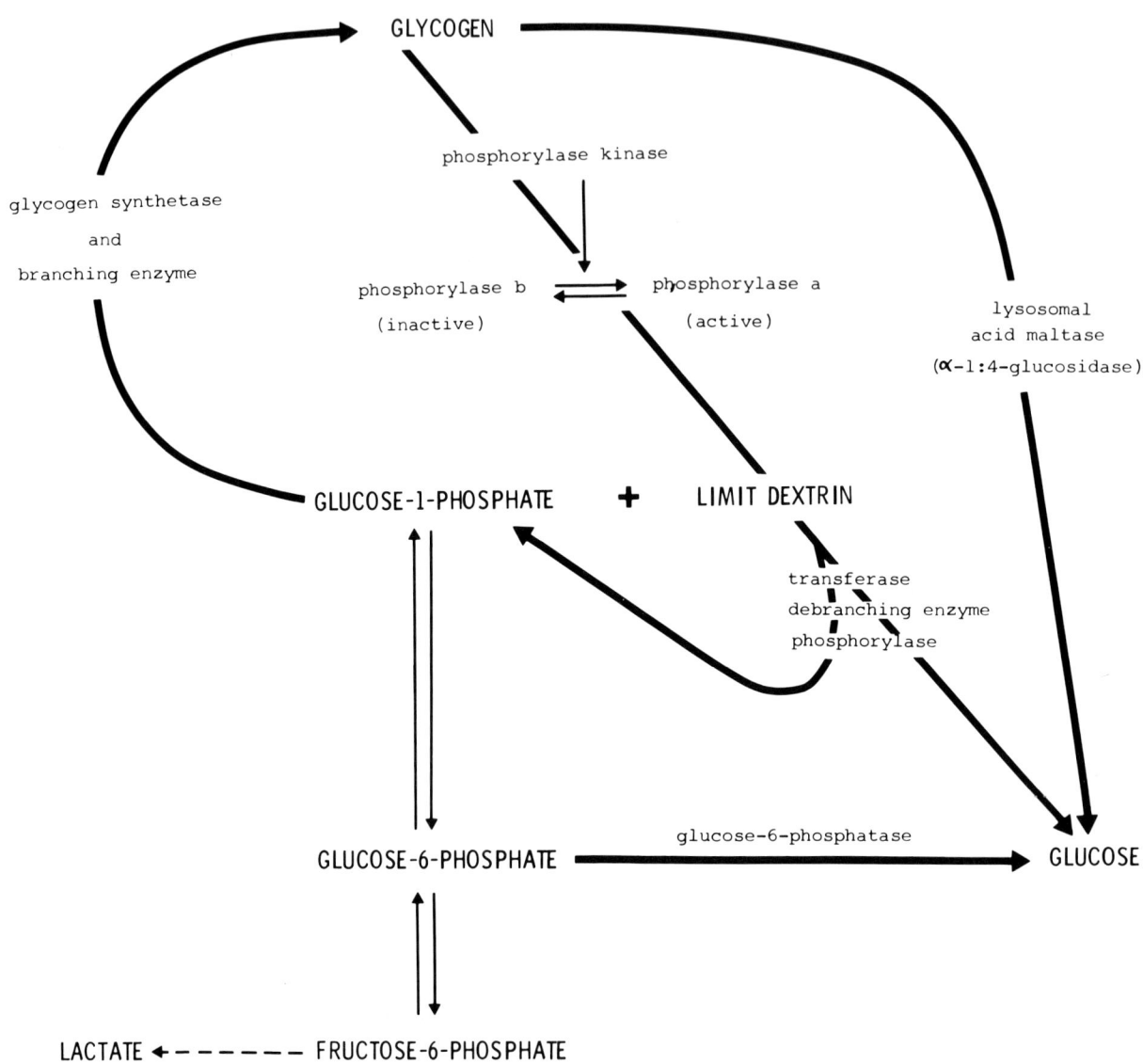

Fig. 16.1   Pathways of glycogen metabolism

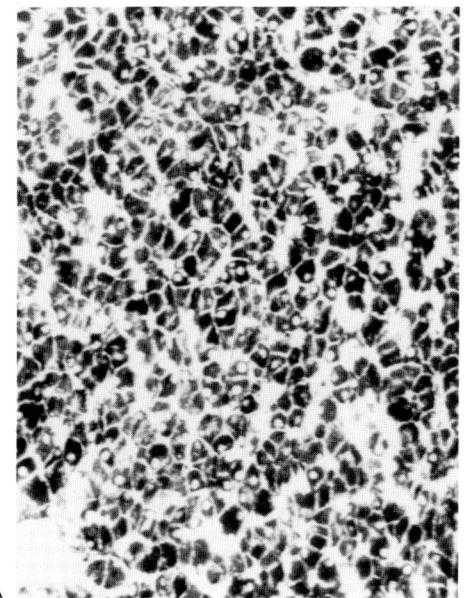 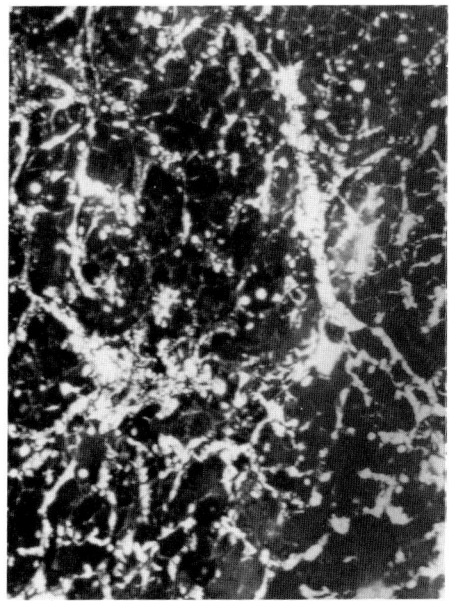

Fig. 16.2 Cryostat sections of liver stained with the protected PAS method (cell PAS) to demonstrate glycogen. A normal liver on the left contains 6% glycogen; on the right a biopsy from a patient with type 3 GSD contains 12% glycogen. Routine sections would show much less glycogen in the patient with type 3 GSD due to the greater solubility of the glycogen in this disorder. × 140

specific phosphatases. In type 1A GSD no activity can be detected in hepatocytes (Fig. 16.3), but normal activity is shown in type 1B GSD because the cells are permeable in frozen sections and the glucose-6-phosphate translocase is not necessary. Type 1B GSD can thus only be diagnosed by biochemical assay of glucose-6-phosphatase activity in *freshly unfrozen* tissue in the presence and absence of a suitable detergent (Triton X 100), or by assay of fresh unfrozen tissue in comparison with frozen thawed tissue (Igarashi et al 1979, Beaudet et al 1980).

Glucose-6-phosphatase is also normally present in proximal renal tubular epithelium and in the epithelial cells of the small intestine. Jejunal biopsy may be used for diagnosis but care should be taken since other conditions with normal jejunal morphology (fructose intolerance for example) show very low levels of activity which may look like deficiency until compared with a deficient patient when subtle differences can be observed. The activity is present mainly in the supranuclear regions with a marked nucler envelope activity. Alkaline phosphatase also hydrolyses glucose-6-phosphatase but its activity is confined to the brush border and causes no problems in interpretation. No excess glycogen is found in the enterocytes in type 1 GSD (Fig. 16.4).

The glucose-6-phosphatase activity in the hepatocytes of types 3 and 6 is reduced and often confined to the nuclear envelope and the periphery of the cell (see Fig. 16.3C), where the endoplasmic reticulum has been displaced by the glycogen deposition. The low activity should cause no difficulty since total absence is the hallmark of type 1 GSD and low activity is interesting but not significant.

Acid phosphatase activity

Acid phosphatase activity in the form of discrete particulate lysosomal staining pattern is found mainly in the pericanalicular regions of normal liver and in a stronger activity in Kupffer cells and histiocytes in the portal spaces. Normal or slightly decreased activity is found in all types of GSD except type 2 where a marked change in distribution and increase in intensity is found in common with all lysosomal enzyme disorders. The discrete particulate pericanalicular activity is replaced by a strong activity throughout the hepatocyte. Enlarged Kupffer cells are also readily visible. Increased acid phosphatase activity is also found in the vacuoles in skeletal and cardiac muscle from patients with type 2 GSD.

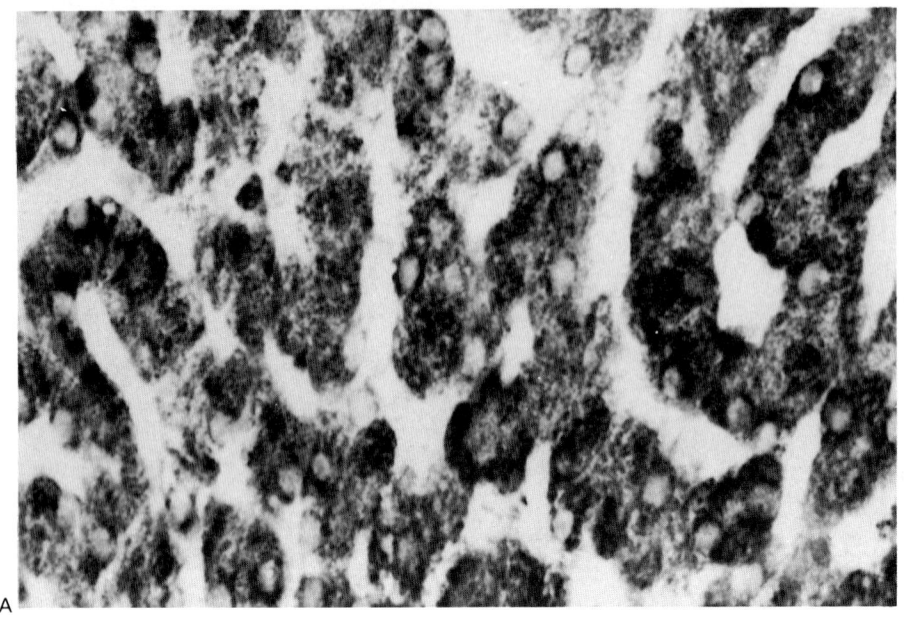

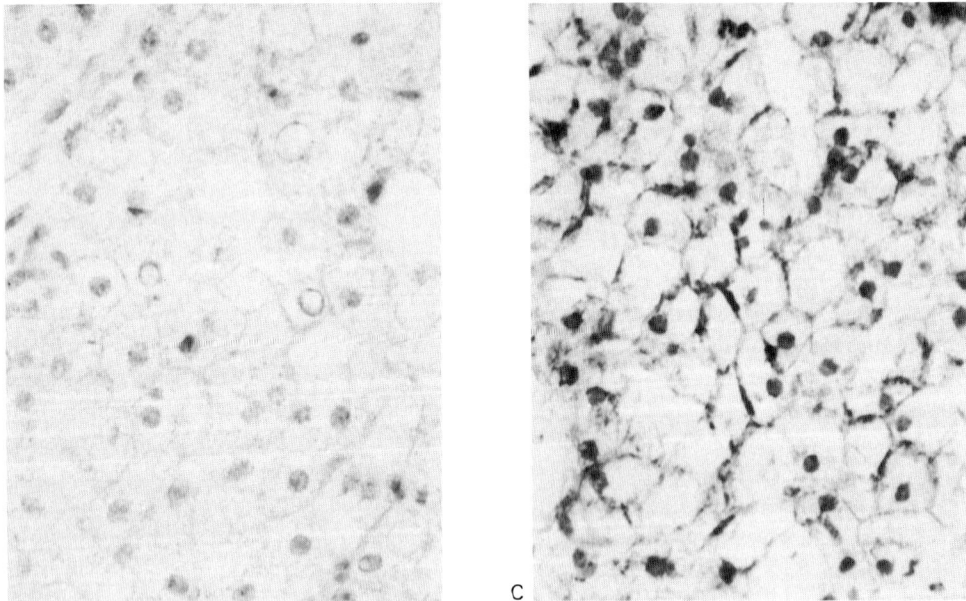

Fig. 16.3 Glucose-6-phosphatase activity in cryostat sections of:
   A. Normal liver. Activity is present in the cytoplasm and in a perinuclear ring. ×600.
   B. Type 1A GSD. No activity is detectable. ×450.
   C. Type 6 GSD. Reduced activity is present and is confined to the periphery of the hepatocyte and around the nucleus. ×350.
A light nuclear counterstain has been added to each preparation.

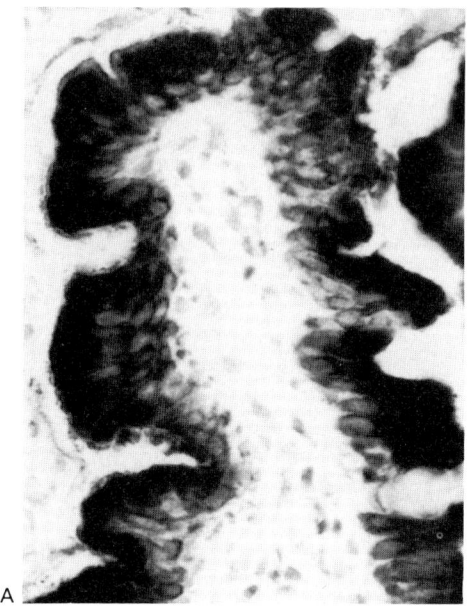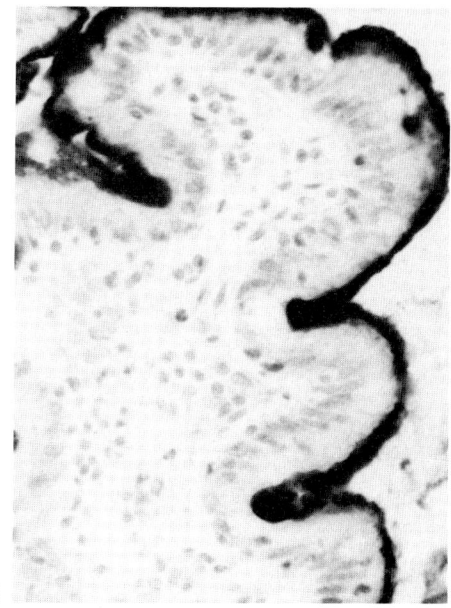

Fig. 16.4 Glucose-6-phosphatase activity in cryostat sections of:
A. Normal jejunum. Glucose-6-phosphatase is present throughout the enterocyte. Perinuclear activity is apparent.
B. Type 1A GSD. No glucose-6-phosphatase activity can be detected within the enterocyte. Alkaline phosphatase hydrolyses the substrate and its activity is present in the brush border.
A light nuclear counterstain has been added. ×490

Phosphorylase activity

Of the many different modifications of the phosphorylase method, none is reliable for the detection of liver phosphorylase deficiency. Although the method will show low staining intensity in cases of phosphorylase deficiency, similar 'deficiency' is also found in phosphorylase kinase deficiency and in debranching enzyme deficiency. A good normal reaction is an indication of normal phosphorylase activity, but a low staining intensity is not of diagnostic significance. Methods for phosphorylase kinase are consequently also unreliable.

Phosphorylase methods for skeletal muscle by contrast are reliable and useful in the diagnosis of McArdle's disease where no activity can be detected in skeletal muscle fibres. Smooth muscle of vessel walls shows activity and acts as an internal control.

Fat

Fatty deposition in hepatocytes is the main feature of GSD I. In type 6A there is a mild to moderate amount of fat, but in other types fat deposition is not found. The fat is mostly triglyceride and is well stained with oil red O.

Glycogen synthetase deficiency

In glycogen synthetase deficiency, for which only two or three cases have been recorded, the glycogen content, as might be expected is very low or even absent, and there is abundant fat. The method for the histochemical demonstration of glycogen synthetase activity produces only a very low amount of reaction product in normal liver, and this coupled with the unpredictability of the glycogen-iodine colour of the phosphorylase reaction makes the detection of glycogen synthetase deficiency almost impossible by histochemical means.

LIPID- AND OTHER STORAGE DISEASES
(see Hers & van Hoof 1973, Stanbury et al 1978)

Wolman's disease

Macroscopically the liver has an orange-yellow glistening appearance due to the very high fat content. The hepatic parenchymal cells are filled with lipid which stains strongly with oil red O and Sudan black. The Kupffer cells and foamy histiocytes in the sinusoids and portal areas are also strongly

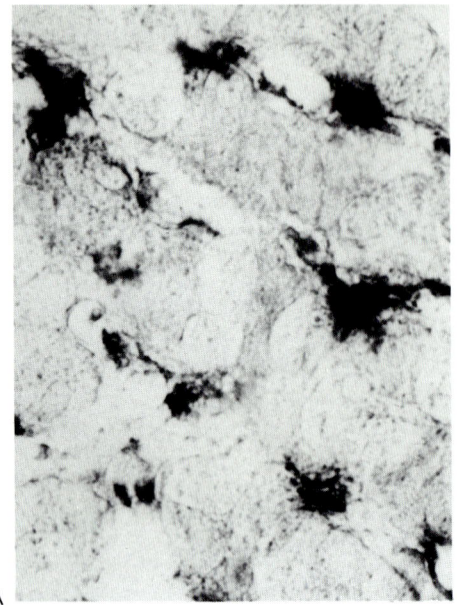
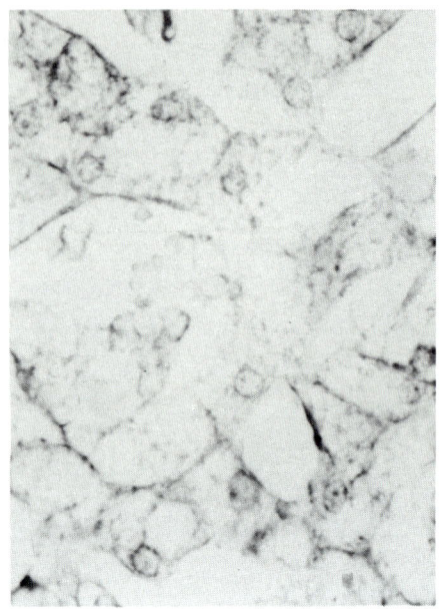

A   B

Fig. 16.5   Acid esterase activity in frozen sections of formol-calcium/gum sucrose fixed liver biopsies. On the left, a normal liver shows activity in Kupffer cells mainly, with some lesser activity in the hepatocytes. On the right no activity can be detected in a liver biopsy from a patient with Wolman's disease. ×950

sudanophilic and in routine sections it is often possible to see that some of the lipid is in the form of a crystalline inclusion (see Chapter 4). Examination of frozen or cryostat sections in polarised light also shows crystalline lipid inclusions. Stains for cholesterol (and esters) are strongly positive in hepatocytes and macrophages. It is not possible to differentiate between cholesterol and its esters by the use of digitonin unless the primary fixation includes digitonin. In sections (frozen or cryostat) the rate of formation of cholesterol digitonide is slower than the rate of solubility of the cholesterol in the solvent and thus although the test in Wolman's disease would appear to indicate the presence of cholesterol esters the same result would be obtained if cholesterol alone was present. Adams' method for triglyceride esters is positive in hepatocytes. The macrophages and enlarged Kupffer cells contain free fatty acids which give a positive reaction with Holczinger's method, and Cain's Nile blue sulphate shows red hepatocytes (neutral fat) with purple macrophages indicating free fatty acids and neutral fat. The acid phosphatase reaction shows the macrophages with the crystalline and lipid inclusions outlined by the reaction product; hepatocytes also show a strong reaction. In frozen sections of formol-calcium/gum sucrose-processed

tissue incubated in the presence of 10^{-4}M E600 to inhibit non-specific esterases, the absence of acid esterase is readily demonstrated with 1-naphthyl acetate as substrate. In normal liver, acid esterase can be shown mainly in Kupffer cells with lesser activity in hepatocytes (Lake & Patrick 1970) (Fig. 16.5).

Gaucher's disease

Gaucher cells with their content of glucocerebroside are PAS-positive in routine and frozen sections. The cells are present as enlarged Kupffer cells in the sinusoids and in the portal areas, and are more numerous in the infantile form of the disease. Hepatocytes are not affected, and this is best demonstrated in acid phosphatase preparations where the Gaucher cells are strongly stained in contrast with the normal pericanalicular distribution in hepatocytes (Fig. 16.6). Other lysosomal storage disorders with visceral involvement affect hepatocytes and this is reflected by a strong diffuse acid phosphatase reaction. The histochemical methods for β-glucosidase (using the substituted indoxyl substrate, or the 6-bromo-2-naphthyl-β glucoside or 1-naphthyl-β-glucosidase) are not sufficiently sensitive to discriminate between the isoenzymes of β-

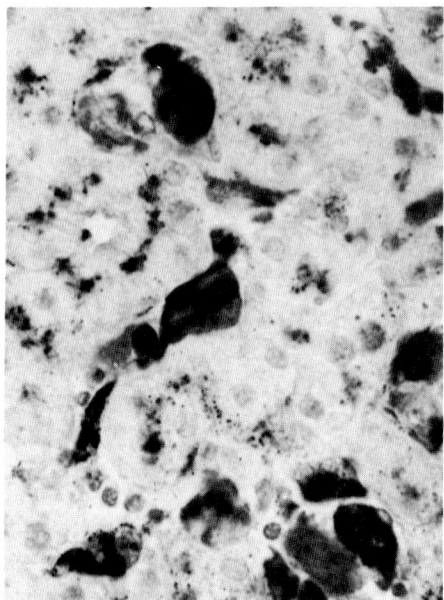

Fig. 16.6 Gaucher's disease; cryostat section of a liver biopsy stained to demonstrate acid phosphatase activity. Gaucher cells and Kupffer cells show strong activity while hepatocytes show normal pericanalicular activity. A light nuclear counterstain has been added. ×450

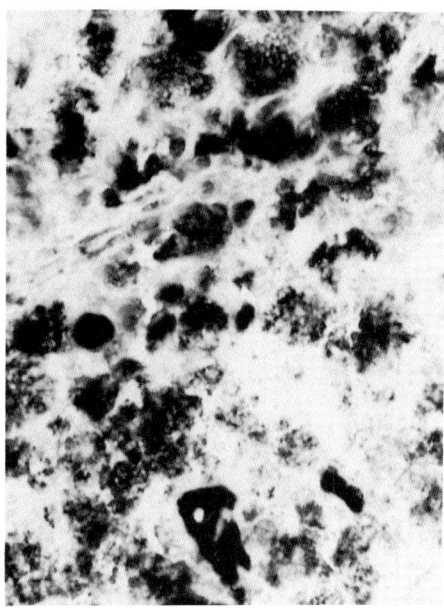

Fig. 16.7 Niemann-Pick disease Type A; cryostat section of a liver biopsy stained to demonstrate acid phosphatase activity. Large foamy Niemann-Pick cells and hepatocytes show strong activity, in contrast with Gaucher's disease where hepatocytes are normal. A light nuclear counterstain has been added. ×470

glucosidase and their differing pH optima, and thus cannot be used in the diagnosis of Gaucher's disease where only one isoenzyme is deficient.

Niemann-Pick disease types A and B

In contrast with Gaucher's disease the hepatocytes are involved in the storage process. Large foamy Kupffer cells and hepatocytes are filled with sphingomyelin and cholesterol, and a variable amount of ganglioside. Sudan black and methods for cholesterol (Schultz, PAN) are positive, and in sections from which glycogen has been removed the foamy cells are weakly positive. The acid haematin method for sphingomyelin is not entirely reliable for two reasons. Firstly, liver normally contains sphingomyelin and secondly the stain appears to depend to a certain extent on the composition of the fatty acid side chain. In several instances the expected blue colour has not been found — instead the Niemann-Pick cells have been yellow-brown. After staining with Sudan black sphingomyelin often will show a red birefringence in polarised light. The acid phosphatase reaction shows a characteristic foamy appearance of the hepatocytes and Niemann-Pick cells with the periphery of the

vacuoles being outlined by the reaction product. There is no staining method available for the demonstration of sphingomyelinase activity.

Niemann-Pick Group C (Ophthalmoplegic lipidosis)

This condition, known in the past by a variety of names (Niemann-Pick type C, Niemann-Pick type D, subacute Niemann-Pick disease, juvenile dystonic lipidosis, neurovisceral storage disease with vertical supranuclear ophthalmoplegia, etc.), does not store sphingomyelin and there is probably no connection with classic Niemann-Pick disease (Neville et al 1973). Routine sections of liver in those presenting early with neonatal hepatitis show changes of neonatal hepatitis with bile pigment retention and giant cell change. As in most hepatitic liver samples, PAS-positive macrophages are present but in Niemann-Pick Group C most Kupffer cells are enlarged and foamy and palely PAS-positive (Fig. 16.8). This change is best seen after digestion of the glycogen, but may be difficult to distinguish from other causes of neonatal hepatitis. In frozen or cryostat sections the foamy cells are sudanophilic. In

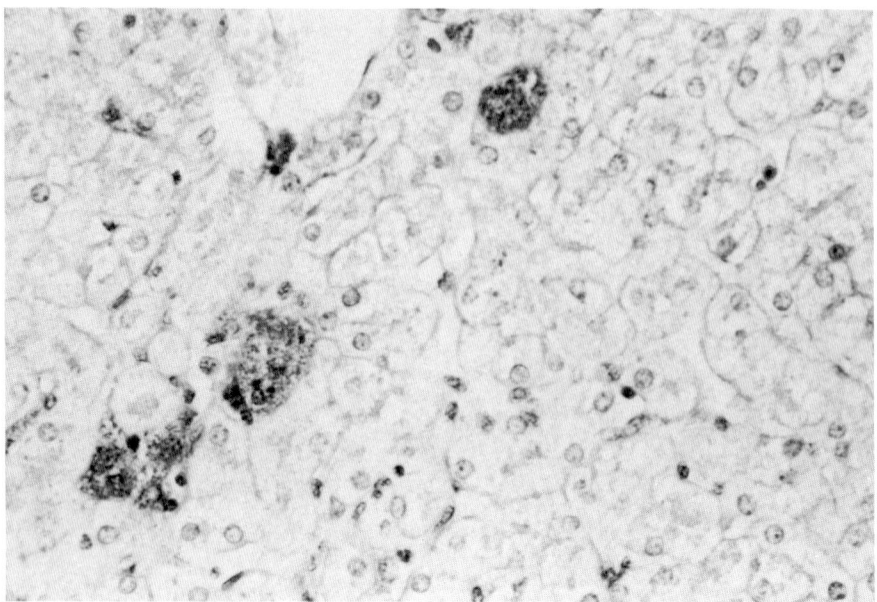

Fig. 16.8 Niemann-Pick disease Group C (Ophthalmoplegic lipidosis). Routine section of a liver biopsy stained with PAS after amylase digestion. Four large storage cells and several Kupffer cells stain positively. Hepatocytes appear to be normal. ×360

cryostat sections, first coated with celloidin then digested with saliva and stained with PAS (cell. dig. PAS), the strong staining of the foamy storage cells and Kupffer cells is clearly seen. Feyrter's thionin shows rose-pink metachromasia in the storage cells and hepatocytes. The storage cells show a strong acid phosphatase reaction as do the hepatocytes, but these are normally quite strong in liver damage.

In those patients presenting with unexplained hepatosplenomegaly at any age routine sections of liver appear quite normal. However, PAS after digestion of glycogen shows positive staining of storage cells and Kupffer cells, which are sudanophilic in frozen or cryostat sections. The cell. dig. PAS method is also strong in these cells but is much reduced in fixed material, indicating a water-soluble component (Lake 1977). Although the hepatocytes appear to be normal they show strong acid phosphatase activity, indicating the lysosomal nature of the disorder which is confirmed by electron microscopy (Fig. 16.9). No enzyme defect is known for this disease.

Farber's disease

Routine sections show a normal liver and although ceramide is present in excess it does not stain to any great extent with lipid dyes or methods. No storage cells are apparent, but as in Niemann-Pick Group C the acid phosphatase reaction shows strong staining in hepatocytes.

G_{M2}-gangliosidosis

There is usually no evidence of storage of ganglioside G_{M2} or its asialo derivative in Tay-Sachs disease and related disorders. In occasional cases, however, the Kupffer cells may show strong PAS-positivity with pale sudanophilia in frozen or cryostat sections.

Cystinosis

Although cystine is relatively insoluble, the volume of fluid used is sufficient to ensure that most cystine is removed from the tissue. All that remains is a foamy appearance in Kupffer cells resembling those found in a variety of other storage disorders. Snap-frozen tissue, sectioned in a cryostat and examined under polarised light, is required to show the crystal habit and will preserve all the cystine within the section. If preferred, the cryostat section can be stained with alcoholic basic fuchsin (1% basic fuchsin in 70% alcohol) to show nuclei and general structure. If the concentration of cystine is high enough, the Woollaston test (Patrick & Lake 1968) is positive.

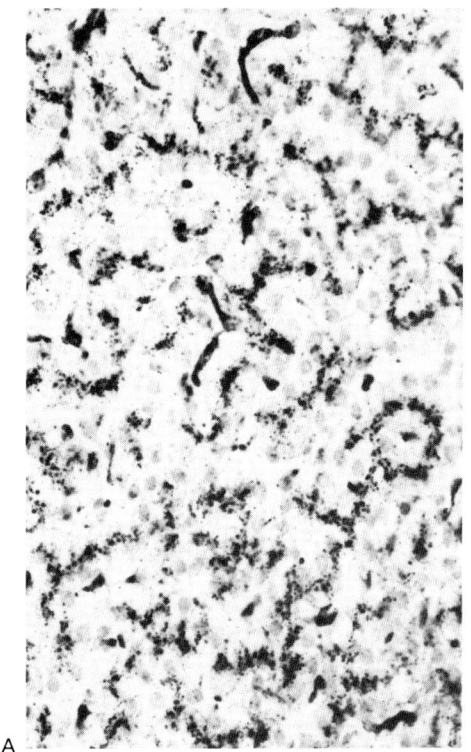 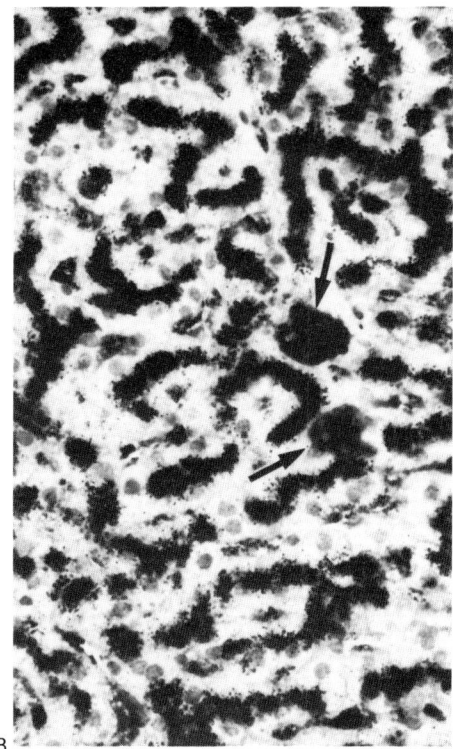

Fig. 16.9 Cryostat sections of liver stained for acid phosphatase activity (30 min incubation). On the left is a normal liver showing Kupffer cells and pericanalicular activity in hepatocytes. On the right is a section of a liver biopsy from a patient with ophthalmoplegic lipidosis showing intense activity in hepatocytes. Two storage cells are evident (arrows). Histologically the liver appeared normal. A light nuclear counterstain has been added. × 300

Under the influence of concentrated hydrochloric acid, cystine dissolves and reprecipitates as cystine hydrochloride with change of crystal shape (see Fig. 18.7).

Metachromatic leucodystrophy

Deposition of sulphatide is not seen in hepatocytes or Kupffer cells. However, sulphatides are found stored in bile duct epithelium where they show brown metachromasia with cresyl fast violet or toluidine blue in frozen or cryostat sections. In some cases the bile duct epithelium can be involved to such an extent that the gall bladder may show papillomatous change with marked sulphatide deposition in epithelium and in macrophages in the subepithelial connective tissue.

G_{M1}-gangliosidosis

Type 1 infantile

There are differences in liver involvement in the two forms of G_{M1}-gangliosidosis. Type 1 (infantile) shows

marked vacuolation of hepatocytes and foamy enlarged Kupffer cells and appears at first sight rather like a fatty liver. No fat can be demonstrated and the vacuoles contain a water soluble substance which has so far defied histochemical demonstration. The substance identified as:

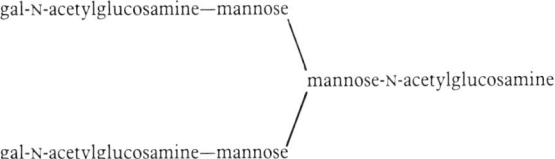

should be PAS-positive but its solubility appears to prevent its demonstration. Although the patients present as a mucopolysaccharidosis no mucopolysaccharide can be demonstrated in the vacuoles by any method.

At least one other storage product has been identified in liver and this contains sialic acid residues in addition to galactose and glucosamine. This suggests that it may be similar to the compound mentioned above and may account for the presence of an alcian blue-positive reaction in the enlarged foamy Kupffer cells seen in sections of routinely processed wax-embedded tissue. These cells are also PAS-positive.

The β-galactosidase deficiency can be detected reliably in cryostat sections of fresh tissue using the substituted indoxyl method (Lake 1974). Normal or unaffected tissue shows marked activity after overnight incubation but no activity can be demonstrated in either hepatocytes or histiocytes in patients with G_{M1}-gangliosidosis of type 1 or type 2 (Fig. 16.10).

Type 2 juvenile

The general hepatic architecture is undisturbed and even though Kupffer cells are known to show storage by electron microscopy no firm evidence can be obtained by light microscopy. Occasional large foamy histiocytes, present in sinusoids, show a moderate PAS reaction in routine sections. These large histiocytes show strong acid phosphatase activity in contrast with normal activity in hepatocytes and Kupffer cells. The demonstration of the enzyme defect by the indoxyl method mentioned above is reliable.

Mucopolysaccharidosis

The several types of mucopolysaccharidosis all show deposition of mucopolysaccharide in the liver. The type of mucopolysaccharide deposited varies from type to type but since the histochemical methods are sufficiently imprecise for individual identification, only the presence of an acidic mucopolysaccharide can be shown. These substances are extremely water-soluble and are totally lost on fixation, and even using cryostat sections of snap-frozen specimens localisation is poor. The only satisfactory method which is reasonably sensitive is that of Haust & Landing (1961) using fixation of the section in a tetrahydrofuran/acetone mixture. Staining of the mucopolysaccharide is based on the metachromasia induced with toluidine blue in a 25% acetone solution. Localisation to individual cells is often impossible although the general impression of hepatocyte or Kupffer cell storage is evident. Normal tissue shows no metachromasia, apart from the occasional mast cell, the granules of which are clearly defined.

In addition to the storage of mucopolysaccharide, foamy cells storing a ganglioside-like substance can be found in routine sections in portal areas and also scattered throughout the lobule. These cells are PAS-

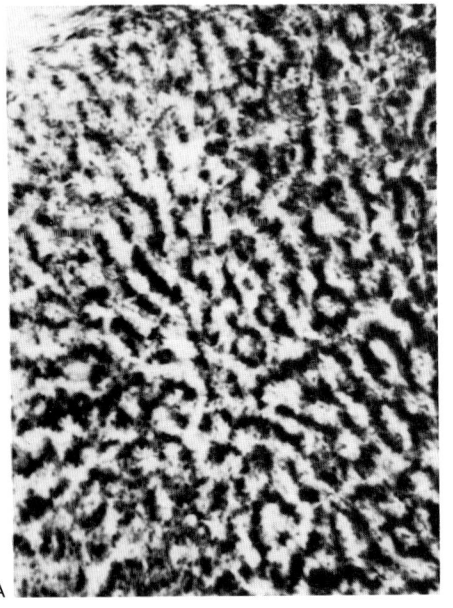
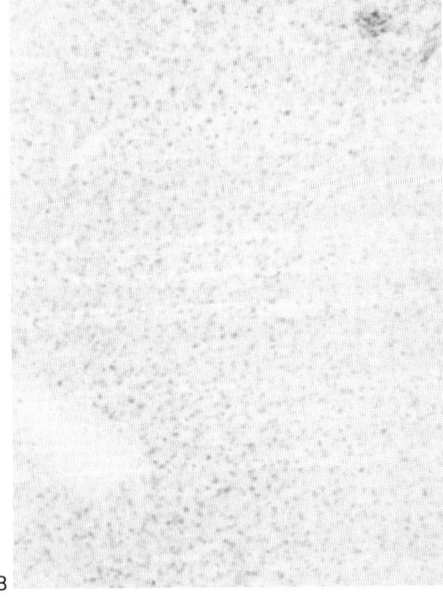

Fig. 16.10 Cryostat sections of liver stained for β-galactosidase activity. Neutral red has been added as a nuclear counterstain. On the left is a normal liver. On the right is a liver biopsy from a patient with G_{M1}-gangliosidosis type 2, in which no activity can be demonstrated. ×110

positive and resist amylase digestion. Their sialic acid content imparts colloidal iron positivity which becomes negative after sialidase digestion.

Fucosidosis

The storage of oligosaccharides, glycoproteins and glycolipids containing a terminal α-linked fucose is found in hepatocytes and foamy Kupffer cells. The epithelium of bile ducts is particularly affected. The substances involved are generally of low molecular weight and very soluble, and are thus difficult to demonstrate without adequate protection. In frozen sections the stored material is variably PAS-positive (or negative where it has been extracted) and is not sudanophilic. The variability of staining from one cell type to another is explained by the differing types of fucose-containing material stored. There are at least two types of fucosidosis and the liver involvement may be different in the differing forms.

Mannosidosis

Routine sections of liver may appear entirely within normal limits, and no evidence of PAS positivity can be seen. The soluble nature of the mannose-containing storage material makes it difficult to detect (as in the peripheral blood lymphocytes). The substances involved are deposited in hepatocytes, Kupffer cells and bile duct epithelium. A reaction for acid phosphase activity should show the lysosomal nature of the disorder, and pick out the enlarged Kupffer cells. The methods for α-mannosidase activity are not sufficiently sensitive to distinguish between the various isoenzymes of α-mannosidase and do not allow deficient activity to be demonstrated.

Aspartylglucosaminuria

There is variable storage of aspartylglucosamine in hepatocytes and marked vacuolation of Kupffer cells. The substance is soluble and does not stain with any method. Lipofuscin deposits may be present in hepatocytes and Kupffer cells.

Zellweger's cerebro-hepato-renal syndrome

Apart from the changes of polymicrogyria in the brain, and microcysts in the kidney cortex there is an

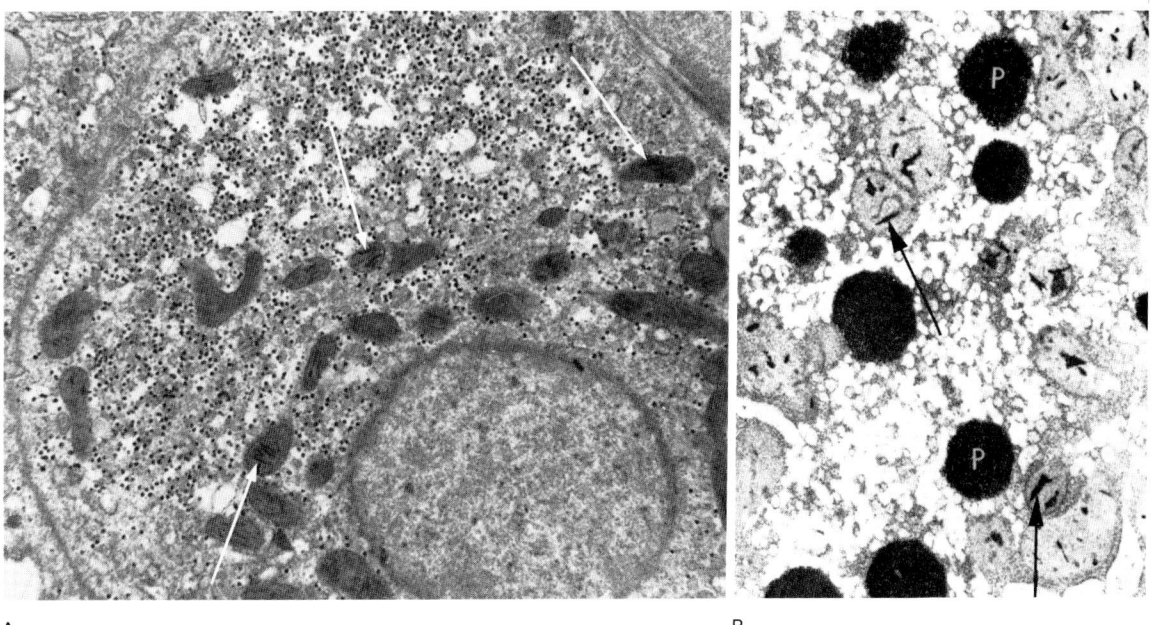

A

B

Fig. 16.11 Comparison of liver sections from an infant with Zellweger's syndrome (A) and normal neonatal liver (B) after incubation in a cytochemical medium to demonstrate the peroxidatic activity of catalase. In the control peroxisomes (P) are darkly stained. Mitochondrial cristae are also reactive, reflecting the presence of cytochrome oxidase. In patients with Zellweger's syndrome peroxisomes cannot be seen. Mitochondrial cristae (arrows) are reactive. Photographs courtesy of Dr S. Goldfischer

absence of peroxisomes in the liver and kidney tubules. The liver is not otherwise abnormal. Peroxisomes, unpredictably shown in cryostat sections of unfixed human tissue, are better demonstrated by the method described by Roels & Goldfischer (1979) on lightly fixed tissue blocks (Fig. 16.11). Electron histochemistry is helpful for confirmation of the absence of organelles with peroxidatic activity. It is important to include a control 'normal' at the same time.

Nephrosialidosis

Few cases of this variant of absent sialidase activity have been recorded (Le Sec et al 1978). The liver sinusoids are filled with foamy storage cells without apparent involvement of the hepatocytes (Fig. 16.12). The stored sialyl-compounds are likely to be extremely water-soluble and vigorous protective treatments (snap-frozen tissue, celloidinised sections or freeze drying) will be necessary.

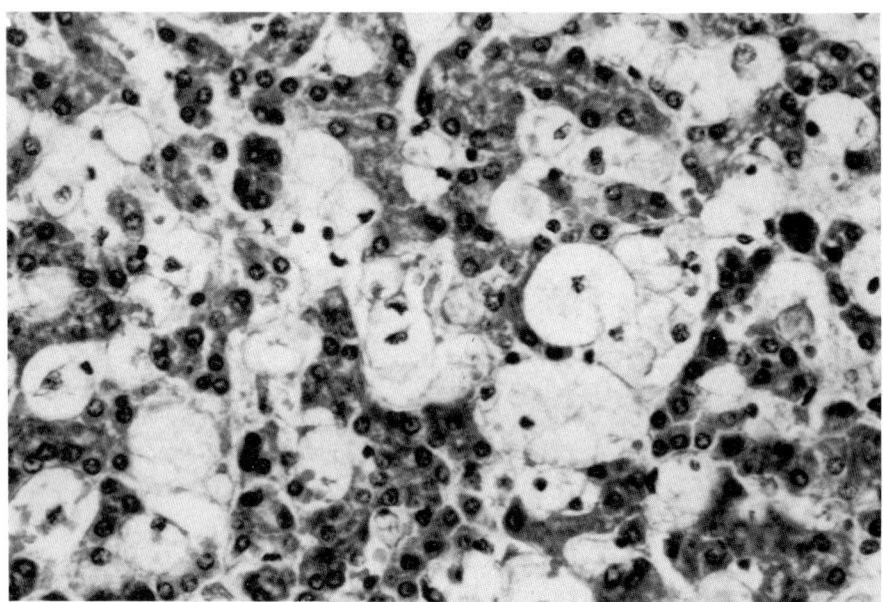

Fig. 16.12 Nephrosialidosis. Section of liver stained with H & E. Large foamy storage cells are present among apparently normal hepatocytes. × 390

REFERENCES

Beaudet A L, Anderson D C, Michels V V, Arion W J, Lange A J 1980 Neutropenia and impaired neutrophil migration in type 1B glycogen storage disease. Journal of Pediatrics 97: 906–910

Haust M D, Landing B H 1961 Histochemical studies in Hurler's disease. A new method for localization of acid mucopolysaccharide and an analysis of lead acetate 'fixation'. Journal of Histochemistry and Cytochemistry 9: 79–86

Hers H G, VanHoof F (eds) 1973 Lysosomes and storage diseases. Academic Press, New York

Igarashi Y, Otomo H, Narisawa K, Tada K 1979 A new variant of glycogen storage disease type 1: probably due to a defect in the glucose-6-phosphate transport system. Journal of Inherited Metabolic Disease 2: 45–49

Lake B D 1974 An improved method for the detection of β-galactosidase activity, and its application to G$_{M1}$-gangliosidosis and mucopolysaccharidosis. Histochemical Journal 6: 211-218

Lake B D 1977 Histochemical and ultrastructural studies in the diagnosis of inborn errors of metabolism. Records of the Adelaide Children's Hospital 1: 337–345

Lake B D, Patrick A D 1970 Wolman's disease. Deficiency of E-600-resistant acid esterase activity with storage of lipids in lysosomes. Journal of Pediatrics 76: 262–266

Le Sec G, Stanescu R, Lyon G 1978 Un nouveau type de sialidose avec atteinte renale: la nephrosialidose. II étude anatomique. Archives français de Pédiatrie 35: 830–844

Neville B G R, Lake B D, Stephens R, Sanders M 1973 A neurovisceral storage disease with vertical supranuclear ophthalmoplegia and its relationship with Niemann-Pick disease. A report of nine patients. Brain 96: 97–120

Patrick A D, Lake B D 1968 Cystinosis: electron microscopic evidence of lysosomal storage of cystine in lymph nodes. Journal of Chemical Pathology 21: 571–575

Roels F, Goldfischer S 1979 Cytochemistry of human catalase: the demonstration of hepatic and renal peroxisomes by a high temperature method. Journal of Histochemistry and Cytochemistry 27: 1471–1477

Stanbury J B, Wyngaarden J B, Fredrickson D S (eds) 1978 The metabolic basis of inherited disease, 4th edn. McGraw-Hill, New York

17

Kidney: glomerular disease

L. Morel-Maroger & M. Levy

The various patterns of glomerular diseases can be identified and classified using both light and immunofluorescent microscopy as routine in the interpretation of renal biopsies. Immunofluorescence adds another dimension to morphology since it is possible to identify material, within the glomeruli, which may have pathogenetic significance. Electron microscopy (EM) is necessary in a minority of conditions and it is therefore recommended to fix and store a small piece of cortex for an eventual electron microscopic study.

For light microscopy (LM) several fixatives have been used and either Dubosq-Brazil or Smith fixatives are preferred in the authors' department. With Dubosq-Brazil the slides are easy to interpret, but acid fixation does not allow immunochemical procedures. A needle biopsy sample for LM is fixed from 3 to 6 h then transferred to 70° alcohol, further dehydrated and embedded in paraffin wax. Ideally 2 μm sections are best, but 3 μm as a routine procedure are also acceptable. The following stains are performed routinely: H & E, Masson's trichrome, Orcein or Weigert's stain, PAS, and silver impregnation according to the method of Wilder or Jones. In addition, in adult patients, thick sections (5 μm) are stained by Congo red and crystal violet to detect amyloid deposits.

Although there have been several studies showing the multiple advantages of peroxidase conjugates, we still consider immunofluorescence as a good reliable method. Some laboratories claim that they have highly reliable results in renal disease using either peroxidase conjugates or fluorescein-labelled antisera in paraffin wax-embedded material (Sinclair et al 1981). Until further information is available in large series, we prefer, for diagnostic purposes, immuno-fluorescence performed on snap-frozen tissue. The following anti sera should be used: IgA, IgG, IgM, C1q, C3, fibrinogen-fibrin, albumin. In adult patients, anti-light chains kappa and lambda are also important. We generally use fluorescein-labelled antisera, preferring direct immunofluorescence (IF) to indirect immunofluorescence which is more susceptible to non-specific fluorescence.

An accepted nosology of glomerulonephritis (GN) does not as yet exist and many and varied classifications have been proposed during the last decades according to clinical data, aetiologic factors, morphology and pathogenetic mechanisms. Inevitably, any classification will therefore be far from perfect.

We will apply the scheme below.

Idiopathic nephrotic syndrome
Membranous GN
Membranoproliferative GN Type I and Type II
Berger's disease
Proliferative GN with subepithelial deposits
Crescentic GN
Systemic lupus erythematosus
Anaphylactoid purpura (Henoch-Schoenlein)
Polyarteritis nodosa
Wegener's granulomatosis
Goodpasture's syndrome
Glomerular lesions in dysproteinaemia
Light chain systemic deposition
Amyloidosis
Diabetes
Thrombotic microangiopathy.

IDIOPATHIC NEPHROTIC SYNDROME

The most common pattern observed is characterised by the absence of significant glomerular lesions by light microscopy. 'Minimal change' is the widely accepted term for this condition. In some cases there may be an increase in mesangial cells and matrix and this pattern is called diffuse mesangial proliferation. Focal sclerotic lesions may be superimposed on both these patterns (Habib & Kleinknecht 1975).

It is generally accepted that these various patterns (minimal glomerular changes, diffuse mesangial proliferation and focal glomerular sclerosis) ...ay be variants of the same disease. However, this is a matter for discussion: many authors consider glomerular lesions, particularly forms with focal glomerular sclerosis, as a separate entity.

Minimal changes

By LM, glomeruli may be completely normal. Usually, however, mild glomerular changes are present. They include swelling and vacuolisation of epithelial cells and some degree of thickening of the mesangial matrix. A mild and focal mesangial hypercellularity may be noted. In contrast to the paucity of glomerular changes seen by LM, ultrastructural changes are constant, mainly involving podocytes and mesangial stalks. There is effacement of foot processes by swelling and retraction so that an almost continuous cytoplasmic layer covers the glomerular basement membrane. Mesangial alterations include mesangial cell hyperactivity, thickened mesangial matrix and frequently the presence of finely granular deposits located along the internal aspect of the basement membrane. However, IF studies do not, in most cases, reveal significant deposition of immunoglobulins or complement. The presence of minimal deposits of IgM and C3 focally either in the mesangial stalks or as peripheral capillary streaks has been reported and we do not regard this finding as a separate disease.

Diffuse mesangial proliferation (Fig. 17.1)

A clear-cut mesangial proliferation with widely patent capillaries whose walls are of normal thickness may be observed. Electron-dense deposits are seen in the mesangium as for minimal changes and in some instances they bind small amounts of IgM and C3. Some authors have begun to describe a pattern characterised by mesangial proliferation and IgM

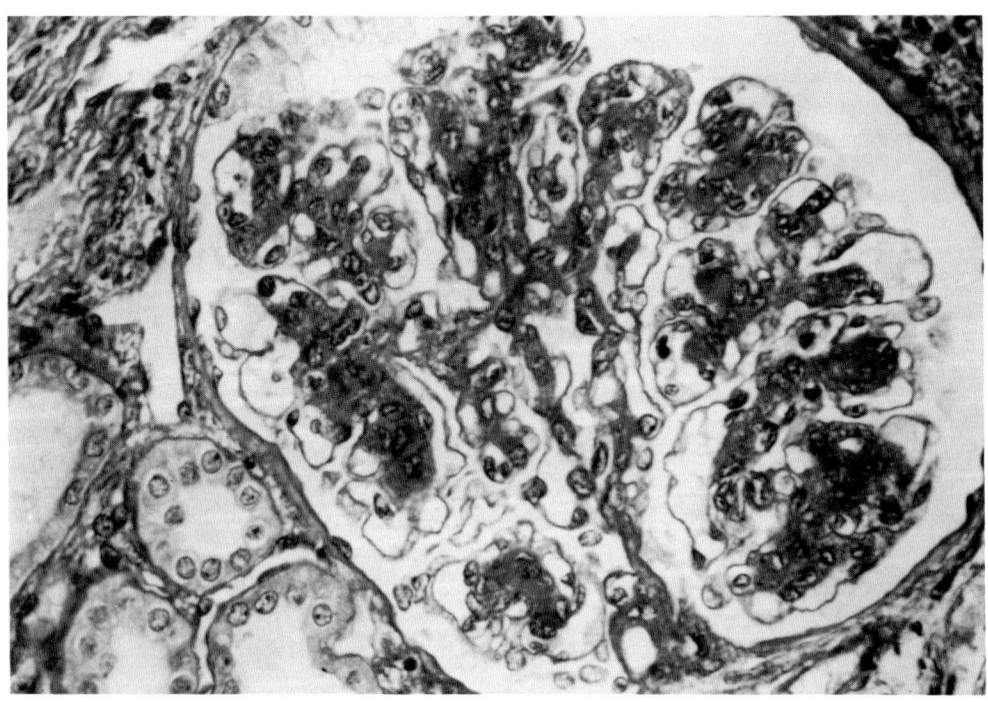

Fig. 17.1 Idiopathic nephrotic syndrome with diffuse mesangial proliferation. Clear-cut mesangial hypertrophy and mesangial cell proliferation. (No deposits are observed by immunofluorescence microscopy.) Trichrome. ×250

deposition, but whether this is a separate entity is a matter for discussion. Repeat biopsies in this group may show normal glomeruli but more often one finds progression to focal glomerular sclerosis with or without persistence of mesangial proliferation.

Focal glomerular sclerosis

Focal glomerular sclerosis is not a specific lesion but it is a sclerosing process which can be seen as a superimposed feature in all kinds of glomerular diseases as well as in other conditions of non-glomerular origin. It encompasses two different patterns. In one the whole of the involved tufts are completely sclerosed (focal global sclerosis) and in the other the focal changes are segmental (focal segmental sclerosis and/or hyalinosis). In both, the uninvolved glomeruli may be normal or may show diffuse mesangial proliferation.

Focal global sclerosis. The coexistence of normal glomeruli and sclerosed glomeruli by light microscopy is characteristic of this type of lesion. The affected glomeruli are irregularly distributed over the renal parenchyma. Only those cases in which at least 15–20% of glomeruli are sclerosed, with sclerosis associated with conspicuous interstitial and tubular damage, should be considered.

Focal segmental sclerosis/hyalinosis (Fig. 17.2). In this variant, juxtamedullary glomeruli are predominantly involved. The lesion usually affects a few capillary loops and either hyaline deposition or sclerosis may predominate. The most characteristic feature of the segmental lesion is an increase in fibrillar material resulting from both collapse of the capillary walls and enlargement of the mesangial matrix, leading to obliteration of the capillary lumen. Hyaline material is often present within the sclerosed areas appearing either as a peripheral rim or as rounded deposits obstructing the lumen. This abnormal material is eosinophilic, strongly PAS-positive, and stains green or red with Masson's trichrome. In addition, foamy endothelial cells as well as lipid inclusions are found in several instances. At the periphery of the sclerotic segments there is in most cases an amorphous 'halo'. Sclerotic segments free in Bowman's space are surrounded by a crown of severely altered podocytes. When the sclerosing lesion is adherent to Bowman's capsule, there are

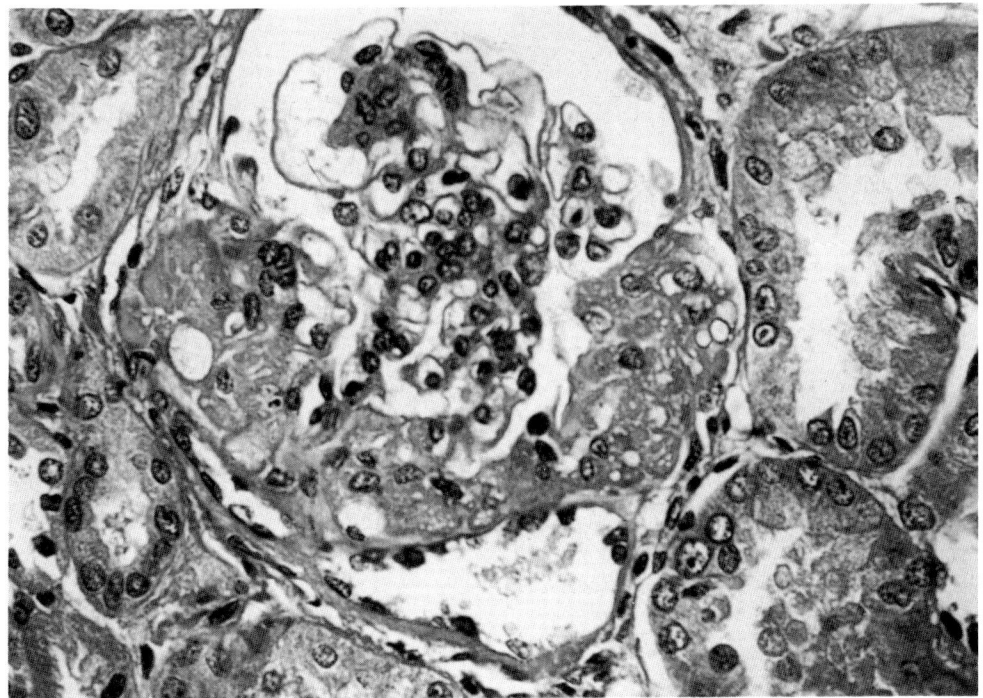

Fig. 17.2 Idiopathic nephrotic syndrome with minimal glomerular changes and focal and segmental hyalinosis. (By immunofluorescence microscopy, IgM, C1q, C3 are trapped within the hyaline deposits.) Trichrome. × 250

adhesions between the collapsed capillary loops covered by the amorphous 'halo' and Bowman's basement membrane.

IF studies do not reveal any significant deposition of immunoglobulins or complement in the majority of glomeruli. In some instances there are small deposits of IgM and C3 in the mesangial stalks. The detection of the segmental hyalinizing lesions is easy since when they are present on the specimen studied, they strongly bind antisera to IgM, C1q and C3. Deposits of IgG and of fibrin are weak and occasional.

MEMBRANOUS GLOMERULONEPHRITIS

Membranous glomerulonephritis (MGN) is the most generally accepted term but various other designations have been used: Ellis type II nephritis, perimembranous, epimembranous or extramembranous glomerulonephritis.

The presence of granular deposits on the epithelial aspect of the glomerular capillary wall is the main pathological finding. In most cases, the changes are diffuse but occasionally the characteristic subepithelial deposits may be found only in some glomerular capillary walls (Bariety et al 1970, Germuth & Rodriguez 1973).

Stage I

By LM the glomeruli appear almost normal but some degree of rigidity of the capillary wall with local thickening may be noted. By EM the deposits are clearly observed. They are small, slightly osmiophilic, and found only on the epithelial side of the basement membrane.

Stage II

The typical image of membranous glomerulonephritis is observed. By LM, the glomerular lesion is characterised by a diffuse and homogeneous thickening of the capillary walls with or without mild mesangial hypercellularity. In most cases, the trichrome stain reveals deposits along the epithelial side of the basement membrane. With silver impregnation these deposits are not argyrophilic but the capillary walls have a comb-like, hatched appearance due to the presence of 'spikes'.

Stage III

By LM, glomeruli show irregular thickening of capillary walls. With silver staining, the capillary walls take the aspect of a mesh. The deposits are easily detected by electron microscopy. They may be arranged in successive layers and surrounded by variable amounts of membrane-like material.

The IF picture is diagnostic showing regular, granular deposits of IgG along the glomerular basement membrane (Fig. 17.3). C3 is sometimes also present in the same pattern and IgA and IgM may be detected in small amounts. C1q and C4 may also be occasionally detected. The IF picture allows easy recognition of the condition, both in the early stages in which the deposits are not yet separated by spikes and in the later stages. Deposition is most often diffuse but may remain segmental.

Membranoproliferative glomerulonephritis

Membranoproliferative glomerulonephritis (MPGN) is a form of chronic diffuse proliferative glomerulonephritis, but many terms have been used: chronic lobular GN, chronic mesangioproliferative GN, persistent hypocomplementaemic GN, mesangiocapillary GN.

As far as the general appearance of the glomeruli is concerned three variants may be described. 'Pure MPGN' is used when the only abnormalities observed are the increase in mesangial cellularity and the thickening of the capillary walls. 'Lobular GN' concerns those cases with a tendency to form centrilobular nodules. In each of these variants superimposed epithelial crescents may be present; hence the term 'MPGN with epithelial crescents'.

Early morphologic observations showed that the problem is still more complex and that based on the location of the glomerular deposits the different variants of MPGN fall into 2 groups: MPGN Type I characterised by the presence of subendothelial deposits, and MPGN Type II marked by the presence of dense intramembranous deposits (Habib et al 1973, Habib et al 1975).

MPGN with subendothelial deposits (Type I MPGN)

The four constant features of Type I MPGN are

1. the proliferation of endocapillary cells;

2. the increase in basement membrane-like material in the axial stalks;

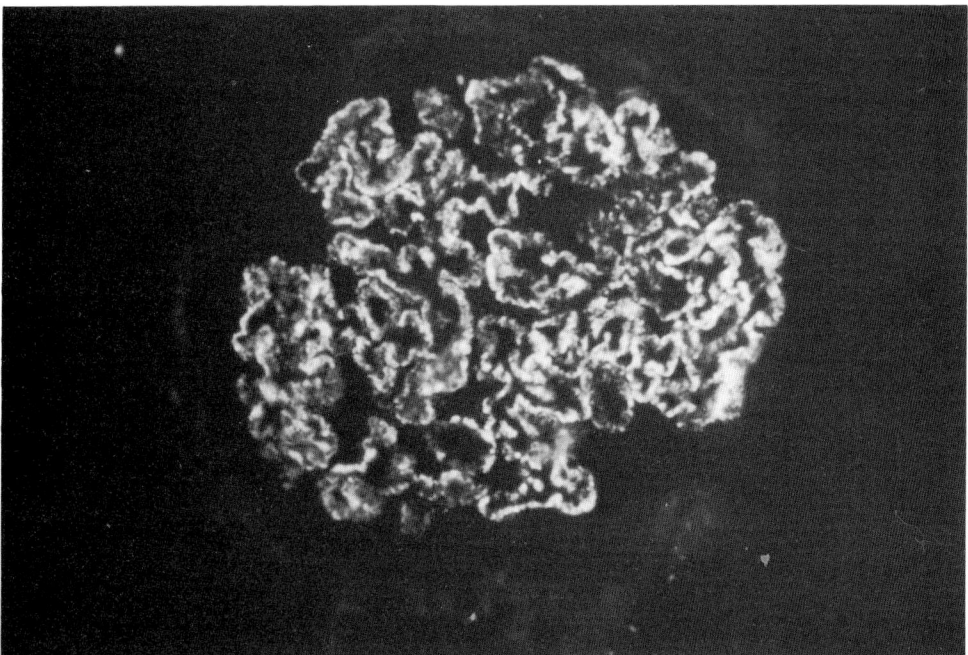

Fig. 17.3 Membranous GN. Immunofluorescence microscopy. Abundant, granular deposits of IgG along capillary loops. × 250

3. the ingrowth or interposition of the mesangium between the lamina densa and the endothelial layer which produce the picture of 'split' capillary wall with 'tram-track' or double contoured basement membrane;

4. the presence of deposits in the subendothelial space and in the mesangial stalks.

By LM, all these features are discernible. However, the peripheral deposits are only visible when very abundant. Several other features are seen in some instances: neutrophils in the capillary lumina, deposits on the epithelial side of the basement membrane having either the appearance of 'humps' or of flat deposits separated by 'spikes' on segments of capillary loops, deposits in the mesangial stalks having the shape of 'worms'. The intensity of the proliferative process, the lobularity and the extent of crescent formation are well appreciated by LM.

By EM, it has been shown that the double contour appearance results from the interposition of mesangial cells between two osmiophilic layers, the normal basement membrane on the external side and the strands of mesangial matrix on the internal side. Finely granular deposits are present along the internal side of the basement membrane and within the mesangial stalks. Subepithelial deposits may also be present. In some cases, subepithelial deposits are contiguous to subendothelial deposits and the interposed basement membrane shows complex alterations (extensive disruption, replication). Recently this pattern has been proposed as a separate entity called MPGN Type III.

By IF, the distribution of the deposits is most often peripheral and no distinction can be made between pure MPGN and MPGN with a lobular pattern. The deposits are always granular, located along the capillary loops where they are more or less diffuse and sometimes present in the mesangial areas. Three patterns may be differentiated on the basis of the presence or absence of Ig and early complement components as well as the presence or absence of mesangial deposits:

1. deposits composed of immunoglobulins, C1q, C3 outlining the periphery of the tuft, sparing the mesangial areas (Fig. 17.4).

2. deposits of immunoglobulin, C1q and C3 outlining the periphery of the tuft associated with deposits of C3 in the mesangial areas.

3. immunoglobulins and early complement components are not detected. Granular deposits of C3 are seen along the capillary walls and frequently in the mesangium.

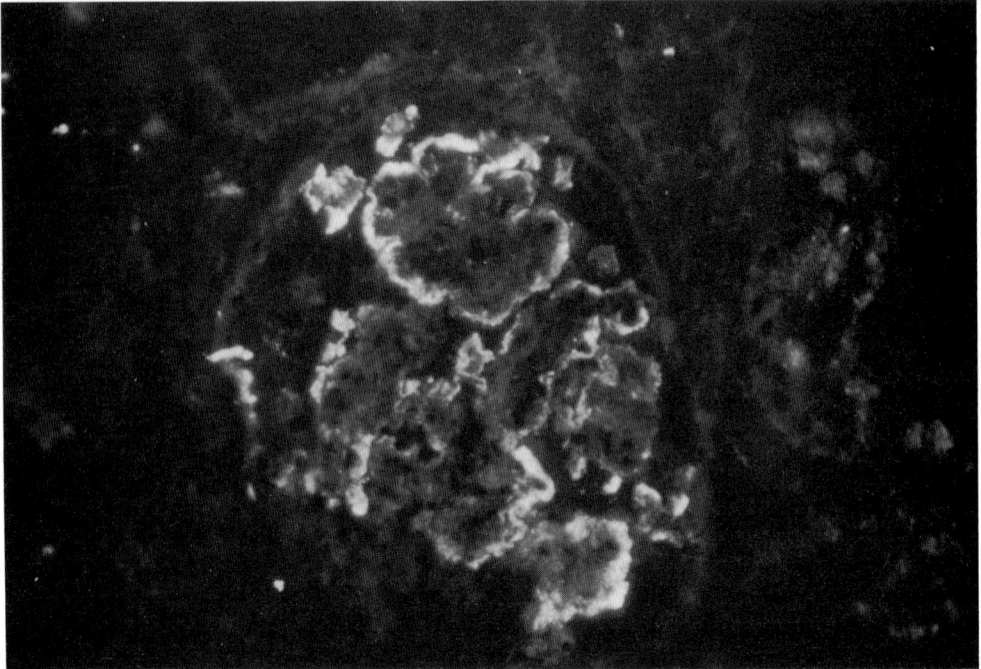

Fig. 17.4 Membranoproliferative glomerulonephritis with subendothelial deposits. Immunofluorescence microscopy. Granular deposits of IgG outlining the periphery of the glomerular tuft. ×250

Contrasting with the constant presence of IgG and the frequent presence of IgM in the first two groups, deposits containing IgA are rarely found and then usually in small amounts.

Properdin may be found in the same location as C3. Fibrin may be detected in Bowman's space if extracapillary proliferation is present and is frequently seen along the capillary walls and in the mesangium.

MPGN with dense intramembranous deposits or Type II MPGN

The characteristic feature of Type II MPGN is the presence of an abnormal material thickening the basement membrane. It is easily identified by LM using oil immersion and the appropriate stains (trichrome, and silver impregnation).

By LM, all the features which characterise Type I MPGN may be seen in Type II. However, mesangial cell proliferation and an increase in mesangial matrix is often less marked than in Type I and may even be absent. The double contour appearance of the capillary walls is not constant. 'Humps' and segmental extramembranous deposits separated by 'spikes' may be seen on the epithelial side of the basement membrane. The amount of neutrophils, the lobularity and the extent of crescent formation is well appreciated by LM and round nodules stained red with the trichrome stain are often found in the mesangial stalks. The distinctive feature of Type II MPGN is the presence of a thickened glomerular basement membrane which is very chromophilic and refractile and has a characteristic 'ribbon-like' appearance. Even in end-stage kidneys, this transformation of the basement membrane is still easily recognisable. This abnormal material is present in Bowman's capsule, in the basement membranes of proximal convoluted tubules, and in the elastic lamina of the arterioles.

By EM, the altered glomerular basement membrane appears thickened and very dark. This feature is due to the presence of strongly electron-dense material located in the midportion of the basement membrane replacing and widening the lamina densa. In the mesangial areas deposits are present in variable amounts in all cases. They form round nodules or masses bulging into the strands of the mesangial matrix. They appear homogeneous and as electron-dense and argyrophilic as the deposits within the basement membrane. Moderately dense and finely granular subepithelial deposits are frequently seen.

By IF, Type II MPGN is characterised by a homogeneous pattern and the almost exclusive fixation of C3. More or less abundant, well-limited round nodules of variable size and intensely stained with anti-serum to C3 are located in the mesangial areas. Deposition of C3 along the glomerular and tubular basement membranes is weaker and appears as a continuous or discontinuous line (Fig. 17.5). In some instances, however, when subepithelial deposits are present, bright granular deposits of C3 seem to be superimposed on the weak linear deposition of C3 on the capillary walls. Small amounts of Ig (IgM most frequently) and, exceptionally, scanty deposits of early complement components, may be observed along capillary walls. Their localization differs from that of C3. Properdin is usually absent. Fibrin is usually found in Bowman's space in cases complicated by crescent formation and may be present along some capillary loops.

Berger's disease

This type of GN, described by Berger & Hinglais (1968), is defined by the presence of diffuse granular deposits of IgA located in the mesangial areas, occasionally extending onto capillary walls. Several terms are currently used: IgA disease, IgA nephropathy, and isolated GN with mesangial IgA deposits. The diagnosis can only be made by the systematic study of renal cortex by IF (Berger et al 1971).

By LM, different patterns are observed but none of them is specific. In some cases, glomerular abnormalities are mild with an enlargement of the mesangial stalks. Round, eosinophilic deposits may be found in the mesangium. Focal GN is a frequent pattern. This type of GN is characterised by the involvement of a limited number of glomeruli, the other glomeruli presenting no abnormality except hypertrophy of the mesangial matrix. In affected glomeruli, a segmental lesion is found, varying from a small, sclerotic adhesion between one or two capillary loops and Bowman's capsule to a larger lesion, consisting of a mixture of segmental mesangial proliferation, excessive matrix, and cellular crescents in Bowman's space. The third pattern, less often encountered, is called endo- and extracapillary GN. In addition to diffuse mesangial hypercellularity of variable intensity involving all glomeruli, the

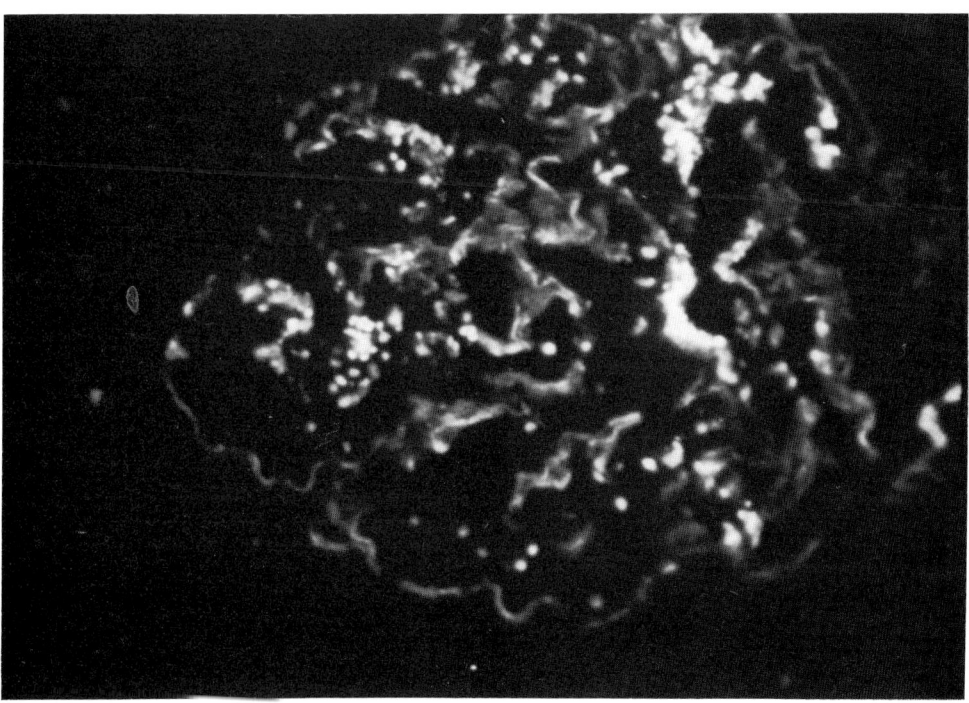

Fig. 17.5   Membranoproliferative glomerulonephritis with dense deposits. Immunofluorescence microscopy. Deposits of C3 in bright nodules located in the mesangium. Small amounts are present along glomerular basement membranes. ×250

presence of cellular or sclerotic crescents is noted. The number of crescents is variable and the percentage of glomeruli involved with crescent formation has to be determined precisely. Frequently, crescent formation affects less than 50% of glomeruli but diffuse crescent formation has been observed in some cases of Berger's disease.

Ultrastructural studies show the presence of finely granular, electron-dense deposits located between the mesangial cells and the mesangial matrix.

Whatever the light microscopic changes may be, the IF pattern is roughly the same, characterised by diffuse mesangial deposits of IgA (Fig. 17.6). However, the importance of the deposits varies from one case to another, from thin filaments to a massive infiltration of the whole mesangium. In addition, subendothelial deposits may be associated with mesangial deposits. Within the IgA deposits the secretory piece has not been identified and the presence of subgroups of IgA (IgA_1 or IgA_2) or both is presently a matter for discussion. IgG deposits are frequently encountered but are far from being constant. IgM is rarely present. Finely granular deposits of C3 and properdin are present in many cases whereas C1q and C4 are not found. Fibrin deposition is not common (20%).

Diffuse proliferative glomerulonephritis with subepithelial deposits or humps

This type of nephritis is usually considered synonymous with acute glomerulonephritis and may or may not follow a streptococcal infection with an acute nephritic syndrome.

Renal biopsies taken early in the course of the disease reveal a diffuse cellular proliferation affecting all glomeruli. Glomerular tufts are enlarged and Bowman's spaces are reduced in size. The capillary lumina are occluded by cells either mesangial and/or endothelial. When numerous neutrophils are found in the capillary lumina, the glomerulonephritis is termed 'exudative'. Recent studies have established that cells of monocyte/phagocyte series may also infiltrate the glomerulus and contribute to glomerular injury. With silver stains, epithelial capillary loops appear normal. Subepithelial, cone-like deposits or 'humps' are usually quite visible, clearly distinct from the glomerular basement membrane to which they are bound. They appear red or green on the trichrome stain. Their number varies from one case to another and from one glomerulus to another. They are not surrounded by silver-stained spikes.

The results of immunofluorescence studies are

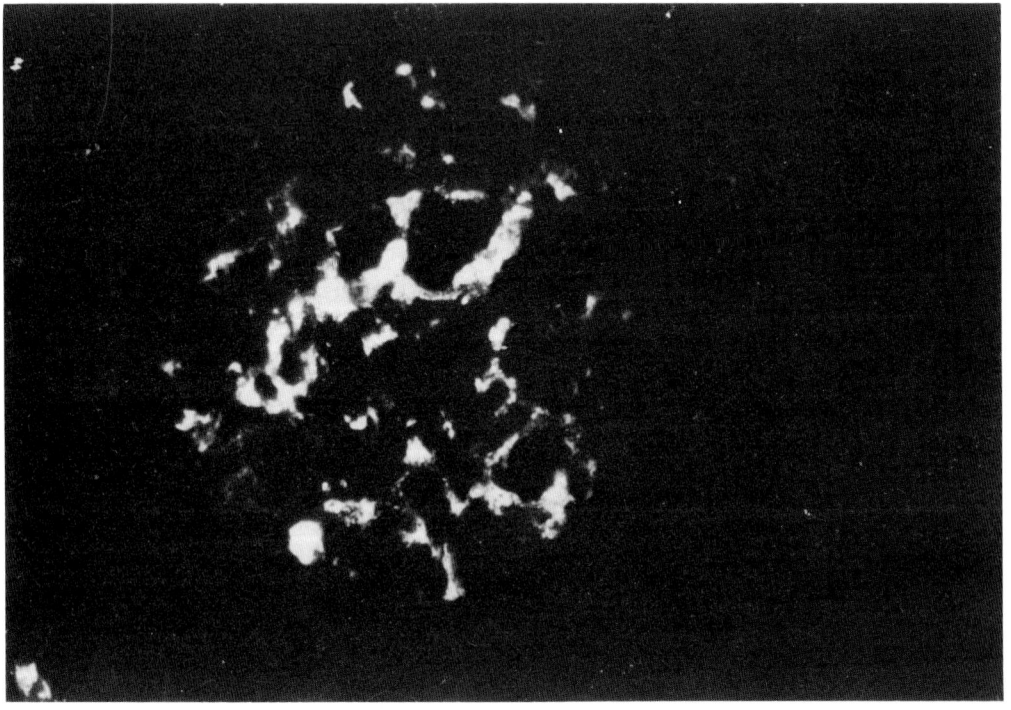

Fig. 17.6 Berger's disease. Immunofluorescence microscopy. Mesangial IgA deposits. ×250

characteristic. C3 is always present and may be the sole immune protein, giving a granular, irregular pattern along the capillary walls and in the mesangium (Fig. 17.7). Less abundant deposits of IgG are often present in the same pattern. Properdin is usually associated with C3. Other immunoglobulins, C1q and C4, are occasionally found (Germuth & Rodriguez 1973).

On late biopsies taken 6 weeks or more after the clinical onset, humps have disappeared, whereas moderate mesangial proliferation may persist. Neutrophils are usually no longer present.

Besides these main lesions, epithelial crescents may be superimposed and part of the prognosis is linked to their extent and percentage.

Diffuse extracapillary or crescentic glomerulonephritis

Extracapillary or crescentic glomerulonephritis is the term used to define a wide variety of glomerular diseases having in common a cellular proliferation in Bowman's space in more than half the glomeruli. In severe cases, virtually every glomerulus may be affected. Because of the difficulties in defining the

limits of this condition, some degree of confusion persists in the literature. This confusion explains the large number of terms proposed: Type I nephritis with a rapidly progressive course, subacute extracapillary GN, malignant GN, acute necrotic GN, and endo- and extracapillary GN.

Traditionally, it was believed that crescents were formed by the proliferation of epithelial cells, probably in response to deposition of fibrin in Bowman's space. It has been shown recently that macrophages migrating within Bowman's space may accumulate, transform into epithelioid and giant cells, and form a substantial proportion of the cells forming the crescent.

By LM, cell proliferation in the Bowman's space is the main lesion. It is composed of a series of cell layers which obliterate the urinary space and compress the glomerular tuft. Some crescents completely surround the tuft while others are segmental. A material showing the characteristic staining of fibrin is seen between the proliferating cells. Secondarily, crescents may become sclerotic. Cellular and sclerotic crescents may coexist in the same biopsy. The tuft is affected by complex alterations associating collapse of capillary loops,

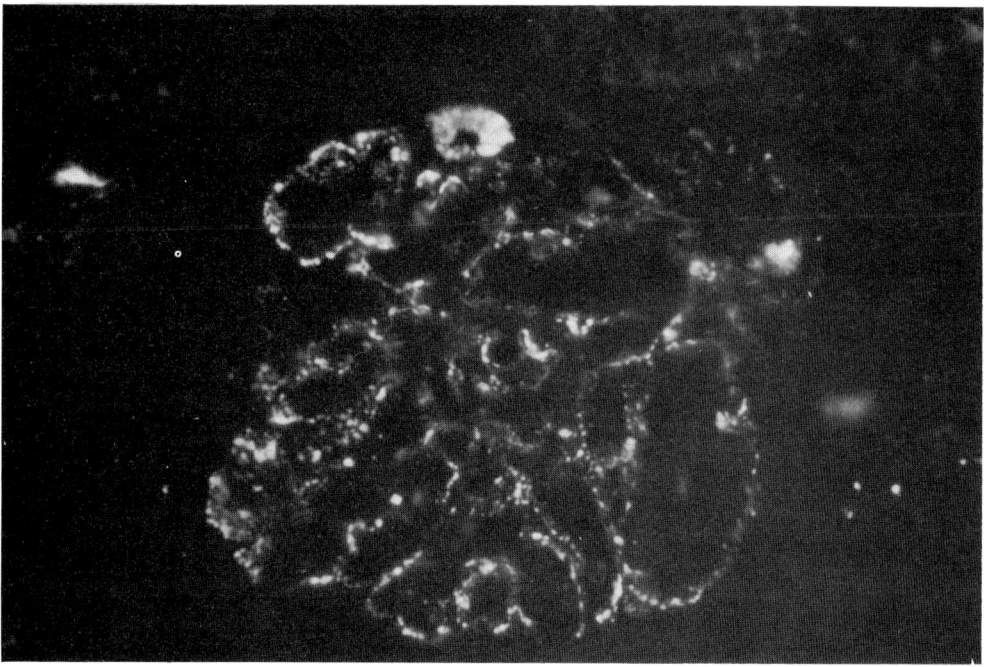

Fig. 17.7 Proliferative glomerulonephritis with subepithelial deposits. Immunofluorescence microscopy. Scattered granular deposits of C3 along the capillary loops. ×250

necrosis, mesangial cell proliferation, polymorphs, and deposits. The glomerular tuft may also be reduced to a small, sclerotic or necrotic mass where structures can no longer be distinguished.

Ultrastructural studies show the proliferating cells and the intercellular fibrin recognised by its specific ultrastructure. Deposits located in variable sites of the tuft may eventually be present. Ruptures or gaps of the glomerular basement membranes may be observed in limited areas of the tuft. They are due to capillary thrombosis or necrosis and allow fibrin, red cells, polymorphs, and cell debris to penetrate into the urinary space.

By IF, the constant finding in early biopsies is the presence of a network of fine flecks containing fibrin. In the tuft itself, IF findings are variable (linear deposits of IgG, granular deposits of immunoglobulin and complement components along the capillary loops and in the mesangium) allowing identification of the pathogenetic mechanisms. In some cases however, deposits cannot be demonstrated in the glomerular tufts.

In conclusion, several causes may trigger mechanisms leading to crescent formation (Figs 17.8 and 17.9). Crescentic GN may appear idiopathic or secondary to various systemic diseases and is observed in diffuse GN with humps, MPGN with subendothelial deposits, MPGN with dense intramembranous deposits, Berger's disease, lupus nephritis, polyarteritis nodosa, Wegener's granulomatosis, mixed cryoglobulinaemia, anaphylactoid purpura and Goodpasture's syndrome.

LUPUS NEPHRITIS

It is well established that the glomerular lesions in systemic lupus erythematosus are heterogeneous and that part of the prognosis of this disease is linked to the type and severity of renal lesions. Renal lesions can be divided into four groups. However, as the lesions are highly pleiomorphic and vary in intensity from one glomerulus to the next, it seems useful to adopt semi-quantitative methods. Some lesions are regarded by most authors as active, i.e. susceptible to treatment, or to worsen, whereas some lesions are fibrotic or inactive. A list of the active lesions has been proposed (Pirani et al 1964). These can be listed as follows: cellular proliferation, nuclear debris,

hematoxyphil bodies, foci of necrosis, hyaline thrombi (which are subendothelial deposits occluding capillary lumena), deposits, often conspicuous, known as wire-loops (well seen on the trichrome stain), crescents, and acute necrotising angiitis. Conversely, subepithelial deposits, or spikes indicating the presence of subepithelial deposits are often seen but represent inactive lesions, together with sclerosis and fibrous crescents.

There is in general a good correlation between IF and LM although both methods are compulsory to assess the activity and severity of lesions (Morel-Maroger et al 1976).

The lesions may be divided into the following groups according to IF:

Minimal changes or normal kidney

Usually there are no deposits by IF, or minimal faint linear deposits of IgG.

Mesangial GN

The deposits made of IgG, C1q, C3 and sometimes IgM, more rarely IgA, are restricted to the mesangial areas. By LM, the lesions are never very severe. The mesangial areas may be diffusely increased with minimal cellular proliferation. Segmental foci of glomerular necrosis with crescents may be present in occasional glomerular segments. Also, nuclear fragments may be found and sometimes they tend to accumulate in the juxtaglomerular apparatus.

Diffuse GN

The lesions, although diffuse, always vary from one glomerulus to the next and within one glomerulus between the different lobules. Cellular proliferation involves mainly mesangial and/or endothelial cells. Neutrophils are frequent. Small cellular crescents are common but extensive crescents are rare. Capillary wall lesions are usually conspicuous and result from the combination of different features: subendothelial eosinophilic deposits known as 'wire-loops', unevenly distributed extramembranous deposits, and sometimes intramembranous deposits. The association of an irregular cellular proliferation with 'wire-loop' and extramembranous deposits is highly

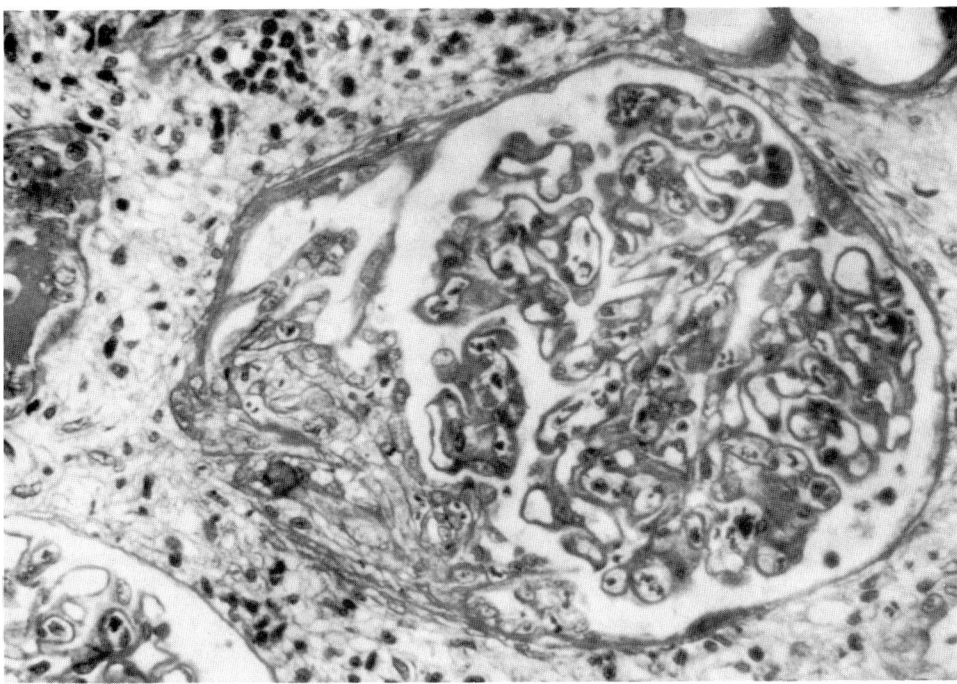

Fig. 17.8 Crescentic glomerulonephritis. Mild mesangial proliferation, neutrophils in the capillary lumina, cellular crescent formation (by IF, granular deposits of C3 were present along the capillary loops). Trichrome. ×250

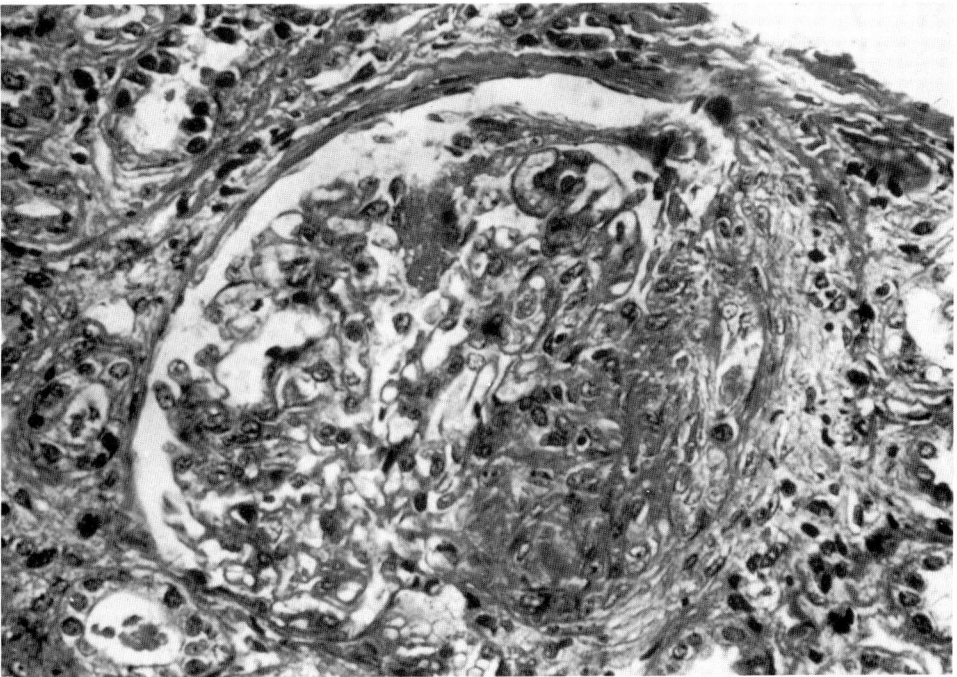

Fig. 17.9 Crescentic glomerulonephritis. Mesangial proliferation. Segmental lesion showing mixture of mesangial cell proliferation, necrosis, crescent formation (by IF, linear deposits of IgG were present along glomerular basement membranes). ×250

suggestive of lupus nephritis. A number of other characteristic active lesions may be present in this form of GN.

By IF, IgG (Fig. 17.10) is almost always detected in a granular diffuse pattern along the GBM. Some deposits are coarse, probably representing the 'wire-loop' whereas some others, smaller and regular, are extramembranous. In addition, granular deposits located in the mesangium are also present. IgA and IgM are focal, and never so abundant as IgG in our experience. C3, C1q, C4 and properdin are detected with a pattern comparable to that of IgG. Granular deposits of IgG, C3 and C1q are frequently found along tubular basement membranes and IgG and/or IgA may be present in the arteriolar walls.

Clinically the majority of patients with this form of GN present with the nephrotic syndrome usually in association with microscopic haematuria and renal failure. When clinical improvement occurs with treatment, it is associated with a decrease of active glomerular lesions without progressive glomerular sclerosis. In our view, serial histological studies remain the most reliable means to follow and monitor therapy in this type of GN.

Extramembranous GN

According to most authors, the term of extramembranous GN in SLE should be restricted to cases in which the deposits are the only detectable lesion. By LM, the appearance is indistinguishable from that of idiopathic extramembranous GN. By IF, early components of complement have been observed (C1q, C4). Provided that the histological criteria are strict, the natural history of this disease is comparable to that of 'primary' extramembranous GN with a long-standing course.

Anaphylactoid purpura nephritis

By LM, the main feature is the presence of crescents (Levy et al 1976). Their appearance is variable: in the early stages, they are purely cellular and contain flecks of fibrin; later, they become sclerotic. In fact, both aspects often coexist in the same biopsy, indicating continuing attacks of glomerular involvement. In the segment of the tuft adjacent to the crescent, one finds a mixture of mesangial proliferation and sclerosis sometimes associated with a segmental double contour of the capillary walls and

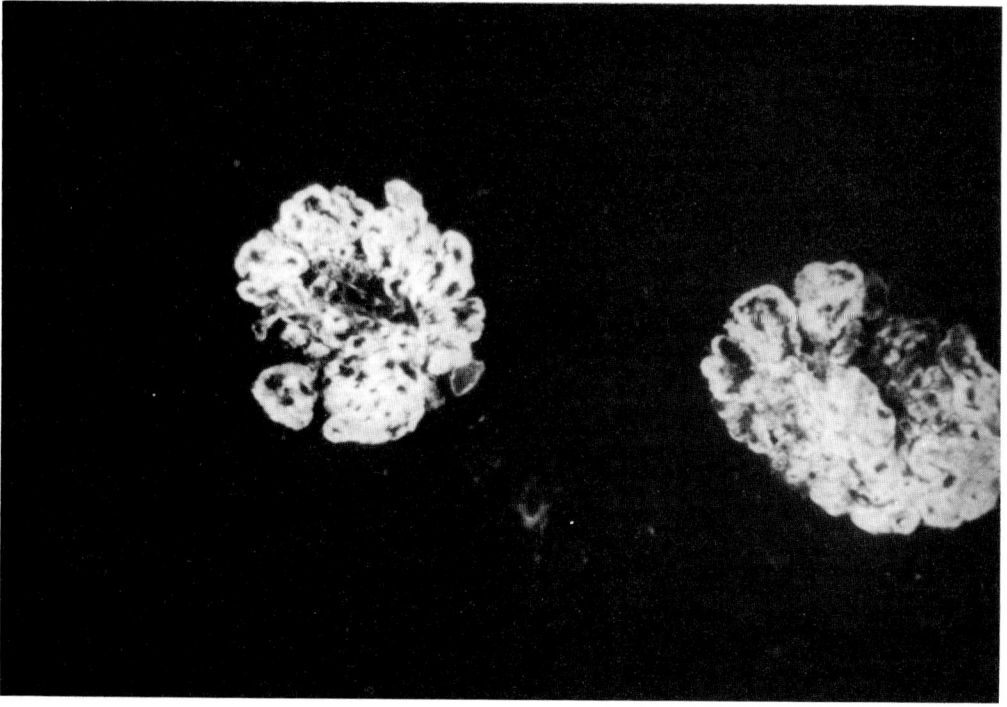

Fig. 17.10 Lupus nephritis. Immunofluorescence microscopy. Abundant granular deposits of IgG along the capillary loops, within the mesangium and in several capillary lumina in a case of diffuse proliferative glomerulonephritis. ×100

in some instances necrosis. The size and the extent of crescents is variable. The percentage of glomeruli involved with crescent formation has to be determined for each case. Prognosis is closely correlated with the percentage of crescents. A histopathologic study demonstrates two types of glomerular lesions characterised by the presence or absence of mesangial proliferation in those glomeruli not involved with crescent formation. In focal and segmental GN (Fig. 17.11), the number of affected glomeruli (i.e. with crescent formation) is variable. Usually the extent is less than 70% and the size never exceeds half of Bowman's space. Crescents may involve 80% or more of the glomeruli and circumferential crescents may be observed.

By IF, granular deposits containing predominantly IgA are observed. In some instances, deposits are purely mesangial, ranging from thin filaments to large clumps. Mesangial and parietal deposits are often simultaneously present. In rare cases, deposits are purely parietal. In fact, these patterns vary from one glomerulus to another, and even within a glomerulus from one portion of the tuft to another. In association with IgA, which is consistently found, deposits may contain IgG and/or IgM. C3 is frequently present. C1q and C4 are not found in the deposits. Fibrin deposits, identical to those observed with anti-serum to IgA, are frequent. Fibrin may also be observed in necrosis of the glomerular tuft and in flecks present in crescents.

Polyarteritis nodosa

The frequency of glomerular lesions in polyarteritis nodosa is difficult to assess, although it is generally recognised that only the so-called microscopic forms are accompanied by glomerular lesions. These lesions vary from small foci of necrosis in the glomeruli to diffuse crescentic GN. Acute necrotising angiitis is the characteristic lesion of the disease. By IF, glomeruli contain fibrinogen both in the necrotic areas and the crescents. However, immunoglobulins and complement are generally not found. Fibrinogen may also be detected in the arterial walls (Churg et al 1979).

Wegener's granulomatosis

As in polyarteritis nodosa, the glomerular lesions vary from crescentic GN, with extensive epithelial proliferation, to mild focal glomerular lesions.

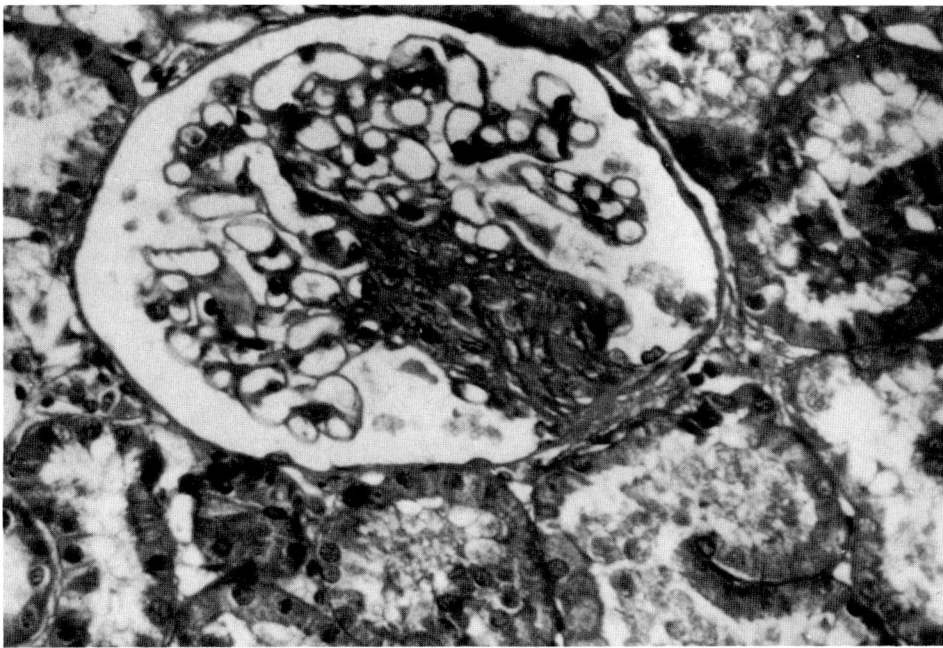

Fig. 17.11 Anaphylactoid purpura. Focal glomerulonephritis. Sclerotic adhesion between capillary loops and Bowman's capsule. Trichrome. × 250

Segmental foci of glomerular sclerosis often coexist with fresh necrotic lesions.

GOODPASTURE'S SYNDROME

In this rare syndrome the characteristic symptomatology consists of a sudden haemoptysis followed by glomerulonephritis. Histologically the renal lesions vary from focal proliferative GN to diffuse crescentic GN with collapse of the glomerular tuft and thickening of GBM. The diagnostic feature is the presence by IF of diffuse smooth linear IgG (Fig. 17.12), sometimes associated with C3 along the GBM. Characteristically circulating antiglomerular basement membrane antibodies are also found. The pattern is the same as that of experimentally induced antiglomerular basement membrane disease (Germuth & Rodriguez 1973).

Anti-GBM antibodies nephritis may occur without lung haemorrhage and in such patients the renal histology is similar to that observed in Goodpasture's syndrome. Clinically most patients present with rapidly progressive GN, although some of them have less severe symptoms, such as proteinuria and haematuria without renal failure.

GLOMERULAR LESIONS IN DYSPROTEINAEMIAS

It is of considerable interest that a correlation has been established between the immunoglobulin deposits and the circulating cryoglobulins or monoclonal para-proteins (Morel-Maroger & Verroust 1974).

The common feature of some lesions occurring in dysproteinaemias is the presence of amorphous intraglomerular 'thrombi' or deposits, which occlude the glomerular capillary lumina. They vary widely in number. Eosinophilic on H & E preparations, they stain by PAS in most instances except in IgG gammopathy in which they are PAS-negative. A conspicuous cellular proliferation is present in mixed essential IgG-IgM cryoglobulinaemias, and in IgG and IgA monoclonal gammopathies. In contrast, in Waldenstrom's disease (Fig. 17.13), we have never observed any increase in the number of glomerular nuclei. Similar 'thrombi' have also been found in a few cases of cryofibrinogenaemia. EM studies have shown, in a few cases, a fibrillar structure of thè deposits as well as crystals in the endothelial cells. In mixed cryoglobulinaemias the GN may be associated with acute necrotising angiitis. By IF, there is a good

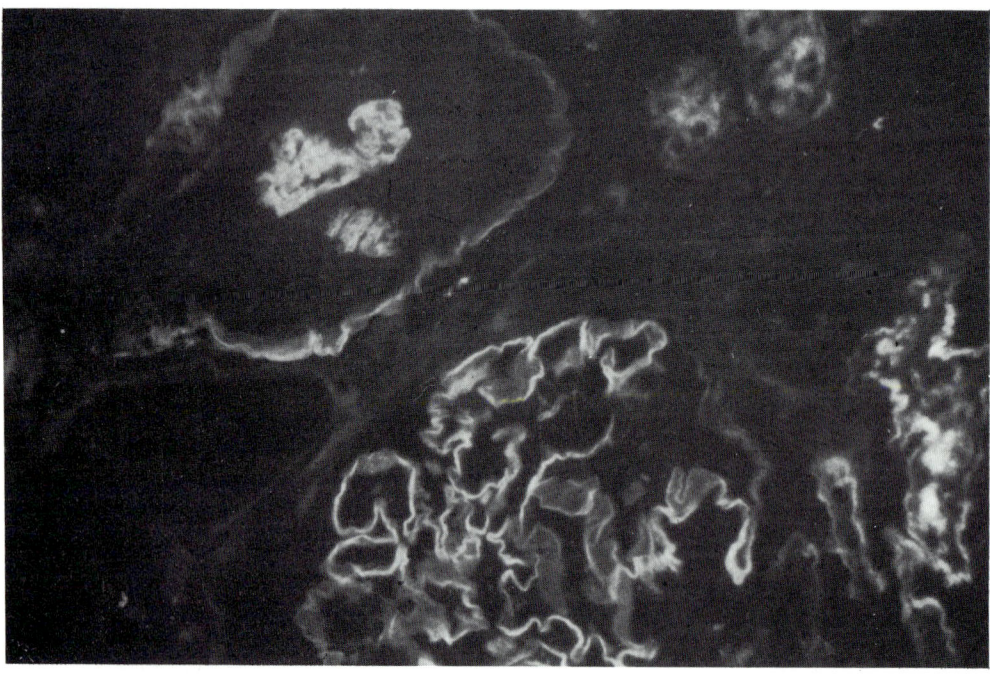

Fig. 17.12   Goodpasture's syndrome. Immunofluorescence microscopy. Bright, linear deposits of IgG along glomerular basement membranes and tubular basement membranes. X250

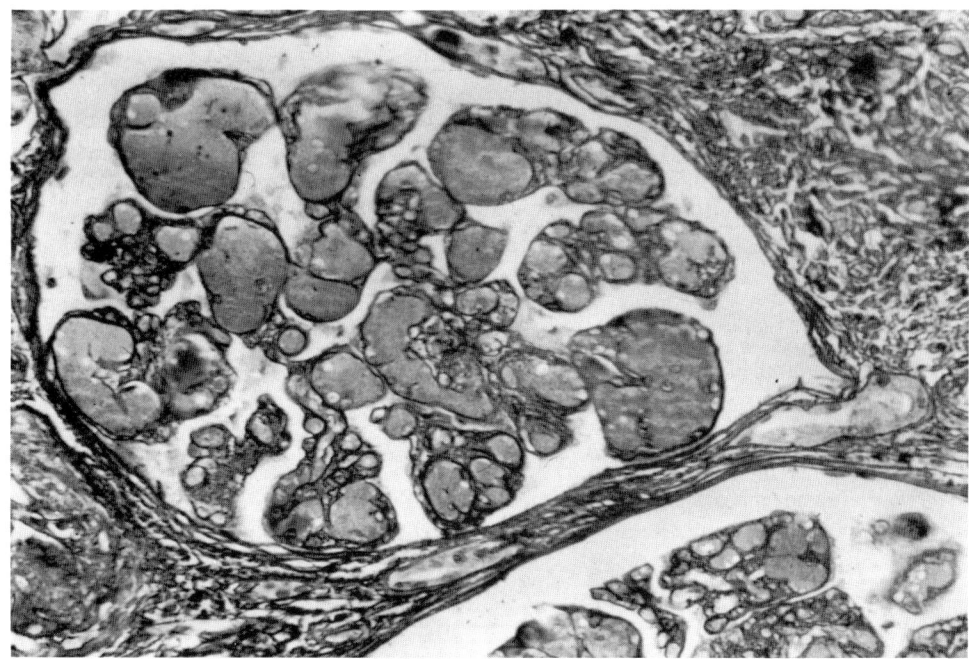

Fig. 17.13 Waldenstrom's syndrome. Occlusive intracapillary deposits invading dilated lumina. Silver stain. ×250

correlation between the circulating abnormal proteins and the deposits.

Clinically, in mixed cryoglobulinaemias and gammopathies, the patients may present with an acute nephritic syndrome and renal failure. With supportive treatment, the lesions may clear. In two of our patients, recurrent and reversible episodes of acute renal failure developed with recurrence of typical lesions. In patients with long-standing cryoglobulinaemia, the clinical presentation varies from a symptomatic proteinuria to nephrotic syndrome with progressive renal failure. On the other hand, patients with Waldenstrom's macroglobulinaemia have very few renal symptoms unless they develop amyloidosis.

Light-chain systemic deposition

Antonovych et al (1973) reported two cases of multiple myeloma with unusual renal lesions; in both cases glomerular and tubular basement membranes were outlined by deposits which were dense and granular by electron microscopy and contained kappa chains detectable by immunofluorescence. Since this initial observation, other cases of similar light-chain deposition nephropathy (often known as Randall's

disease) have been reported; although some occurred in patients with overt multiple myeloma, in others there was no evidence of an underlying plasmacytic malignant disease. The nephropathy is clinically characterised by the association of proteinuria and renal insufficiency. The proteinuria is often heavy, with nephrotic syndrome, although some patients with nodular glomerulosclerosis have only mild proteinuria, or even none. Most patients have renal insufficiency at the time of presentation.

The glomerular and tubular lesions share some common characteristics suggestive of the disease. There is a ribbon-like, refractile appearance of the tubular basement membrane thickening suggestive of the disease, which can in certain cases be differentiated from the non-specific deposits observed in several other types of nephritis with tubular atrophy. The diagnosis, however, is essentially based on IF study, which shows in all cases light chain deposits, most often kappa, delineating the tubular basement membranes. By EM, granular osmiophilic deposits are present along the outer aspect of the tubular basement membranes. This material does not contain fibrillar structures. The glomerular lesions are predominantly mesangial. The most recognisable lesion is a nodular glomerulosclerosis. PAS-positive mesangial nodules are often present (Ganeval et al 1981).

The glomerular lesions can hardly be distinguished from Kimmelstiel-Wilson glomerulosclerosis, or from membranoproliferative glomerulonephritis in its lobular variety, particularly when cellular proliferation is conspicuous. It is noteworthy, however, that the mesangial nodules are not stained by silver impregnation, thus differing from membranoproliferative glomerulonephritis (Fig. 17.14). By IF, the nodules may be stained by the sera against light chain determinants, but this is far from constant. At the EM level the mesangial deposits are dense, non-fibrillar and granular. Since deposits located on the inner aspect of the GBM may also be present, having the same appearance as the TBM deposits along tubular basement membrane.

Amyloidosis

The diagnosis of amyloidosis is usually easy by LM provided that special stains for amyloid are used routinely. However, when deposits are very small, they may be overlooked on LM and EM is then essential. By LM, amyloid appears as an amorphous substance invading the mesangial and subendothelial spaces. The deposits, faintly eosinophilic, are not stained by PAS. The most reliable diagnostic feature is the presence of a green dichroism on preparations stained by Congo red. When stained by crystal violet the amyloid deposits are purple. This stain is better when frozen sections are used. IF gives irreproducible results as it seems that the amyloid substance can trap most of the antisera used. Under the electron microscope amyloid is seen to consist of fine fibrils up to 1 μm in length, with a diameter from 80 to 100 Å. These are arranged in a random array.

Diabetes

Glomerular disease is common in patients suffering from diabetes mellitus. The lesions described by Kimmelstiel & Wilson (1936) as nodular glomerulosclerosis are diagnostic, whereas other varieties of GN are less characteristic. The nodular lesions consist of homogeneous, round, solid areas located in the middle part of glomerular lobules. Stained purple by PAS, these nodules are surrounded by the glomerular nuclei. The other lesions are a diffuse thickening of capillary walls and of the

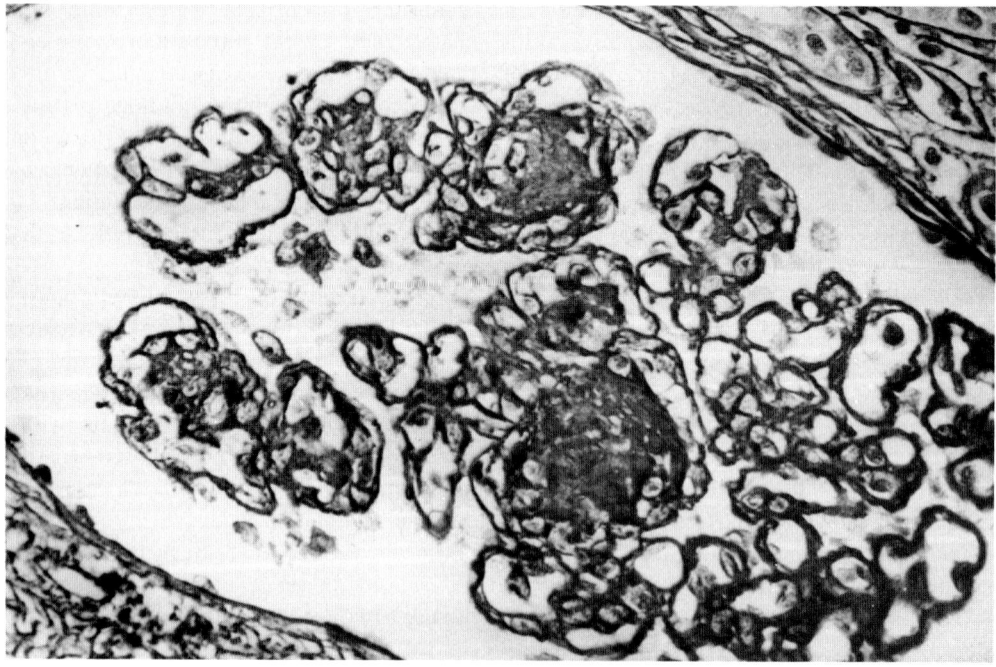

Fig. 17.14 Light chain disease. Centronodular lobules with mild cellular proliferation (by immunofluorescence microscopy, minor deposits of K chains were detected). Trichrome. ×400

mesangial matrix. In some instances, the peripheral capillary walls present aneurysmal dilatations. Eosinophilic deposits may be present along the peripheral thickened capillary walls, sometimes occluding part of the corresponding lumen. Typically these deposits, known as 'fibrin caps', may contain foamy material. Other types of GN have also been reported as a complication of diabetes.

By IF, linear deposits of IgG may be present along the GBM in a few cases but, unlike those found in anti-GBM nephritis, they are usually faint. Anti-GBM antibodies have never been detected in the serum.

THROMBOTIC MICROANGIOPATHY

The haemolytic-uraemic syndrome represents an important cause of acute renal failure. A number of studies have documented the renal changes (Habib et al 1981).

The most characteristic lesion observed is thrombotic microangiopathy. The predominant changes may be glomerular or arterial.

Thrombotic microangiopathy with predominant glomerular involvement

By LM, the striking feature is the variegated pattern of the glomerular involvement. In all cases, a variable percentage of glomeruli show the features of 'thrombotic microangiopathy'. Glomeruli appear enlarged without any mesangial proliferation. The mesangium is hypertrophied and fibrillar. Capillary walls appear thickened with a double contour appearance and, narrowed capillary lumina which may be obstructed by red blood cells or thrombi having the characteristic staining for fibrin. Shrinking of the glomerular tuft at the vascular pole is a frequent finding.

The arteriolar involvement is extremely variable and can be inconspicuous. In some instances, a widened subendothelial space containing a fluffy material leads to the narrowing of the lumen. An associated luminal thrombosis is exceptionally seen. Interlobular arteries appear normal.

By EM, the lesion is characterised by a separation of the endothelial cells from the basement membrane and accumulation of a fluffy material in the subendothelial space giving a double contour appearance to the capillary wall, whereas the basement membrane itself appears unaltered. In association with these lesions, the mesangial matrix has a fibrillar appearance in the absence of mesangial proliferation.

By IF, fibrin is observed on scattered glomeruli, along thickened capillary walls, within a widened mesangium and more intensely on some round thrombi located in the capillary lumina (Fig. 17.15). In addition, in the same glomeruli, deposits containing IgM along some capillary loops as well as granular deposits containing C3 are observed.

Thrombotic microangiopathy with predominant arterial involvement

In the majority of cases, most interlobular arteries are involved and show a nearly complete occlusion of the lumen secondary to the presence of various changes: thrombosis, intimal oedema, intimal proliferation, necrosis of the arterial wall, and thickening of the muscular walls. Many arterioles show a lamellar, onion-skin appearance. Associated with these severe and widespread arterial lesions, most glomeruli appear 'ischaemic' showing a diffuse wrinkling almost always associated with a splitting of capillary basement membranes. Some other glomeruli show a marked dilatation of capillary lumina filled with red blood cells. Occasional glomeruli showing the characteristic features of the first type may be observed.

By IF, fibrin is observed within the lumina of arteries and along some capillary loops on scattered glomeruli.

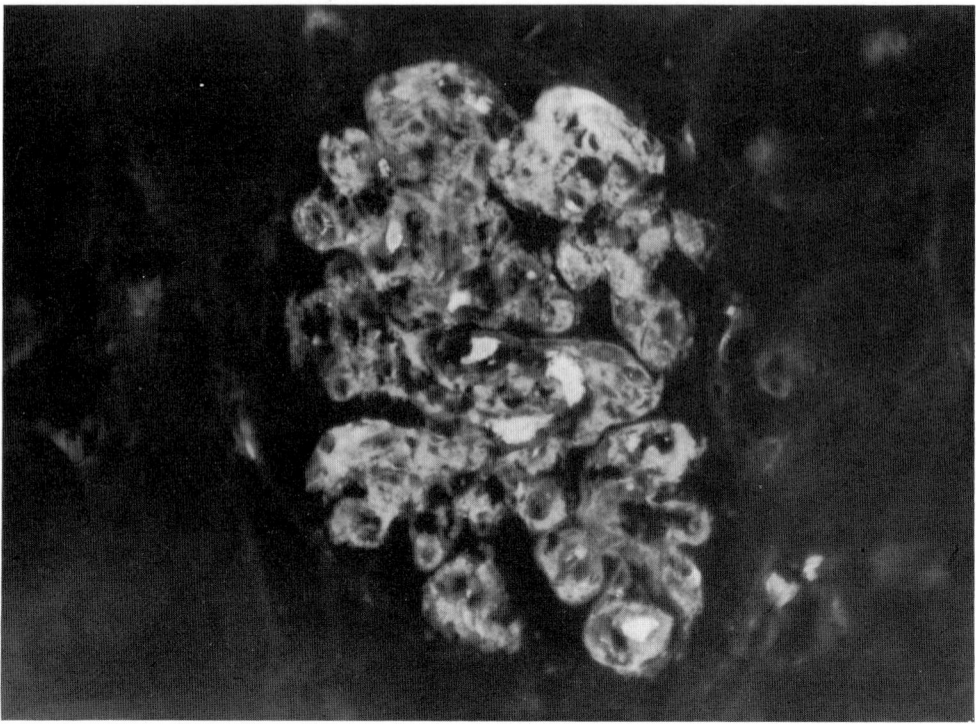

Fig. 17.15   Thrombotic microangiopathy with glomerular involvement. Immunofluorescence microscopy. Deposits of fibrin along thickened capillary loops, widened mesangium and several thrombi. ×250

REFERENCES

Antonovytch T, Lin C, Parrish E, Mostofi K 1973 American Society of Nephrology, abstract 3

Bariety J, Druet P, Lagrue G, Samarcq P, Milliez P 1970 Les glomérulonéphrites extramembraneuses (GEM). Etude morphologique en microscopie optique, électronique et en immunofluorescence. Pathologie et Biologie (Paris) 18: 5

Berger J, Hinglais N 1968 Les dépôts intercapillaires d'IgA-IgG. Journal d'urologie (Paris) 74: 694

Berger J, Yaneva H, Hinglais N 1971 Immunochemistry of glomerulonephritis. In: Hamburger J, Crosnier J, Maxwell M N (eds) Advances in nephrology. Year Book Medical Publishers, Chicago, ch 2, p 11

Churg J, Strauss L, Glabmann S 1979 Necrotizing angiitis. In: Hamburger J, Crosnier J, Grünfeld J P (eds) Nephrology. Wiley-Flammarion, ch 37, p 621

Ganeval D, Mignon F, Preud'homme J L, Noël L N, Morel-Maroger L, Droz D et al 1981 Dépôts de chaînes légères et d'immunoglobulines monoclonales: aspects néphrologiques et hypothèses physiopathologiques. In: Grünfeld J P (ed) Actualités néphrologiques de l'Hôpital Necker. Flammarion Médecine-Sciences, Paris, p 179

Germuth F G, Rodriguez E 1973 Immunopathology of the renal glomerulus. Immune complex deposits and anti-basement membrane disease. Little, Brown and Co, Boston

Habib R, Kleinknecht C 1975 The primary nephrotic syndrome of childhood. In: Sommers S C (ed) Kidney pathology. Decennial, Appleton Century Crofts, New York, p 165

Habib R, Kleinknecht C, Gubler M C, Levy M 1973 Idiopathic membranoproliferative glomerulonephritis in children. Report of 105 cases. Clinical Nephrology 1: 194

Habib R, Levy M, Gagnadoux M F, Broyer M 1981 Le pronostic du syndrome hémolytique et urémique chez l'enfant. In: Grünfeld J P (ed) Actualités néphrologiques de l'Hôpital Necker. Flammarion Médecine-Sciences, Paris, p 245

Habib R, Gubler M C, Loirat C, Ben Maiz H, Levy M 1975 Dense deposit disease: a variant of membranoproliferative glomerulonephritis. Report of 44 cases in children. Kidney International 7: 204

Kimmelstiel P, Wilson C 1936 Intercapillary lesions in glomeruli of kidney. American Journal of Pathology 12. 83

Levy M, Broyer M, Arsan A, Levy-Bentolila D, Habib R 1976 Anaphylactoid purpura nephritis in childhood: natural history and immunopathology. In: Hamburger J, Crosnier J, Maxwell M H (eds) Advances in nephrology. Year Book Medical Publishers, Chicago, ch 9, p 183

Morel-Maroger L, Verroust P 1974 Glomerular lesions in dysproteinemias. Kidney International 5: 249

Morel-Maroger L, Mery J Ph, Droz D, Godin M, Verroust P, Kourilsky O, Richet G 1976 The course of lupus nephritis: contribution of serial biopsies. In: Hamburger J, Crosnier J, Maxwell M H (eds) Advances in nephrology. Year Book Medical Publishers, Chicago, ch 5, p 183

Pirani C, Pollak V E, Schwartz F D 1964 The reproducibility of semi-quantitative analyses of renal histology. Nephron 1: 230

Sinclair R A, Burns Y, Dunnill M S 1981 Immunoperoxidase staining of formalin-fixed, paraffin-embedded, human renal biopsies with a comparison of the peroxidase-antiperoxidase (PAP) and indirect methods. Journal of Clinical Pathology 34: 859

18

Blood, bone marrow, spleen and lymph nodes in metabolic disorders

B. D. Lake

HISTOCHEMISTRY OF BLOOD CELLS IN STORAGE DISORDERS

Many of the metabolic disorders express their enzyme defect in circulating blood cells and in some there is also morphological and histochemical evidence on which a diagnosis can be made (Hansen 1972, Lake 1981a). In some of these disorders the evidence in stained blood films can only point in a particular direction but in others there can be sufficient information to make a definitive diagnosis. Correlation of the findings with clinical information is, as always, most important. The simple screening, by examination of stained blood films, can be of help in the exclusion of certain conditions and can save considerable time and expensive biochemical reagents. It has the added advantage that the techniques are available in the majority of laboratories in contrast with the specialised biochemical methods available only at a handful of laboratories.

Lymphocytes

To most people the presence of vacuolated lymphocytes is a relatively non-specific phenomenon but when the size and distribution of the vacuoles and their staining characteristics are taken into account, together with the clinical presentation, a reasonably accurate diagnosis can be made. Table 18.1 lists a number of conditions and the features found in stained blood films. In the first instance a routinely stained film (Giemsa, Wright, Leishman, etc.) prepared from a fingerprick sample or from anticoagulated blood should be examined. Further examination by the special methods indicated in

Table 18.1 can be made on the spare films prepared at the initial sampling.

Some drugs, particularly those of an amphiphilic nature, are known to induce vacuolation of lymphocytes (Lullman et al 1973) and this should be borne in mind when examining blood films for evidence of a storage disorder.

In the mucopolysaccharidoses the inconstant Reilly bodies cause much confusion. Their presence is variable and their identity is not certain since in the original description it was not clear whether they occurred in circulating blood cells or in bone marrow. They are probably different from Alder granules which occur specifically and constantly in Maroteaux-Lamy disease (MPS VI), mucosulphatidosis and β-glucuronidase deficiency. The presence or absence of Reilly bodies in neutrophils is unhelpful in the differential diagnosis of the mucopolysaccharidoses.

Table 18.1 is not an exhaustive list but includes conditions in which ancillary tests can be helpful. Other disorders in which large, well-defined lymphocytic vacuoles are found include muco-lipidosis I (identical with sialidosis II), and Salla disease. Vacuolated lymphocytes are also mentioned variably in fucosidosis, the cherry red spot — myoclonus syndrome (sialidosis I) and in some patients with Niemann-Pick disease Group C (ophthalmoplegic lipidosis). In none of these are there any specific staining reactions of the contents of the vacuoles and suitable enzyme staining methods are not available.

Ultrastructural abnormalities of lymphocytes can be detected in infantile Batten's disease (deposits of GROD), late infantile Batten's disease (curvilinear bodies) (Lake 1981b) and mucolipidosis IV (membranous cytoplasmic bodies) (Lake et al 1982)

Table 18.1   Histochemical detection of storage disorder in blood cells

Disorder	Lymphocytes (Giemsa stain)	Staining reaction of lymphocytic vacuoles	Enzyme method	Neutrophils	Other Comments
Wolman's disease (including cholesteryl ester storage disease)	small discrete vacuoles 1–6 per cell in most lymphocytes	ORO:SBB	Acid esterase negative (Lake 1971)	Usually not involved; some patients show marked lipid accumulation in neutrophils	Platelets also involved; foam cells in blood are exceedingly rare
Pompe's disease (and other variants of GSD II)	Small discrete vacuoles 1–6 per cell in majority of lymphocytes	cell PAS +ve dig cell PAS −ve	Method for α-glucosidase not suitable or of sufficient sensitivity	Not involved	–
Niemann-Pick Type A	Small discrete vacuoles 1–4 per cell in most lymphocytes	Contents not readily shown	No method available for sphingomyelinase	Not involved	–
Mannosidosis	Numerous small discrete vacuoles in some lymphocytes. Others contain several large, well-defined vacuoles	PAS cell, not digested. Not easily shown because of extreme solubility of mannosides	Method for α-mannosidase not selective enough or of sufficient sensitivity	Not involved	No metachromasia
I-cell disease (mucolipidosis II)	Large bold vacuoles	Vacuoles appear empty	Defect not amenable to histochemical methods	Not involved	No metachromasia
G_{MI}-gangliosidosis type 1	Large bold vacuoles in most lymphocytes	Vacuoles appear empty	β-galactosidase activity not detectable (Lake 1974)	No storage; β-galactosidase activity not detectable	Some residual or different β-galactosidase may be seen in eosinophils. No metachromasia. Platelets also β-galactosidase deficient
G_{MI}-gangliosidosis type 2	No vacuoles	–	ditto	ditto	
Juvenile Batten's disease	Several large bold vacuoles in up to 30% of lymphocytes in the tail of the film	Vacuoles appear empty	Defect not known	Not affected	–
Aspartyl-glucosaminuria	Several large bold vacuoles in some lymphocytes; not frequent Reddish inclusions may be present in the vacuoles	Vacuoles appear empty	Defect not amenable to histochemical methods	Not affected	–
Mucopoly saccharidosis IH (Hurler)	Occasional vacuoles	Metachromatic inclusions present generally in <5% of lymphocytes	None available	–	Gasser cells may be seen (vacuolated lymphocytes with basophilic inclusions)

Disorder	Lymphocytes (Giemsa stain)	Staining reaction of lymphocytic vacuoles	Enzyme method	Neutrophils	Other Comments
Mucopolysaccharidosis (contd)					
IS (Scheie)	Occasional vacuoles	Metachromatic inclusions in <5% of lymphocytes	None available		
IH/S (Hurler-Scheie compound)	5–10% of cells contain vacuoles	Metachromatic inclusions in <5% of lymphocytes	None available		
2 (Hunter)	Occasional vacuoles	Metachromatic inclusions generally in <20% of lymphocytes	None available		
3 (Sanfilippo)	Occasional vacuoles	Metachromatic inclusions generally in >20% lymphocytes	None available		Type A most common in UK. Type C most common in Scandinavia
4 Morquio	Vacuoles present in significant number of lymphocytes	No metachromasia detectable	None available	Occasional, rare basophilic inclusions	
6 (Maroteaux-Lamy)	Vacuoles present in significant numbers of lymphocytes	Metachromatic inclusions present in lymphocytes	None available	Alder granulation present in all neutrophils. The granulation is metachromatic and shows birefringence (Haust & Landing method 1961) in contrast to toxic granulation	
7 β-glucuronidase deficiency			β-glucuronidase activity in lymphocytes and neutrophils is deficient	Alder granulation present	
Mucosulphatidosis			Sulphatases A, B and C. Not tested histochemically.	Alder granulation present	Sulphatides excreted in urine are metachromatic as in metachromatic leucodystrophy

even though no abnormality can be detected by light microscopy.

Neutrophils

Abnormalities of the neutrophil are less common in metabolic disorders but occasionally vacuolated neutrophils are found in carnitine deficiency. The contents of the vacuoles stain with oil red O. Lipid inclusions in neutrophils may also be present in some cases of Wolman's disease. Care should be taken to ensure that the acute toxic changes producing vacuolation of neutrophils are not mistaken for either of the above disorders.

Myeloperoxidase deficiency is readily detected by any of the methods for demonstration of peroxidase activity. Although neutrophils normally contain much peroxidase activity they are negative in

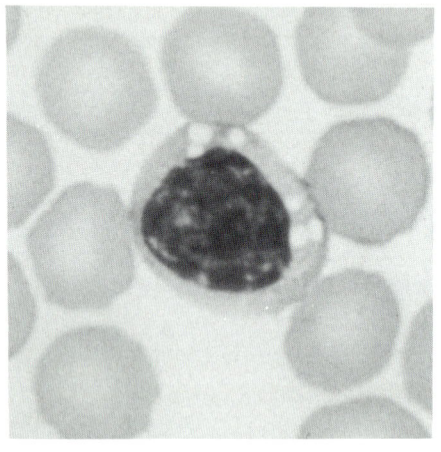
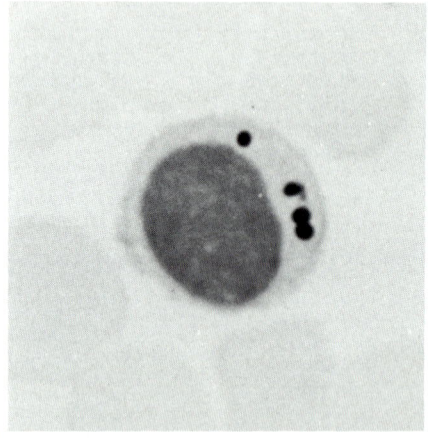

A   B

Fig. 18.1   Pompe's disease (glycogen storage disease type 2); lymphocytes in a blood film stained on the left with a routine haematological method (May-Grunwald-Giemsa) and on the right with the protected PAS method. The cytoplasmic vacuoles seen in the routine film are full of glycogen. × 2100

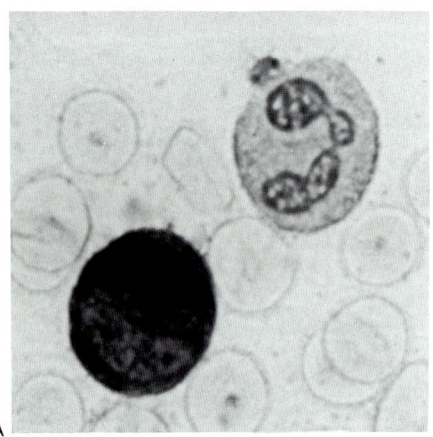
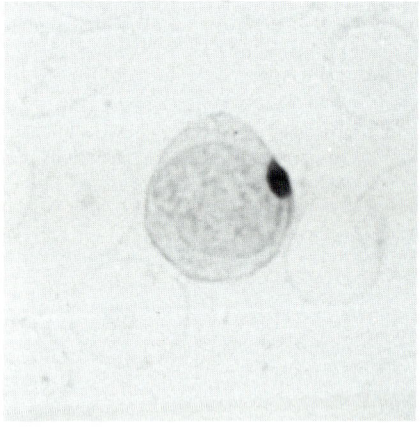

A   B

Fig. 18.2   (A) Normal blood film stained to show acid esterase activity. No E600 has been added and the non-specific esterase activity of monocytes serves as a control for the method. Most lymphocytes (B) show a strong single spot of activity which is not present in Wolman's disease. Monocytes, however are still strongly stained. × 1800

myeloperoxidase deficiency while eosinophils maintain their strong reaction.

In a screening programme (Kitahara et al 1981), one patient with a total deficiency was readily detected. However a large number of partially deficient patients were also found. Their partial deficiency was not clearly defined by staining methods but was apparent in quantitative assays.

BONE MARROW IN STORAGE DISORDERS

In storage disorders affecting the reticuloendothelial (mononuclear phagocyte, MP) system, examination of bone marrow films is an important diagnostic procedure. The differential diagnosis in an infant, young child or adult with hepatosplenomegaly will include storage disorders, infection or malignancy. Study of bone marrow films will exclude storage conditions if an adequate sample of marrow contains no storage cells. Films should be examined in preference to routine sections of wax-embedded trephine or clotted samples of bone marrow, because processing will extract the characteristic storage material and thus prevent an accurate diagnosis.

Although morphology of the affected histiocytes is best assessed in films stained by Giemsa or similar routine haematological methods, the exact nature of the storage cell can only be determined by further staining methods. It should be remembered that the presence of 'foam' cells in bone marrow is not necessarily indicative of a storage disorder since they may be acquired, as in the hyperlipidaemias and chronic myeloid leukaemia, or drug-induced.

In very general terms there are two types of 'foam' cell which may be found in bone marrow samples. The first is the Gaucher type of cell with its fibrillar appearance and the second is the Niemann-Pick type of cell which has foamy rather than fibrillar cytoplasm and is the type found in most storage disorders.

Gaucher's disease

The Gaucher cell is usually described as having cytoplasm 'like wrinkled tissue paper', and illustrations show this feature. However in bone marrow films and tissue sections the characteristic appearance is not seen in every Gaucher cell, and an extended search may be necessary before the 'typical' Gaucher cell is found. The majority of the cells show features ranging from the typical form to almost vacuolar. The glucocerebroside stored within the Gaucher cell gives a mild PAS positivity and stains

only pale grey with Sudan black. The fibrillar nature of the cell is best shown in marrow films which have been fixed before staining with Sudan black. The cells show strong acid phosphatase activity with β-glycerophosphate as substrate, and an acid phosphatase reaction will help to decide whether storage cells (or histiocytes) are present in a marrow film by examination with a low power objective (× 10) (Fig. 18.3).

Megakaryocytes show some acid phosphatase activity but this is often localised in one area of the cell and is finely punctate whereas histiocytes and storage cells show strong diffuse activity throughout the cell. Pseudo-Gaucher cells may occur in chronic myeloid leukaemia, thalassemia or in any condition where there is an overload on the MP system due to excessive turnover of blood cells. The numbers of pseudo-Gaucher cells are small and it should be possible to distinguish these 'acquired' storage conditions from Gaucher's disease by the range of morphology with some histiocytes still containing ingested WBCs.

All forms of Gaucher's disease have similar cells in the bone marrow and it is not possible to differentiate infantile from adult on morphology or staining characteristics. The histochemical method for the detection of β-glucosidase activity is not reliable for

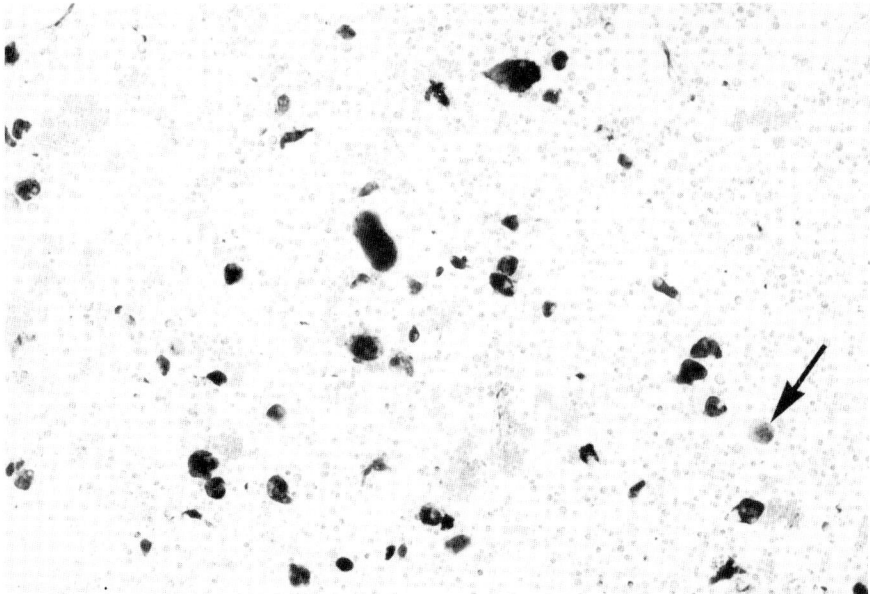

Fig. 18.3 Gaucher's disease; bone marrow film stained for acid phosphatase activity (Gomori). Low-power survey shows numerous storage cells with a variety of shapes and sizes. A megakaryocyte is present (arrow) with only low activity. A light nuclear counterstain has been added. × 90

the diagnosis of Gaucher's disease because, of the several isoenzymes of β-glucosidase, only one is deficient, and with the limitations of the capture reaction, differentiation of pH optima is not possible. It should also be noted that patients with the adult form of Gaucher's disease have a greater incidence of malignancy with myeloma, myeloid leukaemia and lymphoma occurring in 10% of those who died in one series of 200 (Lee 1981).

Niemann-Pick disease types A and B

In the infantile form of Niemann-Pick disease (type A) the marrow cells are foamy with the vacuoles being reasonably uniform in each cell. Only rarely are there any densely-staining inclusions (nuclear debris) or ingested RBCs. Sphingomyelin is stored together with cholesterol and a variable amount of ganglioside, and this combination gives rise to a rather pale blue staining with Sudan black which in most instances shows red birefringence in polarised light, due to sphingomyelin. The acid haematein reaction is not sufficiently reliable in marrow films to be of use in the diagnosis. Stains for cholesterol are also positive (Schultz type or PAN). The variable ganglioside content results in a variable PAS reaction. The foamy nature of the Niemann-Pick cells is readily appreciated in an acid phosphatase reaction when the vacuoles are outlined by the reaction product.

The juvenile and adult forms of the disease can present from 18 months of age onwards (past the age at which neurological involvement is grossly evident in the infantile form) with hepatosplenomegaly, abnormal pulmonary function and coagulation defects. In the older patients the most striking storage cell in the marrow is the 'sea-blue histiocyte' which is profuse in number and whose cytoplasm is filled with blue stained granules in routine haematological stains. The granules are autofluorescent, PAS-positive, stain grey-black with Sudan black and show acid phosphatase positivity. In a marrow with cells which match these criteria the most likely diagnosis is adult Niemann-Pick disease. Fewer cells, and those with sparse blue granules, represent the normally functioning histiocyte, perhaps in response to a variety of other conditions (Long et al 1977). In adult Niemann-Pick disease a few foamy storage cells may also be found.

Niemann-Pick disease Group C (Ophthalmoplegic lipidosis)

Crocker & Farber (1958) in their series of patients with Niemann-Pick disease included a number of patients from Nova Scotia and some with what has been called the sub-acute type of Niemann-Pick disease, later known as Niemann-Pick disease type D and type C respectively. The patients in these two subgroups do not show deficiency of sphingomyelinase, neither do they show accumulation of sphingomyelin in the liver or brain, both of which are grossly affected by storage. The clinical presentations vary widely from neonatal hepatitis to unexplained splenomegaly in an otherwise well child of up to 7 years. All develop ataxia, dementia and a vertical supranuclear ophthalmoplegia later in the disease and it is this latter phenomenon which has led to the term 'ophthalmoplegic lipidosis' (At a recent symposium in Prague 1982, it was decided to retain the term Niemann-Pick disease Group C, until the chemistry is clarified). The pathology is the same in all, the ultrastructure is distinct and different from other storage conditions (Neville et al 1973) and no basis for the disease has yet been identified.

Marrow films show an abundance of large foamy storage cells which may be multinucleate. There is variation in the sizes of the cytoplasmic vacuoles within the cell and it is common to find one or more small, densely-staining inclusions (probably nuclear debris) in the routine staining method. The cells do not stain with Sudan black but show PAS-positivity, the colour of which often seems to be a little redder than the colour of the other marrow cells. The acid phosphatase reaction is strong. In the older patients occasional, rare cells with numerous *pale* blue cytoplasmic granules may be found. These are *not* sea-blue histiocytes.

Even in the absence of neurological deterioration, neuronal storage can be demonstrated in rectal biopsies several years before the onset of dementia. In my experience of over 50 patients there has only been one who had the characteristic bone marrow cells but whose neurons were entirely normal. She is still neurologically normal after 10 years and we would expect her to remain so.

With this exception in mind it is important to be certain of neurological involvement and a suction rectal biopsy should be examined to confirm neuronal storage (see Ch. 7) before the gloomy prognosis is given.

G_{M1}-gangliosidosis

The storage cells present in G_{M1}-gangliosidosis are different in the two known types. In type 1 — presenting early and similarly to patients with Hurler's disease (MPS I) — the numerous storage cells are large and foamy. No inclusions are present and it is not possible to demonstrate any storage material within the vacuoles. The cells do however show acid phosphatase activity. In type 2 — presenting in the late infantile/juvenile age range but without features of a mucopolysaccharidosis and usually with hepatosplenomegaly — the storage cells are few in number. In the routine Giemsa stain the cells have a 'sky-blue' colour and resemble, to a limited extent, Gaucher cells in their morphology. They are PAS-positive, greyish with Sudan black and are positive for acid phosphatase activity. The activity of β-galactosidase is deficient in both types 1 and 2 and this deficiency is readily shown by the indoxyl method (Appendix 5).

Wolman's disease and cholesteryl ester storage disease

Both of these disorders are caused by a deficiency of

acid esterase activity and should be regarded as the infantile and juvenile/adult forms of the same disease. The foamy storage cells present in the marrow contain 'neutral fat' in the vacuoles and stain strongly with oil red O or Sudan black (Fig. 18.4). The neutral fat is a mixture of triglyceride and cholesteryl esters, and stains to demonstrate cholesterol (free or as ester) are also positive. Cain's Nile blue method imparts a deep blue/purple colour to these cells due to their content of free fatty-acids — though why free fatty acids originate in macrophages deficient in acid-esterase activity is not clear. No other storage disorder has cells with this characteristic staining reaction. The acid phosphatase reaction is positive.

Although acid esterase activity is easily demonstrated in blood films without the addition of the non-specific esterase inhibitor E600, its addition is necessary in marrow films because of the widespread distribution of many fatty acid ester hydrolases which would otherwise interfere. In the infantile and juvenile presentations the cells are numerous but in those presenting in the adult age range the storage cells are less common and not readily recognisable in the routine Giemsa stain. The cells may be large and angular and contain neutral fat, the fat vacuoles being diffusely scattered through the cytoplasm. The appearance may be difficult to differentiate from acquired neutral fat storage seen in the hyperlipidaemias.

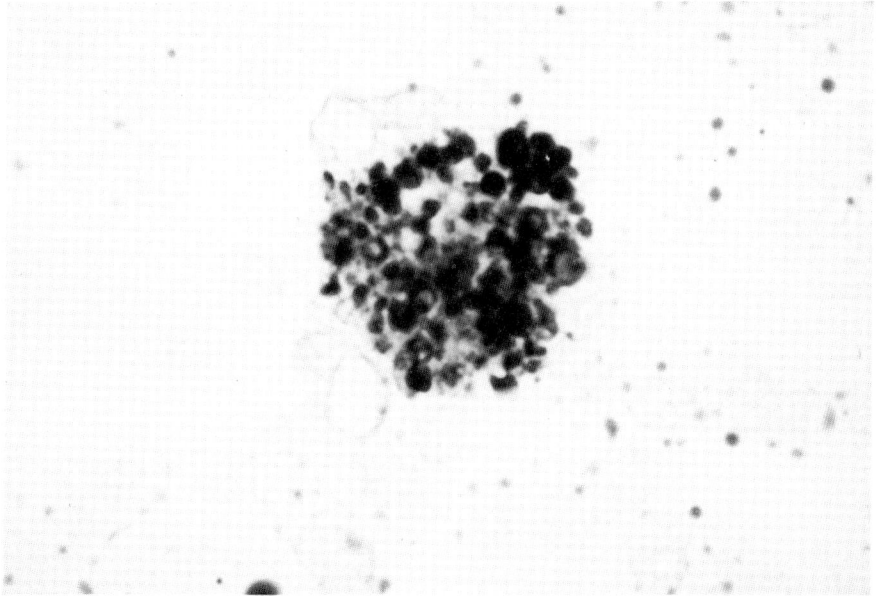

Fig. 18.4 Wolman's disease; bone marrow film stained for fat with oil red O. A large foamy cell is filled with neutral fat and stains strongly. A very similar appearance is found with Nile blue sulphate. ×1200

Cystinosis

The large macrophages containing the cystine crystals are quite fragile and in taking the aspirate and in making the films the cells can become ruptured leaving scattered cystine crystals over the slide. Thus the films should be made with great care. Although cystine is relatively insoluble, aqueous reagents should be avoided since the large volumes used relative to the small concentration of cystine will dissolve some or all of the crystals leaving at best a distorted crystal shape. In routine Giemsa-stained films scattered vacuolated macrophages may be found containing the occasional rounded crystal. The cystine crystals are adquately preserved in films fixed in ethanol, and stained for 5 min in 1% basic fuchsin in 70% ethanol before dehydration, clearing and mounting. Examination with polarised light will show the characteristic rectangular-shaped birefringent crystals. The hexagonal form (end on) shows no birefringence. The concentration of cystine in individual cells is insufficient for chemical confirmation and the Woollaston test (see p. 201) will be negative. A more sensitive but impermanent technique for detection of cystine in marrow aspirates is to place one drop of the aspirate on a slide, add a cover slip and view in polarised light. Many more macrophages can be seen with this technique. It should be borne in mind that dirt, dust and glass chips are also birefringent and the crystal habit should be carefully considered before making the diagnosis.

Mucopolysaccharidoses

Diagnosis of the mucopolysaccharidoses by examination of marrow aspirates is not a usual exercise since the diagnosis is more easily and accurately made on blood and urine samples. However in some cases, particularly in the Sanfilippo type (MPS 3) where the patients do not always have the characteristic physical features, a bone marrow aspirate may be taken in the investigation of hepatosplenomegaly. In routine Giemsa-stained preparations overt storage cells are not seen in MPS 3, but around the marrow fragments occasional histiocytes with scattered fine basophilic granules can be found. The metachromatic character of these granules (Plate 14) is best shown with the Haust and Landing method (Appendix 3) after fixation in tetrahydrofuran-acetone, which will show more histiocytes containing acid mucopolysaccharide than otherwise suspected. In Maroteaux-Lamy syndrome (MPS 6) all granulocytes show Alder granulation which is strongly metachromatic after staining with Haust and Landing's method. In addition to the Alder granulation, occasional large intensely basophilic and metachromatic cells (probably osteoclasts) can be found. Too few studies on the variation between the marrow changes in the various types of mucopolysaccharidoses have been made for diagnostic points to have emerged but it would be expected that all types would show histiocytes containing an acid mucopolysaccharide, and this is best shown with Haust and Landing's method (1961). The deficiency of β-glucuronidase in MPS 7 can be shown using naphthol AS-BI-β-glucuronide as substrate (Peterson et al 1981). Histochemical methods are not available for the detection of the defects in the other mucopolysaccharidoses.

Sialidosis I (cherry red-spot myoclonus syndrome)

These patients show no hepatosplenomegaly but storage cells are found in the bone marrow, albeit in small numbers. In Giemsa preparations the foamy storage cells have large vacuoles and the contents stain a blue-grey colour. The stored substance is strongly PAS-positive and negative with Sudan black. The periphery of the vacuoles is strongly positive for acid phosphatase activity. The cells are probably derived from plasma cells since the stored material (oligosaccharides with sialyl terminals) is very similar to those bodies erroneously attributed to Russell (Lendrum 1981).

Mannosidosis

The bone marrow in patients with mannosidosis contains numerous large foamy histiocytes with particularly well-defined vacuoles. Plasma cells also show marked, very well-defined vacuoles, the vacuoles being separated from each other by a rim of cytoplasm. The stored material is extremely water-soluble but with adequate protective measures (e.g.

celloidinisation) it should be PAS-positive. The patients present similarly to those with a mucopolysaccharidosis but a negative reaction with the toluidine blue method of Haust and Landing will distinguish mannosidosis from mucopolysaccharidosis. The histochemical method for demonstration of α-mannosidase activity is not sufficiently sensitive or of sufficient selectivity to distinguish between the various isoenzymes of α-mannosidase.

The sea-blue histiocyte syndrome

Considerable confusion has been caused by the use of this term since there is no single cause for the occurrence of histiocytes containing deep-blue granules. Profusion of these cells with their cytoplasm packed with granules suggests adult Niemann-Pick disease, while lesser numbers of this type of cell are found in lecithin-cholesterol acyltransferase (LCAT) deficiency. Smaller numbers, with fewer granules or diffusely blue granules, may be seen in thalassaemia, juvenile Batten's disease, chronic myeloid leukaemia or may represent busy histiocytes responding to an increased cell turnover. A careful search in most marrow samples will reveal an occasional histiocyte with some blue cytoplasmic granules, and these are of no real significance. Since there is no specificity, the term sea-blue histiocyte should not be used.

Acquired storage cells

In many situations where there is an excess of a particular substance, bone marrow aspirates will show loaded histiocytes. All forms of hyperlipidaemia, and Tangier disease, show large lipidladen histiocytes staining positively with oil red O or Sudan black. Foamy histiocytes may be induced in cytotoxic therapy or in treatment with drugs designed to decrease lipid biosynthesis.

Histiocytosis X

Large foamy histiocytes which contain no demonstrable substance in the vacuoles occur only late in histiocytosis X and at that time they may be abundant and give the impression of a storage disease.

The strong acid phosphatase reaction does not delineate the vacuoles but has a punctate localisation in the cytoplasm which helps to distinguish these cells from those of the storage disorders (see also p. 202).

SPLEEN AND LYMPH NODES IN STORAGE DISORDERS

Involvement of cells of the MP system results in evidence of the storage disorder in lymph nodes, spleen and other lymphoreticular structures. The morphological and staining characteristics of the storage cell are essentially the same whether they occur in the bone marrow, liver, spleen or lymph nodes. In some disorders, however (e.g. *mannosidosis, mucopolysaccharidosis type 4*), the evidence of storage may be minimal and missed in routine sections. If tonsils and adenoids are removed from patients with *mucopolysaccharidosis type 4 (Morquio)* not only is routine histology unhelpful but cryostat sections also do not show convincing mucopolysaccharide deposition. Thin (1 μm) sections of resin-embedded tissue may reveal a few foamy cells but electron microscopy is necessary for the detection of the storage cells and the membrane-bound empty vacuoles they contain. In *Hunter's disease (MPS type 2)* the adenoids show numerous foamy storage cells in the sub-epithelial areas and evidence of endothelial cell involvement is present throughout the tissue (Fig. 18.5).

In *Gaucher's disease* there is massive infiltration of storage cells in the spleen, which can weigh up to 3 kg in a child of under 10 years. The Gaucher cells, arising from the inability to degrade glucocerebroside derived from effete red and white blood cells, distort the architecture and the Malpighian follicles are not usually seen on the cut surface. The Gaucher cells are arranged characteristically in 'nests', the centres of which often appear empty and necrotic (Fig. 18.6).

Sphingomyelin and cholesterol accumulate in foamy storage cells in the spleen in a variety of conditions ranging non-specifically from *idiopathic thrombocytopenic purpura* (ITP) (Hill et al 1973) to specific deposition in *Niemann-Pick* disease. Some patients with *Niemann Pick disease Group C* also accumulate sphingomyelin and cholesterol in

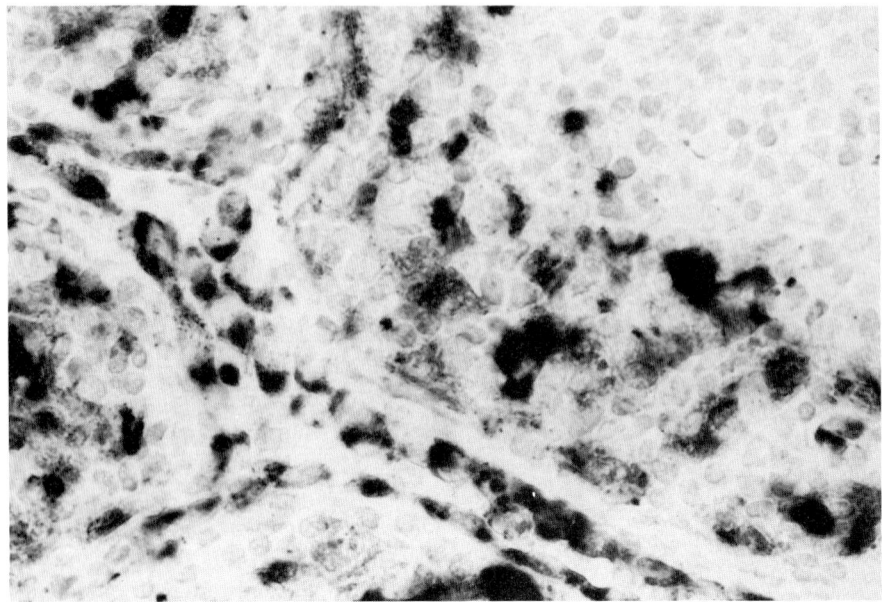

Fig. 18.5 Hunter's syndrome (MPS type 2); Cryostat section of adenoid stained to show acid phosphatase activity (Gomori). Edothelial cells, normally negative, are positive and numerous large foamy cells are readily identified. A light nuclear counterstain has been added. ×600

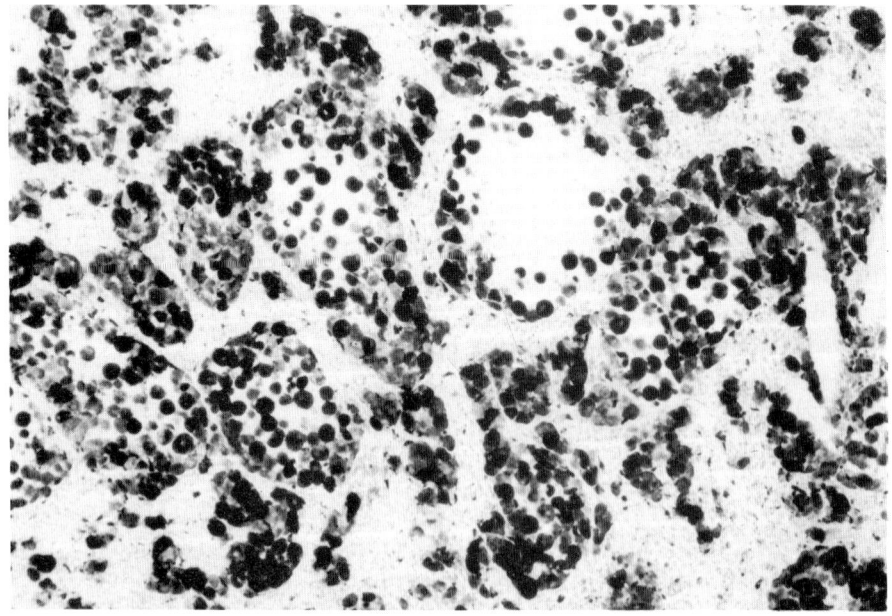

Fig. 18.6 Gaucher's disease; cryostat section of spleen stained for acid phosphatase activity (Gomori). Nests of strongly positive Gaucher cells are present, some of which have necrotic centres. A light nuclear counterstain has been added. × 90

numerous foamy cells in the spleen but although storage cells are present in the liver, the sphingomyelin content of that organ is not increased (Neville et al 1973). Thus spleen should not be the only tissue examined in the diagnosis of a visceral storage disorder.

In ITP the lipid-laden foamy cells are scattered and should not give rise to confusion with a genuine storage condition.

The spleen in *Niemann-Pick disease* (type A & B) is filled with foamy cells and there is loss of normal architecture. In contrast with conditions simulating Niemann-Pick disease, usually only those with sphingomyelinase deficiency show additional storage in the sinusoidal lining cells. This deposition which can be shown in frozen sections by Sudan black (Plate 15) and by its autofluorescence (Plate 16) has very similar properties to that of the 'sea-blue histiocytes' found in the marrow and spleen of patients with adult Niemann-Pick disease. The Niemann-Pick cells filling the spleen, lymph nodes, thymus and liver stain with Sudan black (usually giving red birefringence in polarised light) and with acid haematein, and show variable PAS-positivity associated with a variable ganglioside content. The associated cholesterol can be demonstrated with the Schultz or PAN methods.

Although the storage of sphingomyelin and cholesterol in foam cells in the spleen of some patients with Niemann-Pick disease Group C resembles that of Types A & B the lack of involvement of the sinusoidal lining cells may serve to differentiate the two conditions. Those patients with Group C who do not accumulate sphingomyelin in the spleen show massive infiltration of foamy cells which in frozen sections do not stain with Sudan black. The storage cells are PAS-positive (particularly in protected sections) and show rose-pink metachromasia with Feyrter's thionin. Their acid phosphatase activity is strong.

The precautions to prevent cystine from being removed from spleen and lymph nodes in *cystinosis* are the same as those described for liver and bone marrow, and cryostat sections are essential for the diagnosis and demonstration of cystine (Patrick & Lake 1968). In both spleen and lymph node the concentration of cystine is high enough for the Woollaston test to be positive (Fig. 18.7). The change of crystal shape in converting the free amino acid to the hydrochloride is characteristic of cystine.

OTHER CONDITIONS

Reactive lymph nodes and those involved in lymphoma and other histiocytic disorders may be difficult to interpret in routine sections and thin

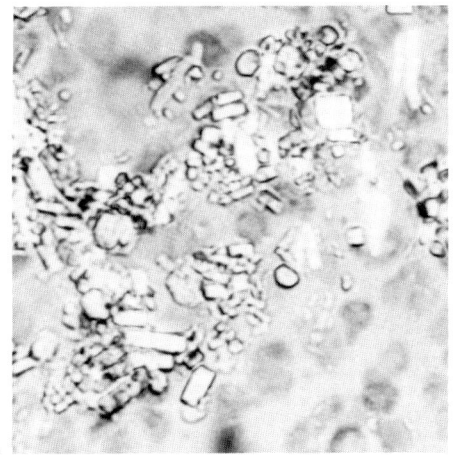 

A   B

Fig. 18.7 Cystinosis; cryostat section of spleen. A. The section has been stained with alcoholic basic fuchsin and photographed in partially polarised light. The characteristic crystal shapes are readily identified. × 825. A serial section (B) has been treated with concentrated hydrochloric acid by allowing the acid to be drawn, by capillary action under a cover-slip placed over the air-dried section, and photographed in fully polarised light. The change of crystal shape from rectangular and hexagonal to fan-shaped needles is specific to cystine and constitutes the Woollaston test. × 170

(1 μm) sections of resin-embedded tissue are usually preferred. Cryostat sections of snap-frozen tissue may be helpful in determining whether the infiltrating cells are histiocytic (acid phosphatase, antibodies to α_1-antichymotrypsin) or lymphocytic (surface membrane immunoglobulins, OKT3 monoclonal antibody). In lymphocytic proliferations it will be important to determine whether there is mono- or polyclonal immunoglobulin production (see Ch. 20).

It has been reported (Tubbs et al 1980) that the histiocytes of malignant histiocytosis are strongly positive to antibodies to lysozyme and this would be consistent with the rapid proliferation of histiocytic cells. Other histiocytic markers are necessary since lysozyme is a characteristic of cells of the granulocytic series and massive extramedullary haemopoiesis can appear 'histiocytic' if lysozyme alone is used as a marker.

Histiocytosis X (eosinophilic granuloma, Letterer-Siwe, Hand-Schüller-Christian disease)

The lesions in this spectrum of disorders have been classified as benign or malignant (Bokkerink & deVaan 1980) depending on the degree of infiltration and types of cell present. Malignant lesions are characterised by infiltration of large rounded histiocytes with relative preservation of the architecture of the organ. Benign lesions show variable degrees of infiltration by a mixture of histiocytes, eosinophils and lymphocytes; giant cells are common and there is loss of normal architecture of the tissue.

In the skin the infiltrating histiocytes appear in the upper dermis. Among the rounded histiocytes occasional irregular-shaped stellate cells may be found. In lymph nodes both rounded and stellate cells are present. These cells and the multinucleate giant cells have slightly different staining characteristics which are best determined in cryostat sections of snap-frozen tissue (Elleder 1982; Thomas et al 1982) (Fig. 18.8). The staining properties are summarised in Table 18.2. The large rounded histiocytes have properties and ultrastructural features (Birbeck granules) which relate them to the Langerhans cell normally found in the epidermis. α-mannosidase activity, demonstrated by the semipermeable membrane technique (Elleder et al 1977, Lake 1982) is present in both Langerhans cells of the epidermis and in the large rounded histiocytes, and this seems the most effective marker for confirmation of the

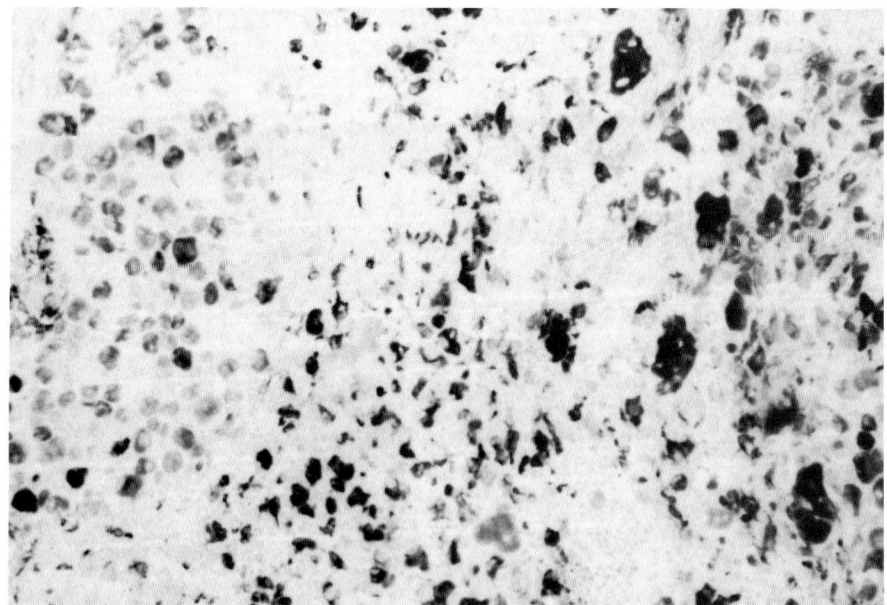

Fig. 18.8 Histiocytosis X; cryostat section of lymph node stained for non-specific esterase activity. Groups of rounded histiocytes (left), angular stellate histiocytes (centre) and large multinucleate histiocytes (right), with varying intensities of reaction are evident. × 140

Table 18.2 Staining reactions of the histiocytes in Histiocytosis X in comparison with Langerhans cells. HTA-1 antigen can be demonstrated with monoclonal antibodies OKT6 or NA1/34

Cell	Reaction: Acid Phosphatase	Non-specific esterase	α-mannosidase	ATPase	Naphthyl-amidase	Ia-like antigen	HTA-1 antigen
Large rounded histiocytes	+	+	+ + +	+/+ +	+	+	+
Irregular stellate histiocytes	+ +	+ +	−	+	+	+	+
Giant multinucleate cells	+/+ + +	+ + +	−,+		±	±	−
Langerhans cells	−	−	+ + +	+ +	+	+	+

diagnosis. The giant cells contain only residual amounts of demonstrable lysozyme which appears to be derived from ingested granulocytes or histiocytes (Figs 18.9, 18.10).

The large lipid-laden histiocytes sometimes seen contain cholesteryl esters, and will stain positively with oil red O and with the Schultz or PAN methods for cholesterol.

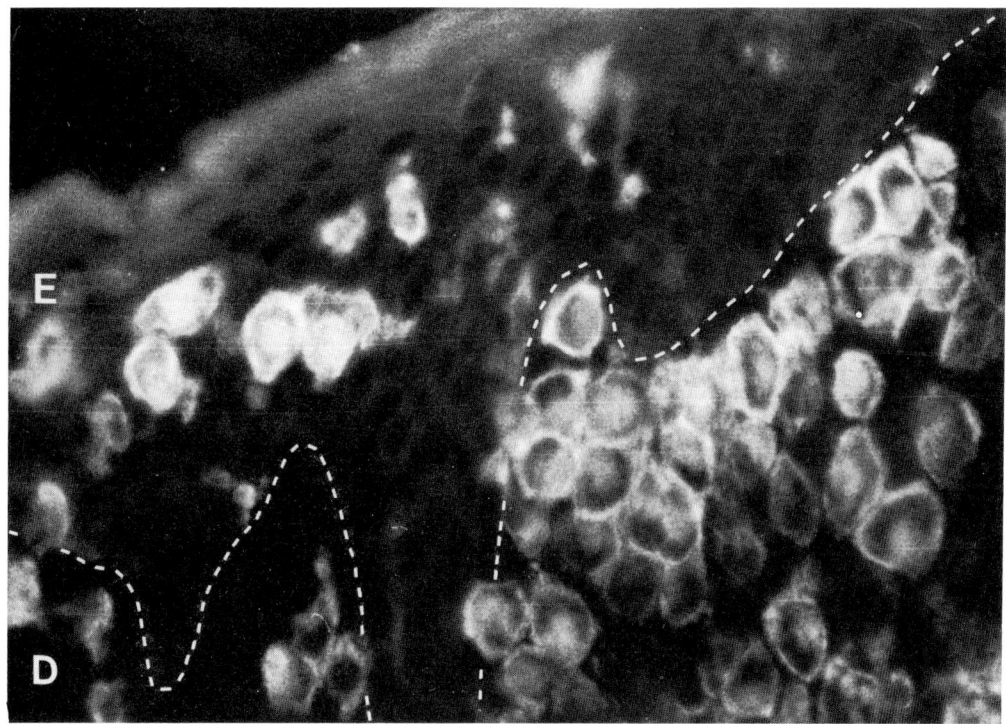

Fig. 18.9 Histiocytosis X; skin biopsy, 6 μm cryostat section, indirect IF method. Fluorescein-labelled monoclonal antibody to human thymocyte antigen (HTA–1) reacts with large round or ovoid cells infiltratiang the dermis (D) and epidermis (E). In normal skin, epidermal Langerhans cells have the HTA–1$^+$, HLA–DR$^+$ phenotype but morphologically similar cells in the normal dermis are HTA–1$^-$. (Photograph by courtesy of Dr J. A. Thomas and Dr G. Janossy)

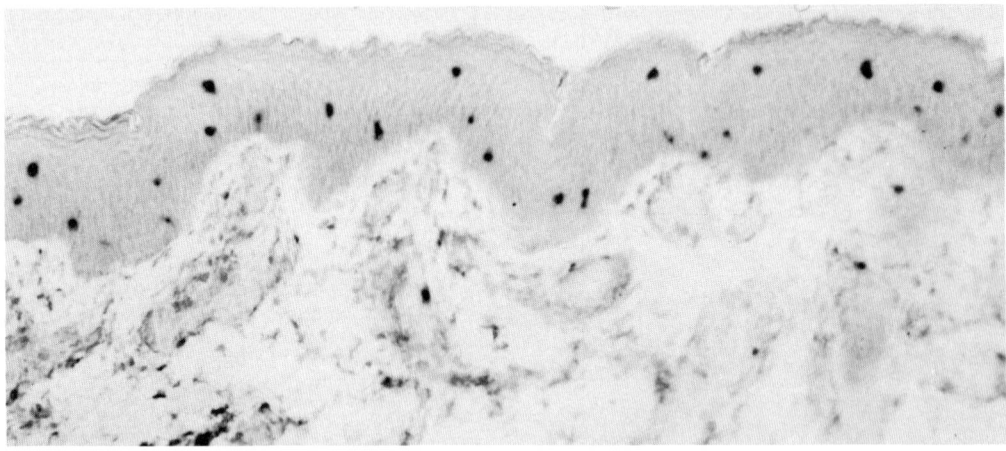

Fig. 18.10A   Skin biopsy. Cryostat section stained for α-mannosidase activity using the semipermeable membrane technique. Langerhans cells in the epidermis are strongly stained. Mast cells, around the vessels in the upper dermis show weaker diffuse activity in this case of mastocytosis. A light nuclear counterstain has been added. × 180

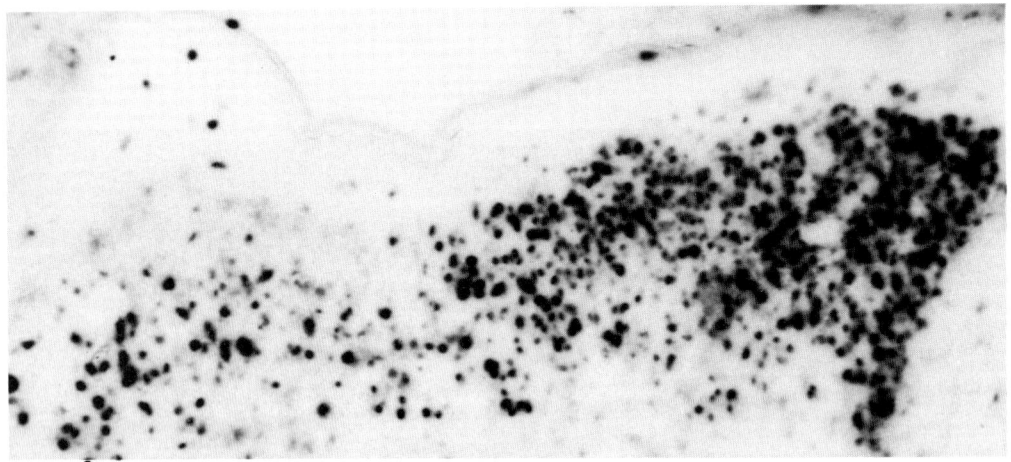

Fig. 18.10B   Histiocytosis X. Cryostat sections of a skin biopsy stained for α-mannosidase activity using the semipermeable membrane technique. The mass of infiltrating histiocytes just under the epidermis show very strong activity. Langerhans cells are markedly reduced in number. × 180

REFERENCES

Bokkerink J P M, deVaan G A M 1980 Histocytosis X. European Journal of Pediatrics 135: 129–146

Crocker A C, Farber S 1958 Niemann-Pick disease: a review of 18 patients. Medicine 37: 1–95

Elleder M, Povysil C, Rozkovcová J, Cihula J 1977 Alpha-D-mannosidase activity in histiocytosis X. Virchows Archiv B. Cell Path 26: 139

Elleder M 1982 Enzyme histochemistry in histiocytosis X. Proceedings of the Royal Microscopical Society 17: 132

Hansen H G 1972 Hematologic studies in mucopolysaccharidoses and mucolipidoses. Birth Defects 8: 15–127

Haust M D, Landing B H 1961 Histochemical studies in Hurler's disease. A new method for localization of acid mucopolysaccharide, and an analysis of lead acetate 'fixation'. Journal of Histochemistry and Cytochemistry 9: 79–86

Hill J M, Speer R J, Gedikoglu A 1963 Secondary lipidosis of the spleen associated with thrombocytopenia and other blood dyscrasias treated with steroids. American Journal of Clinical Pathology 39: 607–615

Kitahara M, Eyre H J, Simonian Y, Atkin C L, Hasstedt S J 1981 Hereditary myeloperoxidase deficiency. Blood 57: 888–893

Lake B D 1971 Wolman's disease. Histochemical detection of the enzyme deficiency in blood films. Journal of Clinical Pathology 24: 617–620

Lake B D 1974 An improved method for the detection of β-galactosidase activity, and its application to G_{M1}-gangliosidosis and mucopolysaccharidosis. Histochemical Journal 6: 211-218

Lake B D 1981a Metabolic disorders. General considerations. In: Berry C L (ed) Paediatric pathology. Springer-Verlag, Berlin, ch 14

Lake B D 1981b Blood and bone marrow biopsy as an aid to diagnosis in the cerebral lipidoses. In: Rose F C (ed) Metabolic disorders of the nervous system. Pitman, London, ch 13

Lake B D 1982 Semipermeable membrane techniques. Histochemical Journal 14: 697

Lake B D, Milla P J, Taylor D S I, Young E P 1982 A mild variant of mucolipidosis type 4 (ML4). In: Berman E, Merin S, Maumenee I (eds) Genetics in ophthalmology. A R Liss, New York

Lee R E 1981 The high incidence of malignant tumours in adults with Gaucher's disease. Laboratory Investigation 44: 37A (abstract)

Lendrum A C 1981 Misapplication of Russell's name (letter). Journal of Clinical Pathology 34: 689

Long R G, Lake B D, Pettit J E, Scheuer P J, Sherlock S 1977 Adult Niemann-Pick disease. Its relationship to the syndrome of the sea-blue histiocyte. American Journal of Medicine 62: 627-635

Lüllmann H, Lüllmann-Rauch R, Wassermann O 1973 Drug-induced phospholipidosis. German Medical Monthly 3: 128-135 (Translated from Deutsche Medizinische Wochenschrift 98: 1616-1623)

Muir H, Mittwoch V, Bitter T 1963 The diagnostic value of isolated urinary mucopolysaccharides and of lymphocytic inclusions in gargoylism. Archives of Disease in Childhood 38: 358-363

Neville B G R, Lake B D, Stephens R, Sanders M 1973 A neurovisceral storage disease with vertical supranuclear ophthalmoplegia and its relationship with Niemann-Pick's disease. A report of nine patients. Brain 96: 97-120

Patrick A D, Lake B D 1968 Cystinosis: electron microscopic evidence of lysosomal storage of cystine in lymph nodes. Journal of Clinical Pathology 21: 571-575

Peterson L, Nelson A, Parkin J 1981 Mucopolysaccharidosis type VII. A morphologic, cytochemical and ultrastructural study of the blood and bone marrow. Laboratory Investigation 44: 5p (abstract)

Thomas J A, Janossy G, Chilsoi M, Pritchard J, Pincott J R 1982 Combined immunological and histochemical analysis of skin and lymph node lesions in histiocytosis X. Journal of Clinical Pathology 35: 327-337

Tubbs R R, Sheibani K, Sebek B A, Savage R A 1980 Malignant histiocytosis. Ultrastructural and immunocytochemical characterization. Archives of Pathology and Laboratory Medicine 104: 26-29

19

The leukaemias

H. Smith

The leukaemias may be regarded as monoclonal accumulations of haemopoietic cells. Conventional Romanowsky staining of films of blood and marrow is the most convenient means of recognising abnormal accumulation, and in some leukaemias, myeloid especially, may be sufficient in itself for detection of heterogeneity and characterisation of the leukaemia. In other leukaemias, lymphoid especially, heterogeneity not evident, or not convincingly evident, in Romanowsky stained films is revealed by other characteristics such as surface markers, cytogenetics, cytochemistry, ultrastructure and response to specific treatment schedules. Thus precision in identification is increased by the use of information obtained from a variety of procedures.

This chapter considers the light microscopic cytochemistry of normal and leukaemic populations of haemopoietic cells (lymphomas and myelomas are considered in Ch. 20; the value for cell identification of the ultrastructural demonstration of enzymes is considered by Breton-Gorius et al 1981). Emphasis is given to those procedures which, by adding significantly to the information obtainable from Romanowsky staining, are of proven value in diagnosis. The chemistry of most of the procedures considered here is discussed in Hayhoe & Quaglino (1980).

INTERPRETATION OF RESULTS

In the interpretation of results of cytochemical procedures, the following points deserve note:

1. The mode of preparation of the tissue affects results. For uniformity of results, and the most faithful indication of conditions in vivo, the use of films made directly from blood or marrow aspirate without anticoagulant is recommended; and the results given here apply, unless otherwise stated, to material prepared in this way. In cytocentrifuged preparations, the distribution and apparent content of cell components are often altered, and caution is needed in their interpretation.

2. The age of the material affects results. As a general rule, the fresher this is, the more consistent the results. While some components such as glycogen retain stability for weeks, the optimal demonstration of others such as acid mucopolysaccharides requires treatment of the material within hours.

3. The reactions should be read as soon as practicable after preparation.

4. The method used should not impede recognition of the leukaemic population. In the methods described here, the leukaemic population is readily identifiable if the observer is familiar with its morphology. As an aid to identification, differential counts may be needed on both the Romanowsky and cytochemical preparation.

5. While semi-quantitative scoring methods are of proven value for some procedures (e.g. neutrophil alkaline phosphatase), for most, the disposition of reaction product is as significant as, and often more significant than its intensity. The description of reactions therefore should note not only strength but also morphology of positive areas (diffuse or granular), distribution in the cell, and proportion of cells with positive reaction. Examination of a minimum of 100 cells is recommended.

6. The proportion of leukaemic cells with cytochemistry regarded as typical for that leukaemia varies from patient to patient, and from time to time in the one patient; variability may be a reflection of differences in intrinsic characteristics, as well as in metabolic activity or degree of 'stimulation' or

Table 19.1 Cytochemistry of normal haemopoietic cells

	Myeloperoxidase	SBB	PAS	Acid phosphatase	ANAE	CAE	β-glucuronidase	Neutral red[1]
lymphocytes	neg	neg	10–40% with fine to coarse granules; rarely heavy clumps; T and B cells similar	T: majority pos, paranuclear granules; B: variable proportion pos, weak scattered granules	T: dense dots and blocks; B: majority neg; fluoride-resistant	neg	T: paranuclear granules and blocks; B: 10–80% pos, scattered granules	0 or a few fine to medium scattered granules
plasma cells	neg	neg	neg to faint diffuse	intense	moderate	neg	strong diffuse + granular	
monocytes	50–100% with scattered granules or aggregates; immature > mature	most pos; fine to coarse scattered granules; immature > mature	70–100% with fine to coarse granules on diffuse background	weak to moderate, diffuse + coarse deposits	75–100% pos, strong diffuse + granular, often coarse aggregates; fluoride-sensitive[2]	scattered granules in occasional cells[2]	85–100%, weak diffuse, occasional granules	numerous chunks scattered in otherwise clear cytoplasm
marrow macrophages	neg	neg	neg to mild diffuse	intense, diffuse, often with granules	intense diffuse + granular	neg	intense diffuse + granular	
neutrophils	pos from promyelocyte on; usually packed; green-black	pos from promyelocyte on; strong localized + heavy overall	granular + diffuse; mature > immature; faint diffuse + occasional granules in myeloblasts	neg to faint, predominately diffuse, immature in occasional myelocytes[2] > mature	few scattered granules	intense granular + diffuse, from promyelocyte on[2]	faint to mod, diffuse from promyelocyte on	numerous fine granules with occasional coarse aggregates

Table 19.1   *(contd)*

	Myeloperoxidase	SBB	PAS	Acid phosphatase	ANAE	CAE	β-glucuronidase	Neutral red[1]
eosinophils	pos from promyelocyte on; core of specific granules neg	pos from promyelocyte on; core of specific granules neg	diffuse pos in intergranular cytoplasm; some myelocyte granules pos	mild to mod, predominately diffuse	occasional granules in rare myelocytes	neg	faint to moderate diffuse + granular	
basophils	neg or minority granules pos; green-black	most neg; some granules may have red metachromasia	intense coarse granules on neg or patchy diffuse background	mod-intense diffuse often + coarse granules	neg	neg	mild-moderate diffuse	
mast cells[3]	neg	neg	mild to moderate intense diffuse	intense	neg	intense	intense	
erythroblasts	neg	neg	neg	coarse paranuclear granules	neg	neg	weak diffuse in earlier cells	
megakaryocytes	neg	scattered fine granules in occasional cells	diffuse + granular often + large dense deposits	intense diffuse, often + coarse deposits	strong predominantly diffuse	neg or faint diffuse	weak diffuse	
platelets	neg	neg	strong	strong	strong	neg	neg to faint	

[1] Ficoll-treated suspension, Collins (1981); [2] Some monocytes and myelocytes may stain for ANAE + CAE in the dual esterase procedure; [3] Intense purple metachromasia with toluidine blue

SBB = sudan black B; PAS = periodic acid-Schiff; ANAE = α-naphthyl acetate esterase; CAE = chloroacetate esterase; neg = negative; pos = positive

Table 19.2 Cytochemistry of lymphocytic leukaemias: blastic (acute)[1]

	non-T non-B[2]	T	B
myeloperoxidase	neg	neg	neg
SBB	neg	neg	neg
acid phosphatase	weak-mod scattered granules in about $\frac{1}{2}$ pts; T pattern in minority; rest neg	mod-strong coarse granules as paranuclear clumps in about $\frac{3}{4}$ pts; rest: scattered granules	neg or weak sparse granules or patches in occasional cells
β-glucuronidase	majority pts: fine to coarse scattered granules often + diffuse patches; T pattern in about 20% pts	mod-strong coarse granules, paranuclear in about 60% pts; rest: scattered granules, often + diffuse patches	neg or sparse granules
PAS	strong coarse granules + blocks on neg background in at least some and often many cells; occasional pts: fine scattered granules; neg in all cells in < 5% pts	as for non-T non-B	neg
ANAE	neg or weak fine granules scattered; fluoride-resistant	coarse granules, often to one side of nucleus; usually minority cells only; fluoride-resistant	neg
CAE	neg	neg	neg
ORO	coarse droplets in occasional pts		some vacuoles pos
neutral red	see Table 19.1		
Niagara sky blue 6B (Kass 1980)	absent blue, purple or turquoise granules		

inclusions: round to oval, to approx. 7 μm size, staining (Romanowsky) orange-pink of varying intensity from faint to brick-red; proportion pos for PAS (Fig. 19.2), acid phosphatase, β-glucuronidase and toluidine blue metachromasia; ultrastructure: membrane-enclosed collections of 40–100 nm particles (Smith 1978); for discussion of other types of inclusion see Yanagihara et al (1980)

short list of preferences: 1. myeloperoxidase or SBB; 2. neutral red; 3. PAS; 4. acid phosphatase

[1] On FAB criteria, acute leukaemias that appear undifferentiated on Romanowsky stain are classified as lymphocytic if < 3% of blast cells are positive for myeloperoxidase or SBB; however, a proportion of leukaemias of this type have neutral red uptake and surface markers indicative of myeloid differentiation; the cytochemistry of these is variable, some having lymphoid rather than myeloid features

[2] Lymphoblastic crises of CML are almost invariably non-T non-B

SBB = sudan black B; PAS = periodic acid-Schiff; ANAE = α-naphthyl acetate esterase; CAE = chloroacetate esterase; ORO = oil red O; mod = moderate; pts = patients; neg = negative; pos = positive

maturity of the cells (Hayhoe & Quaglino 1980, Cawley & Burns 1980).

7. The more procedures used, the more likely is a correct identification to be made; rarely will leukaemic cells be negative for all of a series of procedures (Glick et al 1980).

8. The cytochemistry of a leukaemic population cannot be predicted from the characteristics of its putative normal counterpart. Cytochemical characteristics of haemopoietic cells in the normal are summarised in Table 19.1. The heterogeneity of lymphoid cells is understated (see Cawley & Burns 1980), and only those subtypes which are accepted as having distinctive cytochemistry and which may have leukaemic counterparts have been noted. The tabulation also omits a distinctive minor population of lymphocytes (Smith & Collins 1977) which may have a chronic leukaemic counterpart.

9. The interpretation of cytochemistry may conflict with identification based on other procedures; for example, in occasional blastic leukaemias, the morphology in the Romanowsky stain is L2, the

ANAE strong, diffuse and fluoride-sensitive ('monocytic'), and the karyotype Philadelphia-positive (not described in monocytic leukaemia). In the acute lymphocytic leukaemias especially, the application of immunologic markers is revealing heterogeneity in lineage commitment, and confirming the long-held suspicion that not all cells with the morphology and cytochemistry of lymphoblasts are necessarily committed to lymphocyte lineage; some are almost certainly erythroblastic or megakaryoblastic in nature.

10. Cytochemistry is not of value in distinguishing true leukaemias from transient, pseudo-leukaemic proliferations of immature leucocytes, such as the myeloblastosis of some Down's syndrome neonates, and the promyelocytic hyperplasia or lymphoblastosis of some virus infections.

CYTOCHEMICAL CHARACTERISTICS OF THE MORE COMMON LEUKAEMIAS

The cytochemical characteristics of the more common leukaemias are summarised in Tables 19.2

to 19.5, and some are illustrated in Figures 19.1–19.6. Characteristics have been given only for the visible expanded population, and not for accompanying groups, which, though they might appear normal on Romanowsky stain, may have abnormal cytochemistry, e.g. deficiency of myeloperoxidase and glycogen in mature neutrophils in myeloblastic leukaemias (Catovsky et al 1972). The more common inclusions detectable by light microscopy in leukaemic cells are also considered; and, for each major group of leukaemias, a short list of preferred procedures is suggested as a guide for those occasions when material is scarce on account of difficulty in aspiration or other circumstances (slides may also be split lengthwise for separate procedures). Gaps in tables indicate that we have no useful experience of those procedures.

Lymphocytic leukaemias (Tables 19.2, 19.3, Figs 19.1–19.3)

Acute leukaemias have been identified as lymphocytic in accordance with French-American-British criteria ('FAB', Bennett et al 1976), i.e. absence of myeloid

Table 19.3 Cytochemistry of lymphocytic leukaemias: chronic

	CLL	hairy cell	prolymphocytic
myeloperoxidase	neg	neg	neg
SBB	neg	neg	neg
acid phosphatase	most pts neg; in some pts minority cells pos; T type: strong in most cells	most pts intense, tartrate-resistant; occas. pts: weak or neg or partly or completely inhibited by tartrate	neg to weak diffuse or granular
β-glucuronidase	most pts: scattered granules; rare pts neg; T type: strong in most cells, single block or scattered small granules	neg	neg or scattered granules
PAS	neg in most pts	strong diffuse + granular in the one cell; often stronger at periphery	variable, often neg
ANAE	neg in most pts; in minority, scattered granules or hairy cell pattern	faint diffuse often in crescent; NaF-resistant	B type: most pts neg; T type: single prominent granule in most cells
CAE	neg	neg	neg
neutral red	'lymphoid' (Table 19.1)		'lymphoid' (Table 19.1)
alkaline phosphatase	pos in some apparent CLL		

inclusions: in some apparent CLL (more likely immunocytic lymphoma), rod-shaped crystals of monoclonal immunoglobulin in occas. cells; do not stain with Romanowsky, or for myeloperoxidase, SBB, PAS or phosphatases

short list of preferences: cytochemistry important in hairy cell leukaemia, but of no great diagnostic value in CLL or prolymphocytic leukaemia; preferences: 1. acid phosphatase; 2. peroxidase; 3. ANAE

SBB = sudan black B; PAS = periodic acid-Schiff; ANAE = α-naphthyl acetate esterase; CAE = chloroacetate esterase; CLL = chronic lymphocytic leukaemia; pts = patients; neg = negative; pos = positive

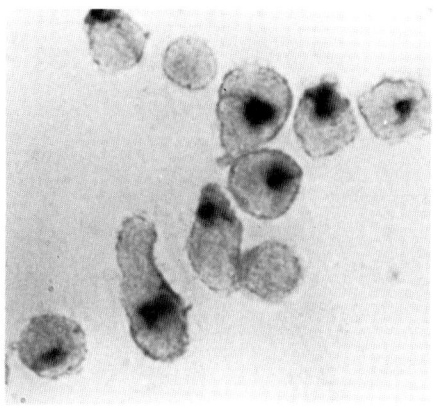
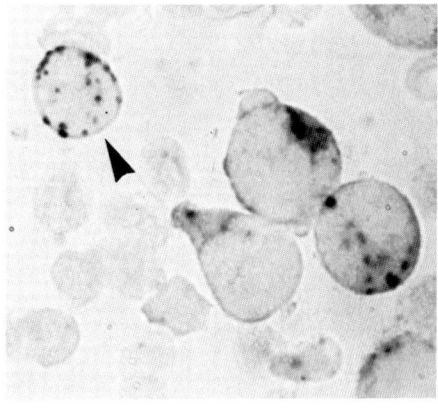

A   B

Fig. 19.1 Acid phosphatase (A) and β-glucuronidase (B) in bone marrow lymphoblasts from a 2-year-old girl with receptor-silent (sheep erythrocytes, HTA–1, Ia, Ig) ALL and mediastinal enlargement. The acid phosphatase has a compact paranuclear ('T' type) distribution; the β-glucuronidase has a compact paranuclear distribution in only one of the group of three lymphoblasts; arrow shows normal lymphocyte. × 1300

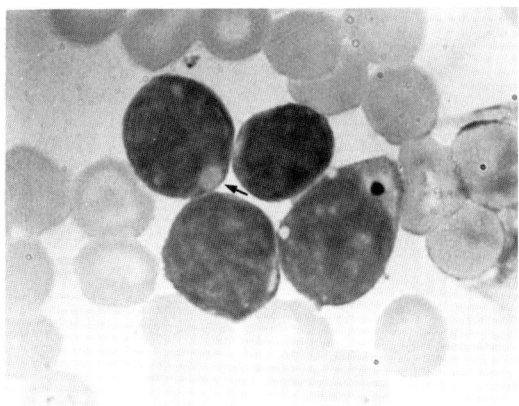
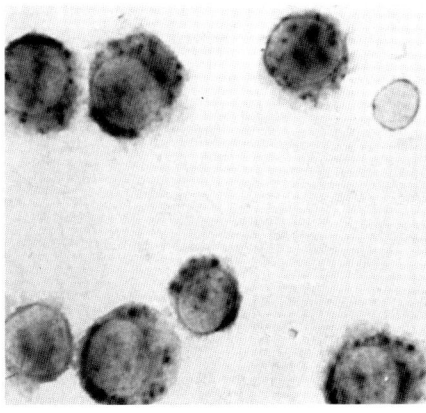

Fig. 19.2 PAS stain, bone marrow, Ph' positive ALL, receptor silent except for Ia. One lymphoblast contains a strongly positive block, another (arrow) a spherical inclusion staining with moderate intensity (compare with clear vacuoles in cell at right). × 1300

Fig. 19.3 Acid phosphatase, bone marrow, hairy cell leukaemia. The field contains six hairy cells with strong activity which was tartrate-resistant, and a normal lymphocyte. × 1000

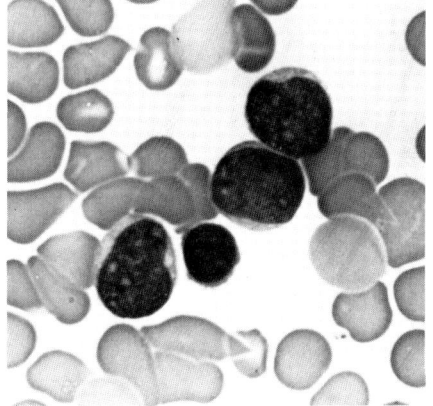
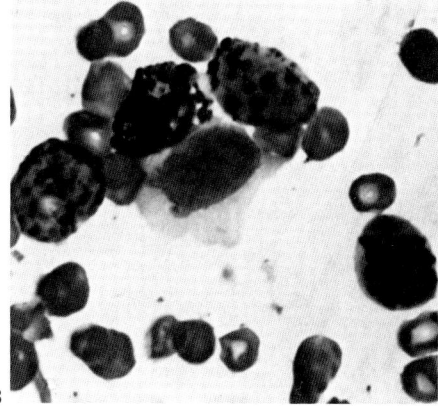

A   B

Fig. 19.4 May-Grünwald-Giemsa (A) and myeloperoxidase (B) stains, bone marrow, AML M1. The myeloperoxidase stain is strongly positive even though no granulation was evident in the Romanowsky stain. The field in both A and B contains a lymphoid cell. × 1000

Table 19.4 Cytochemistry of acute myeloid leukaemias

	M1[1,2]	M2	M3	M4 (monocyte) component	M5	M6 (erythroblast) component
myeloperoxidase	pos in 3 or more % blasts	pos	strong pos	variable % cells pos, usually minority; fine granules	neg or variable numbers of cells pos, fine scattered granules, % lower in type 5a	neg
SBB	variable % cells pos,	pos	strong pos	neg or fine scattered granules	neg or fine granules; % pos cells lower in type 5a	neg
acid phosphatase	most pts neg or faint diffuse; occasional. pts, super-imposed strong scattered granules	neg or faint diffuse	neg or faint diffuse	majority cells pos, medium to strong	mild-mod diffuse, often stronger paranuclear; in occasional. pts, scattered granules	strong paranuclear unipolar
β-glucuronidase	mild-mod diffuse, often + fine granules	neg to weak diffuse	weak diffuse, often + granules	faint to strong often + fine granules	faint-mod diffuse; may be stronger paranuclear	
PAS	neg or faint diffuse; + fine granules in minority of cells	mild diffuse often + fine granules	neg or faint diffuse	usually neg; in occasional pts fine to coarse granules on diffuse background	usually neg; in some pts mild diffuse, often + fine or rarely coarse granules	usually strong; granular or diffuse
ANAE	often pos, fluoride-resistant	often pos, fluoride-resistant	rare pts pos	fluoride-sensitive	usually strong; fluoride-sensitive	strong
CAE	neg or faint in minority of cells	mild to mod	intense packed granular	usually neg	neg or faint pos in minority of cells.	pos in rare cells
dual esterase				cells pos for ANAE or CAE; in some pts some cells pos for both	occasional cells pos for both	
neutral red	'myeloid' pattern (see Table 19.1)			'monocytic' pattern (see Table 19.1)		

Niagara sky blue purple granules in leukaemic myeloblasts and monocytes 6B (Kass 1980)

inclusions: 1. Auer rods mainly in M 1–3; pos myeloperoxidase, SBB, acid phosphatase; PAS: weak-mod pos; CAE pos in some cases, neg for ANAE, β-glucuronidase
2. Charcot-Leyden-like crystals in myeloblasts, granulocytes and marrow macrophages in M1, 2, 3; neg for peroxidase, SBB, PAS, acid phosphatase, CAE, ANAE, β-glucuronidase

short list of preferences: 1. peroxidase or SBB; 2. neutral red; 3. PAS

[1] On FAB criteria, blastic leukaemias with evidence of myeloid maturation on Romanowsky stain, but negative for myeloperoxidase or SBB, are classified as lymphocytic (L2). However, in a proportion, other cytochemical characteristics most closely resemble those of M 1–M 3 leukaemias
[2] Blastic crises of CML may be lymphoblastic or myeloblastic

SBB = sudan black B; PAS = periodic acid-Schiff ANAE = α-naphthyl acetate esterase; CAE = chloroacetate esterase; pts = patients; neg = negative; pos = positive; mod = moderate; CML = chronic myeloid leukaemia

differentiation in Romanowsky stained films, and the presence in < 3% of blast cells of myeloperoxidase activity. However, in the subdivision of these leukaemias, markers have been used in preference to the FAB L1-3 types, since markers appear to have greater precision in characterisation of cell phenotype and in predicting optimal treatment. As stated above, 'lymphoblastic' leukaemias as defined by FAB criteria include cases which markers show are not committed to lymphocyte differentiation; these may have typical 'lymphoid' cytochemistry, e.g. PAS, in addition to appearing lymphoid in Romanowsky morphology and in the absence of myeloperoxidase.

Three types of chronic leukaemia are tabulated, the usage of 'prolymphocytic' comforming to that of Galton et al (1974). The difference in composition of the short lists of preferred procedures for acute v. chronic leukaemias is due to the fact that identification as lymphocytic is more of a problem in acute than in chronic leukaemias.

Myeloid leukaemias (Tables 19.4, 19.5, Figs 19.4–19.6)

The acute leukaemias have been classified by FAB

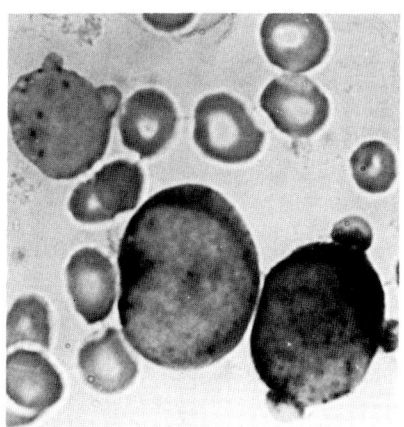

Fig. 19.5 β-glucuronidase, bone marrow, AML M1. Two myeloblasts have a diffuse and granular reaction. A lymphocyte contains dots on a negative background. × 1300

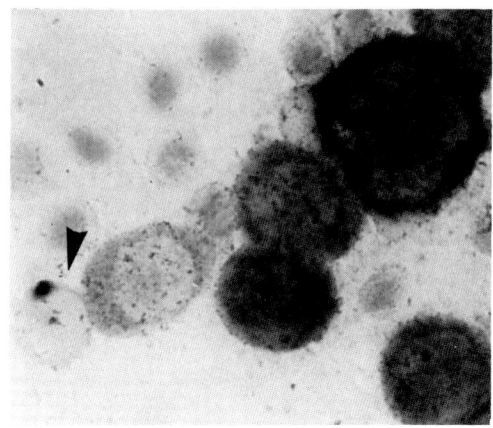

Fig. 19.6 α-naphthyl acetate esterase, bone marrow, AML M5. The field contains four monocytes with positive reaction which was fluoride-sensitive, and a lymphocyte (arrow) with a single block. × 1000

Table 19.5 Cytochemistry of chronic myeloid leukaemias

Most procedures give no additional diagnostic information. Procedures of value or giving unusual results are:

In CML:
1. NAP: low values; also in Philadelphia-negative juvenile CML, paroxysmal nocturnal haemoglobinuria, viral infections, hereditary hypophosphatasia, idiopathic thrombocytopenia, and often in AML
2. myeloperoxidase, SBB, CAE may be faint in some mature neutrophils
3. PAS may reveal intense pools, and SBB may show reddish metachromasia in basophils.

In eosinophil leukaemia:
Eosinophils may contain coarse extragranular PAS pos deposits, acid phosphatase may be strongly pos, and CAE may be pos; similar eosinophils may be a significant component also in AML M 4
In mast cell leukaemia:
Toluidine blue metachromasia is confirmatory

In chronic myelomonocytic leukaemia:
Monocyte-like cells often negative for ANAE

CML = chronic myeloid leukaemia; NAP = neutrophil alkaline phosphatase; AML = acute myeloid leukaemia; SBB = sudan black B, CAE = chloroacetate esterase; PAS = periodic acid-Schiff; ANAE = α-naphthyl acetate esterase

criteria; M1 = myeloblastic without maturation, with 3% or more of the blast cells positive for myeloperoxidase or SBB; M2 = myeloblastic with maturation, with >20% of nucleated cells as myelocytes or beyond; M3 = promyelocytic, hypergranular or microgranular variant; M4 = myelomonocytic with >20% monocytoid forms confirmed by ANAE cytochemistry and >10% granulocyte blasts; M5 = monocytic mature or immature; M6 = erythroleukaemia with >30% myeloblasts, and either a minimum of 50% erythroblasts or a minimum of 30% erythroblasts of which at least some are bizarre. In the chronic leukaemias, cytochemistry adds little to the information obtainable from Romanowsky staining, and only those procedures giving results which are unusual or useful in diagnosis have been noted.

CONCLUSION

Because of its usefulness and convenience, cytochemistry is a significant component in the characterisation of the leukaemias. The increasing heterogeneity being detected by an expanding range of discriminatory procedures is likely to uncover new and useful correlations between cytochemistry and other aspects of the phenotype of leukaemic cell populations.

ACKNOWLEDGEMENTS

I am grateful to Professor D. Galton (London), Mr R. Collins, Mrs I.M. Forbes, Dr I. Bunce, Dr P. Woodford and Dr A. Forsythe (Brisbane) for advice and helpful discussion, and Mr B. Stewart (Brisbane) for assistance with photography.

REFERENCES

Bennett J M et al 1976 Proposals for the classification of the acute leukaemias. British Journal of Haematology 33: 451–458

Breton-Gorius J, Gourdin M F, Reyes F 1981 Ultrastructure of the leukemic cell. In: Catovsky D (ed) The leukemic cell. Churchill Livingstone, Edinburgh, ch 4, p 87–128

Catovsky D, Galton D A G, Robinson J 1972 Myeloperoxidase deficient neutrophils in AML. Scandinavian Journal of Haematology 9: 142–148

Cawley J C, Burns G F 1980 The cytochemistry of human lymphoreticular subpopulations. Immunology Today 1: 85–90

Collins R J 1981 Personal communication

Galton D A G, Goldman J M, Wiltshaw E, Catovsky D, Henry K, Goldenberg G J 1974 Prolymphocytic leukaemia. British Journal of Haematology 27: 7–23

Glick A D, Paniker K, Flexner J M, Graber S E, Collins R D 1980 Acute leukemia of adults. Ultrastructural, cytochemical and histologic observations in 100 cases. American Journal of Clinical Pathology 73: 459–470

Hayhoe F G J, Quaglino D 1980 Haematological cytochemistry. Churchill Livingstone, Edinburgh

Kass L 1980 Niagara sky blue 6B — a new stain for granulocytic cells. American Journal of Clinical Pathology 74: 801–803

Smith H 1978 Inclusion bodies containing virus-like particles in acute lymphocytic leukaemia of childhood. Leukemia Research 2: 133–140

Smith H, Collins R J 1977 A population of lymphocytes in human blood distinctive in morphology and other characteristics. Journal of Clinical Pathology 30: 243–249

Yanagihara E T, Naeim F, Gale R P, Austin G, Waisman J 1980 Acute lymphoblastic leukemia with giant intracytoplasmic inclusions. American Journal of Clinical Pathology 74: 345–349

20

Immunohistology of lymphoid tissue

D. Y. Mason

INTRODUCTION

This chapter reviews the immunohistological techniques which may be used to aid the histopathologist in the diagnosis of human lymphoproliferative disorders. At the outset it should be noted that these procedures have been in use for less than 10 years and they are still being improved and their scope expanded. In particular two important technical innovations in this field have only recently been introduced, and their full potential diagnostic value has therefore yet to be realised. These two advances are, firstly, the use of monoclonal antibodies against human tissue constituents (in place of conventional antisera) for immunohistological labelling (Janossy et al 1980, Pizzolo et al 1980, Warnke et al 1980, Naiem et al 1981, Stein et al 1981); and, secondly, the increasing use of immunoperoxidase-stained cryostat sections of lymphoid tissue (in place of immunoperoxidase-stained paraffin wax sections and of immuno-fluorescently stained cryostat sections) (Stein et al 1980).

The reason for emphasising these advances which are only now beginning to be explored in a routine context is that in the author's opinion the role of immunohistological procedures in the field of human lymphoma diagnosis will change radically in the near future. Until now immunohistological techniques have been of relatively little practical value to the routine histopathologist in the evaluation of lymphoid tissue sections. Although many laboratories now possess the facilities for staining a variety of different antigens in lymphoid tissue samples, in the majority of cases this staining neither alters the established diagnosis nor provides the pathologist, who cannot decide between two (or more) different diagnoses, with a solution to his problem. In consequence the application of immunohistological techniques to the study of human lymphoma biopsies is generally seen as a minority interest, of value mainly as a means of research, rather than as an essential diagnostic procedure.

This state of affairs will not last much longer however, since the 'second generation' immuno-peroxidase techniques outlined in this chapter are likely to play an increasingly important role in the diagnosis of human lymphoma biopsies in the future.

PRACTICAL DETAILS OF LYMPHOMA HISTOLOGY

Substances to be detected by immunohistological techniques

Table 20.1 lists the major constituents which can be detected by immunohistological methods in human lymphoid tissue biopsies. It will be apparent from this table that the results obtained depend to a considerable extent on whether the techniques are performed on cryostat sections of unfixed frozen tissue or on paraffin wax embedded material. In the section which follows the technical aspects of preparing both types of tissue for immunohistological analysis are discussed in detail.

Preparation of tissue samples

Frozen tissue

Handling of tissue before freezing. Ideally a lymphoid tissue sample should be sliced into portions no thicker than 5 mm and frozen without delay after surgical removal. However in the author's experience samples of lymphoid tissue may be kept at room temperature (provided they are immersed in saline or tissue culture medium) for as long as 24 h without obvious

Table 20.1 Antigenic constituents detectable in normal/reactive human lymphoid tissue by immunohistological staining

Antigen	Antibody used for detection	Staining results		Comments
		Paraffin wax sections	Cryostat sections	
Immunoglobulin				
IgG	Polyclonal or monoclonal	Numerous cells are present containing cytoplasmic IgG (both plasma cells and scattered cells of non-plasmactyic morphology in germinal centres). IgG is also detectable in cells which have acquired it from their environment — either by passive diffusion or by active uptake (see text and Fig. 20.9)	Germinal centres contain a dense network of IgG representing immune complexes deposited on dendritic reticulum cells. Plasma cell IgG is also visible but often not clearly localised to individual cells, because of diffusion. Collagen frequently carries large amounts of IgG and this fact together with the presence of free extracellular IgG, often causes excessive background staining of this antigen in cryostat sections of lymphoid tissue	The value of IgG as a marker of neoplastic Ig-secreting cells is restricted by the fact that many of these cells produce IgM rather than IgG and also by the frequency of excessive background staining and uptake by cells of extracellular IgG (see columns 3 & 4)
IgM	Polyclonal or monoclonal	Relatively few cells containing cytoplasmic IgM are seen, except in hyper-reactive states	The germinal centre meshwork stains strongly for IgM (Fig. 20.1). Surface membrane IgM is also present on mantle zone lymphocytes around the germinal centres. Plasma cell IgM is relatively inconspicuous, but IgM associated with collagen may be prominent	Many non-Hodgkin's lymphomas carry surface IgM of a single light chain class. About 50% non-Hodgkin's lymphomas also show intracytoplasmic IgM in at least a small, proportion of cells especially when the tumour is a high grade lymphoma (see Table 20.4)
IgA	Polyclonal or monoclonal	IgA-positive plasma cells occur with variable frequency being most plentiful in tissues producing secretory IgA (e.g. gut- and respiratory tract-associated lymphoid tissue)	Occasional cells with surface membrane IgA and positive plasma cells are seen. However, as with other classes of Ig, diffusion reduces the clarity with which individual plasma cells can be visualized	The majority of lymphomas are negative for IgA although immunoblastic lymphomas arising in α chain disease may express cytoplasmic and/or surface α chains.
IgD	Polyclonal or monoclonal	IgD-positive plasma cells are only rarely encountered in reactive lymphoid tissue	Mantle zone lymphocytes stain strongly for surface membrane IgD (Figs 20.1 & 20.2). Germinal centres however, in contrast to IgM and IgG (see above), are negative for IgD	IgD is found on the cell surface in a minority of cases of non-Hodgkin's lymphoma usually of the chronic lymphatic leukaemia/well-differentiated category. IgD may be found intracytoplasmically in IgD myeloma cells (Plate 19)
κ and λ light chains	Polyclonal or monoclonal	The pattern resembles that of IgG	Staining of the meshwork pattern in germinal centres resembles that seen for IgG and IgM. In addition many mantle zone lymphocytes stain for κ and λ chains.	Tumour cells express cell surface or intra-cytoplasmic Ig it is usually of a single light chain class. However, high levels of background staining in

Table 20.1 *(contd)*

Antigen	Antibody used for detection	Staining results		
		Paraffin wax sections	Cryostat sections	Comments
x and λ light chains *(contd)*			As with IgG excessive background staining is frequently encountered	both paraffin and cryostat sections are not infrequently encountered and may obscure monoclonal Ig patterns
J chain	Polyclonal	Moderate numbers of J chain-positive cells are seen in germinal centres. Plasma cells in extrafollicular regions are less frequently J-chain positive except in tissues containing many cells producing IgM or secretory IgA	No information available	If J chain is found within a cell it is likely to be an endogenous product of the cell, since extracellular levels of J chain are low, and diffusion into the cell consequently is a rare occurrence. As a result staining for J chain may be a valuable means of distinguishing between cells which have absorbed Ig from their environment (which will be J chain-negative) and cells which are endogenous Ig producers (J chain-positive). The best reactions for J chain are obtained using trypsinised formalin-fixed sections or non-trypsinised Bouin's fixed tissue
Lysozyme (muramidase)	Polyclonal	Neutrophils stain strongly for lysozyme. Histiocytes may also show staining, which is usually diffuse and weak and is only found in a minority of cells	No information available	Lysozyme may be found in true histiocytic tumours, but its staining intensity is usually weak. The intensity of staining appears to be related to the degree of differentiation. Lysozyme survives fixation with a wide variety of different fixatives, and trypsinisation is not necessary for its detection
α_1- anti-trypsin	Polyclonal	Staining for this constituent appears to be limited to histiocytes/macrophages and to myeloid cells	No information available	α_1- anti-trypsin is a valuable marker of true histiocytic neoplasms (see Table 20.4 plate 17).
HLA-DR (Ia-like antigen	Monoclonal	This antigen survives very poorly in paraffin-embedded tissues	Lymphoid follicles and interdigitating reticulum cells (also referred to as 'dendritic cells') in T cell areas stain strongly. Vascular endothelium may also stain	The majority, if not all, B cell lymphomas are positive for HLA-DR, whereas T cell neoplasms often lack the antigen
T cell antigen	Monoclonal	Paraffin sections are not suitable for staining (because of antigenic denaturation)	T cells are plentiful in extra-follicular regions of lymphoid tissues. T cells are also found scattered through germinal centres (Fig. 20.2)	T cell antigens may be found on Sezary syndrome/mycosis fungoides cells, and also on more primitive neoplasms of T cell origin

Table 20.1   *(contd)*

| Antigen | Antibody used for detection | Staining results | | Comments |
		Paraffin wax sections	Cryostat sections	
T helper cell	Monoclonal	Paraffin sections are not suitable for staining (because of antigenic denaturation)	Many of the T cells in both extrafollicular areas and within germinal centres are helper cells	Most cutaneous lymphomas are of T helper type
T suppressor cells	Monoclonal	–do.–	A minority of extrafollicular T cells stain as suppressor cells. Very few of the germinal centre T cells are of suppressor type	Only rare instances of T suppressor cell lymphomas are encountered
Leucocyte common antigen	Monoclonal	–do.–	Most cells in lymphoid tissue with the exception of vascular and connective tissue structures stain for this antigen (Fig. 20.7)	This antigen is of value in differentiating between lymphomas (which are positive for the antigen) and non-lymphoid tumours (negative staining). For further details see Fig. 20.7, Table 20.4, Pizzolo et al 1980 and Gatter et al 1982

deterioration of immunoreactivity. Furthermore, morphological detail is usually well preserved under these conditions. The feasibility of keeping fresh lymphoid tissue for several hours before freezing is emphasised in order to encourage pathologists to apply immunohistological techniques to biopsy material which cannot be frozen immediately after excision (e.g. because of transport from another hospital).

Freezing procedure. In the author's experience any procedure which rapidly freezes the sample is satiafactory (e.g. immersion in liquid nitrogen, freezing on a cryostat chuck with CO_2 gas, etc.). When using liquid nitrogen the sample may be immersed directly without using a medium such as isopentane. However when dealing with small samples it may be useful to place the tissue in an embedding compound before freezing. A convenient alternative is to immerse the tissue sample in buffer contained in a small plastic tube and then to freeze the sample by immersing the entire tube in liquid nitrogen. At the time of cryostat sectioning the tube is cut away and the frozen buffer containing the tissue sample recovered as a block for cryostat sectioning.

Storage of frozen tissue. Frozen lymphoid tissue may be stored for long periods in liquid nitrogen or at $-70°C$ without deterioration of antigenic reactivity.

However the tissue should be wrapped in foil or placed in a sealed container to minimise desiccation during storage. If the tissue is likely to be removed from storage more than once for sectioning it may be convenient to freeze it initially on a small cork disc to which it is made to adhere by the use of a small amount of embedding medium or normal saline. The disc bearing the tissue sample can then be mounted on a cryostat chuck and subsequently removed without risk of thawing the tissue.

Sectioning of frozen tissue. The frozen tissue is mounted on a cryostat chuck and sectioned at 5 μm. In the author's laboratory sections are picked up on gelatin-coated slides (to minimise the risk of subsequent detachment) but this is by no means essential. Sections should be allowed to dry thoroughly. Contrary to previous impressions, prolonged drying has little or no deleterious effect on antigenic reactivity, and carries the advantage that it enhances the preservation of tissue morphology. This latter point has only been fully appreciated since immunoperoxidase methods have begun to replace immunofluorescent techniques for staining cryostat sections.

Drying can be performed by leaving sections on the bench at room temperature for several hours, and may be assisted by the use of a hair dryer. In the author's laboratory sections are left for 4–18 h in the vacuum chamber of a freeze-drying apparatus to ensure thorough desiccation but this is not essential. Good results may also be obtained if the slides are left in a refrigerator overnight before staining, although it is advisable to wrap them before doing so in order to prevent condensation forming when they are removed from the refrigerator.

Fixation of cryostat sections. The routine fixative used in the author's laboratory is acetone, slides being immersed (after drying — see above) for 10 min at room temperature and then air-dried. Shorter fixation times, which tended to be used in the past, give inferior preservation of tissue morphology without improving antigenic reactivity. After fixation slides may be stained immediately or else stored (see below) for future examination. It is convenient to prepare more sections than are initially required and to store them in case repeat staining needs to be performed.

Storage of cryostat sections. Dried fixed sections are wrapped in aluminium foil and stored at $-20°C$ (or below). Immediately before staining they are removed from the freezer, brought to room temperature and unwrapped. Staining is then performed in the same way as for slides which have not been stored (see below and Appendix 6).

Sections may be stored in a frozen state for many months without evident loss of antigenic reactivity.

Fixed embedded tissue

When embedded tissue is used for immuno-histological labelling the following steps must be considered: fixation, embedding, and the use of proteolytic enzymes to enhance the reactivity of 'masked' antigens.

Fixation of tissue. Initially immunohistological studies of paraffin wax-embedded lymphoma biopsies were performed on stored routine paraffin wax-embedded material, which in the majority of cases had been fixed in formalin. It subsequently became apparent however that this fixative is far from optimal, and intracellular immunoglobulin in particular often stains very poorly in formol saline-fixed tissue (Mason et al 1980). A characteristic artefact produced by this fixative is that Ig reactivity is restricted to a small area lying close to the plasma cell nucleus, corresponding roughly to the location of the Golgi apparatus (Mason & Biberfeld 1980).

Some laboratories responded to the problems associated with formalin fixation by using proteolytic digestion to reveal antigens hidden as a result of fixation (see below), whilst others explored the use of alternative fixation schedules. It now appears that Bouin's fixative and formol sublimate (or B5 fixative) both appear to preserve the reactivity of immunoglobulin in plasma cells and neoplastic cells satisfactorily. Sublimate has the advantage that cell morphology is preserved better than with formol saline, although some lymphoma histologists may respond with the comment that they have become accustomed to the artefacts produced by formalin fixation and are therefore reluctant to adopt an alternative fixative.

Recently Curran and Jones (from Birmingham, U.K.) have reported that formol saline containing 5% acetic acid preserves the antigenic reactivity of immunoglobulins particularly well in lymphoid tissue, and experience from other laboratories (including our own) has tended to confirm these results. However a disconcerting consequence is that immunohistological staining reveals a much more ubiquitous labelling pattern for immunoglobulin than is obtained with other fixatives (e.g. immunoglobulin can be demonstrated within endothelial cells, tingible body macrophages and in a network pattern in germinal centres).

Finally, in considering the optimal choice of fixation for lymphoma biopsies it may be noted that individual constituents vary considerably in their response to different fixatives. Thus lysozyme appears to survive a variety of fixation procedures rather better than does immunoglobulin (Mason & Taylor 1975). J chain however is more susceptible to fixation and is probably best studied in either Bouin's fixed tissue or digested formol saline-fixed tissue (Isaacson 1979). The macrophage marker α_1-anti-trypsin appears to be most readily demonstrated in digested formalin-fixed tissue (Isaacson et al 1981).

Embedding media. Paraffin wax is almost universally employed as embedding medium in immunohistological studies of fixed lymphoma tissue. However it may be noted that satisfactory labelling can be obtained using resin-embedded tissue. Methacrylate resins are difficult to remove and the best results are obtained with resins such as Araldite or Epon, which may be dissolved with methanol or ethanol containing saturating concentrations of sodium hydroxide.

Proteolytic digestion. The difficulties encountered in staining formalin-fixed lymphoma biopsies (see above) led to the use of proteolytic enzymes (either trypsin or pronase) in several laboratories for uncovering 'masked' antigens (Mepham et al 1979). Constituents (notably immunoglobulin, J chain and α_1-anti-trypsin) which may be completely undetectable without this treatment become strongly reactive following digestion. However it is probable that digested formalin-fixed paraffin wax sections do not stain better for immunoglobulin or J chain than do optimally fixed tissue sections (e.g. Bouin's or sublimate fixed — see above) and consequently it is advisable, if a choice of fixation procedures is available, to avoid the need for digestion by choosing an optimal schedule.

It may be noted that proteolytic digestion is a relatively complicated procedure, since the optimal digestion period for each section cannot readily be predicted. It may therefore be necessary as a preliminary step to process several slides in order to find a digestion time which achieves a balance between optimal antigenic reactivity and avoidance of gross proteolytic damage to the tissue section. Whilst this is simple to perform it obviously reduces the attraction of immunohistological staining in the context of a busy routine histopathology laboratory. It should also be noted that digested sections sometimes show a tendency to detach themselves from the slide.

Immunohistological labelling procedures

Once tissue sections (whether cryostat or paraffin wax-embedded) have been prepared for immunohistological staining, the question arises as to which labelling technique is the most appropriate. The choice lies between immunoenzymatic techniques (almost always utilising peroxidase as the antibody label) and immunofluorescent methods. In the past immunoenzymatic staining has tended to be the

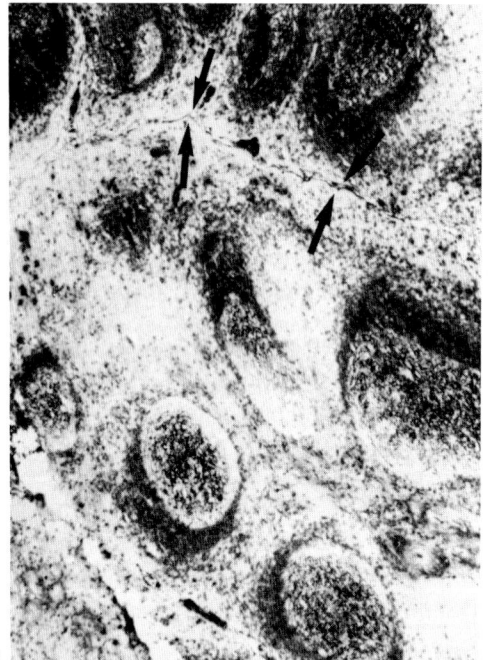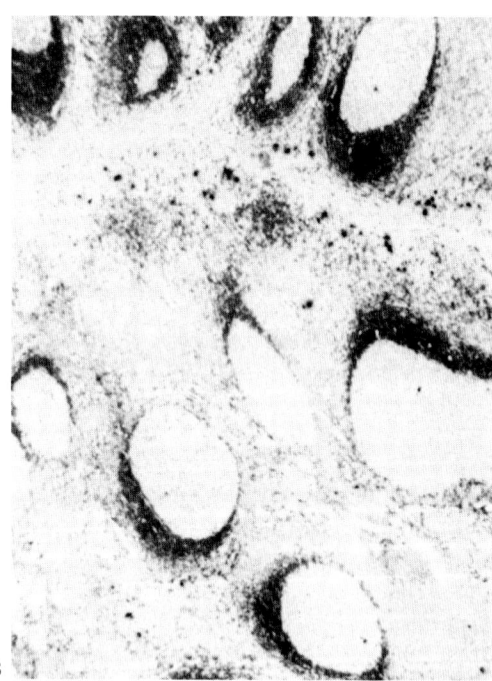

A   B

Fig. 20.1 Immunoperoxidase staining of normal human tonsil (cryostat section) for IgM and IgD. The same area from two adjacent sections is shown. Note that IgM (A) stains the mantle zone of small lymphocytes lying around germinal centres, and also a coarse meshwork of IgM within the germinal centres. This latter staining represents IgM present in immune complexes bound to dendritic reticulum cells within the germinal centres. In contrast IgD staining (B) is restricted to mantle zone lymphocytes, being absent from germinal centres. In addition scattered B cells staining for IgM and IgD are present in the interfollicular areas. Note that there is more background staining in the IgM preparation (A), allowing the position of crypt epithelium to be identified (arrowed). Cryostat sections stained for IgM (and for IgG, kappa and lambda chains) frequently show excessive background staining, particularly in areas containing connective tissue. This is due to the high concentration of these classes of immunoglobulin (relative to IgD) in serum and to their tendency to bind to connective tissue structures. Reproduced by permission from Mason et al (1982) Clinical applications of monoclonal antibodies. Academic Press.

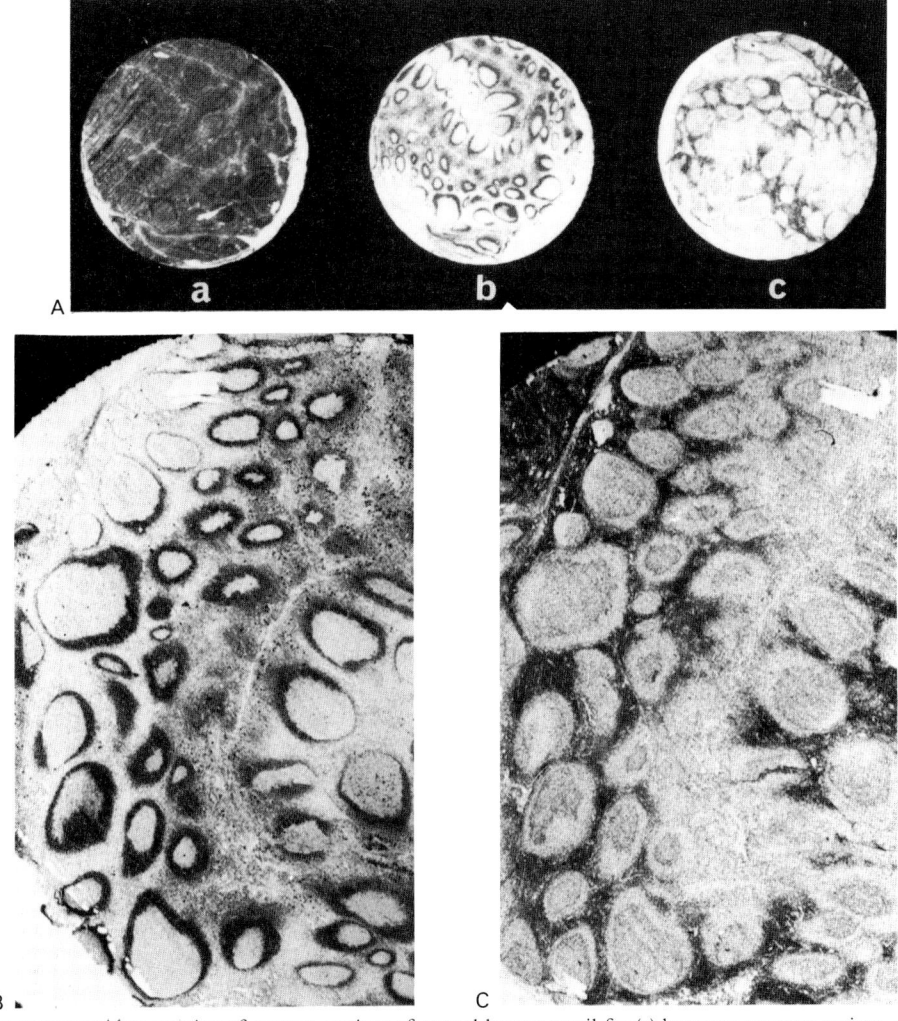

Fig. 20.2 Immunoperoxidase staining of cryostat sections of normal human tonsil for (a) leucocyte-common antigen, (b) IgD and (c) T cells. The illustration (A) shows three adjacent areas on a multi-test slide that has been photographed without a microscope. Slides of this type, on which four or more circular areas are separated by water-repellent masking, are of great convenience when staining multiple sections from one case with a panel of antibodies. Note the clarity with which the different staining patterns revealed by these three antibodies can be differentiated, even at this low-power magnification. In the lower part of the figure higher magnification views are shown of the sections stained for IgD (B) and T cells (C). Reproduced by permission from Naiem et al (1982) Journal of Immunological Methods 50: 145-160

method of choice for labelling paraffin wax sections (Pinkus & Said 1977, Taylor 1978b, Zulman et al 1978, Isaacson et al 1980, Pangalis et al 1980), whilst immunofluorescence has been preferred for labelling cryostat sections (Warnke & Levy 1978, Janossy et al 1980). However recently it has come to be realised that cryostat sections can be labelled equally satisfactorily by immunoperoxidase methods (Stein et al 1980, 1981, Warnke & Levy 1980, Mason et al 1982). Cryostat sections stained in this way offer several advantages over immunofluorescently labelled

sections for the diagnostic histopathologist. Foremost among these is the fact that peroxidase-labelled cryostat sections can be counter-stained with haematoxylin. Furthermore the immunoperoxidase label is permanent (in contrast to immuno-fluorescence). These two advantages of immunoperoxidase labelling are of particular value in the context of lymphoma immunohistology, since lymphoid tissue biopsies are often histologically complex (e.g. both normal and neoplastic areas may be present within the same section) and their interpretation is hence greatly facilitated if tissue morphology can be visualised simultaneously with the immunohistological label, and if labelled sections can be examined for prolonged periods, if necessary by several observers.

In view of the superiority of peroxidase over immunofluorescence in the context of routine lymphoma histology the discussion below is concerned exclusively with the use of this staining procedure.

Immunoperoxidase staining

There are three commonly used techniques for immunoperoxidase labelling (Fig. 20.3). The first two to be introduced historically involve the use of antibodies to which the enzyme has been covalently conjugated. The conjugated antibody is either applied directly to the section, e.g. peroxidase-conjugated-anti-human IgM, or used in a two-stage indirect labelling technique (Fig. 20.3).

The third procedure for immunoperoxidase labelling of tissue sections is the peroxidase:anti-peroxidase (PAP) procedure introduced by Sternberger as a modification of earlier 'unlabelled antibody' immunoperoxidase methods (Fig. 20.3).

Choice of immunoperoxidase labelling procedure

The choice between the three methods outlined in Figure 20.3 may be simplified by excluding the first of these techniques (direct immunoperoxidase staining). This method entails the use of different peroxidase-conjugated antibodies for each antigen studied. Not only is this expensive, but many antisera are not available in a conjugated form. Low sensitivity and excessive background staining are also likely to prove troublesome if direct conjugate methods are used.

The choice therefore lies between the indirect conjugate method and the PAP procedure. It should be appreciated that the best commercial conjugates available today, e.g. peroxidase-conjugated-anti-rabbit or mouse Ig, are considerably better, in terms of strength of reactivity and lack of non-specific background labelling, than they were in the past. In consequence there is little to choose between the two staining procedures if they are performed using optimal reagents. Exaggerated claims of great sensitivity have been made in the past for the PAP procedure relative to indirect immunoperoxidase staining. In reality the PAP technique is probably at most a few times more sensitive than the indirect immunoperoxidase method. In the context of lymphoma diagnosis it should be pointed out that antigens which can only be detected by pushing immunoperoxidase methods to the limit of their sensitivity are not suitable for detection in a routine histopathology laboratory.

IMMUNOPEROXIDASE LABELLING METHODS

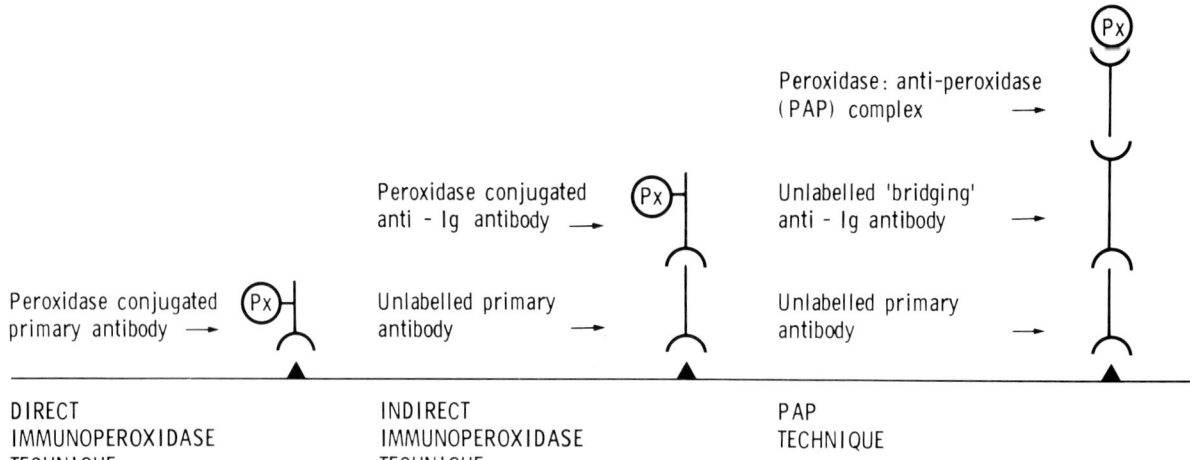

Fig. 20.3 Schematic representation of the three most widely used immunoperoxidase staining procedures.

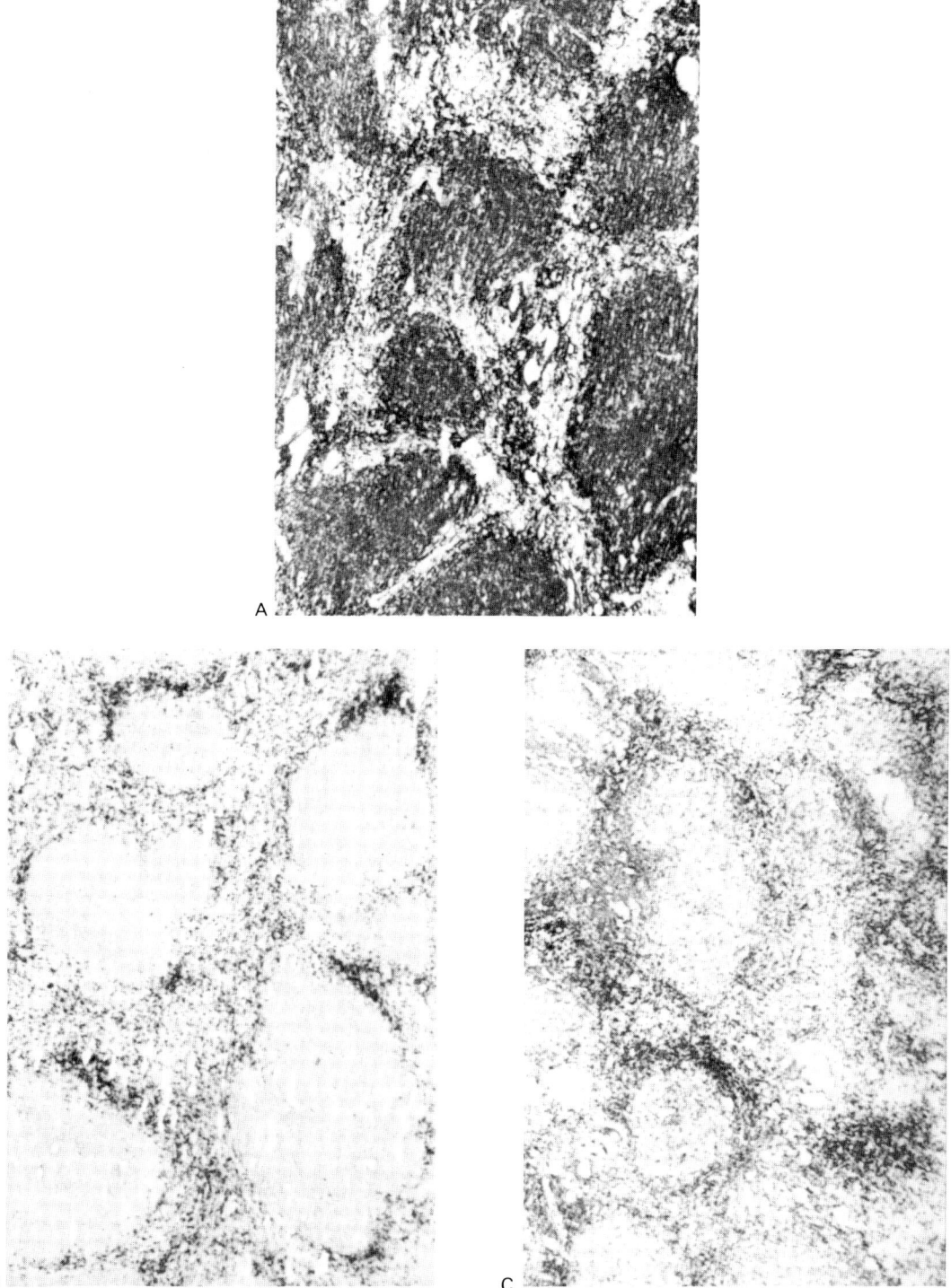

Fig. 20.4 Immunoperoxidase staining of a case of follicular lymphoma (cryostat sections) with monoclonal antibodies against IgM, IgD and T cells. The irregularly-shaped neoplastic follicles show diffuse staining for IgM (A), whilst IgD staining (B) reveals that the mantle zones are thinned and partially destroyed (compare with normal lymphoid follicles in Figs 20.1 and 20.2). T cells (C) are plentiful in the intertollicular areas between the neoplastic follicles and also scattered within the follicles. Reproduced by permission from Mason et al (1982) Clinical applications of monoclonal antibodies. Academic Press.

Table 20.2 Comparative properties of monoclonal and polyclonal antibodies

Antibody type	Origin of antibody	Composition of antibody preparation	Advantages	Disadvantages
Polyclonal	Raised by immunising rabbits, goats, etc. with highly purified antigen	Either whole serum or an immunoglobulin fraction isolated from the serum. Both types of preparation contain antibody specific for the immunising antigen (at a concentration of approx. 1 mg/ml) together with an excess (5–15 mg/ml) of unrelated immunoglobulin	Polyclonal antibodies for immunohistological staining are at present more widely available and more extensively characterised than their monoclonal counterparts	1. The range of antigenic specificities is restricted to those antigens which can be purified in sufficient quantity for immunisation. 2. the excess of unrelated immunoglobulin (see column 3) is an important potential cause of unwanted non-specific staining
Monoclonal	Prepared by immunising mice or rats with antigen-containing preparations (which may be very impure) and then fusing spleen cells from the animal with a cultured myeloma cell line (using polyethylene glycol as fusing agent). Hybrid clones which secrete antibody of the desired specificity are selected and then grown either in tissue culture medium or as ascitic tumours (in mice or rats)	Either tissue culture medium containing approximately $5\mu g$/ml of specific antibody (plus 10% fetal calf serum) or ascitic fluid (or an Ig fraction thereof) containing at least 1 mg/ml of specific antibody, together with a small amount of unrelated immunoglobulin	1. Monoclonal antibodies are homogeneous reagents, all the molecules produced by one individual hybrid cell line have exactly the same antigen-binding site. In consequence they are usually specific for a single antigenic determinant although occasionally binding to closely related antigenic structures may also occur 2. Since little or no unrelated immunoglobulin is present in the antibody preparation (see column 3) non-specific reactivity is minimised 3. Monoclonal antibodies can be raised against new hitherto unrecognised antigens (e.g. those characteristic of T lymphocytes and their sub-sets) as well as against antigens which do lend themselves to the production of antibodies on a large scale (HLA DR) 4. Any individual monoclonal antibody can be produced on a potentially limitless scale, thus facilitating the use by many different laboratories of the same reagent	New monoclonal antibodies directed against unknown antigens may give unexpected results, i.e. binding to apparently unrelated cell populations which presumably share the same antigenic structure. However, in a histopathological context this obstacle can be overcome by testing any individual antibody on a range of different tissue types and thus establishing its operational specificity. Since only one determinant on a molecule is detected there is a theoretical possibility that variant forms of the molecule lacking the relevant determinant will give negative immunohistological staining reactions. However in practice this does not appear to be a serious disadvantage, and can be solved if necessary by using more than one monoclonal antibody directed against the same molecule

Antibodies for immunohistological staining

The primary antibodies used for immunohistological labelling have in the past been produced by immunisation of animals such as rabbits, goats etc. To an increasing extent however the specificities of these polyclonal antibodies are now beginning to be duplicated by monoclonal antibodies, which possess a number of advantages in the context of lymphoma immunohistology (Mason et al 1982). The relative characteristics of these two types of antibodies are set out in Table 20.2. It should be emphasised that in discussing the role of monoclonal antibodies all references are to their use as primary antibodies (see

Fig. 20.3). Antibodies for subsequent stages in the immunoperoxidase procedure are still almost universally prepared by conventional means as polyclonal reagents.

Choice between polyclonal and monoclonal antibodies. There is no single answer to the question of whether monoclonal antibodies should be used in preference to polyclonal reagents for the immunohistological analysis of lymphoma biopsies. In the case of some antigens no such choice arises, since only one type of antibody is available: i.e. T cells and their subsets cannot be detected satisfactorily with any commercially available polyclonal antisera, monoclonal anti-J chain is not available, etc. However in the case of immunoglobulin there are now a number of sources of both monoclonal and polyclonal antibodies suitable for their detection. The choice in this instance will obviously be influenced by practical considerations, e.g. whether techniques based upon polyclonal antisera have already been established in the laboratory. However three practical points,

arising from the author's own experience, may be made in this context.

Firstly, it is possible to obtain monoclonal anti-Ig antibodies which give immunohistological labelling reactions fully comparable in strength to those obtained using conventional antisera. This point is emphasised because of suggestions made in the past that monoclonal antibodies will always give inferior immunohistological labelling reactions because of their ability to recognise only a single determinant on a molecule. In practice this theoretical argument against their use appears to be of no validity, although it must be stressed that monoclonal antibodies vary in their suitability for immunohistological labelling (as do polyclonal antibodies) and some will inevitably give unsatisfactory labelling reactions.

Secondly, monoclonal antibodies tend to give cleaner reactions than conventional antisera (because of their greater purity — Table 20.2). This is not to deny that conventional antisera can give excellent low-background immunohistological reactions, but

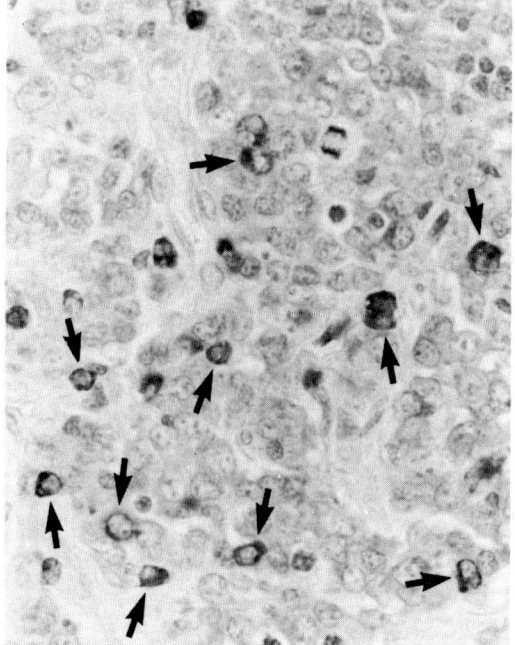
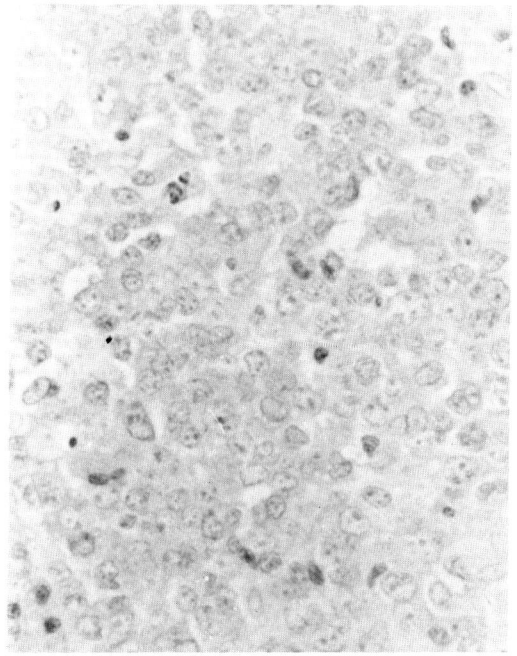

A   B

Fig. 20.5 Immunoperoxidase staining of a germinal centre-derived lymphoma (paraffin wax section) showing the same areas stained for (A) lambda light chains and (B) kappa light chains. Scattered lambda-positive cells are seen (arrowed), whereas kappa staining is negative. Staining for μ chains (not shown) gave a similar appearance to that seen for lambda chains. Note that only a minority of the neoplastic cells contain detectable amounts of immunoglobulin. This pattern of scattered positive cells is characteristic of this type of tumour and contrasts with plasmacytoma (see Plate 19) or high-grade lymphoma (Plate 22) in which a much higher percentage of cells are positive for immunoglobulin. However if the neoplasm shown in these illustrations had been stained in cryostat sections for surface (rather than cytoplasmic) immunoglobulin the majority of cells would have been positive for Ig of a single light chain type (see Fig. 20.6).

rather to indicate that less care is required (in terms of selecting optimal dilutions) when using monoclonal antibody.

Finally, it may be noted that monoclonal antibodies present in ascitic fluid sometimes give background staining which is not observed if the same antibody present in culture supernatant is used. This presumably reflects the presence of non-specific immunoglobulin in ascitic fluid from hybridoma-bearing mice. In consequence, if a choice exists, it is often advisable to use culture supernatant rather than ascitic fluid.

Specificity controls. The question of specificity controls in immunohistological staining is a complex one and a full discussion is beyond the scope of this chapter. Table 20.3 summarises the different types of control which may be used. In the context of the immunohistological staining of lymphoma sections by a routine histopathology laboratory the following points may be made:

1. Staining of control sections in which the primary antiserum has been omitted or in which a non-immune serum or an unrelated antiserum is substituted for the primary antiserum (at the same or higher concentration) should always be performed. It may be noted that when staining a number of different antigens (e.g. kappa chains, lambda chains, IgM etc.) this control is already in effect achieved.

2. Sections of normal lymphoid tissue (against which the reactivity of each antiserum has already been evaluated) should always be included as a positive control. This provides a baseline against which staining of a lymphoma sample can be assessed. It may be noted that human tonsil is particularly suitable for this purpose since it is readily available and contains all the major elements of lymphoid tissue.

3. The possibility of a contaminating antibody in a commercial antiserum will always be a potential risk and the control to eliminate this possibility (incubation with purified antigen — see Table 20.3) is beyond the scope of most routine histopathology laboratories. However a simple but often highly informative procedure which may be of value is to use at least one (and if possible more) alternative antibodies of the same specificity. If three antisera from independent sources all give the same staining patterns this provides compelling evidence that the reaction is specific, since the three reagents are

unlikely to contain the same contaminating antibody. If the reaction patterns are different the antiserum giving the most restricted pattern of reactivity is likely to be specific (unless this limited reaction reflects a low titre antiserum).

4. The risk of false-positive staining, whether due to contaminating antibodies or to non-specific binding of background immunoglobulin, is minimised by the use of monoclonal antibodies.

Storage of reagents and choice of optimal dilutions. These important technical points are dealt with in Appendix 6.

PRACTICAL APPLICATION OF IMMUNOPEROXIDASE STAINING

Present status of lymphoma immunohistology

In this concluding section details are given of the particular contexts in which immunoperoxidase staining may be of value in the diagnosis of human lymphoproliferative diseases. This review may be prefaced by reminding the reader of a point made in the introduction to this chapter, namely that the full potential value of the staining procedures in the field of lymphoma diagnosis is only just now beginning to be realised.

The reasons for the limited role of immuno-peroxidase methods in the past for the diagnosis of lymphoma are two-fold. Firstly most studies were performed on paraffin wax sections rather than on cryostat sections. The argument for using this type of material was that it is much more readily available than cryostat sections. However this argument does not take into account the fact that a very large proportion of the antigenic reactivity in a tissue is rendered undetectable following conventional fixation and embedding procedures. In particular the majority of surface membrane antigens cannot be detected in paraffin wax sections, even when they are readily demonstrable in cryostat sections (see Table 20.1). As the introduction of new monoclonal antibodies (see below) increases the number of antigens detectable in human lymphoid tissue the argument for using cryostat sections rather than paraffin wax sections will become increasingly persuasive.

A second reason for the failure in the past of immunohistological techniques to achieve a routine

Table 20.3 Negative controls for immunohistological staining

Procedure	Aim of control	Significance of unwanted staining
Omission of primary antibody (i.e. substitution with buffer)	To establish that the second stage reagent (and also the third stage reagent when using the PAP procedure) shows no affinity for the tissue section	*Two stage method:* Staining is either due to endogenous peroxidase activity or to binding of the peroxidase conjugate to the section. These two patterns are usually readily distinguished (i.e. red cell or myeloid cell staining is likely to indicate endogenous peroxidase activity etc). If in doubt however a further control should be performed in which substrate alone is applied — any staining which persists under these conditions must represent endogenous peroxidase. Binding of the second stage conjugate may be attributed to: (a) Recognition by the conjugate of human Ig (due to the antigenic cross-reactivity of Ig from different species). This staining will be limited to sites known to contain Ig (e.g. germinal centres when staining cryostat sections, plasma cells in paraffin sections). This type of reactivity can be blocked by the addition of normal human serum to the anti-Ig peroxidase conjugate (see Appendix 6). (b) Non-specific absorption of the conjugate to the section. This will cause diffuse staining over much of the section. (c) A contaminating antibody (e.g. anti-actin etc) in the conjugate. This is relatively uncommon, but may be suspected if there is staining of isolated structures/cell populations in the section. *PAP method:* As noted in the text non-specific staining due to binding of either the second stage (bridging antibody) or the third stage (PAP reagent) is very rarely encountered.
Substitution of the primary antibody by either normal serum or an unrelated antibody (which has no specificity for any antigen in the tissue). In either case the substituting reagent should be from the same species as the primary antibody and the dilution used should be no greater than that of the primary antibody	To assess whether binding of immunoglobulin in the primary antibody is causing non-specific staining. Such binding may occur either because of a contaminating antibody (unrelated in specificity to the putative specificity of the primary antiserum or because of non-specific binding of immunoglobulin molecules of no single specificity	Almost all polyclonal antisera will bind non-specifically to tissue sections and thus lead to unwanted staining if applied at too high a concentration. The level at which this occurs depends on several factors, including the tissue being stained, the sensitivity of the detection system, and the properties of the individual antiserum. Concentrations above 1/20 frequently give rise to such non-specific staining. The limitation of this control is that negative staining does not eliminate the possibility that the reactions of the primary antiserum represent the presence of unwanted contaminating antibody (e.g. anti-immunoglobulin contaminating anti-lysozyme antiserum etc). Evidence on this possibility can only be provided by blocking or immunoabsorption controls (see below)
Blocking of primary antibody by the addition of purified antigen	To demonstrate that positive staining is due to specific antibody (directed against the antigen used for immunisation) and is not due either to a contaminating antibody against an unrelated antigen or to non-specific binding of immunoglobulin molecules in the primary antiserum	Failure to block the reactivity of the primary antiserum by the addition of antigen provides strong evidence that the reactivity of the antiserum is not specific. However it is possible that too little antigen has been added to achieve full blocking. It is recommended that the primary antiserum is diluted to its working concentration and that differing amounts of antigen are pre-incubated with aliquots of this diluted preparation. If antigen is added to undiluted antibody it is much more difficult to block reactivity
Absorption of primary antibody on an immunoabsorbant of purified antigen — followed by the recovery of activity in material eluted (e.g. at low pH) from the immunoabsorbant	–do.–	Failure of the immunoabsorbant to remove reactivity, and the absence of reactivity in the fraction subsequently eluted from the immunoabsorbant after thorough washing strongly suggests that the reactivity of the antiserum is non-specific

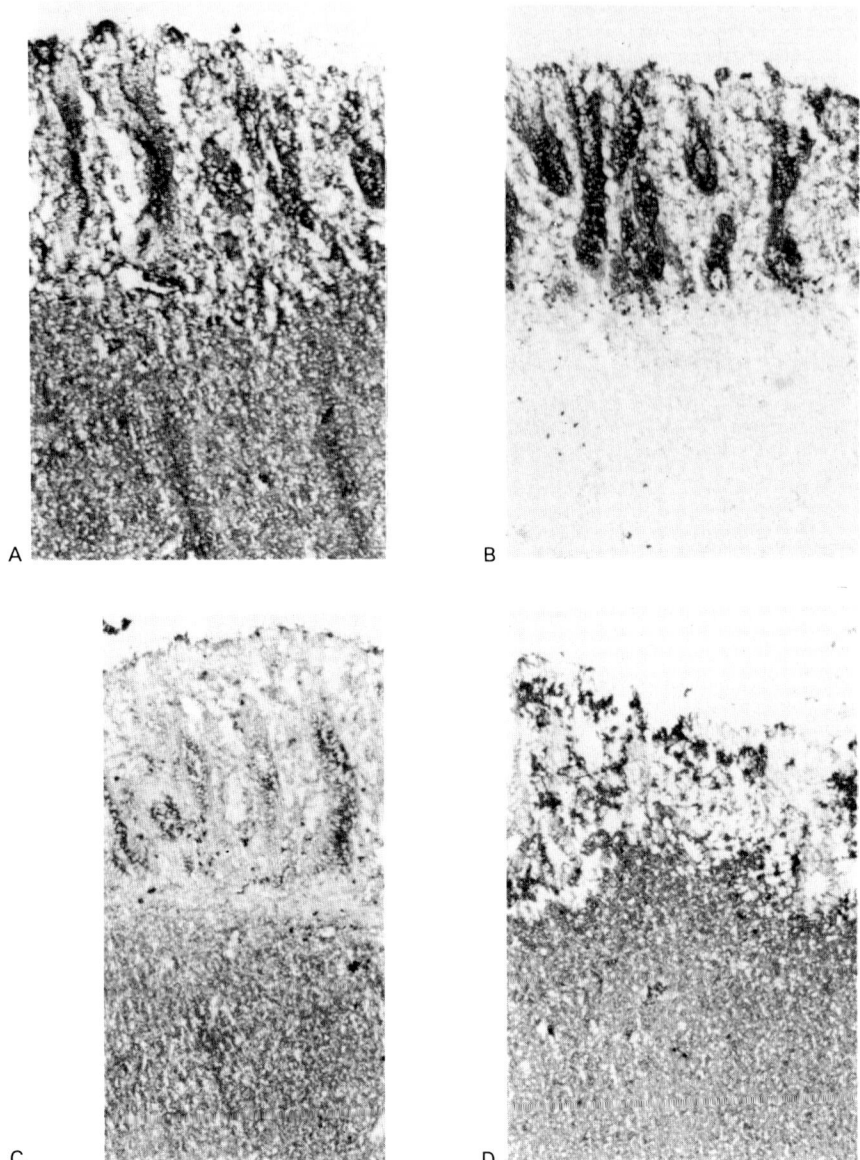

Fig. 20.6 Immunoperoxidase staining of gastro-intestinal lymphoma (cryostat section) with a panel of monoclonal antibodies, revealing positive staining for IgM (A), kappa (B), lambda (C) and for HLA–DR (D). In each illustration epithelium occupies the upper half of the picture whilst the lower portion shows the lymphoma. Note that polyclonal immunoglobulin is present in the epithelium, as indicated by the positive staining for kappa light chains (B). The positive staining for HLA–DR (D) in the epithelium represents staining of resident gut macrophages rather than of tumour cells. Reproduced by permission from Mason et al (1982) Clinical applications of monoclonal antibodies. Academic Press.

diagnostic role in the analysis of human lymphoma lies in the fact that very few antigenic substances have been detectable with existing antibodies. Furthermore it should be pointed out that the majority of these antigenic markers have been subject to a variety of technical limitations, e.g. the antigen is only expressed in a minority of cells of a certain type, or is subject to artefactual or misleading reactivity patterns.

The limited range of antigens studied in the past and the problems associated with their detection may be illustrated by considering each antigen separately:

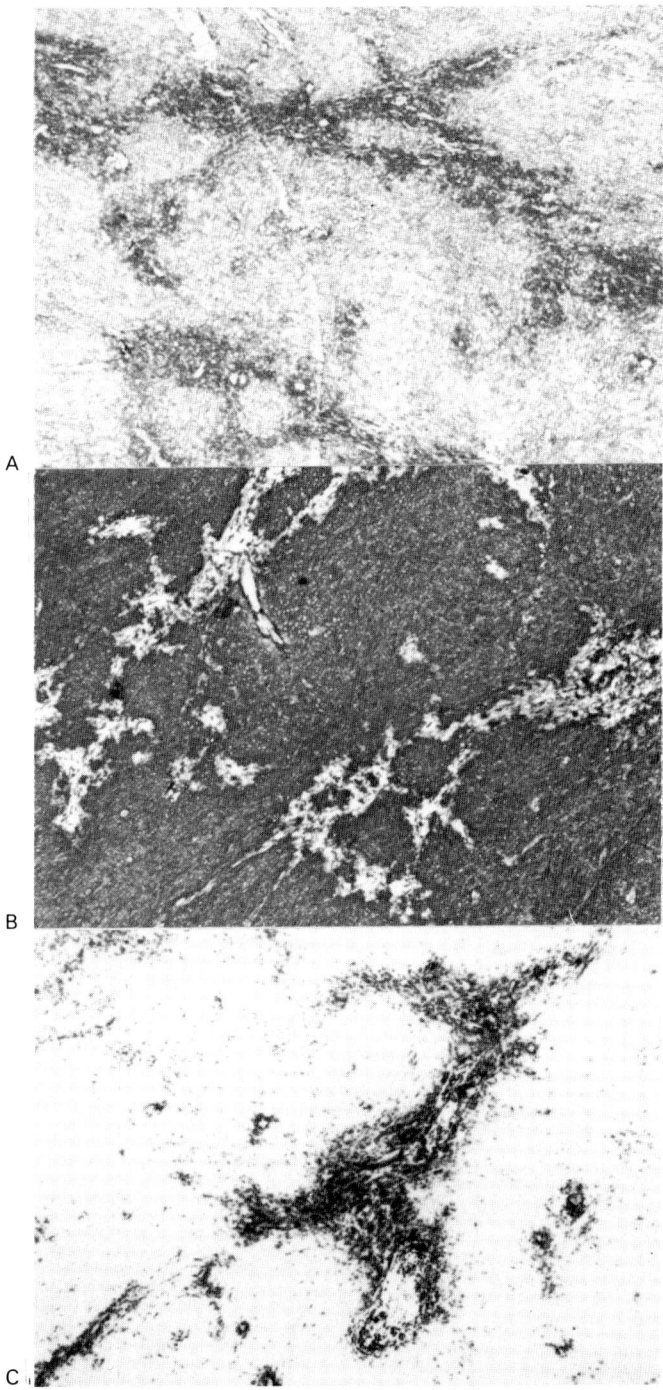

Fig. 20.7 Immunoperoxidase staining of a diffuse high-grade lymphoma (cryostat section) with monoclonal antibodies against leucocyte-common antigen (A), HLA–DR (B) and T cells (C). On routine histological examination the tumour was highly anaplastic and was thought most probably to be a carcinoma. However immunoperoxidase staining revealed its B lymphoid nature. In addition to the reactions shown the tumour stained positively with a monoclonal antibody against B cells and failed to react with any anti-epithelial antibodies. The tumour was negative for surface immunoglobulin, as is often frequently the case in high-grade B cell lymphomas. Note that the tumour consists of sheets of neoplastic cells (labelling for leucocyte-common antigen and for HLA–DR) interspersed with areas of small lymphoid cells (which are strongly positive for leucocyte-common antigen and for the T cell antigen but predominantly negative for HLA–DR). This latter cell population represents infiltrating host lymphocytes. For further details of how monoclonal antibodies may be used to distinguish between different types of anaplastic neoplasms see Gatter et al (1982). Reproduced by permission from Mason et al (1982) Clinical applications of monoclonal antibodies. Academic Press.

Immunoglobulin

Cytoplasmic immunoglobulin is only detectable in the more mature members of the B cell series. Hence diseases such as chronic lymphocytic leukaemia and a substantial proportion of non-Hodgkin's lymphoma will be negative for this constituent (Taylor 1978a). Furthermore the great potential value of kappa/-lambda staining as a marker of monoclonal B cell proliferation is reduced by the fact that these two light chains are ubiquitously present in normal immunoglobulin. As a result true monoclonal reaction patterns are frequently obscured, in both paraffin wax-embedded and cryostat material, by excessive background staining due to serum immunoglobulin.

Lysozyme

Although initially considered of great potential value as a marker of myeloid and histiocytic cells (Mason & Taylor 1975), especially since it survives well in paraffin wax-embedded tissue whatever the fixative, this constituent in practice turns out to be of relatively little value. This is probably attributable to the fact that histiocytic cells rapidly excrete the enzyme after synthesis. In consequence it does not achieve a high enough intracellular concentration to give strong staining. A contrast may be drawn between the weak staining of most histiocytic cells for lysozyme and the strong staining of cells such as polymorphs, Paneth cells and serous salivary gland cells, in all of which the enzyme is packaged within secretion granules. Furthermore neoplastic histiocytes may, because of their less differentiated state, produce smaller quantities of lysozyme (and hence stain even more weakly) than do their non-neoplastic counterparts (Mendelsohn et al 1980).

J chain

This marker (first studied in human neoplastic lymphoid tissue sections by Brandtzaeg and by Isaacson) has the great potential advantage over immunoglobulin as a B cell marker that it is present in low levels in the extracellular environment and in consequence its presence in the cytoplasm of a cell indicates endogenous synthesis (Brandtzaeg & Berdal 1975, Isaacson 1979). It may hence be of value in distinguishing between cells which have acquired Ig from their environment and those in which it is of endogenous origin. However J chain suffers from one of the limitations noted above for cytoplasmic Ig as a B cell marker in that it only appears in more mature B cells (at approximately the same time that Ig synthesis is initiated). Furthermore it is not uncommon to encounter human B cell neoplasms which contain cytoplasmic Ig (synthesised by the cells themselves) without accompanying J chain.

α_1-anti-trypsin and α_1-anti-chymotrypsin

These proteins are of interest in that they appear to be present in many cells of the histiocytic series, and are often detectable in histiocytic tumours in which the lysozyme reaction is negative or equivocal. One disadvantage however is that the reaction may be relatively faint and require optimal technique and some interpretative skill to detect it. Secondly α_1-anti-trypsin is detected most readily in trypsinised formalin-fixed tissue and stains very poorly in non-trypsinised sections of tissue fixed in other fixatives (Isaacson et al 1981).

Impact of monoclonal antibodies on lymphoma immunohistology

The restricted range of antigens discussed above is now steadily being widened as a result of the production of new monoclonal antibodies. At the time of writing only a few of these reagents are commercially available (e.g. anti–HLA DR, anti–T cell antigens etc.). However by a fortunate coincidence monoclonal antibodies possess the inherent advantage, apart from their high specificity and purity (see Table 20.2), of being easy to produce on a large scale, once the antibody-secreting cell line has been established. In consequence there is no major obstacle to the distribution of a monoclonal antibody on a commercial basis to numerous laboratories around the world, and the next few years are likely to see scores of new monoclonal antibodies specific for human lymphoid antigens becoming widely available. Inevitably some of these reagents will prove to be of great diagnostic importance, e.g. antibodies which can distinguish different stages of lymphoid cell maturation, antibodies specific for hystiocytes/macrophages, antibodies allowing neoplastic lymphoid cells to be distinguished from reactive cells, etc.

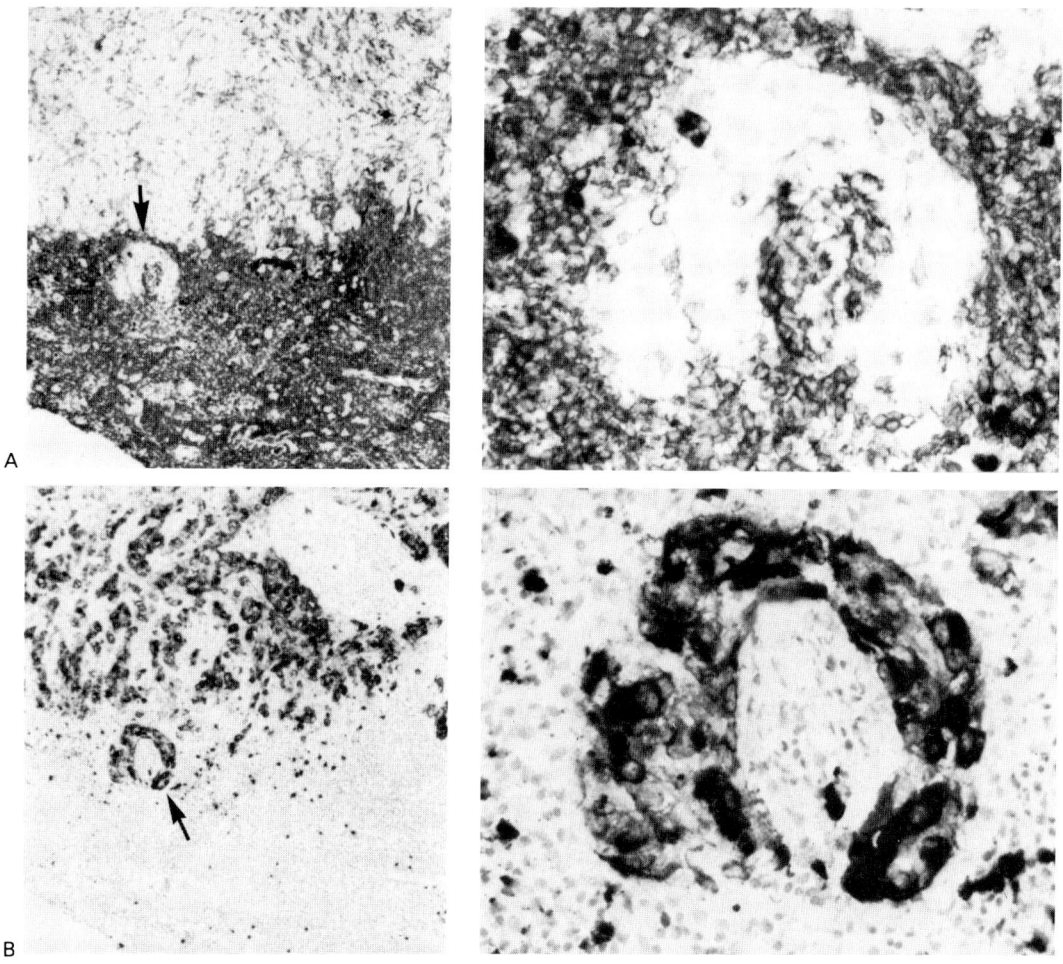

Fig. 20.8   Immunoperoxidase staining of the same area of a lymph node (cryostat section) infiltrated by metastatic adenocarcinoma using monoclonal antibodies against (A) leucocyte-common antigen and (B) epithelial intermediate filaments. The deposits of carcinoma are clearly delineated by virtue of their negative staining for leucocyte-common antigen (A) and their positive reaction with the anti-intermediate filament antibody (B). The higher power views on the right show one of the carcinoma deposits (arrowed on the left) in greater detail. Note that scattered myeloid cells are visible in the sections stained for leucocyte-common antigen because of their content of endogenous peroxidaae. In practice the characteristic appearance of these cells and their low frequency means that their presence does not prevent the interpretation of immunoperoxidase-stained cryostat sections.

IMMUNOHISTOLOGICAL PATTERNS OBTAINED IN NEOPLASTIC LYMPHOID TISSUE BIOPSIES

Table 20.4 summarises the staining patterns which may be encountered in different types of human lymphoproliferative disorders. Further details may be found in the references listed at the end of this chapter.

A few comments may be made to supplement the results summarised in Table 20.4.

Interpretation of staining

The results obtained following immunoperoxidase staining of paraffin sections are often more difficult to interpret than is generally realised. In particular immunoglobulin is frequently found in lymphoid tissue biopsies within the cytoplasm of cells which have not synthesised this material, but rather absorbed it from their environment.

Uptake of immunoglobulin may be non-specific, as demonstrated by the fact that staining for other serum proteins (e.g. albumin) is also positive (Isaacson &

Table 20.4 Immunoperoxidase staining patterns in human neoplastic lymphoid tissue biopsies

Disorders	Paraffin wax sections	Cryostat sections
Non-Hodgkin's lymphoma		
Chronic lymphocytic leukaemia	Usually negative. Crystalline Ig inclusions (of a single light chain class) are occasionally detectable (Plate 18)	Surface Ig (usually IgM and IgD) of a single light chain class is detectable on the surface of the majority of lymphoid cells. Neoplastic cells are also positive for HLA-DR
Plasmacytoma	Neoplastic cells usually contain Ig of a single light and heavy chain class (Plate 19). J chain is often also present	No data
Lymphoplasmacytoid lymphoma	Plasmacytoid cells contain Ig (usually IgM) of a single light chain class. J chain is often also present	No data
Follicular lymphoma	A proportion of cells in some cases contain Ig (usually IgM) of a single light chain class (Plate 20). This constituent may occasionally be present in small granules or as larger inclusions (Plate 18; see also van den Tweel 1978). J chain is also often detectable	Surface IgM of a single light chain class is present on the cells of the neoplastic follicles in at least 50% of cases. IgD is confined to mantle zone cells. This region lies outside the neoplastic germinal centres and is often partially destroyed by the neoplastic process. The neoplastic follicles do not contain the meshwork of polyclonal Ig found in reactive germinal centres. Neoplastic cells are HLA-DR-positive and T cells are often plentiful around the neoplastic follicles (Fig. 20.4)
Diffuse lymphoma of follicle centre cell origin	As for follicular lymphoma (Plate 21 & Fig. 20.5). Giant cells containing IgG kappa lambda may be present	Neoplastic cells usually have surface IgM (of a single light chain class) and are HLA-DR-positive (Fig. 20.6)
Diffuse large cell lymphoma (equivalent to 'histiocytic lymphoma' in Rappaport's classification)	The majority of these tumours are of B cell origin and many stain for cytoplasmic Ig (usually IgM) of a single light chain class (Plate 22). Staining is frequently limited to the perinuclear space. Giant cells containing polyclonal IgG may be found (Mason et al 1981). J chain is detectable in a proportion of cases.	Many cases express surface Ig (most commonly IgM) of a single light chain class. At least 25–30% appear to lack surface Ig (Fig. 20.7). A minority of cases are of T cell origin and stain for T cell antigens. These cases are usually HLA-DR-negative. All other cases of diffuse large cell lymphoma are HLA-DR-positive. Occasionally cases in this category are of true histiocytic origin
True histiocytic lymphoma	α_1- anti-trypsin is usually detectable, most frequently localised as a clump of reactivity close to the nucleus (Plate 17). Lysozyme is present in a small number of cases	No data
Sezary syndrome/ mycosis fungoides	No staining of tumour cells with available antisera	T cell antigens (almost always of 'helper' type) are detectable. HLA-DR staining is negative
Hodgkin's disease	Reed-Sternberg and Hodgkin's cells usually stain for IgG kappa and lambda (Plate 23). α_1- anti-trypsin is also detectable in a proportion of these cells as a small clump of reactivity close to the nucleus	In most cases T cells are plentiful (showing a normal excess of helper cells over suppressor cells). B cell follicles may be present (as indicated by staining for IgM and IgD) but are often partially obliterated. In lymphocyte-predominant Hodgkin's disease however the majority of cells are B cells (IgM and IgD-positive)
Anaplastic carcinoma	Negative with commercially available antisera	Such cases may be distinguished from lymphoma by their positive staining for epithelial markers (e.g. intermediate filaments) and their negative reactions for white cell markers (e.g. Ig and the leucocyte common antigen). HLA-DR is almost always present on lymphoma cells but frequently absent from carcinoma cells (Fig. 20.8)

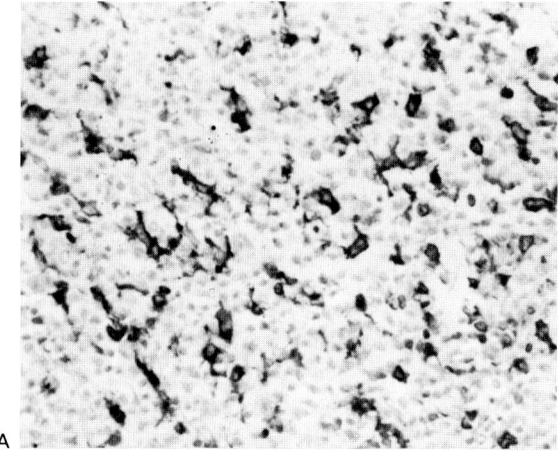

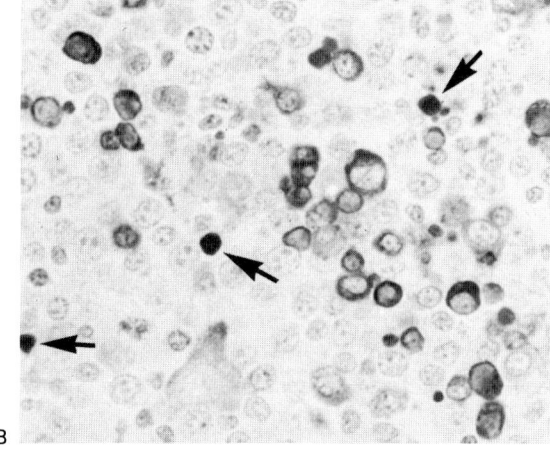

Fig. 20.9   Immunoperoxidase staining of two lymphoma biopsies (paraffin wax sections) for lambda light chains. In each sample scattered, strongly-staining cells are present. However staining for kappa light chains, IgG and albumin gave identical staining patterns to those observed for lambda chains. Furthermore double staining for kappa and lambda light chains (see Plate 23) revealed that both these constituents were present in the same cells. These findings indicate that the positive staining of these lymphoma cells is due to uptake of exogenous serum proteins rather than to endogenous synthesis of immunoglobulin (which would have produced positive staining for a single light chain class and negative labelling for albumin). Staining for immunoglobulin due to this 'leakage' phenomenon is relatively frequently encountered when studying paraffin wax-embedded lymphoma biopsies. Positive cells are often (as in the illustrations) found scattered against a negative background with the result that uptake of exogenous protein is not immediately suspected. However a number of characteristic features, including the presence of extracellular deposits of protein (arrowed) allow the phenomenon to be readily recognised (see Table 20.5).

Wright 1979, Mason et al 1980). This phenomenon is presumed to be due to diffusion of Ig into degenerate cells (Fig. 20.9), although it should be emphasised

that cells showing this appearance usually appear morphologically intact. Table 20.5 lists some of the features which may allow this diffusion artefact to be identified.

Table 20.5   Non-specific uptake of exogenous immunoglobulin by lymphoma cells

Characteristic immunohistochemical features that allow this phenomenon to be recognized:

    1. Cytoplasmic Ig staining is diffuse rather than granular, and may be strongest towards the periphery of the cell
    2. Positive cells may show degenerative changes
    3. Positive cells tend to be found scattered within circumscribed zones in the section, sometimes in relation to vessels. Positive cells are frequently commonest in the central poorly-fixed area of the section
    4. Extracellular clumps of precipitated serum protein are often seen in the vicinity of positive cells (Fig. 20.9)
    5. The intensity of staining for serum proteins is related to their concentration in extracellular fluid (i.e. albumin staining is stronger than that for transferrin) and inversely to their molecular weight (i.e., IgM nonspecific uptake is minimal)
    6. Double labelling for kappa and lambda light chains reveals mixed staining. J chain staining is negative.

An alternative type of staining for cytoplasmic Ig occurs when cells absorb immunoglobulin unaccompanied by other serum proteins. This may be accounted for by Fc-receptor mediated uptake of exogenous immunoglobulin. This pattern is most frequently seen in Reed-Sternberg and Hodgkin's cells (Taylor 1976, Curran & Jones 1978, Poppema et al 1978), (Plate 23) although it may also be seen in giant cells in a variety of types of non-Hodgkin's lymphoma (Mason et al 1981) — see Table 20.4.

The importance of exogenous immunoglobulin uptake should not be underestimated, since it probably accounts for many reports of anomalous staining patterns in human lymphoma in the past (Mason et al 1980). In theory the phenomenon should be suspected because of the fact that the same cell population will stain for IgG kappa and lambda. However it should be remembered that kappa chains are present in serum at a higher concentration than are lambda chains. Furthermore an anti-lambda antiserum may react more strongly with free lambda chains than with lambda chains bound to heavy chains. These two factors, especially when associated with weak staining (see below), may artefactually give rise to an apparently monoclonal pattern of staining in which kappa chains appear to be present in the absence of lambda chains.

Technical quality of immunoperoxidase stained sections

Weak staining. There is considerable variation from laboratory to laboratory in the intensity of peroxidase staining which is obtained when labelling tissue sections by immunohistological techniques. In the author's experience relatively weak staining, amounting to no more than a pale golden brown deposit of diaminobenzidine reaction product, is often accepted as satisfactory. An inevitable danger in these circumstances is that minor variations in technique from section to section may produce artefactually negative reactions. Generally speaking it should be possible to label any antigen present in substantial concentration in a section (e.g. plasma cell immunoglobulin) sufficiently strongly that the reaction product is dark brown in colour. Only under these conditions can neoplastic cells (which will often contain smaller amounts of an antigen than do their normal counterparts) be satisfactorily labelled.

In consequence the immunoperoxidase technique should be optimised on a tissue sample known to contain plentiful numbers of positive cells, and only when strong labelling is obtained should neoplastic biopsies be analysed.

Background staining. The simple rule concerning background staining is that it should be absent. There are exceptions to this; when an antigen is present diffusely through a tissue (e.g. immunoglobulin in inflamed tissue) some degree of background staining may be inevitable. However it should be emphasised that background staining is as great an enemy to the correct interpretation of immunoperoxidase labelling as is weak specific staining (see above). A frequently encountered trap is to identify strongly stained cells scattered in a sea of diffuse weaker background staining as showing positive labelling. However it should be realised that when non-specific background staining occurs (usually due to too high a concentration of primary antiserum) inevitably some structures will be more stongly labelled than others, and there is no justification for assuming that their greater intensity of labelling indicates that they alone contain the antigen under investigation.

In consequence histologists should be wary of interpreting any preparations showing diffuse background staining and should try to eliminate it (i.e. by diluting the primary antiserum further). If background staining cannot be overcome it may be wiser to report that a sample is uninterpretable, rather than to hazard an opinion which may prove to be mistaken.

CONCLUSION

As the preceding pages have illustrated, immunoperoxidase staining of lymphoid tissue biopsies is starting to play an increasingly important role in the diagnosis of human lymphoproliferative disorders. It is probable that the scope of these investigations will greatly enlarge in the next few years. Consequently the present chapter should be taken as an indication of the technical approach to be used in studying lymphoid tissue biopsies, rather than as a definitive account of all the staining patterns which may be obtained.

REFERENCES

Brandtzaeg P, Berdal P 1975 J chain in malignant human immunocytes. Scandinavian Journal of Immunology 4: 403–407
Curran R C, Jones E L 1978 Hodgkin's disease: an immunohistochemical and histological study. Journal of Pathology 125: 39–48
Gatter K C, Abdulaziz Z, Beverley P, Corvalan J R F, Ford C, Lane E B et al 1982 Use of monoclonal antibodies for the histopathological diagnosis of human malignancy. Journal of Clinical Pathology 35: 1253–1267
Isaacson P 1979 Immunochemical demonstration of J chain: a marker of B cell malignancy. Journal of Clinical Pathology 32: 802–807
Isaacson P 1979 Middle East lymphoma and alpha-chain disease. An immunohistochemical study. American Journal of Surgical Pathology 3: 431–437

Isaacson P, Wright D H 1979 Anomalous staining patterns in immunohistologic studies of malignant lymphomas. Journal of Histochemistry and Cytochemistry 27: 1197–1202
Isaacson P, Jones D B, Millward-Sadler G H, Judd M A, Payne S 1981 Alpha-1-anti-trypsin in human macrophages. Journal of Clinical Pathology 34: 982–990
Isaacson P, Wright D H, Judd M A, Jones D B, Payne S V 1980 The nature of immunoglobulin-containing cells in malignant lymphoma: an immunoperoxidase study. Journal of Histochemistry and Cytochemistry 28: 761–770
Janossy G, Thomas J A, Pizzolo G, Mattingley S, McLaughlin J, Habeshaw J et al 1980 Immunohistological diagnosis of lymphoproliferative diseases using selected combinations of antisera and monoclonal antibodies. British Journal of Cancer 42: 224–230
Mason D Y, Sammons R E 1978 Alkaline phosphatase and peroxidase for double immunoenzymatic labelling of cellular constituents. Journal of Clinical Pathology 31: 454–460

Mason D Y, Biberfeld P 1980 Technical aspects of lymphoma immunohistology. Journal of Histochemistry and Cytochemistry 28: 731–745

Mason D Y, Bell J I, Christensson B, Biberfeld P 1980 An immunohistological study of human lymphoma. Clinical and Experimental Immunology 40: 235–248

Mason D Y, Stein H, Naiem M, Abdulaziz Z 1981 Immunohistological analysis of human lymphoid tissue by double immunoenzymatic labelling. Journal of Cancer Research and Clinical Oncology 101: 13–22

Mason D Y, Naiem M, Abdulaziz Z, Gatter K, Nash J R G, Stein H 1982 Immunohistological applications of monoclonal antibodies. In: Fabre J W, McMichael A J (eds) Clinical applications of monoclonal antibodies. Academic Press p.585

Mason D Y, Taylor C R 1975 The distribution of muramidase (lysozyme) in human tissues. Journal of Clinical Pathology 28: 124–130

Mendelsohn G, Eggleston J C, Mann R B 1980 Relationship of lysozyme (muramidase) to histiocytic differentiation in malignant histiocytosis. Cancer 45: 273–279

Mepham B L, Frater W, Mitchell B S 1979 The use of proteolytic enzymes to improve immunoglobulin staining by the PAP technique. Histochemical Journal 11: 345–354

Naiem M, Gerdes J, Abdulaziz Z, Nash J R G, Stein H, Mason D Y 1981 Production of monoclonal antibodies for the immunohistological analysis of human lymphoma. In: Knapp W (ed) Leukaemia markers. Academic Press, New York, p 117

Pangalis G A, Nathwani B N, Rappaport H 1980 Detection of cytoplasmic immunoglobulin in well-differentiated lympho-proliferative diseases by the immunoperoxidase method. Cancer 45: 1334–1339

Pinkus G S, Said J W 1977 Specific identification of intra-cellular immunoglobulin in paraffin sections of multiple myeloma and macroglobulinaemia using an immuno-peroxidase technique. American Journal of Pathology 87: 47–58

Pizzolo G, Sloane J, Beverley P, Thomas J A, Bradstock K F, Mattingley S et al 1980 Differential diagnosis of malignant lymphoma and non-lymphoid tumours using monoclonal anti-leucocyte antibody. Cancer 46: 2640–2647

Poppema S, Elema J D, Halie M R 1978 The significance of intracytoplasmic proteins in Reed-Sternberg cells. Cancer 42: 1793–1799

Stein H, Bonk A, Tolksdorf G, Lennert K, Rodt H, Gerdes J 1980 Immunohistologic analysis of the organisation of normal lymphoid tissue and non-Hodgkin's lymphomas. Journal of Histochemistry and Cytochemistry 28: 746–760

Stein H, Mason D Y, Gerdes J, Ziegler A, Naiem M, Wernet P et al 1981 Immunohistology of B cell lymphomas. In: Knapp (ed) Leukaemia markers. Academic Press, p 99

Taylor C R 1976 An immunohistological study of follicular lymphoma, reticulum cell sarcoma and Hodgkin's diseaae. European Journal of Cancer 12: 61–75

Taylor C R 1978a Immunocytochemical methods in the study of lymphoma and related conditions. Journal of Histochemistry and Cytochemistry 26: 496–512

Taylor C R 1978b Immunoperoxidase techniques. Archives of Pathology and Laboratory Medicine 102: 113–121

van den Tweel J G, Taylor C R, Parker J W, Lukes R J 1978 Immunoglobulin inclusions in non-Hodgkin's lymphomas. American Journal of Clinical Pathology 69: 306–313

Warnke R, Levy R 1978 Immunopathology of follicular lymphomas. New England Journal of Medicine 298: 481–486

Warnke R, Levy R 1980 Detection of T and B cell antigens with hybridoma monoclonal antibodies: a biotin-avidin-horseradish peroxidase method. Journal of Histochemistry and Cytochemistry 28: 771–776

Warnke R, Miller R, Grogan T, Pederson M, Dilley J, Levy R 1980 Immunologic phenotype in 30 patients with diffuse large cell lymphoma. New England Journal of Medicine 303: 293–299

Zulman J, Jaffe R, Talal N 1978 Evidence that the malignant lymphoma of Sjogren's syndrome is a monoclonal B cell neoplasm. New England Journal of Medicine 299: 1215–1220

21

Breast tumours

M. G. Ormerod & J. P. Sloane

INTRODUCTION

There are only a limited number of aspects of breast pathology to which currently available histochemical techniques can usefully be applied and unfortunately these do not include such major problems as the distinction of in situ carcinoma from epithelial hyperplasias and the prediction of prognosis in individual breast carcinomas. However, there are a few circumstances in which histochemistry may be of help, such as in the distinction of anaplastic breast carcinoma from non-epithelial lesions like malignant lymphoma and in ascertaining local origin of a carcinoma in the breast in the absence of an in situ component. Certain methods may also facilitate the sub-typing of breast carcinomas but unfortunately this usually appears to have more functional than prognostic significance. Apart from mucin stains, there is also little to assist in the identification of metastatic breast carcinoma.

Immunocytochemical stains have been investigated extensively in recent years but the early promise of many methods has not been fulfilled. Antisera to a variety of substances have been used to stain sections of breast carcinomas and the reagents we describe fall broadly into four categories: milk-specific proteins which it was hoped would serve as tissue-specific markers but turn out to have little value; steroid hormone receptors whose cytochemical detection would have clear application but for which the available reagents are unsatisfactory; epithelial components such as Thompson-Friedenreich antigen, carcinoembryonic antigen (CEA) and epithelial membrane antigen (EMA) and the contractile proteins (myosin and actin) of myoepithelial cells. Antisera to epithelial components, especially EMA, may be of value in distinguishing carcinomas from non-epithelial lesions and also in the identification of metastatic breast carcinoma. They may also be of some assistance in identifying breast as the primary site of the tumour.

HISTOCHEMICAL STAINS

Mucin

Conventional mucin stains such as alcian blue pH 2.5-periodic acid Schiff (AB-PAS) reveal the presence of both neutral and acidic mucins in breast carcinomas as well as in the epithelial cells of the normal breast and in those of fibrocystic mastopathy. The acid mucin is usually sialidase-susceptible and hyaluronidase-resistant (Spicer et al 1962). In some breast carcinomas, mucin may be demonstrated in the form of intracytoplasmic globules; some of these are small and multiple but others are single and large enough to fill half the cell (Gad & Azzopardi 1975). In AB-PAS stains, the large globules often exhibit a blue rim with a magenta or purple dot in the centre producing a bull's-eye appearance (Fig. 21.1). Such vacuoles are most frequently found in in situ and infiltrating lobular carcinomas and almost certainly correspond to the intracytoplasmic lumina described ultrastructurally and immunohistochemically (Battifora 1975, Sloane & Ormerod 1981).

Mucin stains may be of value in distinguishing poorly differentiated carcinomas in the breast from tumours of different histogenesis, particularly malignant lymphomas. Although the bull's-eye vacuoles have been reported in carcinomas of other organs such as the stomach and lung they nevertheless appear to occur in the breast carcinomas more frequently than in those from other primary sites and may thus be of some value in determining the origin of metastatic lesions.

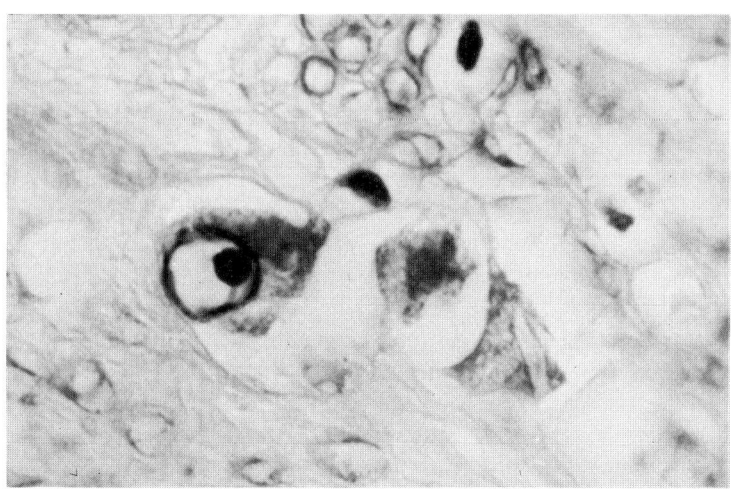

Fig. 21.1 Three metastatic breast carcinoma cells infiltrating the uterine cervix stained by the alcian-blue-periodic acid-Schiff method. One of the cells contains a large purple bull's eye globule which occupies about half the cell. (×1260)

Patterns of mucin staining may also assist in identifying the less common variants of breast carcinoma. In colloid carcinomas, the lakes of mucinous secretion stain uniformly for hyaluronidase-resistant, sialidase-susceptible acid mucin. In the rare adenoid cystic carcinoma, there is a biphasic appearance with AB positive stromal 'cysts' and PAS-positive ductal structures; this helps to distinguish it from morphologically similar carcinomas with a 'basaloid' cell pattern and from cribriform intraduct carcinomas (Anthony & James 1975).

The histiocytoid carcinomas usually secrete hyaluronidase-resistant AB-positive mucin which can be used to distinguish them from granular cell myoblastoma and the various histiocytic lesions with which they may be confused (Hood et al 1973). Lipid-rich carcinomas, on the other hand, do not contain mucin but stain strongly for neutral lipid (Aboumrad et al 1963, Ramos & Taylor 1974).

Elastic tissue

Conventional elastin stains reveal a cuff of fibrillar elastic tissue around the main ducts of the normal breast which is absent from the terminal ducts and lobules. Larger ducts have more elastin than smaller ones and the material tends to disappear on involution. Increased amounts of elastic tissue (elastosis) have been reported in up to 86% of breast carcinomas (Azzopardi & Laurini 1974) mostly around ducts and veins but also occasionally in a diffuse distribution in the stroma. It is much less prominent in medullary and colloid carcinomas. Elastosis is not specific for carcinoma and may also occur in association with benign breast lesions. However, elastosis is very rarely associated with metastatic carcinoma in the breast and elastin stains may thus be of value in determining whether a breast carcinoma is a primary or secondary tumour. Unfortunately, significant elastosis is uncommon in metastatic deposits from breast carcinoma and is thus of only limited value in determining the primary site of a metastatic carcinoma; metastases in certain sites such as gut and pancreas are said to exhibit elastosis more frequently than deposits in other tissues such as liver and bone (Azzopardi 1979). It has been reported that significant elastosis is associated with low histological tumour grade but claims that it is, by itself, associated with an improved prognosis have been disputed. There have been several reports that marked elastosis is related to oestrogen receptor positivity and response to hormone therapy. The rarity of elastosis in male breast carcinoma and the disappearance of elastin in the normal involuted breast certainly suggest that hormonal factors may be involved in elastin production.

Silver stains

It has been recognised recently that a proportion of breast carcinomas contain intracytoplasmic granules which stain with argyrophil methods such as Bodian and Grimelius (Fig. 21.2) but not with argentaffin techniques such as Masson-Fontana. In one study of 21 breast carcinomas (Taxy et al 1981), 11 exhibited focal argyrophilia and 5 showed zones where the growth pattern resembled that of a carcinoid tumour. The lesions exhibited either an intraduct or an intralobular component and some were oestrogen receptor-positive. Electron microscopy showed the presence of intracytoplasmic granules of variable morphology; some were greater than 600 nm and resembled lysosomes and others were less than 300 nm and resembled typical neurosecretory granules with a double membrane (Fig. 21.2, insert). Intermediate forms also existed. In another series (Capella et al 1980) approximately one third of mucoid carcinomas of the breast contained argyrophilic granules. Occasional argyrophilic cells have been reported in normal breast ducts.

Argyrophilia and dense core granules are, of course, usually associated with polypeptide hormone production but as no such substance has yet been identified in the breast, their significance in these lesions remains uncertain. From a practical point of view it is important to be aware of such tumours in order not to confuse them with metastatic carcinoids from other sites such as the gut or lung; they appear to behave as ordinary breast carcinomas.

Colonisation of breast carcinoma by melanocytes has also been reported (Azzopardi & Eusebi 1977) and appears to be a relatively common phenomenon in tumours which infiltrate the skin; such tumours are distinguished from the carcinoid-like lesions by their argentaffin positivity. As pigment may be present in the carcinoma cells as well as the melanocytes, confusion with malignant melanoma may arise. However, the mixture of tumour cells and dendritic melanocytes and the restriction of pigmentation to those tumour cells in the upper dermis should serve to make the distinction clear.

ENZYME HISTOCHEMICAL METHODS

These methods are not generally applicable to the diagnosis of breast tumours as frozen sections are required and they rarely contribute any information of significant diagnostic value. However, stains for alkaline phosphatase activity may be useful in identifying myoepithelial cells which are usually strongly positive in contrast to the negative breast epithelial cells. In intraduct carcinomas, a positive rim of alkaline phosphatase-positive myoepithelial cells can often be seen around negative carcinoma cells (Fig. 21.3). Breast carcinomas generally exhibit stronger staining for acid phosphatase (Fig. 21.4) and glucose-6-phosphate dehydrogenase than normal breast epithelial cells (Livni & Laufer 1975, Elias et al 1980).

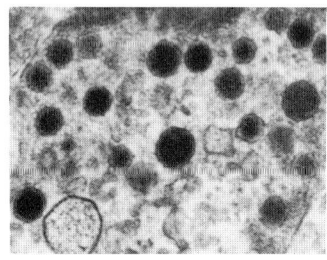
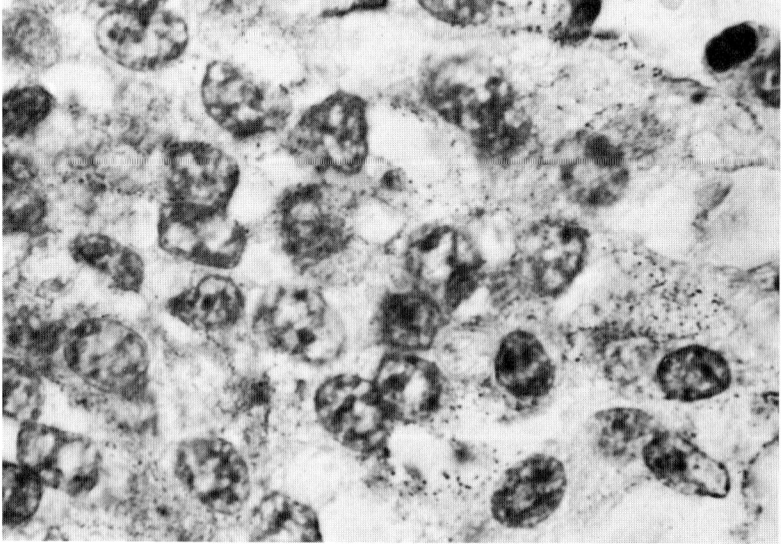

Fig. 21.2 Breast carcinoma stained by the Grimelius method for neurosecretory granules (× 1260). Inset is part of an electron micrograph of the same tumour showing that the granules are characteristic double membrane bound dense core vesicles. (× 35 100)

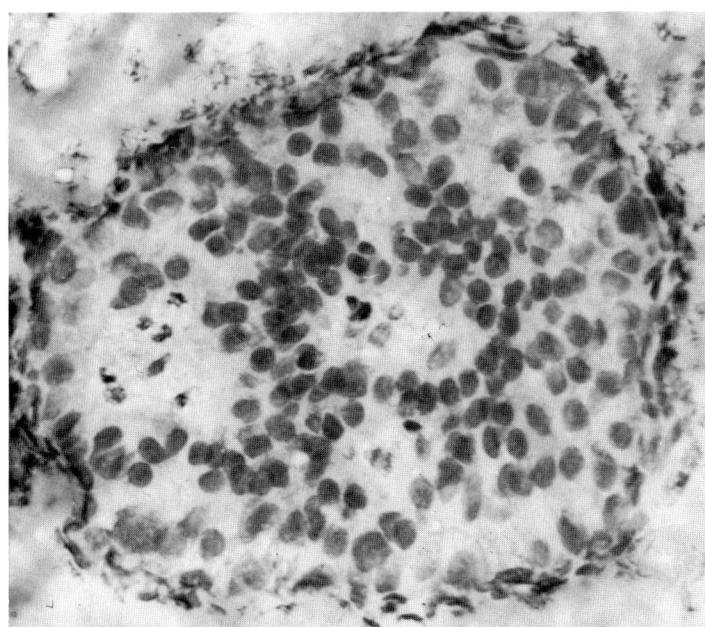

Fig. 21.3 Intraduct carcinoma stained for alkaline phosphatase by a diazo coupling method using naphthol AS:BI phosphoric acid and brentamine fase red R. The malignant cells are surrounded by a thin rim of myoepithelium. (Haemalum counterstain, ×315)

IMMUNOHISTOCHEMICAL STAINS

Milk proteins

It was hoped that milk proteins might serve either as tissue-specific markers for mammary carcinoma or have prognostic significance. Despite early results which showed some promise it can be concluded that carcinomas do not produce significant amounts of milk-specific proteins and that methods for their detection are consequently of little value. The two proteins which have been studied most extensively are α-lactalbumin (α-LA) and casein.

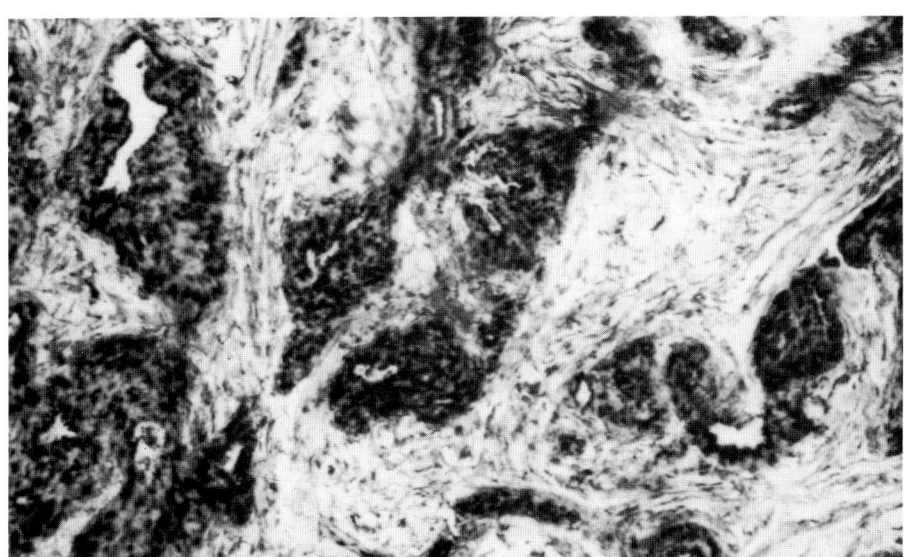

Fig. 21.4 Infiltrating breast carcinoma stained for acid phosphatase by diazo coupling method using naphthol AS:BI phosphoric acid and hexazonium pararosaniline. The epithelial islands are strongly positive and there is also weak patchy staining of the stroma. (Haemalum counterstain, ×126)

Although Walker (1979) found that an antiserum to α-lactalbumin stained approximately 50% of the breast carcinomas studied we were able to detect αLA only in sections of pregnant or lactating breast or in small lactational lesions occurring in non-pregnant patients. Breast carcinomas were negative even when they arose in lactating breasts (Bailey et al 1981). In experiments using αLA cDNA probes, Hall et al (1981) have shown that although αLA transcripts are present in normal mammary tissue in pregnancy and lactation, they cannot be detected in human breast tumour tissue. They did, however, demonstrate a peptide sharing antigenic determinants with human αLA in some breast tumours. Thus the balance of evidence at present seems to suggest that human breast carcinomas do not produce αLA but may secrete an αLA-like molecule which cross-reacts with many antisera raised to αLA. Further work is needed to characterise this molecule and to assess its value in the histological diagnosis and evaluation of human breast carcinoma.

Antisera to casein may react with breast carcinomas as well as with other non-mammary tissues (Pich et al 1976, Fortt et al 1979). The tissue distribution of the antigen is very similar to that reported for epithelial membrane antigen (EMA) (compare description of 'casein' in Pich et al 1976 with that of EMA in Sloane & Ormerod 1981). Using a radioimmunoassay for EMA we have shown that casein preparations can contain small amounts of EMA as an impurity (Ormerod, Bussolati, Steele and Sloane, in preparation). The EMA activity can be separated from casein electrophoretically. We have also found that the immunocytochemical activity can be removed from certain anti-casein sera by absorption with purified EMA; conversely, absorption of anti-EMA sera with some casein samples also removed activity. We suggest that many anti-casein sera contain antibodies to EMA and this gives rise to the observed immunocytochemical patterns.

Oestrogen receptors

Approximately two-thirds of all mammary carcinomas contain oestrogen receptors and about half of these tumours respond to endocrine therapy. The highest response rates (about 80%) are found in patients whose tumours contain both oestrogen- and progesterone-receptors while less than 10% of those which are receptor-negative respond. Receptor activity is estimated by measuring the binding of tritiated steroid to receptors in the cytosol of the tumour. A histochemical method for visualising steroid receptors would be valuable; it would demonstrate the distribution of cells which contain receptors as well as offering a rapid method of analysis which would be economical in tissue.

Several methods have been published. They rely on either an antibody to detect the reaction of oestradiol with the section or on the conjugation of oestradiol to peroxidase either directly or indirectly. These methods have been adversely criticised — the main grounds for criticism have been summarised by Chamness et al (1980).

One of the problems associated with all methods is that of fixing the receptors in a frozen section of the tissue. The receptors are labile and it has not been demonstrated unequivocally that they survive fixation or which is the best fixative to use. In unfixed sections, the soluble receptor is liable to be extracted by the aqueous incubation media (McCarty et al 1980).

Another major problem is that, as well as the high affinity receptors, there are binding sites for oestradiol which have high capacity and low affinity. The concentrations of oestradiol or oestradiol conjugate used in all the reported methods are far in excess of that needed to saturate the high affinity sites. It has been suggested that the receptor itself is not visualised but that hormone-specific binding to low affinity sites is detected (Mercer et al 1980).

At present, one may conclude that high-affinity oestrogen receptors have yet to be demonstrated unequivocally in tissue sections.

Thomson-Friedenreich antigen and the peanut lectin receptor

The Thomson-Friedenreich or T-antigen is a precursor of the MN blood group antigens. Both M and N antigens can be converted to T by removal of neuraminic acid. The immuno-dominant determinant of the T antigen is the terminal sequence, β-D-galactosyl-1 → 3-N-acetyl-D-galactosamine. This disaccharide is also the binding site of the peanut (*Arachis hypogaea*) lectin (PNA). Using serological methods, Springer et al (1975) showed that normal and neoplastic mammary glands carried M- and N-active substances. The precursor T antigen was only exposed in malignant tissue. Several authors have since shown that PNA receptors are expressed patchily on tissue sections of normal human breast (Stegner et al 1981, and references therein). The intensity of staining is increased by prior treatment of the section with neuraminidase. The PNA receptors are on the luminal membrane and on the secreted intraductal products. They are not found on the adjacent cell membranes or in the cytoplasm.

In tumours, PNA receptors are found on the surfaces and secretions of any lumen but they are also present, in many tumours, on the adjacent cell membranes and in the cytoplasm (see Stegner et al 1981, and references therein). In this respect, the distribution of those receptors is similar to that of EMA. However, the PNA receptors are frequently absent from poorly differentiated mammary carcinomas.

Similar results to the above have been obtained using a rabbit anti-T serum (Stegner et al 1981). Using human anti-sera, Howard & Taylor (1979) failed to detect T antigen on normal tissue and benign lesions but obtained staining on all 13 breast tumours studied including two metastatic deposits. This system needs further study.

Caution needs to be exercised in comparing results using PNA with those using antisera since their reactions are not necessarily identical. Although Stegner et al (1981) observed no differences in breast, they found that their rabbit antisera stained vascular endothelia and blood cells while PNA did not. No attempt to use PNA or anti-T sera to detect micrometastases has been reported.

Carcinoembryonic antigen (CEA)

Although many mammary carcinomas express CEA, the distribution is frequently uneven and patchy. Shousha & Lyssiotis (1978) stained sections from 74 carcinomas and 43 benign lesions of the breast; 66% of the carcinomas were positive for CEA. The most interesting claim was that all tumour tissue found in lymphatics and in metastases in lymph nodes contained CEA. (Using our anti-serum this is not the experience in our laboratory.) They also found a significant relationship between CEA-positivity of the primary and the presence of lymph node metastases.

However, a similar study by Walker (1980) failed to confirm this point. The possible prognostic significance of CEA in breast tumours needs further investigation.

Epithelial membrane antigen

This antigen is a cell surface component which can be stained in conventional formalin-fixed paraffin wax-embedded sections by antisera raised to milk fat globule membranes (Sloane & Ormerod 1981). It is confined to luminal and surface membranes (and to a lesser extent to the cytoplasm) of epithelial tissues in the normal state; increased staining may be observed in many neoplasms as well as some non-neoplastic states. The staining of tumours is related both to their histogenesis and degree of differentiation being found only in lesions of surface epithelial or mesothelial origin and being more consistently present in well and moderately differentiated neoplasms although a significant proportion of anaplastic carcinomas are positive.

Although EMA is not specific for breast and breast carcinoma, it has, like other epithelial cell markers, a number of useful applications in the diagnosis of breast tumours. It can be used to distinguish carcinomas from non-epithelial tumours of the breast; this is especially useful with malignant lymphomas and with breast carcinomas exhibiting spindle cell metaplasia which may be mistaken for spindle cell sarcomas. Intracytoplasmic lumina may also be identified in many breast carcinomas. These structures have been previously described ultra-structurally and although they may occur in carcinomas other than those of breast origin, they seem to occur with greater frequency in breast tumours and may thus have some value in identifying breast as the primary site in metastatic lesions.

Antisera to EMA can be used to identify small deposits of metastatic carcinoma, which would normally go unrecognised, in sites such as lymph node, liver, marrow, adrenal, ovary and skin where the normal tissues do not express the antigen (Sloane et al 1980). We have found that the most useful applications are in identifying metastatic cells in sections of bone marrow and skin (see Fig. 21.5) and in detecting single carcinoma cells in aspirates of marrow (Plate 24) (Dearnaley et al 1981). Finally, EMA is expressed with greater consistency in breast

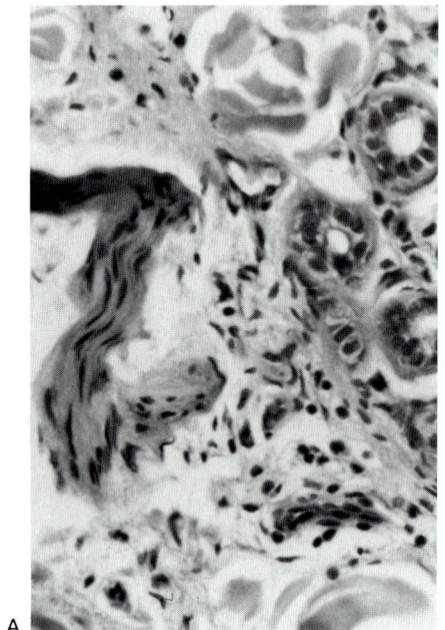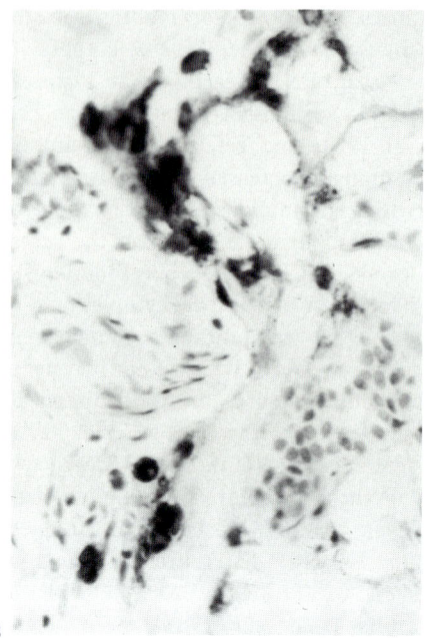

A B

Fig. 21.5 Skin biopsy from a patient with breast carcinoma. (A) H & E stain showing dermal infiltration between a sweat gland and a nerve by cells with small densely staining nuclei and copious cytoplasm. (B) Immunocytochemical method for epithelial membrane antigen showing the epithelial nature of the infiltrating cells. (×315)

carcinomas than in those arising from other organs and, indeed, no negatives were found in a series of 50 cases studied (Sloane & Ormerod 1981). A negative stain would thus tend to indicate that a carcinoma in the breast was not of local origin.

Contractile proteins

Myoepithelial cells contain high concentrations of contractile proteins and antisera to either actin or myosin can be used to identify those cells in the breast. A weak reaction with the apical surface of the normal epithelial cells is also observed. Many carcinomas react with antisera, the proteins being concentrated at the cell surface (Gabbiani et al 1978, Macartney et al 1979a, Bussolati et al 1980). These stains can be helpful in distinguishing tubular carcinoma, in which there are usually no myoepithelial cells surrounding the tubules, from benign lesions (Eusebi et al 1979). Visualisation of the myoepithelial cells in in situ carcinoma could allow recognition of the early steps in invasion in which the tumour breaks the line of myoepithelial cells surrounding the duct. Fixing and embedding tissue frequently destroys the antigenic groups on myosin

and actin and this limits the potential value of these antisera. This is not invariably true; Bussolati et al (1980) have obtained a rabbit antiserum against denatured chicken gizzard actin which reacts with human actin in formalin-fixed tissue. Another method of solving this problem is to abandon the immunocytochemical method and use the tannic acid-phosphomoblybdic acid stain for myoid fibrils (Macartney et al 1979b), although, in our experience, this method is not as sensitive (Fig. 21.6).

Mammary tumour virus-related protein

Spiegelman and co-workers (1980) have raised antisera against a glycoprotein (gp52) from mouse mammary tumour virus and have found that this antiserum cross-reacts with a protein found in human breast cancers. Using an immunoperoxidase technique on paraffin wax-embedded material they found positive staining in 212 of the 447 cases of human mammary carcinoma examined. 119 benign lesions were negative as were 9 normal and 9 lactating glands. Only 1 out of 107 non-mammary malignancies was positive. The group is attempting to isolate the human antigen responsible and to raise a

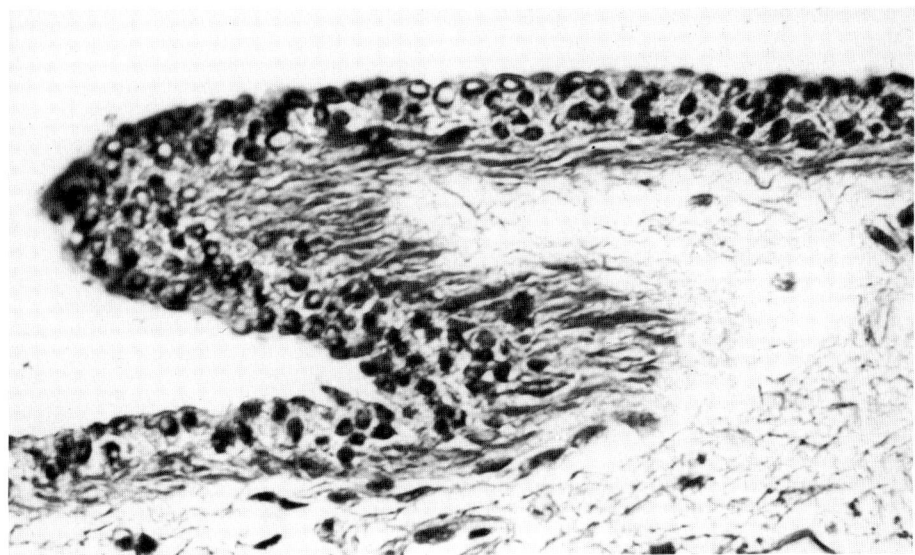

Fig. 21.6 Major breast ducts stained by the phospho-tungstic acid haemotoxylin method (PTAH) for myofibrils showing a well-developed myoepithelial layer (×315).

monoclonal antibody which potentially would be of considerable value.

FUTURE DEVELOPMENTS

Monoclonal antibodies to membranes derived from lactating breast tissue and from mammary carcinoma are being produced by several laboratories. These antibodies can be used for immunocytochemical staining and some of them show an apparent specificity for carcinomas. Their use in diagnostic histopathology has yet to be explored fully but it seems probable that monoclonal antibodies will extend the range of useful immunocytochemical stains.

REFERENCES

Aboumrad M H, Horn R C, Fine G 1963 Lipid-secreting mammary carcinoma: report of a case associated with Paget's disease of the nipple. Cancer 16: 521–525

Anthony P P, James P D 1975 Adenoid cystic carcinoma of the breast: prevalence, diagnostic criteria, and histogenesis. Journal of Clinical Pathology 28: 647–655

Azzopardi J G 1979 Problems in breast pathology. W B Saunders, Philadelphia, p 387

Azzopardi J G, Eusebi V 1977 Melanocyte colonization and pigmentation of breast carcinoma. Histopathology 1: 21–30

Azzopardi J G, Laurini R N 1974 Elastosis in breast cancer. Cancer 33: 174–183

Bailey A J, Sloane J P, Trickey B A, Ormerod M G 1982 An immunocytochemical study of α-lactalbumin in human breast tissue. Journal of Pathology 137: 13–23

Battifora H 1975 Intracytoplasmic lumina in breast carcinoma. Archives of Pathology 99: 614–617

Bussolati G, Alfani V, Weber K, Osborn M 1980 Immunocytochemical detection of actin on fixed and embedded tissues: its potential use in routine pathology. Journal of Histochemistry and Cytochemistry 28: 169–173

Capella C, Eusebi V, Mann B, Azzopardi J G 1980 Endocrine differentiation in mucoid carcinoma of the breast. Histopathology 4: 613–630

Chamness G C, Mercer W D, McGuire W L 1980 Are histochemical methods for estrogen receptor valid? Journal of Histochemistry and Cytochemistry 28: 792–798

Dearnaley D P, Sloane J P, Ormerod M G, Steele K, Coombes R C, Clink H McD et al 1981 Increased detection of mammary carcinoma cells in marrow smears using antisera to epithelial membrane antigen. British Journal of Cancer 44: 85–90

Elias E A, Elias R A, Bijlsma P J, Tazelaar D J 1980 The enzyme histochemistry of metastasizing basal cell carcinoma of the skin. Journal of Pathology 131: 235–241

Eusebi V, Betts C M, Bussolati G 1979 Tubular carcinoma: a variant of secretory breast carcinoma. Histopathology 3: 407–419

Fortt R W, Gibbs A R, Williams D, Hansen J, Williams I 1979 The identification of 'casein' in human breast cancer. Histopathology 3: 395–406

Gabbiani G, Csank-Brassert J, Schneeberger J-C, Kapanci Y, Trenchev P, Holborow J 1976 Contractile proteins in human cancer cells: immunofluorescent and electron microscopic study. American Journal of Pathology 83: 457–474

Gad A, Azzopardi J G 1975 Lobular carcinoma of the breast: a special variant of mucin-secreting carcinoma. Journal of Clinical Pathology 28: 711–716

Hall L, Craig R K, Davies M S, Ralphs D N L, Campbell P N 1981 α-Lactalbumin is not a marker of human hormone-dependent breast cancer. Nature 290: 602–604

Hood C I, Font R L, Zimmerman L E 1973 Metastatic mammary carcinoma in the eyelid with histiocytoid appearance. Cancer 31: 793–800

Howard D R, Taylor C R 1979 A method for distinguishing benign from malignant breast lesions utilising antibody present in human sera. Cancer 43: 2279–2287

Livni N, Laufer A 1975 Histochemical studies of human breast tumors: activity of alkaline phosphatase, acid phosphatase and glucose-6-phosphate dehydrogenase. Pathology and Microbiology 42: 159–170

Macartney J C, Roxburgh J, Curran R C 1979 Intracellular filaments in human cancer cells: a histological study. Journal of Pathology 129: 13–20

Macartney J C, Trevithick M A, Kricka L, Curran R C 1979 Identification of myosin in human epithelial cancers with immunofluorescence. Laboratory Investigation 41: 437

McCarty K S Jnr Woodard BH, Nichols DE, Wilkinson W, McCarty KS Snr 1980 Comparison of biochemical and histochemical techniques for oestrogen receptor analyses in mammary carcinoma Cancer 46: 2842–2845

Mercer W D, Lippman M E, Wahl T M, Carlson C A, Wahl D A, Lezotte D et al 1980 The use of immunocytochemical techniques for the detection of steroid hormones in breast cancer cells. Cancer 46: 2859–2868

Pich A, Bussolati G, Carbonara A 1976 Immunocytochemical detection of casein and casein-like proteins in human tissues. Journal of Histochemistry and Cytochemistry 24: 940–947

Ramos C V, Taylor H B 1974 Lipid-rich carcinoma of the breast: a clinicopathologic analysis of 13 examples. Cancer 33: 812–819

Shousha S, Lyssiotis T, Godfrey V M, Schener P J 1979 Carcinoembryonic antigen in breast cancer tissue: a useful prognostic indicator. British Medical Journal 1: 777–779

Sloane J P, Ormerod M G 1981 Distribution of epithelial membrane antigen in normal and neoplastic tissues and its value in diagnostic tumor pathology. Cancer 47: 1786–1795

Sloane J P, Ormerod M G, Imrie S F, Coombes R C 1980 The use of antisera to epithelial membrane antigen in detecting micrometastases in histological sections. British Journal of Cancer 42: 393–398

Spicer S S, Neubecker R D, Warren L, Henson J G 1962 Epithelial mucins in lesions of the human breast. Journal of the National Cancer Institute 29: 963–975

Spiegelman S, Keydor I, Mesa-Tejada R, Ohro T, Ramanarayanan M, Nayak R et al 1980 Possible diagnostic implications of a mammary tumour virus-related protein in human breast cancer. Cancer 46: 879–892

Springer G F, Desai P R, Banatwala I 1975 Brief communication: blood group MN antigens and precursors in normal and malignant breast glandular tissue. Journal of the National Cancer Institute 54: 335–339

Stegner H E, Fischer K, Poschmann A 1981 Immunohistochemical localization of Thomsen-Friedenreich antigen in normal and malignant breast tissue using peroxidase-antiperoxidase technique. Tumor Diagnostik 3: 127–130

Taxy J B, Tischler A S, Insalaco S J, Battifora H 1981 'Carcinoid' tumor of the breast: a variant of conventional breast cancer? Human Pathology 12: 170–179

Walker R A 1979 The demonstration of α-lactalbumin in human breast carcinomas. Journal of Pathology 129: 37–42

Walker R A 1980 Demonstration of carcinoembryonic antigen in human breast carcinomas by the immunoperoxidase technique. Journal of Clinical Pathology 33: 356–360

22

Soft tissue tumours

D. H. Mackenzie & M. Isabel Filipe

The term 'soft tissue tumour' is not very scientific. Nevertheless it is reasonably well understood and refers to masses occurring in soft tissues excluding lymphomas and tumours of the skin. Most of these tumours will be of mesenchymal or, less often, neuroectodermal origin but epithelial neoplasms arising from adnexial structures and occasional metastases will enter the differential diagnosis. The spectrum of behaviour of soft tissue tumours is so wide that accurate identification of any given neoplasm is vital if the correct treatment is to be instituted. Histochemistry can play an important if somewhat limited role in the diagnosis of these neoplasms and can be relevant in the following situations.

TUMOURS CONTAINING MUCOPOLYSACCHARIDES

Focal accumulation of mucopolysaccharide material can be seen in many abnormal conditions of skin and soft tissues and cannot be considered here. However, some soft tissue tumours owe their bulk to production of mucopolysaccharides and, in some of these, histochemistry can help. The mucosubstances produced are usually the same as those produced by the tissue of origin (Dobrogorski & Braunstein 1963) and the most important substances are hyaluronic acid, chondroitin 4- and 6-sulphates and keratan sulphate. The so-called myxoid tumours of soft tissues are listed as follows: the benign varieties include the myxoma, myxoid tumours of nerve sheath origin, myxoid chondroma, myxoid lipoma and lipoblastoma (lipoblastomatosis); the malignant tumours to be considered are myxoid liposarcoma,

myxoid malignant fibrous histiocytoma, extraskeletal myxoid chondrosarcoma and myxoid fibrosarcoma.

The histochemical differentiation of mucopolysaccharides can be achieved by staining with alcian blue at pH 2.5 and with alcian blue containing increasing concentrations of $MgCl_2$ (CEC method) (Scott & Dorling 1965). Further identification is obtained by pretreatment of the sections with hyaluronidase and chondroitinases (CHase) ABC.

A. Alcian blue pH 2.5 (AB) with and without pretreatment with hyaluronidase

Sulphated and non-sulphated acid mucosubstances are stained by this method. Testicular hyaluronidase depolymerises hyaluronic acid as well as chondroitin 4- and 6-sulphates whilst streptomyces hyaluronidase removes hyaluronic acid only.

B. Alcian blue pH 5.6 + $MgCl_2$ (CEC method)

Hyaluronic acid has a low CEC and the AB staining is abolished at concentrations above 0.1–0.25M $MgCl_2$, whilst chondroitin sulphates (CS) and keratan sulphate (KS) have a higher CEC and the staining is retained at 0.55M and 1.0M $MgCl_2$ respectively. Additional use of testicular hyaluronidase and CHase ABC may be helpful in equivocal CEC results to distinguish chondroitin sulphates from keratan sulphate. CHase digests chondroitin 4– and 6-sulphate, has a partial effect on dermatan sulphate and hyaluronic acid and no effect on heparin and keratan sulphate (Yamada 1974, Kindblom & Karlsson 1977).

Unfortunately in the majority of such neoplasms the mucosubstance is hyaluronic acid and histochemistry cannot separate one such neoplasm from another. When cartilagenous differentiation is present however, even poorly differentiated neoplasms tend to contain chondroitin 4- and 6-sulphate as in fetal cartilage, while better-differentiated ones may also contain keratan sulphate as in adult cartilage. These may explain the variation in the effect of hyaluronidase and CHases ABC on the alcian blue staining. In the well-differentiated cases positive alcian blue staining at pH 2.5 will be at least partially retained after pretreatment with CHases ABC. In addition the CEC will also be variable. In the presence of increasing concentrations of $MgCl_2$ in a solution of alcian blue at pH 5.6, staining will be retained at $0.55M$ $MgCl_2$ and sometiems at $1.0M$ $MgCl_2$ indicative of keratan sulphate (Kindblom &

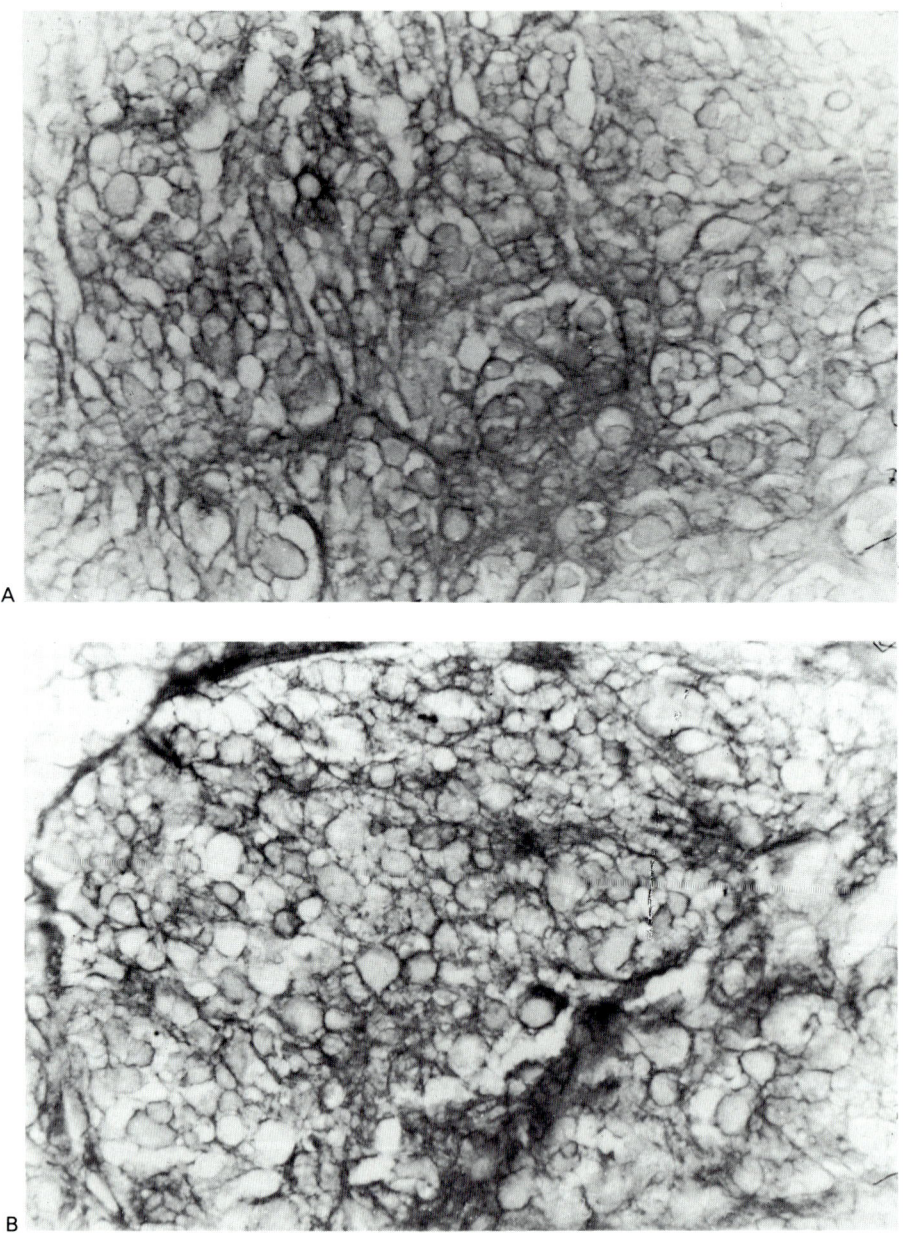

Fig. 22.1 Extraskeletal myxoid chondrosarcoma. A. The mucopolysaccharide material shows positive AB staining at pH 2.5. B. The stain is partially retained by prior treatment with hyaluronidase. ×125

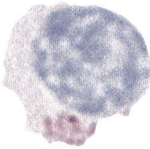

Plate 14 Mucopolysaccharidosis; blood film stained with toluidine blue (Muir et al) showing a collection of metachromatic granules in the cytoplasm of a lymphocyte. ×2000

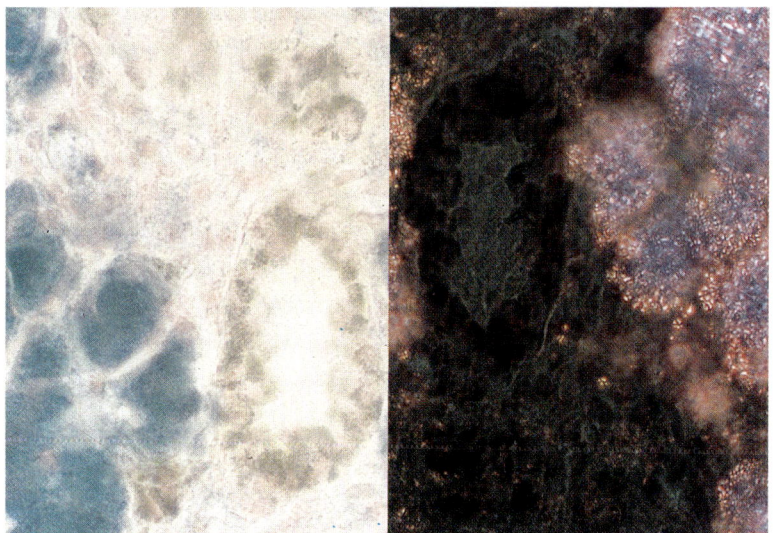

Plate 15 Niemann-Pick disease; cryostat section of spleen stained with Sudan black and carmalum, showing the large foamy Niemann-Pick cells staining blue and the endothelial cells which contain a greyish-black granular storage substance. In polarised light only the Niemann-Pick cells show red birefringence. ×468

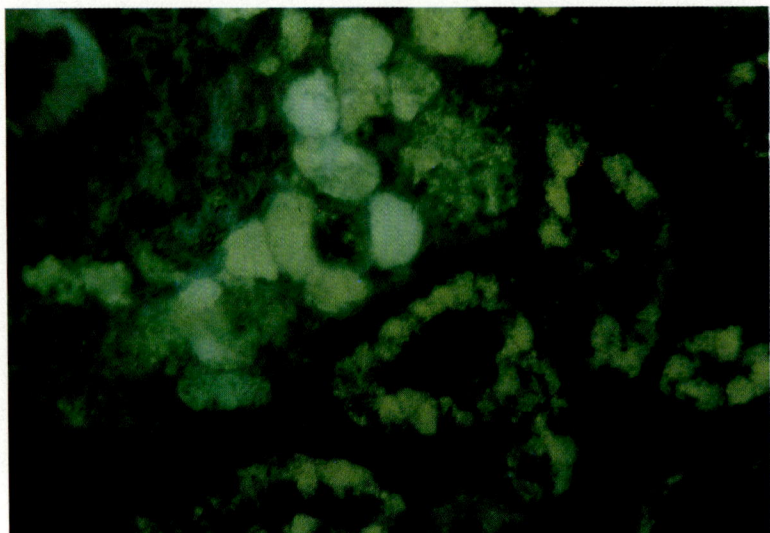

Plate 16 Niemann-Pick disease; unstained cryostat section photographed to show autofluorescence in the Niemann-Pick cells and in the endothelial cells. Dark-ground conditions. ×468

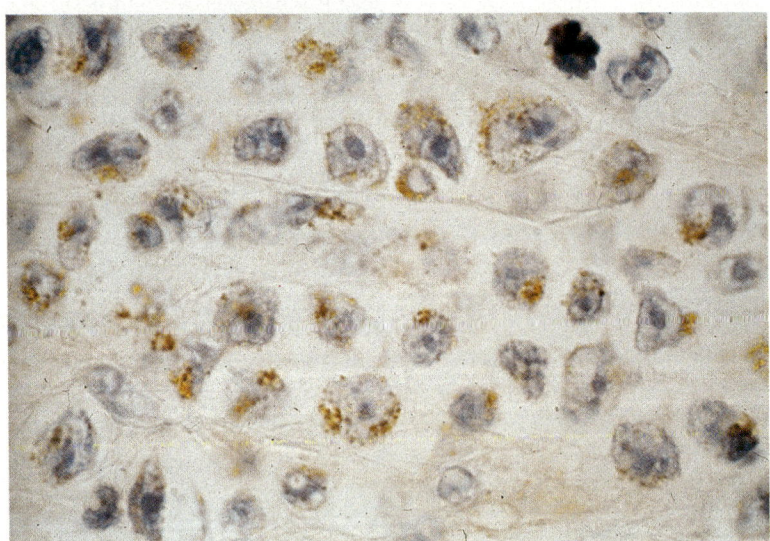

Plate 17 Immunoperoxidase staining of a true histiocytic lymphoma for a_1-anti-trypsin (paraffin wax section). Note the characteristic granular deposits of this constituent in many of the neoplastic cells, frequently localised to one zone close to the nucleus. This antigen is a useful marker of histiocytic cells and is best revealed in paraffin wax sections following treatment with trypsin. Illustration kindly provided by Dr P. Isaacson.

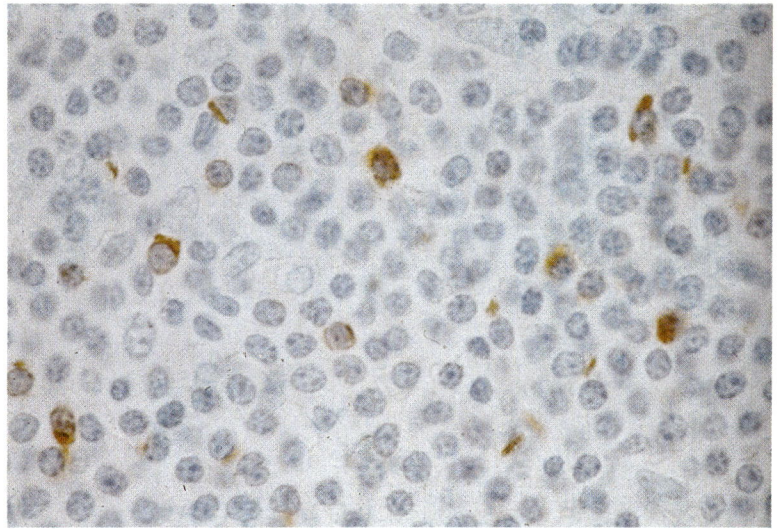

A

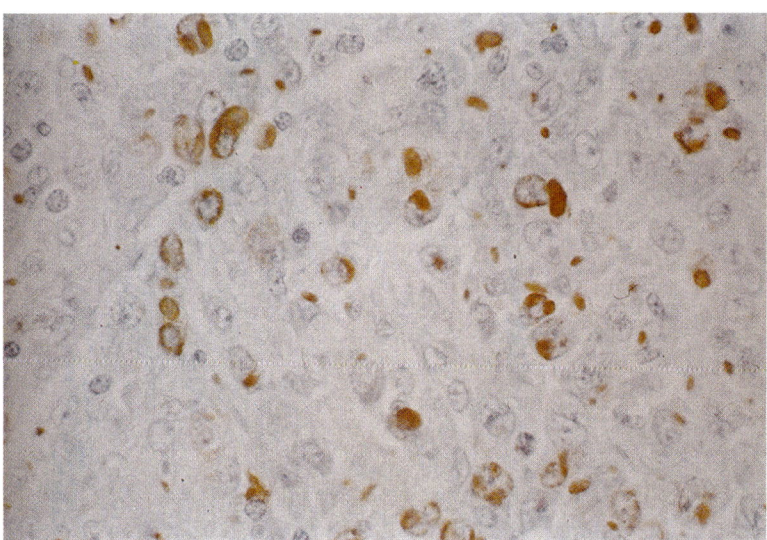

B

Plate 18 Immunoperoxidase staining of paraffin wax sections from two cases of lymphoma in which the neoplastic cells contained immunoglobulin-positive inclusion bodies. In one sample (A), from a case of chronic lymphocytic leukaemia, the inclusions are small and rod-shaped in outline whereas in the other case (B), a high grade intestinal lymphoma, the inclusions are larger and more rounded in appearance. In other cases inclusions may take the form of multiple small granules. The significance of Ig-positive cytoplasmic inclusions is that they are indicative of endogenous production by the neoplastic cell, since immunoglobulin absorbed from the cell's environment (see Fig. 20.9 and Plate 23) never adopts this form.

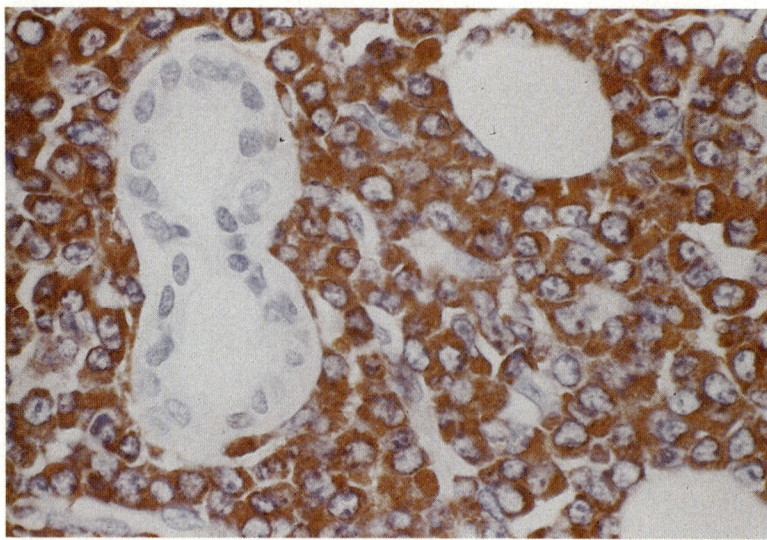

Plate 19 Immunoperoxidase staining of a cutaneous deposit of neoplastic cells in a case of IgD myeloma. The section is of paraffin wax-embedded tissue and has been stained for IgD. Note that the majority of neoplastic cells contain immunoglobulin, whilst a skin appendage is negative.

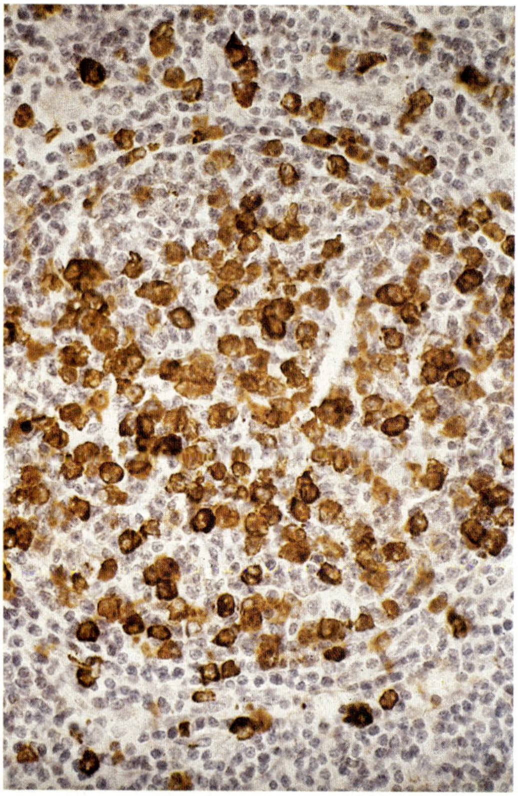

A 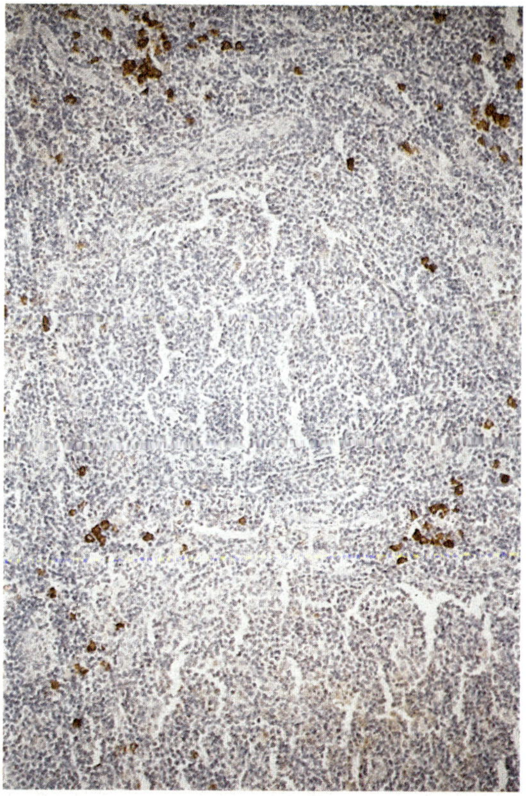   B

Plate 20 Immunoperoxidase staining of follicular lymphoma (paraffin wax section) for (A) lambda and (B) kappa light chains. Note the presence of many neoplastic cells within the follicle staining for lambda chains but not for kappa chains. A few scattered kappa-positive cells are present in the extrafollicular tissue. Illustration kindly provided by Dr P. Isaacson.

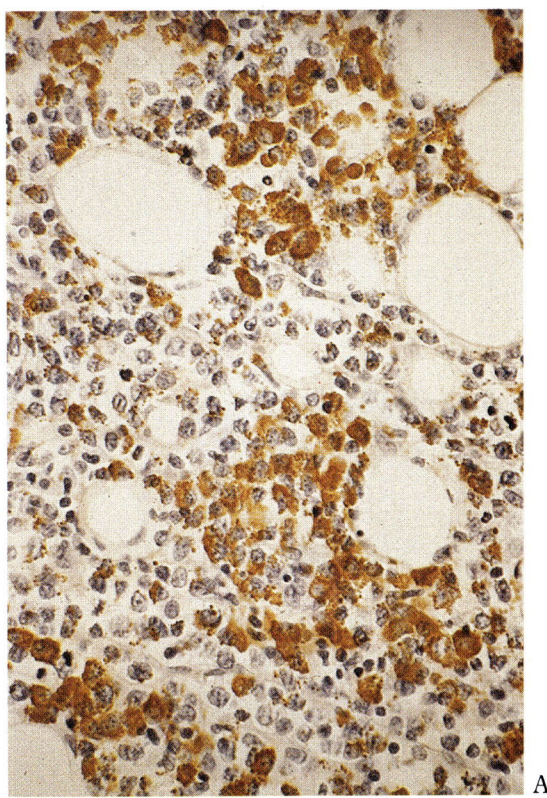

A

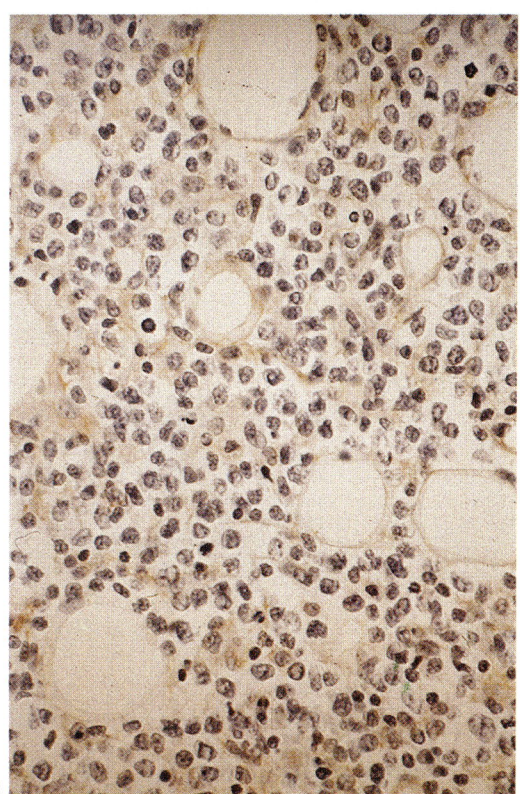

B

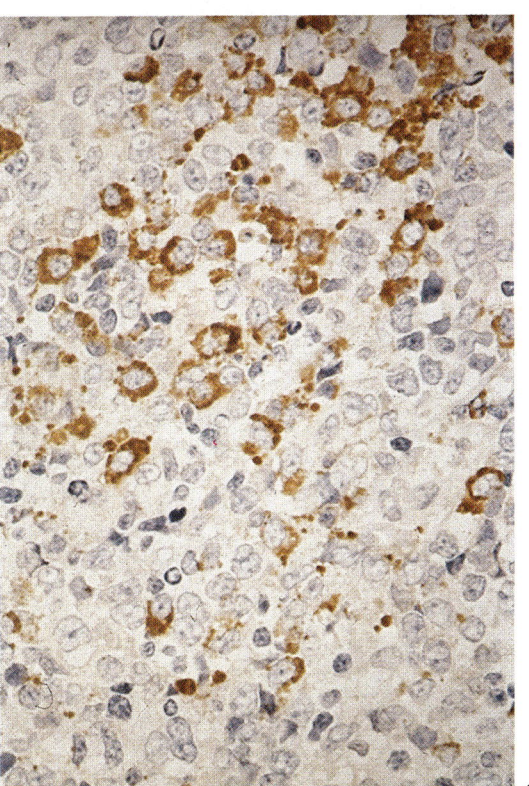

C

Plate 21 Immunoperoxidase staining of a germinal centre-derived neoplasm (paraffin wax section) for (A) kappa chains, (B) lambda chains and (C) J chains. Clusters of neoplastic cells containing kappa and J chain are present. Illustration kindly provided by Dr P. Isaacson.

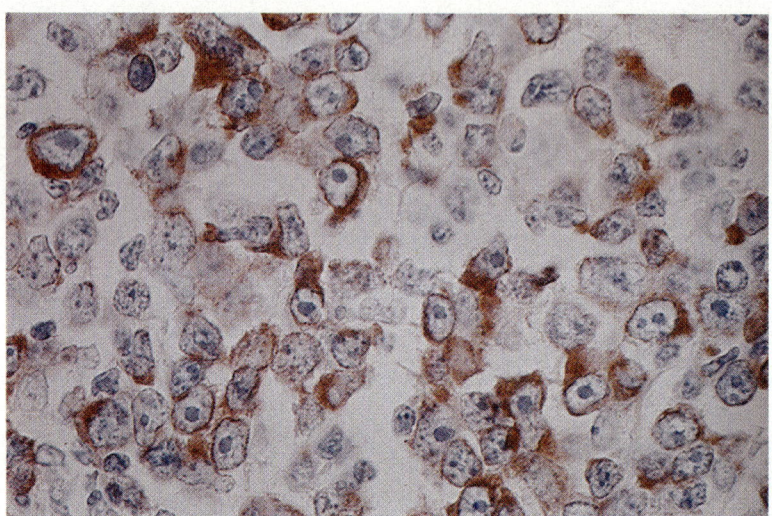

Plate 22 Immunoperoxidase staining of a high-grade lymphoma (paraffin wax section) showing that many of the tumour cells contain immunoglobulin of a single light chain class. Illustration kindly provided by Professor H. Stein.

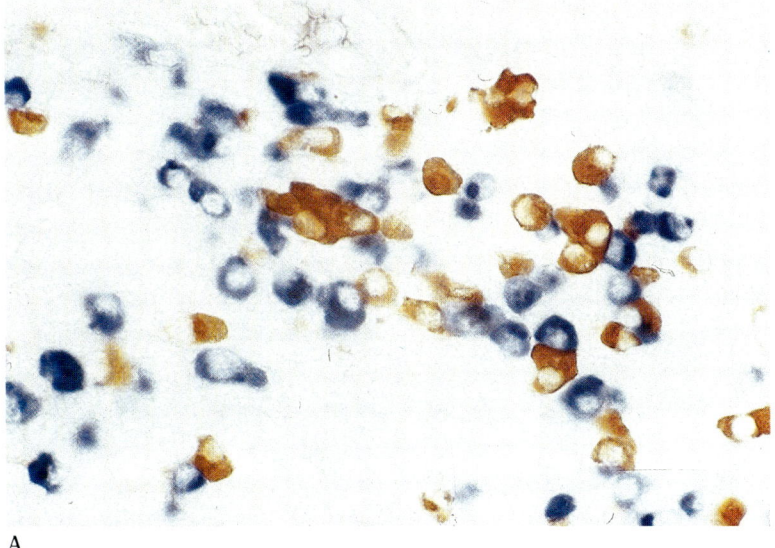

A

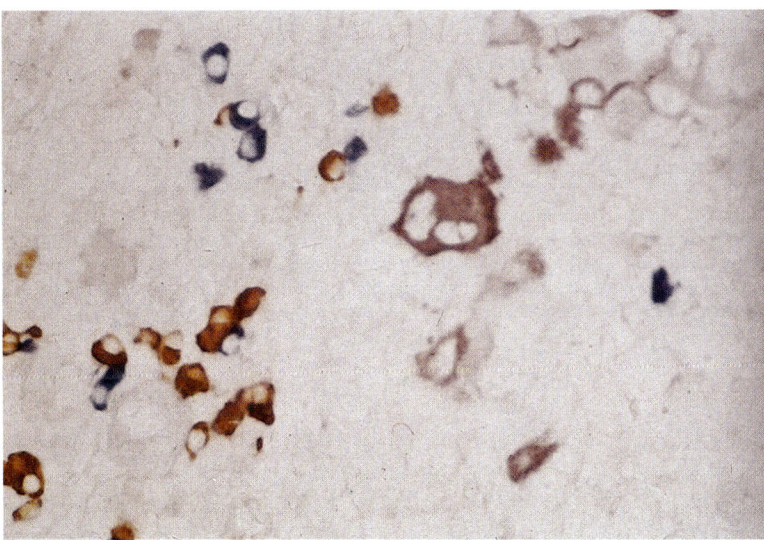

B

Plate 23 Double immunoenzymatic staining of paraffin wax sections from (A) normal tonsil and (B) Hodgkin's disease. This technique (details of which may be found in Mason & Sammons 1978 and Mason et al 1981) utilises alkaline phosphatase as a second label in conjunction with peroxidase. The two enzymes are revealed using substrates which give contrasting colour reactions. Both sections have been stained for kappa and lambda light chains. In the reactive tissue (A) plasma cells are labelled for one or other light chain type whereas in the case of Hodgkin's disease a Reed-Sternberg cell, showing a mixed labelling reaction, is seen, indicating that both types of light chain are present within its cytoplasm. This is accounted for by the presence of polyclonal serum IgG within the cell. A few plasma cells staining for only one light chain class are also seen in this illustration.

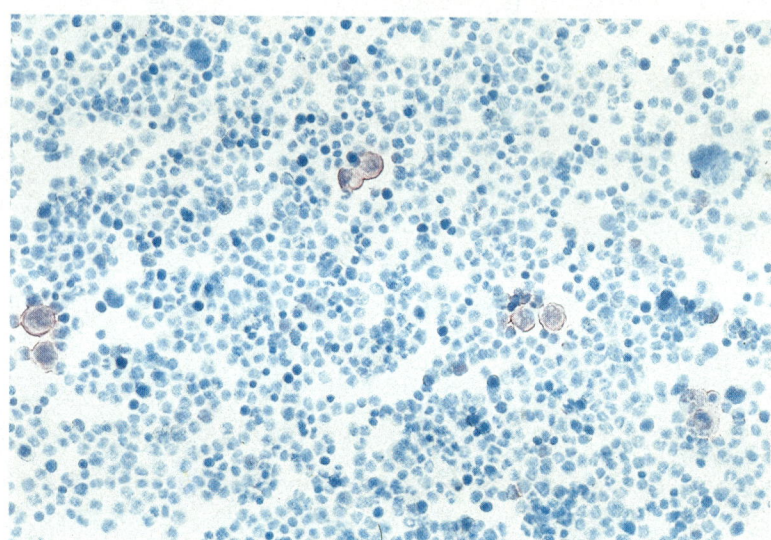

Plate 24 An indirect immuno-alkaline phosphatase method for epithelial membrane antigen on a smear made from an aspirate of bone marrow from a patient with carcinoma of the breast. EMA-positive cells are clearly demonstrated. These are presumptively metastatic carcinoma cells. ×3600. Photograph supplied by Dr D.P. Dearnaley, Ludwig Institute for Cancer Research.

Angervall 1975, Allen 1980, Mackenzie 1981). These methods are particularly valuable in distinguishing such entities as the extraskeletal myxoid chondrosarcoma (Fig. 22.1) from the myxoid liposarcoma (Fig. 22.2) (Table 22.1).

One must be aware that other lesions not derived from cartilage may contain sulphated glycosaminoglycans e.g. the myxoid tumour of nerve sheath origin, nodular fasciitis. In the interpretation of the results the staining of collagen which contains CS-4 and -6, should not be confused with the staining of the mucopolysaccharide material produced by the tumours.

The colloid-type of carcinoma metastasizing in soft tissue can be differentiated from myxoid tumours with D-PAS method for epithelial mucins. Under

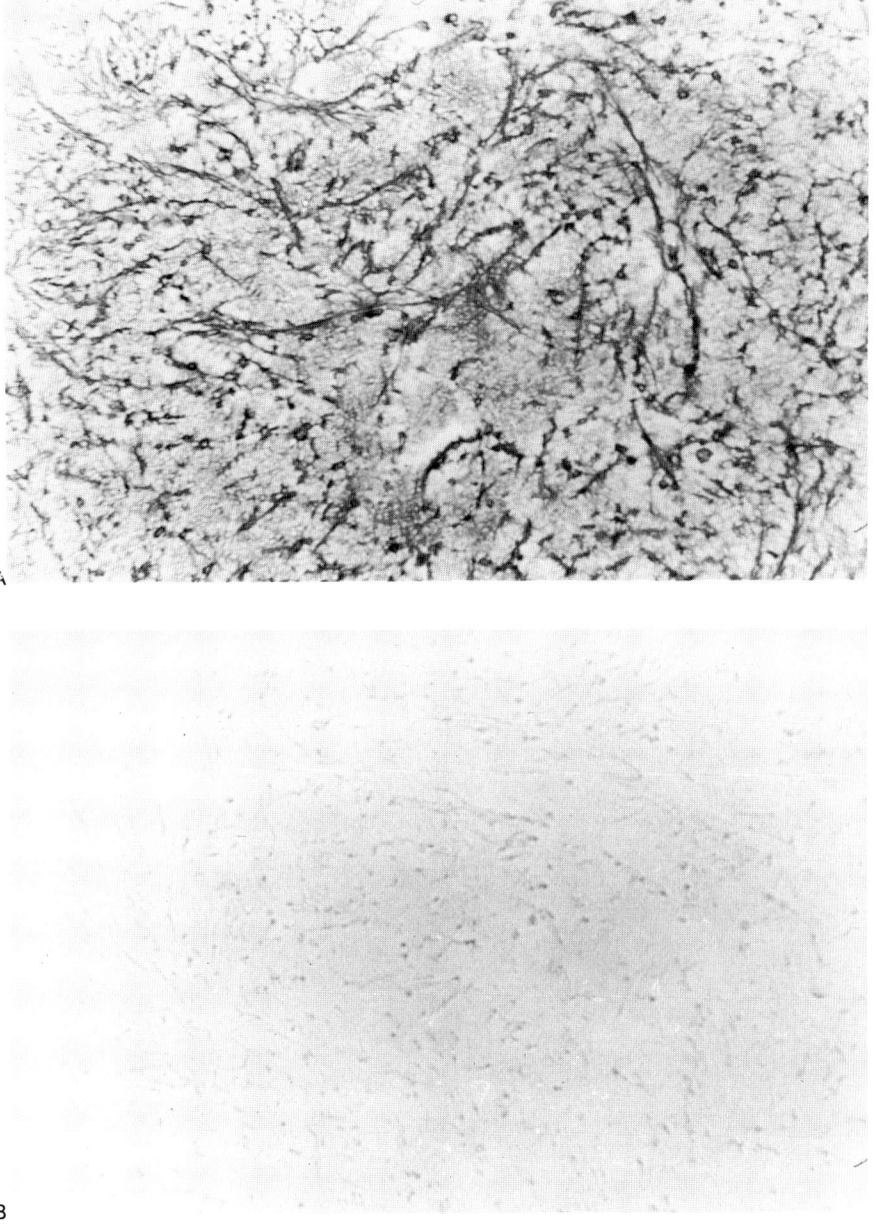

A

B

Fig. 22.2 Myxoid liposarcoma. A. The mucopolysaccharide material stains with AB pH 2.5. B. The stain is totally abolished by prior treatment with hyaluronidase. ×75

Table 22.1 Histochemical characterisation of myxoid and chondromatous tumours

	Alcian blue pH 2.5	Alcian blue pH 2.5 + hyaluronidase	CEC (AB + MgCl$_2$)	Type of mucosubstance
Myxoid tumours without chondromatous element Myxoma Myxoid lipoma Lipoblastoma Myxoid liposarcoma Myxoid MFH Myxoid fibrosarcoma	+	–	<0.25M MgCl$_2$	Hyaluronic acid
Myxoid tumours with chondromatous element				
Extra-skeletal myxoid chondrosarcoma	+	± (variable)	>0.55M MgCl$_2$	Keratan sulphate CS-4 and 6
Benign chondromatous tumours	+	± (variable)	>0.55M MgCl$_2$	Keratan sulphate CS-4 and 6
Chondrosarcomas Well differentiated	+	± (variable)	<0.55M MgCl$_2$	Keratan sulphate CS-4 and 6
Poorly differentiated	+	–	>0.55M MgCl$_2$	CS-4 and 6

standard staining conditions, hyaluronic acid and CS-4 and -6 are negative with D-PAS.

TUMOURS CONTAINING FAT

Stains for fat are of very limited value. Many soft tissue neoplasms give a positive reaction particularly in degenerate areas while some liposarcomas, particularly the myxoid type, show minimal lipogenesis. Fat stains may be of value in the diagnosis of the rare round cell type of liposarcoma while the failure of lipoblast-like cells in malignant fibrous histiocytomas to stain for fat may also be helpful.

TUMOURS CONTAINING GLYCOGEN

Small amounts of glycogen are a common finding in many soft tissue tumours. Occasionally, particularly with biopsy material, the demonstration of glycogen can help in the elucidation of malignant round cell tumours. Glycogen will usually be present in embryonal and alveolar rhabdomyosarcomas and in Ewing's tumour of soft tissues. Glycogen will not be found in lymphomas and is exceedingly rare in neuroblastomas.

ENZYME MARKERS FOR TUMOURS

Studies of alkaline and acid phosphatase are useful in differentiating osteoblastic from chondromatous and fibroblastic lesions in soft tissues (Sanerkin 1980, Sanerkin & Jeffree 1980).

Alkaline phosphatase activity*

This is strong in osteoblasts in both benign and malignant lesions (Figs. 22.3 and 22.4). This may be of value in biopsy material where osteoid tissue may

*Neutral phosphatase: *some methods* for alkaline phosphatase also demonstrate a neutral phosphatase which is present in osteoclasts. Therefore a control section for neutral phosphatase activity should also be included.

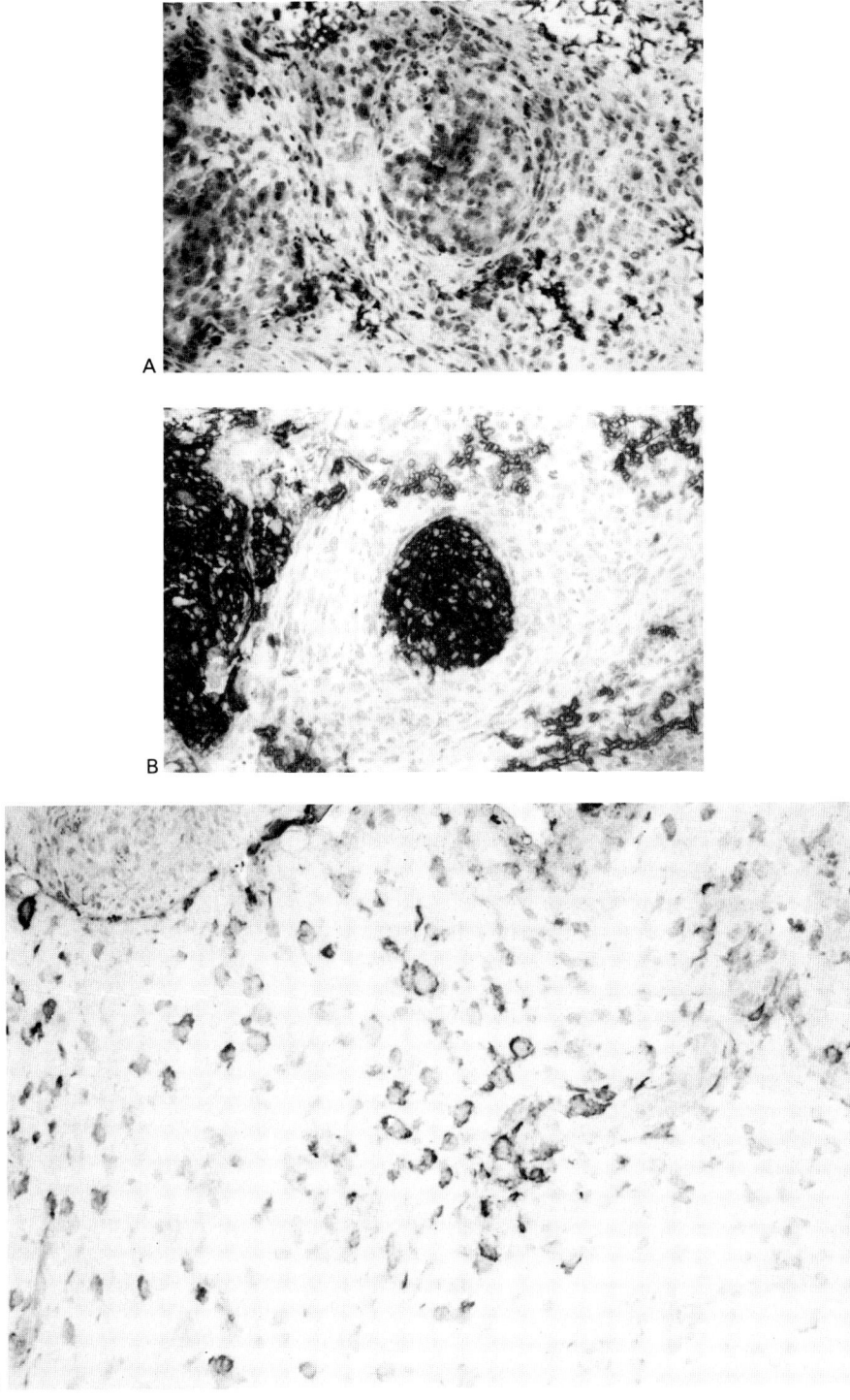

Fig. 22.3 Osteogenic sarcoma. A. On H & E, nests of malignant cells are surrounded by a stroma without osteoid formation. In the serial section (B) the osteoblastic nature of the malignant cells is demonstrated by strong alkaline phosphatase activity. C. Periosteal connective tissue infiltrated by small and scanty tumour cells clearly identified by a positive alkaline phosphatase reaction. Cryostat sections.

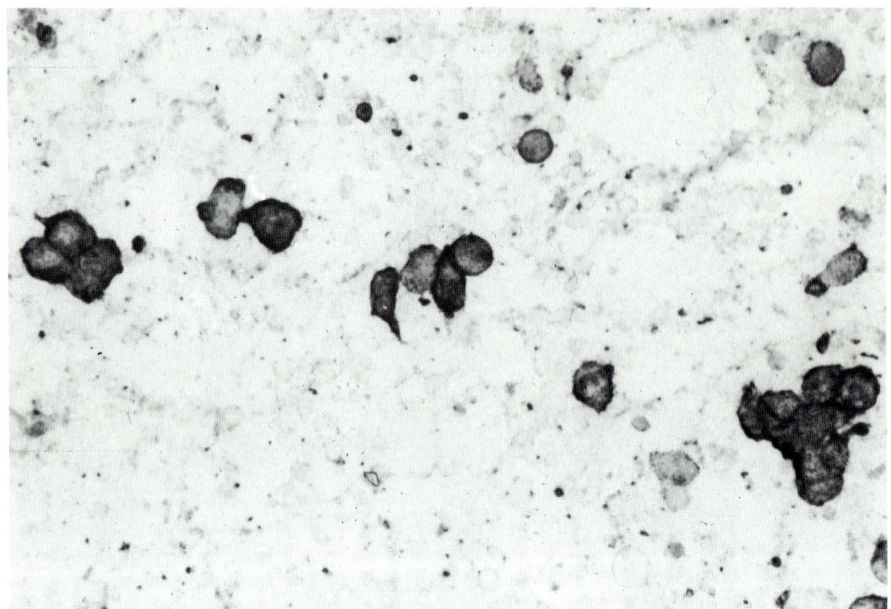

Fig. 22.4   Imprint of bone fragment with osteogenic sarcoma. Malignant cells show strong alkaline phosphatase activity.

be difficult to identify. Such enzyme activity is negative in malignant tumours containing fibroblasts while, in chondrosarcomas, alkaline phosphatase activity is either absent or very variable.

Alkaline phosphatase is also present in endothelial cells and adipocytes and therefore may be found in angiosarcomas and liposarcomas. These entities, however, are not likely to be confused with osteosarcomas.

Acid phosphatase activity

Acid phosphatase activity is almost ubiquitous. Amongst the many cells containing it are histiocytes and osteoclasts. In osteosarcomas with large numbers of giant cells the use of both alkaline phosphatase and acid phosphatase methods may be useful in discriminating between tumour giant cells which are alkaline phosphatase-positive and osteoclasts which are acid phosphatase-positive but negative for alkaline phosphatase.

Enzyme studies have also been carried out on a number of round cell tumours of soft tissue (Jeffree 1974). The degree of overlap, however, renders the method of limited value.

IMMUNOHISTOCHEMISTRY

Antisera to CEA, EMA and prekeratin, present in epithelial cells and in their neoplastic counterparts but absent from non-epithelial tumours, may be of value in deciding the epithelial or non-epithelial origin of an undifferentiated tumour (Goldenberg 1976, Sloane et al 1981; Sieinski et al 1981).

The demonstration of myoglobin in rhabdomyosarcomas and its absence in malignant fibrous histiocytomas, other soft tissue tumours and in lymphomas may prove a useful diagnostic tool (Brooks 1981, Kindblom et al 1981). The often-difficult diagnosis of embryonal and alveolar rhabdomyosarcoma may be helped by the use of antisera to a range of cell products such as myoglobin, myosin, creatine kinase BB and particularly creatine kinase MM (Tsokos et al 1981, Kindblom et al 1981).

Lysosyme and α_1-antichymotrypsin have been used as markers for histiocytic tumour cells. Immunohistochemistry studies have shown their value in distinguishing malignant fibrous histiocytoma (MFH) and malignant histiocytosis (MH) from other malignant soft tissue tumours and round cell tumours respectively. False-negative and false-positive reactions may occur and caution should be taken in the interpretation of the results (Meister & Nathrath 1981).

The endothelial cell origins of angiosarcoma and Kaposi's sarcoma are identified by the presence of factor VIII related antigen.

These are certainly fields to be explored.

Miscellaneous

Alveolar soft part sarcoma

Many alveolar soft part sarcomas contain PAS-positive, diastase-resistant crystalline bodies which probably represent a protein-carbohydrate complex. These bodies are unique to this tumour and may be helpful in the diagnosis of slightly atypical examples.

Synovial sarcoma

The pseudoglandular spaces of biphasic synovial sarcomas contain a PAS-positive diastase-resistant material. This feature is seldom helpful in diagnosis.

Digital fibrous tumours of infancy

This particular variant of the fibromatosis family is unique in having intracytoplasmic rounded inclusions. These stain red with phloxine tartrazine. Their nature is uncertain but no good evidence of a viral origin exists.

Chordoma

Occasionally it may be necessary to distinguish a chordoma from a chondrosarcoma. Crawford (1958) states that the distinction is achieved with PTAH and reticulin stains. With both these techniques chordomatous matrix remains unstained while cartilaginous substance is strongly impregnated. Alcian blue pH 2.5 and toluidine blue are also useful staining methods for the identification of the physaliferous cells.

REFERENCES

Allen P W 1980 The myxoid tumours of soft tissue. In: Pathology annual. Appleton-Century Crofts, New York, p 133

Brooks J J 1981 Myoglobin immunohistochemistry of soft tissue tumours: identification of rhabdomyosarcoma in patients over 30 years of age. Laboratory Investigation 44: 6A (abstract)

Crawford T 1958 The staining reactions of chordomas. Journal of Clinical Pathology 11: 110

Dobrogorski O J, Braunstein H 1963 Histochemical study of staining lipid, glycogen and mucin in human neoplasms. American Journal of Clinical Pathology 40: 435–443

Goldenberg D M 1976 Oncofetal and other tumour-associated antigens of the human digestive system. In: Morson B C (ed) Current topics in pathology. Springer-Verlag, Heidelberg

Jeffree G M 1974 Enzymes of round cell tumours in bone and soft tissue: a histochemical survey. Journal of Pathology 113: 101

Kindblom L G, Angervall L 1975 Histochemical characterization of mucosubstances in bone and soft tissue tumours. Cancer 36: 985–994

Kindblom L G, Karlsson K 1977 Differential staining of glycosaminoglycans, utilizing bacterial chondroitinase and chondrosulphatase. Acta pathologica et microbiologica scandinavica section A 85: 665–670

Kindblom L G, Seidal T, Karlsson K 1981 Immuno-histochemical localization of myoglobin in human muscle tissue and embryonal and alveolar rhabdomyosarcoma. Acta pathologica et microbiologica scandinavica section A 89: 271–285

Mackenzie D H 1981 The myxoid tumours of somatic soft tissues. American Journal of Surgical Pathology 5: 443–457

Meister P, Nathrath W 1981 Immunohistochemical characterization of histiocytic tumours. Diagnostic Histopathology 4: 79–87

Sanerkin N G 1980 Definition of osteosarcoma, chondrosarcoma and fibrosarcoma of bone. Cancer 46: 178–185

Sanerkin N G, Jeffree G M 1980 Cytology of bone tumours. John Wright, Bristol

Scott J E, Dorling J 1965 Differential staining of acid glycosaminoglycans (mucopolysaccharides) by alcian blue in salt solutions. Histochemistry 5: 221–223

Sieinski W, Dorsett B, Joachim H L 1981 Identification of prekeratin by immunofluorescence staining in the differential diagnosis of tumours. Human Pathology 12: 452–457

Sloane J P, Ormerod M G 1981 Distribution of epithelial membrane antigen in normal and neoplastic tissues and its value in diagnostic pathology. Cancer 47: 1786–1795

Tsokos M, Zweig M, Howard R, Bowling M C, Costa J 1981 Immunocytochemical markers in embryonal and alveolar rhabdomyosarcoma: a diagnostic aid. Laboratory Investigation 44: 68A (abstract)

Yamada K 1974 The effect of digestion with chondroitinases upon certain histochemical reactions of mucopolysaccharide-containing tissues. Journal Histochemistry and Cytochemistry 22: 266–275

Immunohistological techniques for the diagnosis of ovarian and testicular neoplasms

C. R. Taylor & N. E. Warner

Surgical pathologists are generally of a conservative disposition, ever guided by tradition and lore, rarely given to unbridled speculation. Yet, it appears that restraint has been cast aside as pathologists have attempted to incorporate the manifold expressions of testicular and ovarian neoplasms into the general schemata of tumour classification.

> 'Tumors are classified like normal tissues on a histologic basis — the type of cell is the one important element in every tumor. From it the tumor should be named.' (F.B. Mallory 1923)

This concept is disarmingly simple, but presupposes that one can confidently identify and reliably distinguish the various types of normal cells. Such a supposition is not justifiable with reference to recognition of the various germ cell and stromal cell derivatives found in the ovary and testis. Willis, in *Pathology of tumours,* (Willis 1948) prefaces the chapter pertaining to ovarian neoplasms with the following statement:

> 'Several factors have combined to cause confusion in their (ovarian tumours) classification and nomenclature, namely (1) uncertainties as to the histogenetic relationships of some of the tissues of the normal ovary, (2) uncertainties as to the derivation of particular tumours from particular ovarian tissues, (3) practical difficulties in some cases of distinguishing growths of ovarian origin from those of parovarian or tubal origin, and (4) the frequent mistaking of secondary for primary growths in the ovaries.'

In his chapter on tumours of the testis, Willis encounters a similar veil of confusion. Critical review of the literature reveals that with the passing of time one particular concept may attain popularity and become so widely accepted that it is mistaken for fact. This has the desirable effect of inhibiting the

formulation of further hypotheses, but at the price of sometimes suppressing the truth when it finally threatens to emerge.

Much as in the area of malignant lymphomas, whenever many minds devote themselves to extensive studies of a particular group of neoplasms, differences of opinion develop and find expression in the form of diverse classifications varying in concept, nomenclature and biological relevance (Taylor 1978a). Within the study of testicular and ovarian neoplasms, these difficulties are compounded by the lack of a firm scientific basis for the recognition and distinction of individual normal cells, and their neoplastic counterparts. It is often forgotten that normally proliferating cells of different cell lines may appear remarkably similar; tumours composed of a high proportion of proliferating cells may appear similar on this basis alone. Often pathologists distinguish such tumours not by the majority components, for these are quite similar, but by a minority population displaying some degree of differentiation along one or another cell line. In the final analysis, such distinctions, based as they are on subtle subjective morphological judgments, may be very difficult to make. To state the matter bluntly, the recognition and classification of many of the tumours of the ovary and testis depend solely upon morphologic interpretation; interpretation is a matter of opinion, and opinions often differ.

CLASSIFICATION OF OVARIAN AND TESTICULAR TUMOURS

The basic classifications of ovarian and testicular neoplasms used within this chapter have been adopted on the basis of convenience and popularity.

The adoption of these classifications is not intended to convey support for their scientific rectitude, or any preference for these schemata over those championed by other investigators. The classification of testicular neoplasms (Table 23.1) is based on that used by Mostofi & Price in the Armed Forces Institute of Pathology (AFIP) 'Atlas' (Mostofi & Price 1973). This in turn was founded upon the conceptual view advanced by Friedman and Moore that embryonal carcinomas, teratomas and choriocarcinomas all originate from the pluripotent germ cells. Approximately 93% of the AFIP cases can then be considered to be of germ cell origin.

Table 23.1 Classification of testicular neoplasms (Modified from Mostofi & Price 1973)

1. Germ cell tumours
    a. Single histologic type
    Seminoma
    Embryonal carcinoma
    Choriocarcinoma
    Teratoma
    b. More than one histologic type
    Any combinations of types within (a)
2. Gonadal stromal tumours
    Undifferentiated
    Leydig cell
    Sertoli (granulosa-theca) cell
    Combinations of these
3. Mixed germ cell and gonadal stromal tumours
4. Metastatic and other

Editorial Note: The British Testicular Tumour Panel (BTTP) classifies germ cell tumours as follows
1. Seminoma
2. Malignant teratoma
    a. differentiated
    b. intermediate
    c. undifferentiated
    d. trophoblastic
and recognises overlap between these groups.

The simplistic classification given for ovarian neoplasms (Table 23.2) is based upon the 1973 World Health Organisation recommendations (Sero et al 1973), somewhat condensed for the purposes of this discussion.

Clearly many parallels exist between tumours of the ovary and tumours of the testis, as might be expected considering that the cellular composition of these two organs is identical in embryologic terms and differs only to the extent that the cells are subject to differing hormonal environments during embryologic differentiation and subsequent maturation. It will be

Table 23.2 Classification of ovarian neoplasms (Modified from WHO, Sero et al 1973). (Bracket entries are relatively rare; they are included to maintain the parallel with Table 23.1)

1. Germ cell tumours
    Dysgerminoma (seminoma-like)
    (Embryonal carcinoma)
    Choriocarcinoma
    Teratoma
    Endodermal sinus (yolk sac) tumour
    (Mixtures of the above)
2. Gonadal stromal tumours
    (Undifferentiated)
    Sertoli-Leydig cell (androblastoma)
    Granulosa-theca cell
3. Mixed germ cell and gonadal stromal tumours
4. Tumours of ovarian surface epithelium
    Serous cystadenoma and carcinoma
    Mucinous cystadenoma and carcinoma
    Endometrioid carcinoma
    Undifferentiated carcinoma
5. Metastatic and other

shown that immunohistologic techniques have given new insight into the origins and interrelations of these tumours.

THE ADVENT OF IMMUNOHISTOLOGICAL METHODS

'The histopathologic diagnosis of neoplasia is at times hedged with uncertainty, owing in part to conceptual difficulties regarding the nature and origin of the neoplastic cells and also to a continuing lack of alternative methods of cell identification with which to validate the morphologic criteria employed.' (Taylor 1978a)

Orthodox histological staining methods are designed to introduce contrast into the cellular and extracellular elements of a tissue section, thus facilitating the identification and distinction of the various cell types. Ordinary histological stains are mostly based upon dyes (e.g. haematoxylin and eosin) that show varying degrees of affinity for different subcellular components. More refined histochemical stains, often termed 'special stains,' are designed to exploit the chemical characteristics of the molecules that make up the various cell types, and are based in principle upon the occurrence of specific chemical reactions between the stain and particular cell components. Histochemical methods have proved extremely valuable in some areas of pathology, but have had limited application in the realms of testicular and ovarian neoplasms, for the normal cells of the ovary and testis do not show uniquely characteristic histochemical patterns. In addition, many histochemical stains depend upon preservation of the active sites of enzymes in order to 'catalyse' the production of a colour change within the staining reagent, thereby producing a specific coloured reaction product. This requirement for preservation of the activity of enzymes precludes, in many instances, the use of routinely processed formalin-fixed paraffin wax-embedded tissues.

The development of immunohistologic methods, depending upon the specificity of an antibody marker for a particular cell or tissue antigen, offers the prospect of developing an alternative form of 'special stain' for use in routinely processed tissues. In practice, various types of horseradish peroxidase-labelled antibody procedures have been increasingly employed since 1974, when it was shown that these methods are able to demonstrate at least some antigens within routinely processed tissues (Taylor 1974).

There are three limitations of immunoperoxidase methods for use in routine surgical pathology:

1. The range of stains is limited by the availability of specific antisera

2. Performance of the stain in the surgical pathology laboratory is dependent upon preservation of the immunoreactivity of the antigen in question through the processes of fixation, dehydration, paraffin wax embedding, etc.

3. As with any special staining method, pathologists must employ controls and must learn to interpret the observed patterns of staining.

PRACTICAL ASPECTS OF IMMUNOPEROXIDASE STAINING OF 'ROUTINE TISSUES'

The most commonly employed variant of the basic immunoperoxidase procedure is the PAP (peroxidase-anti-peroxidase) method (Fig. 23.1); it is both sensitive and reproducible, and the basic reagents are available commercially. More recently, immuno-histology staining kits have been introduced by two manufacturers (Dako Corporation and Immulok): Kits currently available and of relevance to the present discussion are listed in Table 23.3.

For many pathologists the advantage of employing

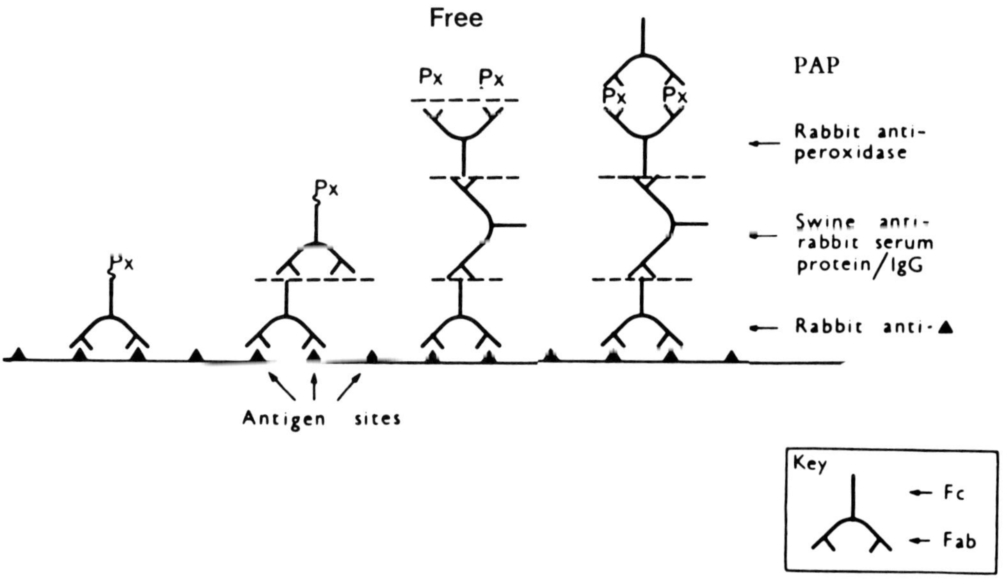

Fig. 23.1 Immunoperoxidase techniques. The direct and indirect conjugate procedures, shown on the left, are analogous to the corresponding immunofluorescence methods. The enzyme bridge method (second right) and the PAP method (peroxidase-anti-peroxidase method, extreme right) were devised to avoid the problems of chemical conjugation of horseradish peroxidase (Px) to antibody.

Table 23.3 Commercially available 'immunostains'*

	Dako®	Immulok®
Primary tumours	Human chorionic gonadotrophin	Human chorionic gonadotrophin
	Pregnancy-specific beta-1-glycoprotein	Pregnancy-specific beta-1-glycoprotein
	Human placental lactogen	
		Oestradiol
		Testosterone
		Alpha fetoprotein
Metastases	Carcinoembryonic antigen	Carcinoembryonic antigen
	Prostate-specific acid phosphatase†	Prostate-specific acid phosphatase†

* Kits listed pertain particularly to the study of ovary and testis; more than 30 different immunostains are available for other purposes (Taylor & Kledzik 1981), including stains for immunoglobulin light and heavy chains (Dako®, Immulok®), TSH, FSH, GH, ACTH, prolactin, insulin, glucagon, VIP, somatostatin, gastrin, glial fibrillar, acid protein, factor VIII, haemoglobin A, lysozyme, ferritin, transferrin, lactoferrin (Immulok®) and hepatitis B surface antigen

† Prostate-specific epithelial antigen and prostate-specific acid phosphatase are distinct antigens; both are virtually restricted in occurrence (in postnatal life) to prostatic epithelium

Dako Corporation, 22 N. Milpas, Santa Barbara, CA 93103
Immulok, 1019 Mark Avenue, Carpinteria, CA 93013

kits, containing pre-titred matched reagents plus all of the necessary substrate chromogen systems, may be considerable in terms of routine laboratory practice. Those who prefer not to use the commercially available kits may purchase individual reagents from a wide variety of commercial sources or may prepare their own reagents, subject of course to the necessity for strict specificity control. Even with the use of commercial kits, the pathologists should introduce specificity controls in the form of tissue sections known to contain the antigen in question ('positive' control), or known to lack it ('negative' control). The pattern and intensity of staining observed in the 'test' section may then be judged against the pattern and intensity of staining observed in the 'positive and negative controls.'

For the staining of paraffin wax sections the procedure employed is basically as follows. Sections first are deparaffinised through xylol and the graduated alcohols. The sections in phosphate buffered saline or Tris buffer are then incubated in the primary antibody (Fig. 23.1), having specificity

for the antigen under investigation, for periods of 10 min to 1 h according to the titre of activity of the antibody utilised. Sections are then washed twice in buffer and incubated in the bridge or linking antibody, which serves to link the primary antibody to the PAP reagent that will be added next. Following incubation for 10–30 min, sections are again washed twice in buffer prior to the addition of the PAP reagent for 30 min. In general terms the PAP reagent and the primary antibody must contain immunoglobulins derived from the same species (or from species possessing immunoglobulins that show a strong cross-reactivity, both binding with the bridge antibody). The procedure is set out in diagrammatic form in Figure 23.1. Two alternate procedures, the direct and indirect conjugate methods also are depicted; these have the advantage of simplicity but are considered by many to be less sensitive, thereby requiring the use of precious specific primary antibodies at higher concentration.

If the tissue under study contains large amounts of 'endogenous peroxidase' activity, then this should be inactivated by pretreatment of the sections in 0.3% hydrogen peroxide in methanol for 20 min, otherwise the reaction of the endogenous peroxidase with the chromogen may mask the sites of localisation of the horseradish peroxidase label. Nonspecific binding of the reagents (antibodies) to charged components within tissue sections may also cause problems in interpretation, but may be successfully combatted by pretreatment of sections with diluted serum of the bridge antibody species (Taylor 1978b).

The chromogen most commonly used in the investigative work is diaminobenzidine, with a hydrogen peroxide substrate. This produces a crisp brown reaction product that is alcohol-fast and will not be leached out during dehydration steps preparatory to mounting the section in a permanent mounting medium. The commercial kits include amino ethyl carbazole (AEC) as the chromogen, for this is a noncarcinogen. The disadvantage of AEC is that it is alcohol-soluble, necessitating use of a water-miscible mounting medium. The principles of performance of immunoperoxidase techniques, the wide range of immunostains available and their practical usage have been considered in greater detail in a number of reviews (Taylor 1978b, Sternberger 1979, Mukai & Rosai 1980, DeLellis 1981, Taylor & Kledzik 1981).

IMMUNOSTAINING OF OVARIAN AND TESTICULAR TUMOURS

Table 23.4 lists the principal types of testicular and ovarian neoplasms and the various immunostains that may be employed to assist in their discrimination from one another. It must be emphasised that in devising immunohistological staining methods, the pathologist is not casting aside morphological criteria, but rather is taking advantage of the remarkable specific staining offered by immunological techniques in order to validate, refine, and redefine, if necessary, the morphological criteria upon which diagnosis is traditionally based.

Observed patterns of staining — germ cell tumours

The seminoma/dysgerminoma group

This represents the most commonly encountered malignant tumour of germ cell origin. So-called 'pure' examples of this group of tumours do not show staining either for HCG or for alpha fetoprotein (AFP), either in the majority of the tumour cells or in the associated tumour giant cells that may be present (Kurman & Scardino 1981). However, some cases show the presence of large multinucleated giant cells that do give positive immunostaining for HCG, but not for alpha fetoprotein; these cells are thought to represent isolated examples of syncytiotrophoblastic differentiation within a tumour that otherwise is pure seminoma. Kurman points out that prior to immunohistological staining these cells had often been termed pseudotrophoblast, and their relationship to functional true syncytiotrophoblastic elements had not been appreciated. Seminomas may also show the presence of a minority of cells giving low intensity staining reactions with anti-oestradiol and anti-testosterone antisera, most likely representing the presence of the steroid hormone bound in vivo by active receptors or binding globulins.

Embryonal carcinomas

These frequently contain syncytiotrophoblastic elements, emphasising the close biological relationship of these two germ cell-derived neoplasms. The embryonal carcinoma cells per se typically do not stain for HCG. However, syncytiotrophoblastic elements commonly present within these tumours do show a strong positive reaction. On the other hand, alpha fetoprotein (AFP) staining is never observed within the syncytiotrophoblastic elements, but may often be seen within an apparently pure embryonal carcinoma, that apparently has functional differentiation even in the absence of histological yolk sac differentiation (Kurman & Scardino 1981).

Choriocarcinomas

Choriocarcinomas, representing a malignant counterpart of the trophoblast, typically produce large amounts of HCG and stain strongly with anti-HCG sera (Fig. 23.2). The normal trophoblast also produces beta-1-pregnancy-specific glycoprotein (SP1); choriocarcinoma typically mimics this production. It should be emphasised that while the pattern of staining of choriocarcinomas for both HCG and SP1 is very striking and very characteristic, the mere presence of staining with either of these antisera does not of itself define the tumour as of trophoblastic origin. For example, as many as 50% of cases of breast carcinoma have been observed to stain for HCG or for SP1 (Horne et al 1977), or for both, and HCG staining has been reported less often in other tumours (Taylor & Kledzik 1981).

Teratomas

These do not show positive staining for either HCG or AFP unless, within the teratoma, there is differentiation either towards trophoblastic elements or yolk sac elements. In such instances the differentiated elements show the pattern of staining typical of trophoblast or yolk sac respectively. It should be remembered that teratomas may occasionally produce functional epithelial and glandular elements and that these may be stained for the appropriate cell product (e.g. thyroglobulin within 'teratomatous thyroid,' or carcinoembryonic antigen within teratomatous gastro-intestinal type epithelium).

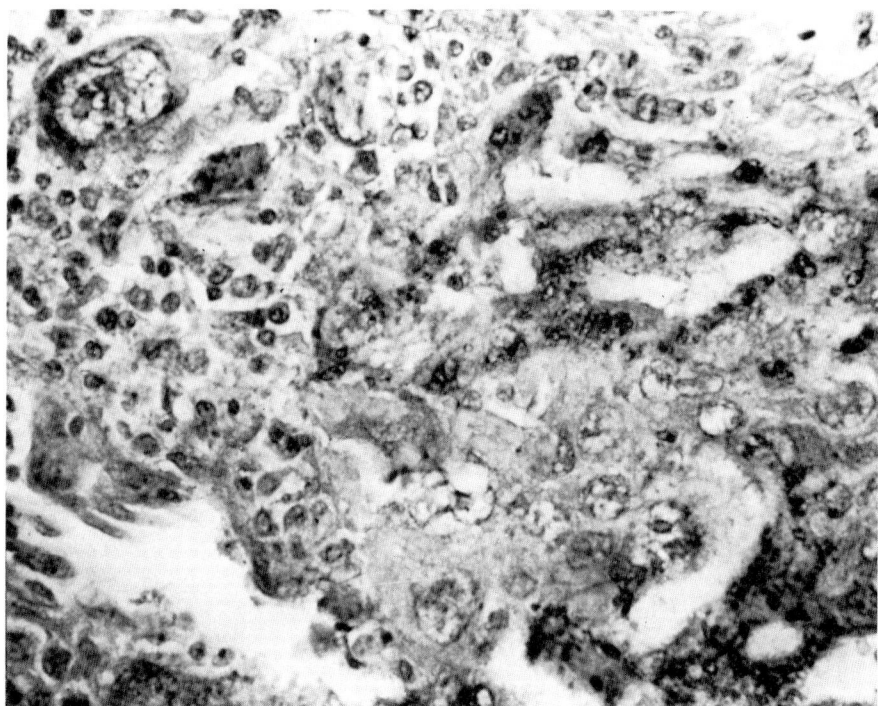

Fig. 23.2   Choriocarcinoma presenting in the lung of a young girl, immunostained for human chorionic gonadotrophin by the PAP method. Positive staining appears grey-black in the cytoplasm of tumour cells, including giant cells. Paraffin wax section, counterstain haematoxylin, ×500

Endodermal sinus (yolk sac) tumours

These tumours show a very typical pattern of staining regardless of the fine histological subtypes. It is important to recognize that positivity for AFP may be distributed throughout the tumour or may occur only focally (Fig. 23.3). Typically morphological elements in the tumour that resemble yolk sacs show intense positive staining. However, in other areas of the tumour, or in other tumours that show no evidence of morphological yolk sac differentiation, it often is possible to demonstrate variable amounts of staining for AFP. Histologically it has been shown that AFP staining correlates with intracellular and extracellular hyaline eosinophilic globules identifiable in routine H & E sections. Many examples of tumours showing yolk sac differentiation will also stain for other liver cell products, such as transferrin and alpha-1-antitrypsin. Positive staining for alpha fetoprotein can be used to identify an endodermal sinus or yolk sac tumour, even when there is little evidence of morphological yolk sac differentiation.

Observed patterns of staining — gonadal stromal tumours

Immunostaining for the different steroid hormones has shed light upon the histogenesis and the histological differentiation of hormonally active tumours within the ovary and testis. Prior to this the nomenclature and classification of these tumours has largely been dependent upon morphological features coupled with clinical status and evidence of endocrine manifestations (Kurman et al 1978, 1979, 1981, Taylor et al 1978). The fact that these neoplasms often present variable patterns of hormone production and that the tumours themselves may display diverse histological patterns has led to considerable difficulties in developing a consistent working hypothesis relating individual cell types to the production of particular hormones. The feasibility of preparing antisera against the different steroid hormones and of applying these antisera to the demonstration of these hormones in formalin paraffin wax sections promises to shed new light on this controversy.

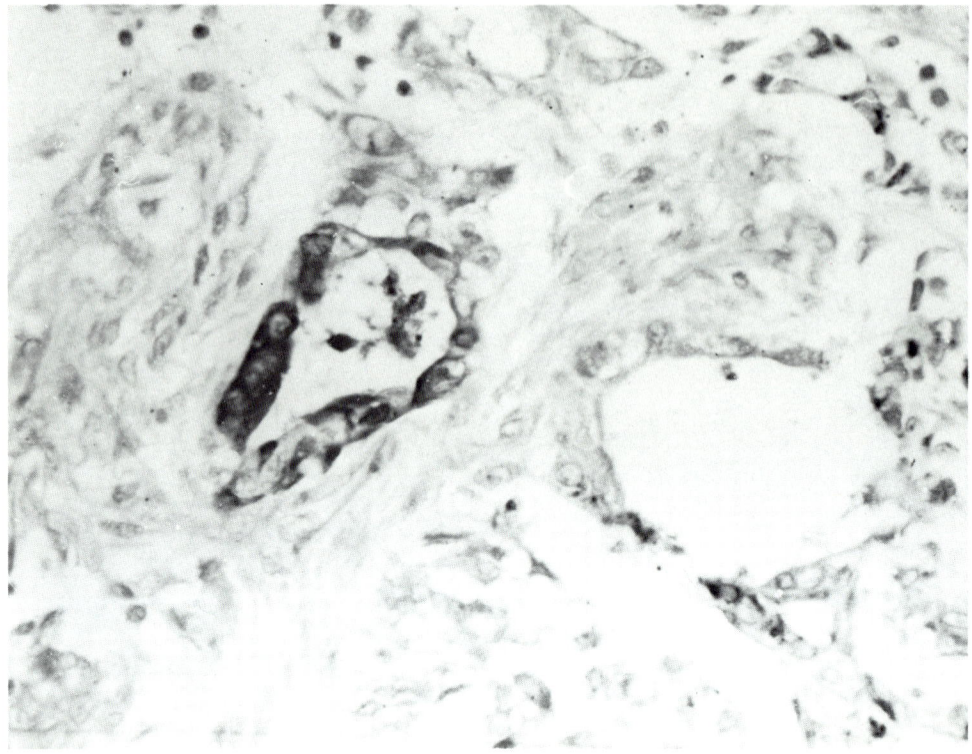

Fig. 23.3 Endodermal sinus (yolk sac) carcinoma immunostained for alpha-fetoprotein by the PAP method. Positive black staining is visible in a recognisable 'yolk sac' and in isolated scattered cells elsewhere in the tumour. Paraffin wax section, counterstain haematoxylin, ×720

Immunohistologic techniques have proved remarkably successful, with some reservations, as discussed below. In general terms, staining for a particular steroid hormone within cell cytoplasm is interpreted as evidence of intracellular synthesis although, of course, the possibility of binding to various intracellular binding proteins or receptors is a real one. It has for example been recognised that Sertoli cells contain receptors for various androgens, while granulosa cells apparently contain oestrogen-binding proteins (Christensen et al 1977, Schreiber et al 1976). In addition, when using antisera to steroid hormones, one must be aware of the possibility of cross-reactivity between the different antisera. In practical terms, this can be controlled to some degree by assessing the specificity of the antisera, one against another, in radioimmunoassay systems; with regard to immunohistologic studies, it is important also to include biological control, as for example in the intense staining of normal interstitial Leydig cells in testis with anti-testosterone serum, and the lesser degree of staining or absence of staining with anti-oestradiol serum. Another potential problem is that in order to generate antisera against the steroid hormones, the hormones are first linked to some sort of carrier molecule. Bovine serum albumin (BSA) has often been used for this purpose; the resulting antiserum will thus contain antibodies against the steroid moiety, and also against the bovine serum albumin. These anti-albumin antibodies will show varying degrees of cross-reactivity with human albumin and will cause immunohistological staining that may be confused with specific staining for anti-oestradiol. For these reasons, antisera that are to be used for immunohistological studies must, if coupled with BSA, be extensively absorbed with human albumin. A better course is to link the steroid hormone with keyhole limpet haemocyanin as a carrier, thus avoiding the possibility of cross-reacting antisera.

The different patterns of staining anticipated among gonadal stromal tumours are illustrated in Table 23.4.

Table 23.4 'Immunostains' of ovarian and testicular neoplasms

Tumour category	HCG	SP1	AFP	Oestra-diol	Testos-terone	Proges-terone	CEA	Breast antigens[1]	Other
Seminoma	(+)[2]			(+)[3]	(+)[3]				
Embryonal carcinoma	+ +	+	+						
Choriocarcinoma	+ + +	+ +							
Teratoma									
with yolk sac differentiation			+ +						
with trophoblastic differentiation	+ + +	+ +							
with other secretory differentiation[4]							(+)[4]	(+)	+[4]
Endodermal sinus (yolk sac)	+	+	+ + +						
Leydig cell				+	+ + +	+			
Sertoli				+[3]	+	+			
Androblastoma (Sertoli-Leydig)				+[3]	+	+			
Granulosa-theca:									
Granulosa				+ +	+	+			
Luteinized theca				+ +	+	+ +			
Undifferentiated gonadal stromal				+ +	+ +	−			
Mucinous cystadenocarcinoma							+		
Metastic and direct spread									
Colon (GI tract)							+ + +		
Breast	+[5]	+[5]		+[2,6]		+[3]		+ +	
Other	+[5]	+[5]	+[7]	(+)[6]					(+)[8]

1 Breast-associated antigens include 'T antigen', mammary epithelial membrane antigen, and a variety of other antigens identifiable by use of monoclonal antibodies
2 Semiquantitative scoring, + to + + +; not every case, nor every cell of positive cases, will be positive. Bracketed entries may or may not show positivity
3 Positivity of some cell elements, such as Sertoli cells or even breast epithelial cells, for oestradiol or other steroid hormones, may reflect receptor binding of the hormone rather than intrinsic production
4 Rare teratomas containing 'thyroid' elements will show immunostaining for thyroglobulin; teratomas with 'gastro-intestinal' type epithelium may give a reaction for CEA; other epithelial elements may show other staining patterns according to product
5 Non-trophoblastic tumours sometimes contain HCG or SP1, e.g. carcinoma of breast, for reasons unknown
6 Non-ovarian/testicular tumours may show positivity, presumably also due to 'receptor binding of steroid hormone'; breast particularly is often positive for oestradiol (Taylor et al 1981); prostate is sometimes positive for testosterone
7 Primary hepatocellular carcinomas typically stain strongly for AFP
8 Hormonally active tumours of many types may produce secondary deposits in the ovary, and rarely the testis; immunostains are available against many of the hormones (Immulok®), and for many other cell products (Taylor & Kledzik 1981)

Leydig cell tumours and Sertoli-Leydig cell tumours (androblastomas)

Characteristically these show intense staining for testosterone in the recognisable Leydig cell elements. Leydig cells may also show positive staining with antiserum to oestradiol and less often with antiserum to progesterone. Occasionally the recognisable Sertoli cells, forming more or less well organised tubules, also show staining for testosterone and/or oestradiol.

Granulosa-theca cell tumours

Quite commonly these tumours are associated with clinical manifestations of hormone production. Classically oestrogen synthesis has been considered to be confined to theca cells while the luteinised granulosa cells have been held responsible for progesterone production. These rigid definitions of cell type correlated to hormone production do not appear to hold true according to the evidence of immunoperoxidase staining. Granulosa cells have been observed to show positive staining not only for oestradiol, but also for progesterone and even for testosterone. Similarly luteinised theca cells contain progesterone, but also may show staining for oestradiol or testosterone.

These findings challenge the time-honoured concept that individual morphological cell types are responsible for, and restricted to, the production of single hormones; theca cells for oestrogen, luteinised granulosa cells for progesterone, and Leydig cells for testosterone. Immunoperoxidase studies suggest that the neoplastic counterparts of these cells certainly do not recognise these restrictions; there is increasing evidence that normal cells also break the accepted rules (Kurman et al 1981). Clearly immuno-peroxidase studies, staining for these specific steroid hormones, may be expected to contribute to our understanding of the biology of these neoplasms. Such studies can be performed upon tissue biopsies in order to gain a more realistic assessment of the potential hormonal manifestations of individual tumours in individual patients, for it is clear that morphology alone is not a reliable indicator of the type of hormone produced by any individual morphologic type of neoplasm.

Undifferentiated gonadal stromal cell tumours

There has been another finding of real interest amongst the gonadal stromal tumours, namely the use of immunostaining in the recognition and differential diagnosis of these undifferentiated gonadal stromal tumours. Morphologically such tumours consist of highly undifferentiated 'primitive' appearing cells, reflecting the high proportion of cells in the S-(synthesis) phase of cell cycle. These neoplasms bear a close resemblance to other undifferentiated neoplasms that may be encountered in ovary and testis, and in our experience have sometimes been misdiagnosed as reticulum cell sarcoma (histiocytic lymphoma, or in more modern terminology immunoblastic sarcoma) of the testis. The staining of such tumours for steroid hormones, specifically the demonstration of testosterone and/or oestradiol in the cytoplasm of the neoplastic cells, provides support for the diagnosis of gonadal stromal neoplasm (Fig. 23.4) (Maurer et al 1980).

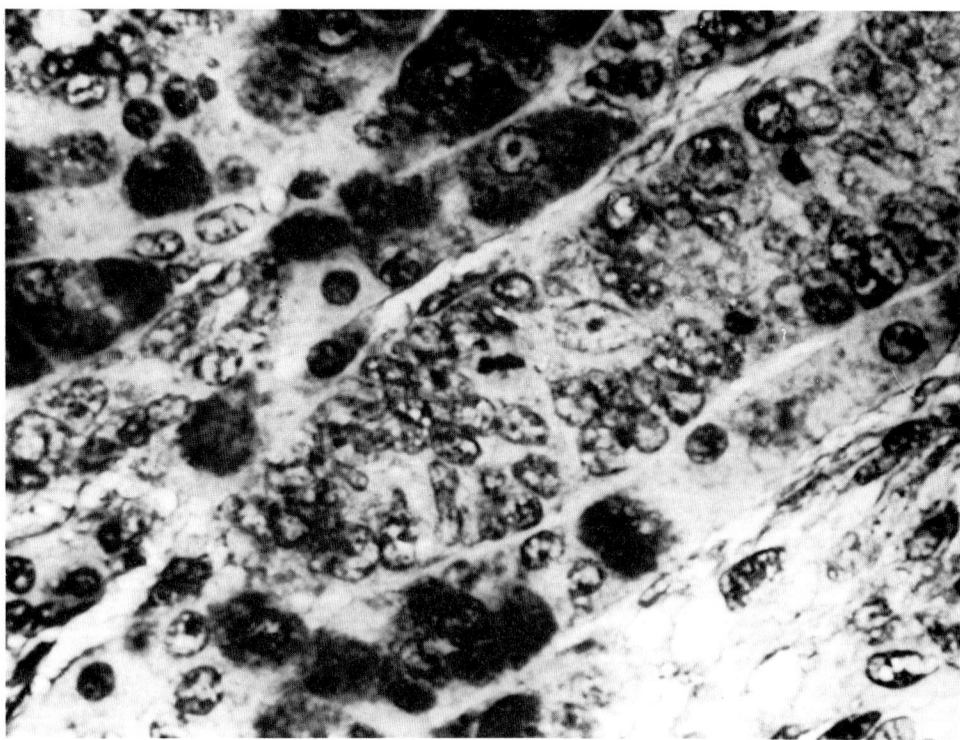

Fig. 23.4 Gonadal stromal cell tumour, presenting in the pelvis of an otherwise normal male, immunostained for oestradiol. Positive intense black staining is seen in cells having some resemblance to Leydig cells; Sertoli-like tubules showed scattered light staining. Antiserum to testosterone gave an even more intense staining pattern. Paraffin wax section, counterstain haematoxylin, ×720. This case was reported by Maurer et al (1980)

Metastatic tumours of ovary and testis

Metastatic neoplasms may present difficulties in differential diagnosis of ovarian tumours. This is less often a problem with reference to the testis, except perhaps for some of the malignant lymphomas.

Whereas a positive immunoperoxidase stain for HCG, alpha fetoprotein, or one of the steroid hormones may be of real value in establishing a diagnosis of primary neoplasm of the gonad, a negative stain is of much less value; even in the presence of procedural controls and biological 'positive' and 'negative' controls, one can never entirely exclude a false-negative result due to technical reasons. In addition, tumours of a lineage with the potential for producing a particular cell product do not necessarily produce that product in amounts detectable by the techniques described. Thus a negative immunoperoxidase stain cannot be given too much weight in considering a differential diagnosis of primary versus metastatic tumour within the testis or ovary. At the present time the range of antisera available, with specificity against antigens associated with other tumours that might occur as metastases within the gonads, is quite limited. Nonetheless new reagents are becoming available almost daily. Antisera against carcinoembryonic antigen can be useful in recognising metastases from a colonic primary; with the reservation that primary cystadenocarcinomas of the ovary may be carcinoembryonic antigen-positive, mucinous carcinomas more so than their serous counterparts (Heald et al 1979). Antisera against prostatic acid phosphatase or prostatic epithelial antigen may assist in the recognition of lymph nodal metastases as of prostatic origin versus a primary testicular neoplasm (Taylor & Kledzik 1981); and antisera against breast-associated antigens may be helpful in recognising poorly differentiated breast carcinoma metastatic to the ovary (Imam & Tokes 1981).

CONCLUSION

It has been written that 'the importance of immunohistologic methods in surgical pathology can scarcely be overestimated' (Bosman & Kruseman 1979). Immunohistologic techniques applicable to paraffin wax sections are only 7 years old, but already the variety of specific immunostains available is enormous, many of which can be applied quite routinely in diagnostic and investigative laboratories (Table 23.5) (Taylor & Kledzik 1981). It is tempting to speculate that this represents the beginning of a new era in diagnostic pathology in which the pathologist will learn to rely not only upon traditional histologic methods, but also upon specific immunostains for cell and tumour recognition. With regard to neoplasms of the testis and ovary, a beginning has been made, but it is only a beginning.

Table 23.5 Antigens for which immunostains are routinely available in University of Southern California Immunopathology Laboratories

On paraffin wax sections

Immunoglobulin chains kappa, lambda, G, A, M, D & J	Growth hormone ACTH
secretory piece	TSH
Lysozyme (muramidase)	Prolactin
Ferritin	FSH
Lactoferrin	Thyroglobulin
Transferrin	Calcitonin
Factor VIII	Parathormone
Haemoglobin A	Insulin
Haemoglobin F	Glucagon
Alpha-1-antitrypsin	Somatostatin
Alpha fetoprotein	Vasoactive intestinal polypeptide
CEA	Gastrin
Prostatic acid phosphatase	Testosterone
Prostate-specific epithelial antigen	Oestradiol
Glial fibrillary acid protein	Progesterone
Mammary epithelial membrane antigen	HCG
Myoglobin	Pregnancy-specific glycoprotein
Hepatitis B surface antigen	Human placental lactogen
Herpes virus antigen	

On cryostat sections
1. All of above
2. Surface immunoglobulin light and heavy chains
3. T lymphocyte antigen, PAN T, helper, suppressor, 'pre-T' etc.
4. Ia antigens

Kits for the staining of almost all of these antigens in paraffin wax sections are available from Immulok, 1019 Mark Avenue, Carpinteria, CA 93013, USA (UK Distributor: LKB Inc., Croydon, Surrey) and Dako Corporation, 22 N. Milpas, Santa Barbara, CA 93103, USA (UK distributor: Mercia Diagnostics, Watford, Herts.)

REFERENCES

Bosman F T, Kruseman A C N 1979 Clinical applications of the enzyme-labeled antibody method. Immunoperoxidase methods in diagnostic pathology. Journal of Histochemistry and Cytochemistry 27: 1140–1147

Christensen A K, Wisner J R, Orth J 1977 Preliminary observations on the localization of androgen and FSH receptors in the rat seminiferous tubule, studied by autoradiography at the light microscope level. In: Troen P, Nankin H R (eds) The testis in normal and infertile men. Raven Press, New York, p 153

DeLellis R A (ed) 1981 Diagnostic immunohistochemistry. Masson Monographs in Diagnostic Pathology. Masson, New York

Heald J, Buckley C H, Fox H 1979 An immunohistochemical study of the distribution of CEA in epithelial tumours of the ovary. Journal of Clinical Pathology 32: 910–926

Horne C H W, Reid I N, Milne G D 1976 Prognostic significance of inappropriate production of pregnancy proteins by breast cancer. Lancet 2: 279–282

Imam A, Tokes Z A 1981 Immunoperoxidase localization of a glycoprotein on plasma membrane of secretory epithelium from human breast. Journal of Histochemistry and Cytochemistry 29: 581–584

Kurman R J, Scardino P T 1981 Alpha-fetoprotein and human chorionic gonadotropin in ovarian and testicular germ cell tumors. In: DeLellis R A (ed) Diagnostic immunohistochemistry. Masson Monographs in Diagnostic Pathology. Masson, New York, ch 17, p 277

Kurman R J, Goebelsmann U, Taylor C R 1979 An immunohistological study of steroid localization in granulosa-theca tumors of the ovary. Cancer 43: 2377–2384

Kurman R J, Goebelsmann U, Taylor C R 1981 Steroid hormones in functional ovarian tumours. In: DeLellis R A (ed) Diagnostic immunohistochemistry. Masson Monographs in Diagnostic Pathology. Masson, New York, ch 8, p 137

Kurman R J, Andrade D, Goebelsmann U, Taylor C R 1978 An immunohistological study of steroid localization in Sertoli-Leydig tumors of the ovary and testis. Cancer 42: 1772–1783

Mallory F B 1923 Principles of pathologic histology. Saunders, Philadelphia

Maurer R, Taylor C R, Schmucki O, Hedinger C E 1980 Extratesticular gonadal stromal tumor in the pelvis: a case report with immunoperoxidase findings. Cancer 44: 985–990

Mostofi F K, Price E B Jr 1973 Tumors of the male genital system. Atlas of tumor pathology, fascicle 7, series 2. Armed Forces Institute of Pathology, Washington DC

Mukai K, Rosai J 1980 Applications of immunoperoxidase techniques in surgical pathology. In: Fenoglio C M, Wolff M (eds) Progress in surgical pathology. Masson, New York

Schreiber J R, Reid R, Ross G T 1976 A receptor like testosterone binding protein in ovaries from immature female rats. Endocrinology 98: 1206–1213

Sero S F et al 1973 International histological classification of tumors (benign). Histologic typing of ovarian tumors. World Health Organization, Geneva

Sternberger L A 1979 Immunocytochemistry, 2nd edn. John Wiley & Sons, New York

Taylor C R 1974 The nature of Reed-Sternberg cells and other malignant cells. Lancet 2: 802–807

Taylor C R 1978a Classification of lymphomas: 'new thinking' on old thoughts. Archives of Pathology and Laboratory Medicine 102: 549–554

Taylor C R 1978b Immunoperoxidase techniques: theoretical and practical aspects. Archives of Pathology and Laboratory Medicine 102: 113–121

Taylor C R, Kledzik G 1981 Immunohistologic techniques in surgical pathology — a spectrum of 'new' special stains. Human Pathology 12: 590–596

Taylor C R, Kurman R J, Warner N E 1978 The potential value of immunohistological techniques in the classification of ovarian and testicular tumors. Human Pathology 9: 417–427

Taylor C R, Cooper C L, Kurman R J, Goebelsmann U, Markland F S Jr 1981 Detection of estrogen receptor in breast and endometrial carcinoma by the immunoperoxidase technique. Cancer 47: 2634–2640

Willis R A 1948 Pathology of tumours. Butterworths, London

24

Prostate

A. C. Jöbsis

INTRODUCTION

There are two major diagnostic problems in tumour pathology of the prostate in which histochemistry can provide helpful discriminative data: (a) Is the prostatic epithelial proliferation malignant? and (b) Could the carcinoma encountered be of prostatic origin? At present, no reliable methods are available for predicting the chances of malignant transformation or the sensitivity of a prostatic carcinoma to hormonal treatment. Problem (b) can be resolved by investigating the presence of the prostate associated isoenzyme: prostatic acid phosphatase (PRAP), whereas arguments for the solution of problem (a) may result from mucin histochemistry. Mucin is used as a synonym for mucosubstances, mucus and mucopolysaccharides.

MUCIN HISTOCHEMISTRY

With regard to the differential diagnosis between adenocarcinoma and epithelial hyperplasia, no reliable data can be obtained by the practicing pathologist using investigations based on tumour-associated antigens or loss of blood group antigens. Although cytophotometric determination of DNA may offer a solution for this differential diagnostic problem, as it does in many other organs, the technique will not be discussed here since it is outside the scope of this book. The prostate is one of the few organs in which mucin histochemistry may yield discriminative results in the investigation of glandular-like proliferations suspected of malignancy. The potential value of some of the mucin stains as a diagnostic tool in prostatic pathology was described in

the 1960s by, among others, Hukill & Vidone (1967). Their observations were essentially substantiated by the advances in mucin histochemistry in general as reviewed by Filipe (1979) with regard to gastro-intestinal tumour pathology.

Acid mucins, which are not generally present in non-neoplastic prostatic glands, can be demonstrated in about 50–70% of prostatic adenocarcinomas. These mucins are rarely seen in poorly differentiated adenocarcinomas or undifferentiated prostatic carcinomas. A high content encountered in well-differentiated prostatic carcinomas (small gland type) allows differentiation between malignant and benign in about two out of three questionable cases. The neutral mucins are of little or no discriminative value in the prostate (Hukill & Vidone 1967).

The acid mucins (alcian blue-positive at acidic pH) can be classified histochemically as: (1) sulphate-containing mucins (sulphomucins), and (2) sialic acid-containing mucins (sialomucins), which are neuraminidase (sialidase)-sensitive. The most suitable technique for routine use is the high-iron diamine method (HID) combined with the alcian blue staining carried out at pH 2.5 (AB) (Spicer 1965). The HID/AB stain offers optimal visualization of sulphomucins (brown-black) and sialomucins (blue) in one section of routinely fixed (formaldehyde or Bouin is recommended), paraffin wax-embedded prostatic tissue (Fig. 24.1). Other techniques for acid mucins yield approximately the same results, but are usually more expensive and laborious, less sensitive and more difficult to standardise.

Sulphomucins and sialomucins, either separately or together, are more frequently seen in the small gland type of prostatic carcinomas than in the large gland or the cribriform type. The mucins are usually located in the lumen or on the luminal surface and are only

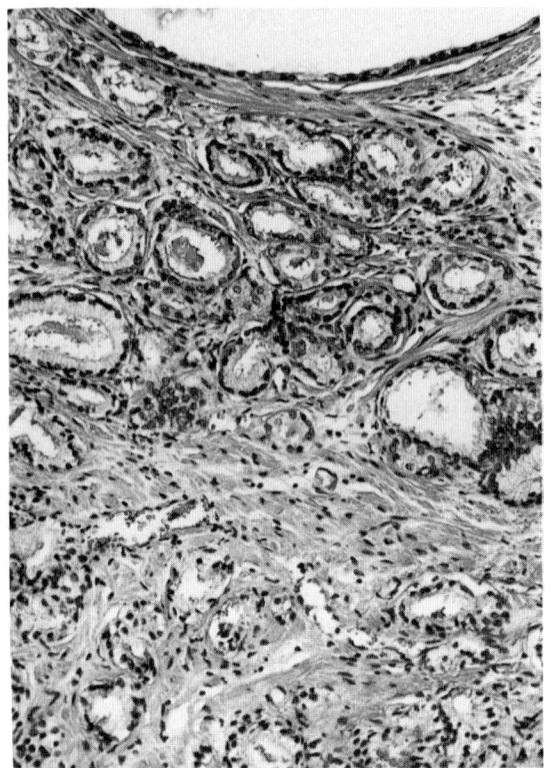

Fig. 24.1 HID/AB staining for acid mucins which are best discernible in the lumen of the gland-like malignant proliferations. Nuclei counterstained. ×126

rarely detectable in the cytoplasm (Fig. 24.2). Colloid prostatic carcinomas stain strongly positive.

Occasionally, a positive HID/AB reaction is found in a non-neoplastic gland. These results, which are in keeping with those reported by others employing the same or analogous methods (Hukill & Vidone 1967, Dollberg 1980) may therefore offer some help in doubtful cases, but cannot in themselves be decisive. In addition, a more intense staining for lipids in neoplastic epithelium (cryostat sections), can be helpful (Braunstein 1964).

The presence of acid mucins which is neither of prognostic value within the group of well-differentiated adenocarcinomas nor of discriminative value with regard to bladder carcinomas (Hukill & Vidone 1967), may increase diagnostic accuracy especially in cases with crush artefacts or tiny fragments (needle biopsy).

ENZYME HISTOCHEMISTRY

The question: 'Does this carcinoma originate in the prostate?' can be answered by investigating the presence of prostatic acid phosphatase (PRAP). This isoenzyme of acid phosphatase is almost exclusively associated with prostatic epithelial cells. Exceptions which have only been encountered in the islets of Langerhans, some insulinomas and some carcinoids, do not essentially diminish the differential diagnostic property of this histochemical feature.

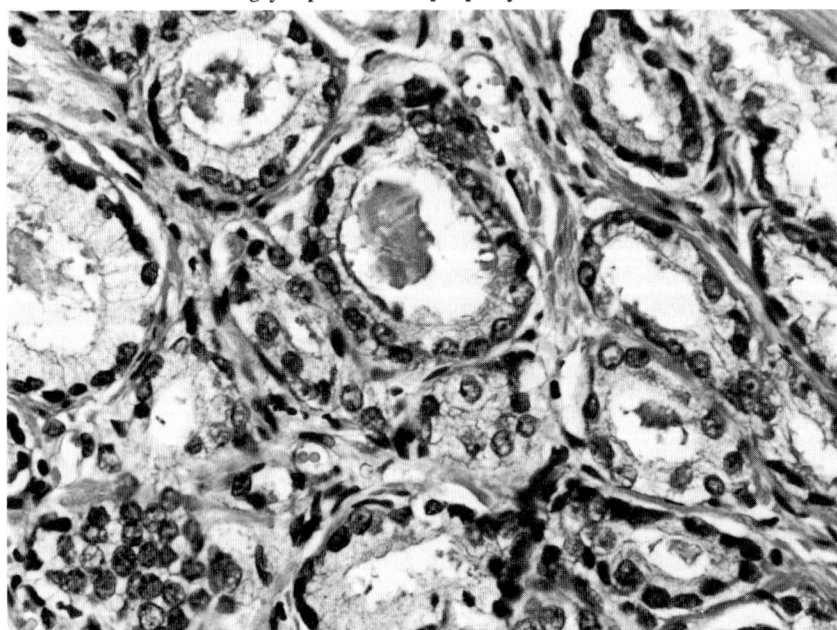

Fig. 24.2 Details of malignant part in Figure 24.1 ×350

PRAP is distinguishable from other acid phosphatase isoenzymes in epithelial cells by its relatively high resistance to formaldehyde (Pearse 1968) and by its almost absolute substrate specificity for phosphorylcholine (Serrano et al 1976). The sensitivity of PRAP for L(+) tartrate is relatively high but does not offer a reliable discriminative histochemical feature. These biochemical aspects can be demonstrated in cryostat sections of fresh (Figs 24.3–24.6) or formaldehyde-fixed tissues. The most

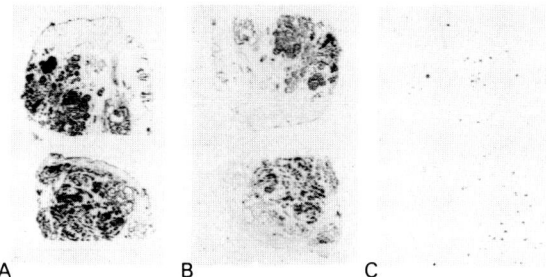

A   B   C

Fig. 24.3 Acid phosphatase activity in three consecutive cryostat sections of unfixed prostate with beta-glycerophosphate as substrate. In A the formalin prefixation was omitted, whereas in B the intensity of staining is reduced slightly due to this prefixation. In C no staining has developed due to the inhibition of the enzyme activity by 10 mmol/l L(+) tartrate. No counterstain. × 1

reliable enzyme histochemical method uses phosphorylcholine as substrate (Appendix 5). Positive staining in serial cryostat sections, one with no fixation and the other after formalin fixation indicates the prostatic nature of the acid phosphatase activity. Although PRAP is relatively insensitive to formalin, the acid phosphatases of epithelial cells of other organs may display, after fixation, some activity which may be demonstrated using substrates such as β-glycerophosphate and α-naphthylphosphate (Pearse 1968) which may cause some problems of interpretation (Figs 24.3–24.4). A section of non-prostatic tissue should therefore be incorporated as a control. If only formalin-fixed tissue is available, PRAP activity may be demonstrated in cryostat sections, preferably with phosphorylcholine as substrate (Fig. 24.5). A negative result under these circumstances does not exclude the prostatic origin of the carcinoma, since some reduction in PRAP activity does occur in formalin-fixed tissue at room temperature. The critical time period depends on the size of the tissue and, of course, the amount of PRAP in the carcinoma cells.

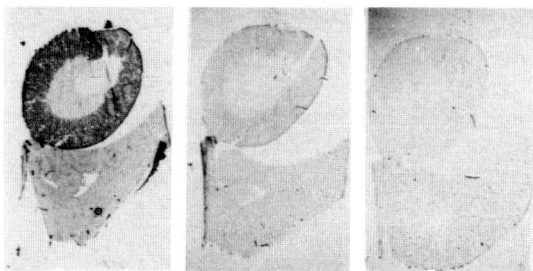

Fig. 24.4 Acid phosphatase activity in three consecutive cryostat sections of a composite block containing unfixed liver and kidney. The incubation circumstances were identical with those for the sections A, B and C shown in Figure 24.3. There is a significant degree of inhibition by means of formalin, so these organs contain little or no PRAP isoenzyme. × 1

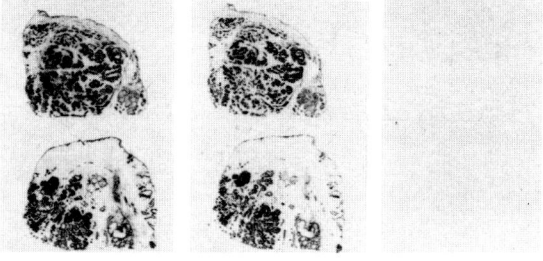

Fig. 24.5 Acid phosphatase activity in three consecutive cryostat sections of the same unfixed prostate as in Figure 24.3 with phosphorylcholine as substrate. The incubation circumstances were otherwise identical with those for the sections A, B and C shown in Figure 24.3. There is no inhibition by means of formalin, so the staining is due to the activity of the PRAP isoenzyme. × 1

Fig. 24.6 No staining of acid phosphatase activity with phosphorylcholine as substrate in this cryostat section of unfixed liver and kidney. The incubation circumstances were identical with those for the section shown in Figure 24.3A. × 1

The enzyme histochemical methods discussed, which can also be applied to cytological smears (imprint or aspiration) (Fig. 24.7), stain the cytoplasm of carcinoma cells less intensely than that of benign epithelial cells. There is no direct relationship between the degree of differentiation and the staining intensity. Papillary adenocarcinomas usually show a weak apical reaction. The staining intensity of carcinomas varies greatly, even in one section. This implies that a negative result in a small fragment of tumour cannot exclude a prostatic origin. Metastatic

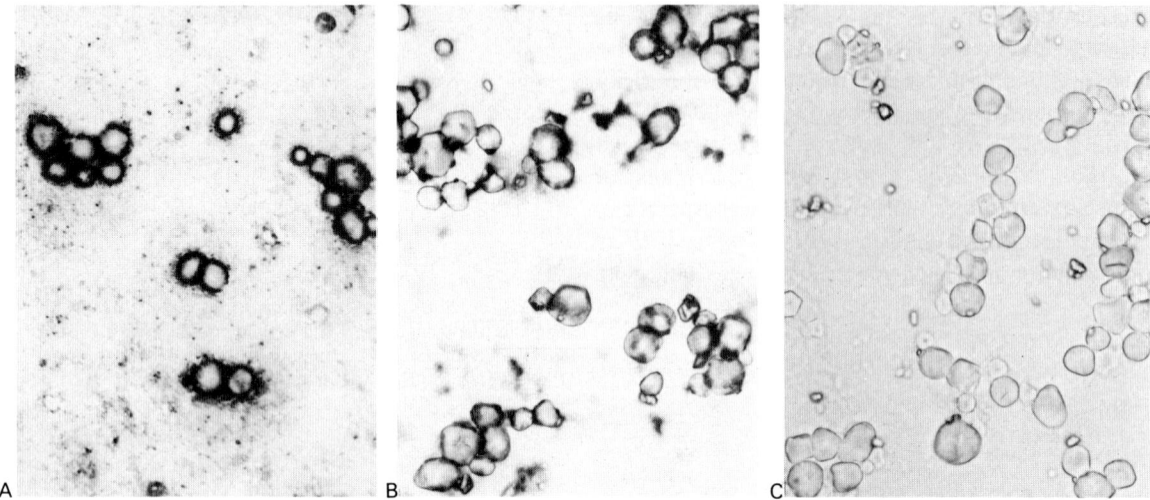

Fig. 24.7 Acid phosphatase activity in three imprint smears of unfixed prostatic tissue with beta-glycerophosphate as substrate. In (A) no formalin prefixation, in (B) formalin prefixation and in (C) the activity was inhibited by 10 mmol/l L(+) tartrate. No counterstain. ×280

foci have the same histochemical properties as the primary tumour. Androgen deprivation or oestrogen therapy reduces the staining intensity. PRAP activity is irreversibly inhibited by alcohol and by most decalcifying solutions. PRAP activity cannot therefore be investigated in paraffin wax sections or in skeletal metastases. These limitations, however, can be circumvented by demonstrating PRAP immunologically.

IMMUNOHISTOCHEMISTRY

With a specific antiserum against human PRAP essentially the same results — both qualitatively and quantitatively — are obtained in sections and smears (Nadji 1980) as with the enzyme histochemical method using phosphorylcholine. The great advantage of the immunological method is that PRAP antigen is not altered by the fixation solutions usually employed, or by decalcifying solutions or embedding procedures. Exceptions are Zenker's solution, the Susa fixation and the almost obsolete decalcification procedure with nitric acid. Small fragments of tumour tissue are another potential source of false-negative results since the staining intensity may vary greatly within one tumour (Fig. 24.8). The staining intensity, which has no predictive value for the effect

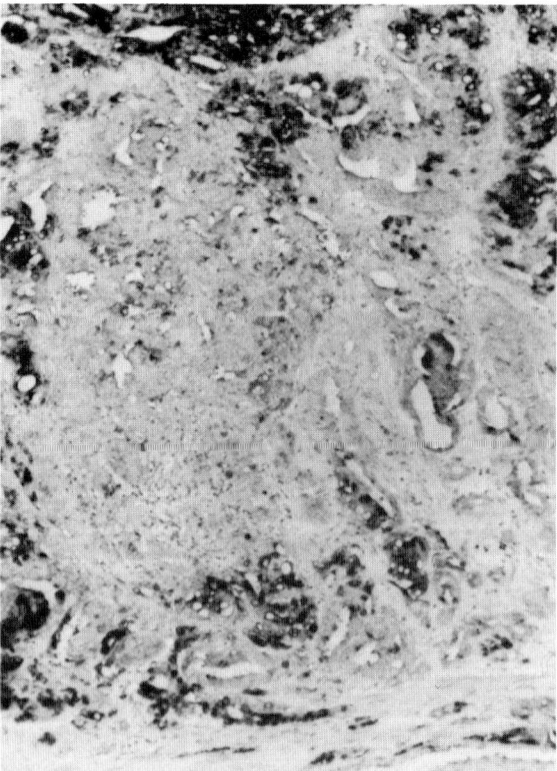

Fig. 24.8 Immunohistochemical demonstration of PRAP in paraffin wax section of prostatic carcinoma. The great variation of staining intensity (peroxidase technique) within one carcinoma is shown. Weak counterstain. ×81

of hormonal treatment, may be so low that it is advisable to repeat the incubation procedure under different conditions. It is then obligatory that the same changes be made in both control incubation procedures. By repeating the procedure under carefully selected conditions several originally negative cases (small fragments of tumour or hormonally influenced tumours) proved positive without obtaining false-positive results.

Non-specificity of the antiserum against human PRAP or inappropriate handling of the antiserum (repeated freezing and thawing for instance) and other technical shortcomings are essentially the only sources for false-positive results in paraffin wax sections (Aumüller et al 1981, Jöbsis et al 1981). Apart from the absorbance techniques which can in general purify a non-specific antiserum into a mono-specific antiserum, the non-specificity problem can be circumvented by means of the mixed aggregation immunocytochemical technique (Jöbsis et al 1981).

The degrees of specificity and sensitivity of the immunohistochemical technique for PRAP in paraffin wax sections were 96% and 98% respectively in a series of 200 proven prostatic tumours and 330 proven non-prostatic tumours (23 different types) and their normal analogues (Jöbsis et al 1981). Metastases of prostatic carcinomas (Fig. 24.9) were positive as were the primaries, whereas 50 bladder carcinomas were negative. Apart from the positive reaction in the islets of Langerhans (beta-cell pattern) and in several insulinomas, a finding which offers the pathologist an unexpected differential diagnostic tool in that area, the anti-PRAP serum discriminates for the prostate origin of carcinomas.

Essentially the same results have been reported for an antiserum against another prostate associated antigen (Nadji et al 1981). The immunohistochemical reaction for this antigen, which is also demonstrable in benign prostatic epithelium, has been claimed to be consistently negative in all tested non-prostatic tumours.

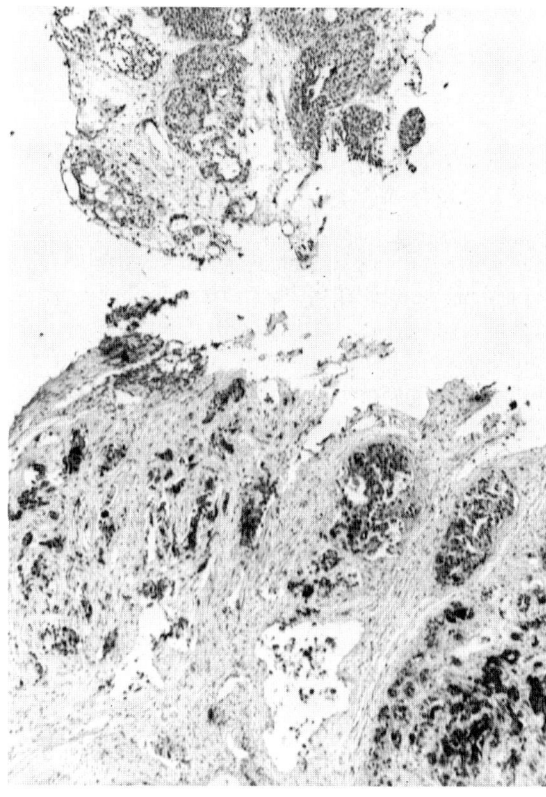

Fig. 24.9 Immunohistochemical demonstration of PRAP in paraffin wax section of a prostatic carcinoma. The primary (big fragment) shows a greater intensity of staining than the secondary (small fragment), which was biopsied after 6 months of hormonal treatment. Weak counterstain. ×50

REFERENCES

Aumüller G, Pohl C, van Etten R L, Seits J 1981 Immunohistochemistry of acid phosphatase in the human prostate: normal and pathologic. Virchows Archiv B 35: 249

Braunstein H 1964 Staining lipid in carcinoma of the prostate gland. American Journal of Clinical Pathology 41: 44

Dollbert L 1980 Early prostatic carcinoma. Human Pathology 11: 688

Filipe M I 1979 Mucins in the human gastro-intestinal epithelium: a review. Investigative Cell Pathology 2: 195

Hukill P B, Vidone R A 1967 Histochemistry of mucins and other polysaccharides in tumours. Laboratory Investigation 16: 395

Jöbsis, A C, de Vries G P, Meijer A E F H, Ploem J S 1981 Prostatic acid phosphatase immunologically detected: Possibilities and limitations with regard to tumour histochemistry. Histochemical Journal 13: 961

Nadji M 1980 The potential value of immunoperoxidase techniques in diagnostic cytology. Acta cytologica 24: 442

Nadji M, Tabei S Z, Castro A, Chu M, Murphy G P, Wang M C et al 1981 Prostatic-specific antigen: an immunohistologic marker for prostatic neoplasms. Cancer 48: 1229

Pearse A G E 1968 Acid phosphatases. Histochemistry, 3rd edn. Churchill, London, ch 16, p 547

Serrano J A, Shannon W A Jr, Sternberger N J, Wasserkrug H L, Serrano A A, Seligman A M 1976 The cytochemical demonstration of prostatic acid phosphatase using a new substrate, phosphorylcholine. Journal of Histochemistry and Cytochemistry 24: 1046

Spicer S S 1965 Diamine methods for differentiating muco-substances histochemically. Journal of Histochemistry and Cytochemistry 13: 211

25

Specific red cell adherence testing (SRCA) in bladder tumours

N. Javadpour

SRCA TESTING IN BLADDER TUMOURS

The major problem in diagnosis and management of a superficial non-invasive bladder cancer has been the lack of criteria to assess its potential invasiveness. The conventional histopathologic examination of the primary or metastatic tumour rarely predicts the potentiality of this cancer to invade. During the last few years it has become apparent that cellular differentiation in certain cancers is reflected in the presence or absence of certain cell surface antigens. The objective here is to update our experience with cell surface antigens as detected by the specific red cell adherence test (SRCA) in superficial and invasive bladder cancer (see Fig. 25.1). (Details of method are given in Appendix 6.)

Surface antigens, which are designated as ABO(H) blood groups, are present on the surface of normal urothelial cells (Fig. 25.2A & B) as well as cells from a variety of other tissues. The development of SCRA to determine the presence or absence of these cell surface antigens in certain cancers has been reported (Davidsohn 1972). Subsequently, DeCenzo et al (1975) applied this test to studies of stage A transitional cell carcinomas of the bladder and found that of the 13 tumours which retained the ABO(H) cell surface antigen, none progressed to invasive stages. In contrast, 8 of the 9 tumours that lost the antigen ultimately did advance to invasion. Reports by our laboratory (Javadpour 1978, Bergman & Javadpour 1978, Emmot & Javadpour 1978, Emmott et al 1981) and others (Lange et al 1978) revealed that

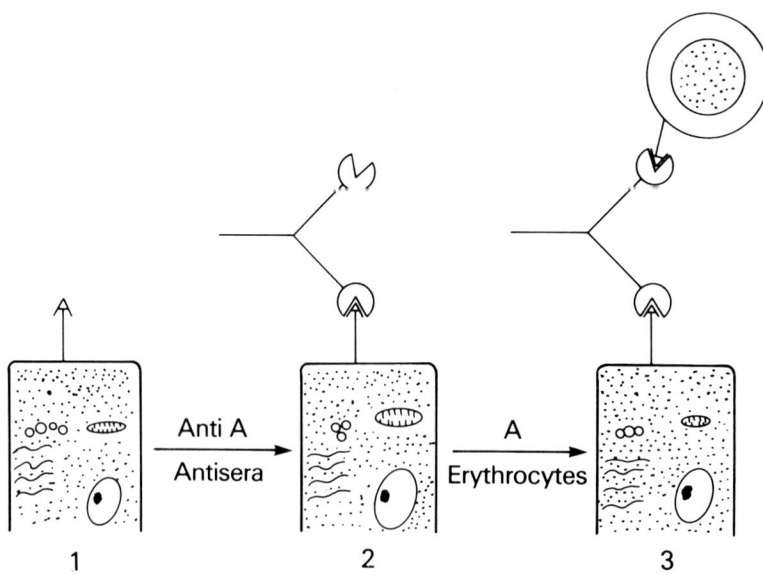

Fig. 25.1 Specific red cell adherence reaction (SRCA)

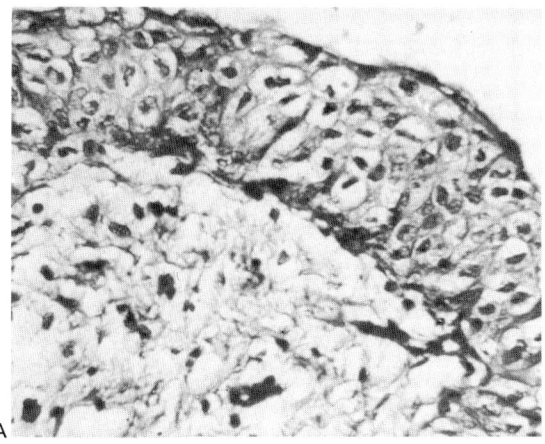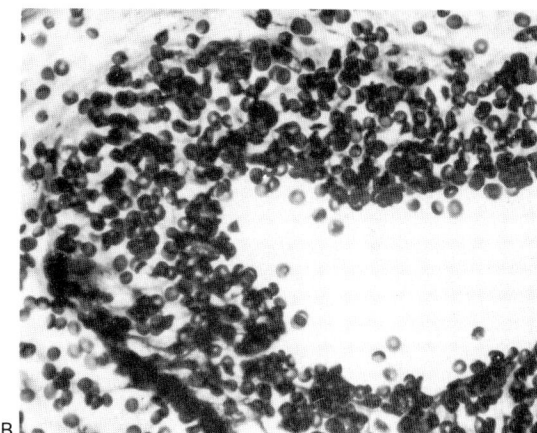

Fig. 25.2 Specific red cell adherence test in normal bladder urothelium, (A) before staining and (B) SRCA test in a patient with O-positive blood group

lesions beyond stage A consistently showed absence of the ABO(H) cell surface antigen and that loss of the antigen could be correlated with change in histologic grade towards a less differentiated tumour.

CORRELATION OF SRCA WITH TUMOUR GRADE

The results in a series of 76 patients are shown in Table 25.1 and suggest that the SRCA test only confirms the potential invasiveness of grade III tumours. The results also suggest that even in the grade I and grade II tumours there is an invasive potential in some, which could not have been predicted by conventional histological methods.

Table 25.1 Correlation of specific red cell adherence with tumour grade

Pathologic grade	Antigen present	Antigen absent
I	7	3
II	15	25
III	0	26

CORRELATION OF SRCA WITH TUMOUR STAGE

Table 25.2 shows the results in 76 patients. None of the patients beyond stage A had retained cell surface antigen. Its presence in 12 out of 28 patients in Stage A suggests that the test is only meaningful in stage A

where pathologic staging cannot predict invasive potential.

Table 25.2 Correlation of specific red cell adherence with tumour stage

Clinical stage	Antigen present	Antigen absent
A	12	16
B_1	0	3
B_2	0	6
C	0	0
D	0	39

CORRELATION OF SRCA WITH PROGNOSIS

In the study of 76 patients with various stages of bladder cancer, the recurrence rate and invasive potential are compared with presence or absence of cell surface antigen (Table 25.3). The figures provide some support for the prognostic capability of the SRCA test.

Table 25.3 Correlation of SRCA with prognosis

	Antigen present	Antigen absent
Initial lesion stage A:		
no recurrence	5	2
Recurrence, no invasion	2	10
Recurrence with invasion or metastasis	1	8
Initial lesion stages B_1 to D: lesion invasive or metastatic	0	48

SRCA TEST IN NON-MALIGNANT LESIONS OF THE BLADDER

It might be expected that inflammatory conditions could also influence the SRCA test. Table 25.4 shows in a limited number of cases that the SRCA test is unaffected in cystitis cystica and cystitis glandularis.

Table 25.4 Specific red cell adherence (SRCA) in non-malignant lesions

Pathology	Patients	SRCA Positive	Negative
Cystitis cystica	8	8	0
Cystitis glandularis	8	8	0
Squamous metaplasia	4	0	4

SRCA AND TUMOUR CHROMOSOMAL ANALYSIS

In a smaller study of 20 patients in whom chromosomal analysis of the tumour was possible, the grade, stage and prognosis of the tumour correlated with chromosomal abnormality. All patients with grade III tumours had chromosomal abnormalities and had lost cell surface antigens. Of 8 patients with Grade II tumours 4 had normal chromosomes and only 3 had retained cell surface antigen.

All patients with stages B–D had abnormal chromosomes and had lost cell surface antigen. In stages 0 and A, half the patients showed abnormal chromosomes and/or loss of cell surface antigen.

Combining chromosomal analysis and the SRCA test may be helpful in stages 0 and A where routine histopathological examination cannot predict anaplasia or invasive potential.

Coon & Weinstein (1981) have recently described the immunoperoxidase detection of ABO(H) antigens in normal and neoplastic urothelium. They found that the SRCA test gave comparable results, but that for H antigen the immunoperoxidase method was superior. Although loss of surface antigens appears to be a reliable guide in bladder cancer, the phenomenon does not hold for other tissues (Slocombe et al 1980).

REFERENCES

Bergman S, Javadpour N 1978 The cell surface antigen A, B or O(H) as an indicator of malignant potential in stage A bladder carcinoma. Journal of Urology 119: 48

Coon J S, Weinstein R S 1981 Detection of ABH tissue isoantigens by immunoperoxidase methods in normal and neoplastic mesothelium. American Journal of Clinical Pathology 76: 163

Davidsohn I 1972 Early immunologic diagnosis and prognosis of carcinoma. American Journal of Clinical Pathology 57: 715

DeCenzo J M, Howard P, Frish A 1975 Antigenic deletion and prognosis of patients with stage A transitional cell bladder carcinoma. Journal of Urology 114: 874

Emmott C, Javadpour N 1978 Correlation of stage and grade with ABO(H) cell surface antigen in bladder cancer. Journal of Urology 119: 31

Emmott R C, Droller M J, Javadpour N 1981 The ABO(H) cell surface antigens in carcinoma in-situ and non-malignant lesions of the bladder. Journal of Urology 125: 32

Javadpour N 1978 Biologic tumor markers in management of testicular and bladder cancer. Urology 12: 177

Lange P H, Limas C, Fraley E E 1978 Tissue blood group antigens and prognosis in low stage transitional cell carcinoma of the bladder. Journal of Urology 119: 52

Slocombe C W, Berry C L, Swettenham K V 1980 The variability of blood group antigens in gastric carcinoma as demonstrated by the immunoperoxidase technique. Virchows Archives of Pathological Anatomy and Histology 387: 289

26

Mesothelioma

M. Isabel Filipe

The differential diagnosis between mesothelioma (malignant) and adenocarcinoma, either primary in the lung or metastatic, is sometimes difficult and a definite diagnosis is not always possible on biopsy material using routine histological methods. Benign proliferative mesothelial lesions may also mimic metastatic adenocarcinoma.

Histochemical and immunocytochemical methods can be used as diagnostic aid, in biopsy material and in smears from effusions (Ch. 29).

METHODS TO DEMONSTRATE HYALURONIC ACID

Mucosubstances containing hyaluronic acid have been demonstrated in mesotheliomas and pleural and peritoneal effusions from patients with mesothelioma but shown to be absent from adenocarcinomas. The latter secrete glycoproteins which do not contain uronic acid.

Preparation of tissues (fixed; paraffin wax-embedded)

Fixation

Fixation of tissues in aqueous solutions such as formol-saline results in some loss of its hyaluronic acid content. 10% formol-calcium is a good alternative and easily available in any laboratory. Other suitable fixatives are alcohol, Zenker's, Helly's, Bouin's or Carnoy's.

Staining

1. D-PAS (diastase/amylase digestion followed by PAS)

2. Alcian blue pH 2.5 with and without pretreatment with hyaluronidase (testicular or streptomyces hyaluronidases)

Instead of alcian blue pH 2.5, other methods to identify hyaluronic acid in tissues can be used such as

3. Hale's colloidal iron with and without pretreatment with hyaluronidase

4. AB-MgCl$_2$(CEC) method.

However, for this particular problem they do not offer any advantage over AB pH 2.5 and they are more time-consuming.

Results and interpretation (see Table 26.1 and Fig. 26.1)

It is accepted that under the standard oxidation conditions (no longer than 10 min at room temperature), the PAS reaction is negative with

Table 26.1 Histochemical differential diagnosis of mesotheliomas and adenocarcinomas

	Mesotheliomas	Adenocarcinomas
D-PAS	−	+
AB pH 2.5	+	+
AB pH 2.5 after hyaluronidase	−	+
CEA	−	+

hyaluronic acid. In contrast, secretory glycoproteins (epithelial mucins) in adenocarcinomas contain substances with 1:2 glycol groups and give a positive D-PAS reaction. Glycoproteins in the majority of adenocarcinomas contain acidic groups, carboxylic and/or sulphated. Because of its acid radicals both hyaluronic acid and glycoproteins stain with AB at pH 2.5. Pre-treatment with hyaluronidase reduces or abolishes the AB staining of hyaluronic acid but has no effect on the glycoproteins.

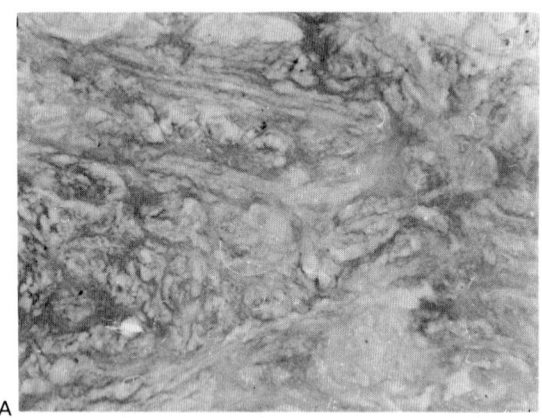
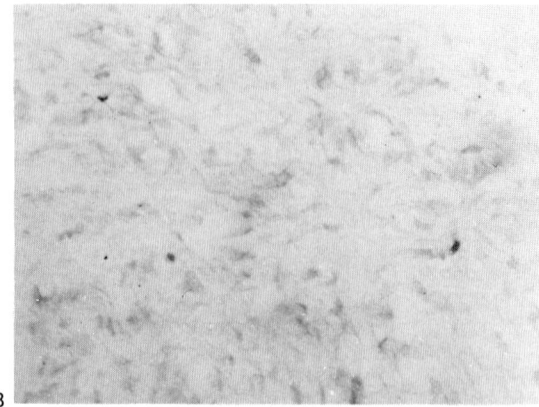

Fig. 26.1 Mesothelioma. The AB pH 2.5 staining in (A) is abolished by prior hyaluronidase digestion (B)

False-negative results may be obtained in mesotheliomas with low content in hyaluronic acid or from which it has been extracted by fixation. However, a positive result will be helpful in the large majority of cases. The accuracy of diagnosis may be improved by the complementary use of other techniques described below.

CARCINOEMBRYONIC ANTIGEN (CEA)

The principle for demonstration of CEA to distinguish mesothelial tumours from lung and gastro-intestinal adenocarcinomas is based on the assumption that the different ontogeny of mesothelium and gut and respiratory epithelia of endodermal origin may confer a different oncofetal antigenic expression.

Methods for immunochemistry

Formol-fixed, paraffin wax-embedded tissues.
  a. Indirect alkaline phosphatase technique or alternatively:
  b. Indirect immunoperoxidase technique.

Interpretation

Although the complete antigenic identity of oncofetal antigen of the lung cancers and those of GI tract is not firmly established the presence of CEA-like material in the lung neoplasias and its absence in tumours of mesothelial origin may be helpful in the differential diagnosis. (Table 26.1).

As in any immunohistochemical staining, it is essential that proper specificity controls are used. Even so, one has to be cautious in the interpretation of the results. Positive and false-negative reactions may occur.

IMMUNOCYTOCHEMISTRY USING ANTI-MESOTHELIAL CELL SERUM

Probable diagnostic value

This method has only recently been described and its value in the differential diagnosis of mesothelioma and other tumours involving the coelomic surfaces has yet to be assessed. However, the results are encouraging and because of its potential use in the specific identification of mesothelial cells, we feel it should be mentioned in this chapter.

In summary, the method uses anti-mesothelial cell serum in an indirect immunofluorescence technique. The results in smears and frozen sections of a variety of solid non-mesothelial tumours suggest that the antiserum is specific for mesothelial cells, benign and malignant, and fails to react with both normal and neoplastic non-mesothelial tissue.

CONCLUSION

In the majority of cases mesotheliomas can be distinguished from adenocarcinomas by a combination of histochemical methods in addition to routine histology and cytology (see Ch. 29).

Each individual method has its limitations.

However, a tumour which presents with the following staining characteristics: negative D-PAS reaction, AB pH 2.5 staining abolished after hyaluronidase digestion and negative CEA, is highly consistent with a diagnosis of mesothelioma rather than adenocarcinoma.

REFERENCES

Abelanet R, Jagueux M, Fondimare A, Roujeau J 1979 Les mésothéliomas pleuraux. Morphologie, histocytochimie, difficultés diagnostiques et problèmes nosologiques. Revue française des maladies respiratoires 7: 243–264

Kannerstein M, Churg J, Magner D 1973 Histochemical studies in the diagnosis of mesothelioma. In: Bogovski P et al (eds) Biological effects of asbestos. Lyon IARC

Kazuyori Y 1973 The effect of digestion with streptomyces hyaluronidase upon certain histochemical reactions of hyaluronic acid-containing tissues. Journal of Histochemistry and Cytochemistry 21: 794–803

Singh G, Whiteside T L, Dekker A 1979 Immunodiagnosis of mesothelioma. Use of antimesothelial cell serum in an indirect immunofluorescence assay. Cancer 43: 2288–2296

Wagner J C, Munday D E, Harington J S 1962 Histochemical demonstration of hyaluronic acid in pleural mesotheliomas. Journal of Pathology and Bacteriology 84: 73–78

Wang N S, Huang S N, Gold P 1979 Absence of carcino-embryonic antigen-like material in mesothelioma. An immuno-histochemical differentiation from other lung cancers. Cancer 44: 937–943

Waxler B, Eisenstein R, Battifora H 1979 Electrophoresis of tissue glycosaminoglycans as an aid in the diagnosis of mesotheliomas. Cancer 44: 221–227

Whitaker D, Shilkin K B 1981 Carcinoembryonic antigen in tissue diagnosis of malignant mesotheliomas. Lancet 1: 1369

27

Neuroendocrine tumours and hyperplasia

A. G. E. Pearse

INTRODUCTION

This chapter, originally designated as endocrine rather than neuroendocrine, now appears under the latter term to signify that the great majority of tumours formerly classed as endocrine should be regarded as derivatives of the diffuse neuroendocrine system.

It might appear logical to begin by considering the hyperplasias of the system but the very subordinate position of the neuroendocrine hyperplasias makes this impractical. Thus the chapter not only begins with the neoplasias but, for the greater part of its course, is concerned with them alone.

NEOPLASIAS OF NEUROENDOCRINE CELLS

Definitions

Tumours of the diffuse neuroendocrine system (DNES) were originally called *apudomas*. This Latin-Greek hybrid was derived by Szijj et al (1969) from the pseudo Latin acronym APUD and the misused Greek termination 'oma' which latter, for pathologists at least, carries the connotation of tumour. The alternative term *neuroendocrinoma* was proposed by Gould (1977) in his review of APUD cell system neoplasms, together with the extended term *neuroendocrine carcinoma* to describe malignant varieties. These variants, as used by Kameya et al (1980) and Gould et al (1980), may well come into general use, replacing the older and less euphonious term apudoma but carrying with them, by implication, possession of all the APUD characteristics described in succeeding sections of this chapter. Thus the terms apudoma and neuro-

endocrinoma are to be regarded as synonymous, allowing for the fact that the former embraces both benign and malignant forms of neuroendocrine tumours and the latter the benign ones only. Leibson (1979) has proposed the term *neurocrine*, in place of the more cumbersome neuroendocrine, to describe secretory nerve cells. Even though this term has sometimes been used to distinguish secretion *between* nerve cells, or between a nerve cell and its target, and even though it is probable that all nerve cells are neurocrine in the sense intended by Leibson, it may come into general acceptance. The natural corollary is to use the word *neurocrinoma* in place of the longer variant.

The apudomas, neuroendocrinomas or neurocrinomas, as they will be called coequally in this chapter, are tumours capable of producing a variety of physiologically active amines and/or a larger number of physiologically active or inactive peptides. Biologically they can be defined as tumours composed of cells derived from APUD cell precursors, and thus 'neuroendocrine-programmed', which can produce one or more of the 35 known APUD peptides, or variations of their correct amino acid sequences, or any of the known or presently unknown prohormones and preprohormones, or fragments thereof. Additionally, or solely, they may synthesise and secrete one or more of the amine hormone/transmitters which are associated with the series.

From the pathological point of view the neuroendocrinomas, and the neuroendocrine carcinomas, are tumours which possess the amine-handling and associated cytochemical and functional qualities of their presumptive APUD cell precursors. They are characterised ultrastructurally by the presence of 'endocrine' or 'neuroendocrine'-type storage granules which can be shown to contain a peptide component with or without a catecholamine or indolalkylamine. Their peptides may or may not be identifiable by immunocytochemistry and their amines may or may not be identifiable by cytochemistry and cytofluorometry.

Types of neuroendocrine tumours

The neuroendocrine tumours can be divided into two classes, those whose status as apudomas is proved, or at least generally accepted for one reason or another, and those for whom an APUD cell origin is postulated, but unproven and in many cases still unaccepted by established pathological opinion.

The list of accepted neurocrinomas includes many categories of tumour which will certainly and sensibly continue to be called by their older names. Among these are melanomas, phaeochromocytomas, carotid body tumours, paragangliomas, neuroblastomas, ganglioneuromas, medullary thyroid carcinomas, pituitary adenomas and pinealomas. Two further categories, in which the older terms are at last giving way to alternatives reflecting our more recent understanding of their nature, are carcinoids (Pearse et al 1974) and islet cell tumours (Creutzfeldt 1975, Larsson 1978).

The principal tumour whose neuroendocrine status is constantly in dispute is the small cell (oat cell) carcinoma of the lung, and this is true despite its commonly acknowledged relationship to the bronchial carcinoid (Bensch et al 1968) and the accepted derivation of the latter from Kultschitsky-type cells in the bronchial mucosa (Bensch et al 1965). In a similar position are the oat cell, or argyrophil cell, carcinomas of various organs, including particularly those of the oesophagus (Tateishi et al 1976, Reid et al 1980) and the uterine cervix (Tateishi et al 1975, Matsuyama et al 1979), and the hitherto infrequently described Merkel cell tumours of the skin (Sidhu et al 1980, Wolf-Peeters et al 1980).

Diagnostic characteristics of neuroendocrine tumours

The original list of APUD cell characteristics (Pearse 1969) appeared in two parts. These are given below as Tables 27.1 and 27.2.

Table 27.1 Cytochemical characteristics of polypeptide hormone-secreting cells of the APUD series

1. Fluorogenic amine content (catecholamine, indolamine or other)
    a. primary
    b. secondary uptake
2. Amine precursor uptake (5-HTP, DOPA)
3. Amino acid decarboxylase
4. High side chain carboxyl or carboxamide content (masked metachromasia)
5. High nonspecific esterase or cholinesterase (or both)
6. High α-glycerophosphate menadione reductase
7. Specific immunofluorescence

5-HTP = 5-hydroxytryptophan
DOPA = L-3, 4-dihydroxyphenylalanine

Table 27.2 Ultrastructural characteristics of polypeptide-secreting APUD cells

1. Low levels of rough endoplasmic reticulum
2. High levels of smooth endoplasmic reticulum as vesicles
3. High content of free ribosomes
4. Electron-dense, fixation-labile mitochondria
5. Prominent microtubules, centrosomes
6. Tendency to produce fine protein microfibrils (especially when neoplastic)
7. Membrane-bound secretion granules, average size 100–200 nm

In Table 27.1 the seventh entry could not and cannot be considered as a characteristic, but only as a prime means of determining which, if any, of the known peptide hormone or prohormone products of the APUD cell series are present in a given tumour. In 10 years of experimentation a further short list of APUD cell characteristics has been added, and these are discussed briefly in the section on molecular markers.

Soon after the appearance of the original lists (Tables 27.1 and 27.2) it became apparent that the cells composing APUD cell tumours possessed all the characteristics of mature APUD cells of normal constitution, often at substantially higher levels. Thus the stage was set for the establishment of a collection of techniques for the identification of apudomas, and these methods can be used either to confirm a clinical diagnosis and the initial histological impression, or for the fuller investigation of material from possible neuroendocrine tumours in patients whose symptomatology lacks diagnostic features (the majority of neuroendocrinoma cases). Most of the features listed in Table 27.2 are to be regarded as ancillary rather than specific.

Although on rare occasions, and especially after

physiological or artificial stimuli, rough endoplasmic reticulum levels in the APUD cells rise sufficiently to equal those of major protein-exporting cells, they normally exhibit a very low level of basophilic material and thus appear as clear cells by light and electron microscopy equally. This fact emphasises their close relationship to the clear cells of Feyrter (1946, 1953), which composed his peripheral endocrine (paracrine) glands and which were considered as the progenitors of a number of tumours of the carcinoid variety.

Identification of apudomas

The methods which can and should be employed for the identification of neuroendocrine tumours are given below in Table 27.3. Mainly composed of

Table 27.3 Identification techniques for neuroendocrine tumours

Histological methods	Lead haematoxylin*
	Masked metachromasia*
	Argyrophilia*
Enzyme cytochemistry	Nonspecific esterase*
	Cholinesterase*
	α-Glycerophosphate dehydrogenase
Amine cytochemistry	Formaldehyde-induced fluorescence*
	Glyoxylic acid*
	Uptake and decarboxylation of amine precursors (APUD-FIF)*
Immunocytochemistry	Specific peptides*
	Marker hydroxylases and transferases
	Neuron-specific enolase
Ultrastructure	Neuroendocrine granule analysis

* These methods are given in full in the Appendices

methods capable of demonstrating the characteristics already outlined above, the list contains three useful paracytochemical techniques and some additional and more specific marker techniques.

Of the histological techniques in Table 27.3, argyrophilia (Grimelius 1968) and lead haematoxylin (Solcia et al 1968) are the most reliable indicators for neuroendocrine status at the light microscope level, always provided that at least some of the tumour cells contain the characteristic storage granules. In the absence of storage of either amine or peptide, the in vitro APUD-FIF method provides a reliable indicator and is easily carried out.

Of the three enzyme techniques in Table 27.3 those for nonspecific esterases and cholinesterases may sometimes afford diagnostic assistance. A positive result with either, or both, may be regarded as complementary to more direct indications obtained by other means.

If a specific peptide has been detected in raised amounts in the patient's serum by radioimmunoassay it will usually be detectable in the tumour cells by immunocytochemistry. If no neuroendocrine peptide can be found, antisera to the two hydroxylases (dopamine β-hydroxylase, DBH, tyrosine hydroxylase, TH) or to the single transferase (phenylethanolamine-N-methyl transferase, PNMT) may give a positive indication of the neural (neuroendocrine) nature of a tumour.

If the tumour contains an endogenous amine this will be revealed by formaldehyde-induced fluorescence (Fig. 27.1). When this gives a positive result, identification of the amine can be achieved by spectrofluorometry combined with ancillary cytochemical techniques. If no endogenous amine is present the cells of the tumour may take up preferentially, and decarboxylate, one or both of the two amine precursors, 5-hydroxytryptophan and

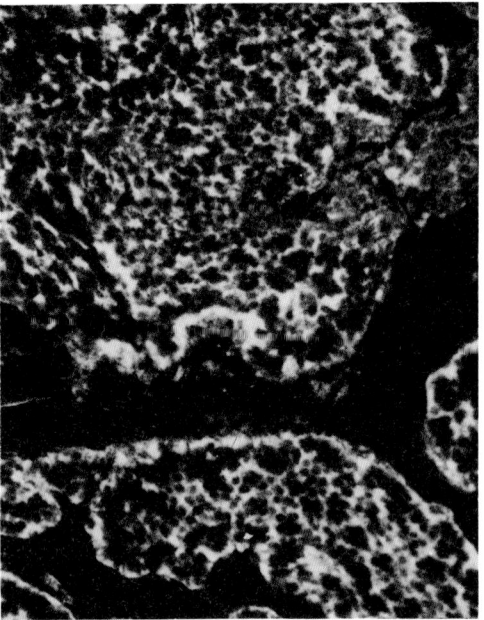

Fig. 27.1 Ileal carcinoid (female aged 45). Freeze-dried formaldehyde vapour-treated section shows FIF with spectral characteristics of 5-hydroxytryptamine. ×470

DOPA. A positive APUD-FIF reaction confirms the presence in the cells of the enzyme dopa decarboxylase.

By far the most convenient and convincing indication of the neuroendocrine status of a tumour is provided by the immunocytochemical demonstration in its cells of the so-called neuron-specific enolase (NSE) (Fig. 27.2) (Fletcher et al 1976, Marangos et al 1978). The gene coding for this isoenzyme, which

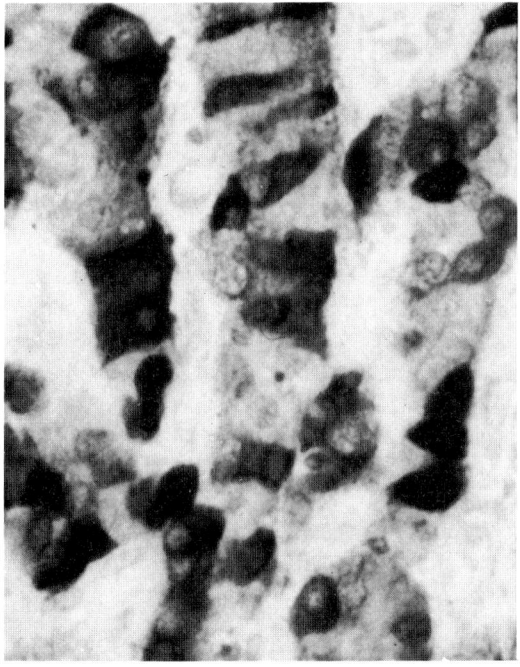

Fig. 27.2 Insulinoma (pancreatic, male aged 53). Freeze-dried, benzoquinone vapour-fixed section. Shows a strong reaction for neuron-specific enolase in a proportion of the tumour cells. PAP method. × 335

occurs in the form of a dimer composed of two γ-subunits, was considered to be expressed solely in neurons until Schmechel et al (1978) showed that it was present in the peripheral and central neuroendocrine cells of the diffuse neuroendocrine system. The sole exception to this rule, so far detected, is the chief cell of the parathyroid gland.

The presence of 'endocrine' or 'neuroendocrine' type granules constitutes the ultrastructural indication that a tumour may be a neuroendocrinoma. In many cases the appearance of the granules will be such that a confident confirmation can be given and, albeit only occasionally, even a tentative identification of the stored peptide. More usually, and especially in

the predominant mixed endocrine tumours, even a complete granule analysis, carried out with an image analysing computer, will require complementary information from other techniques.

Failure to recognise the endocrine nature of a tumour is no longer acceptable. Formerly considered as rare, endocrine (neuroendocrine) tumours composed of APUD cells are now more and more frequently identified, and described in the literature as solitary examples or as a part of some more widely based disorder belonging to the spectrum of multiple endocrine neoplasias (Sizemore et al 1980, Yamaguchi et al 1980).

Molecular markers and neuroendocrine tumours

The demonstration of a specific identifying function or antigen, in or on the surface of tumour cells, has been for many years a principal aim of diagnostic pathology. In the case of neuroendocrine tumours their identification as apudomas by the in vivo demonstration of their acronymous A-P-U-D characteristics constitutes, despite the relatively unspecific nature of the test, a valid molecular marker system. Much more specific, of course, is the immunocytochemical demonstration of NSE, as outlined in the previous section, and there are several other markers for neuronal function which either have been, or will be, used to confirm the neuroendocrine nature of tumours. Among these are the three enzymes of the catecholamine-synthesising pathway, already mentioned in the text (DBH, TH, PNMT), and tryptophan-5-hydroxylase which carries out the first stage synthesis of 5-hydroxytryptamine. Of a similar nature is the enzyme indoleamine 2,3-dioxygenase, demonstrated immuno-cytochemically by Watanabe et al (1979) in several types of APUD cell. Normal and tumour cells synthesising indoleamines, increasingly found to be present in subphysiological or subclinical levels in cells and tumours of the APUD series, will contain also the specific serotonin-binding protein demonstrated by Bernd et al (1979) in the thyroid C cells and shown to be present in serotoninergic neurons in brain and gut.

Histogenesis of neuroendocrine tumours

This problem was very fully considered by Tischler

et al (1977). Reporting their studies on clonal lines of genetically identical neoplastic APUD cells in culture, taken from phaeochromocytomas, oat cell carcinomas, medullary thyroid carcinomas, bronchial carcinoids and neuroblastomas, the authors emphasised the potential pluripotency of such cells and their ability to switch production and morphology towards or away from the neural mode according to the variety of stimulus applied. It is often suggested, at least by omission or inference, that neuroendocrine tumours arise from mature APUD cells. This is not true; they are necessarily the progeny of pluripotent precursor cells whose origin and status alone requires elucidation.

In simple terms the case is as follows. For approximately half the known neuroendocrine tumours their precursor cell is acceptably proposed to be of neural or neuroectodermal origin. A list of such tumours is given in Table 27.4, and only forward transformation is required for their development.

Table 27.4 Apudomas of 'neural' origin

Phaeochromocytoma	Melanoma
Neuroblastoma	Merkeloma
Ganglioneuroblastoma	Pinealoma
Medullary thyroid carcinoma (MCT)	Pituitary adenoma
Thymic carcinoid	

For the other half a common (endodermal) epithelial cell progenitor is usually postulated, which is presumed to give rise equally to non-endocrine or endocrine (neuroendocrine) neoplasms. Tumours having this attribution are listed in Table 27.5.

Table 27.5 Apudomas usually presumed to be of endodermal epithelial origin

Gastro-intestinal carcinoids
Pancreatic islet carcinoids
Bronchial carcinoids
Small cell carcinoma of lung (SCC)
Other small cell carcinomas
Prostatic small cell tumours

The second of the two postulates requires either that neoplastic change arises by dedifferentiation (retrograde differentiation), or that a pluripotential precursor is induced to develop forwards and specifically in a single direction. Retrograde differentiation has never been shown to occur in experimental in vitro studies.

Elucidation of the origin of medullary thyroid carcinoma provides one of the best examples of neurocrine tumour histogenesis. Baylin and co-workers (1976, 1978), studying cases of human C cell neoplasia occurring in females mosaic for the two (A and B) isoenzymes of glucose-6-phosphate dehydrogenase, were able to show that the tumours result from an initial mutation giving rise to multiple clones of precursor cells susceptible to neoplastic change, followed by a second mutation affecting a single clone of either type.

Despite these findings with reference to a single type of neuroendocrine tumour, the histogenesis of the majority remains unproven. We are presently obliged, for lack of concrete information, to accept the fact that half of the tumours (Table 27.4) are certainly of 'neural', or at least ectodermal origin while the other half (Table 27.5) still remain endodermal. Sufficient doubt exists, however, with regard to the inclusion in Table 27.5 of bronchial carcinoids and small cell lung carcinomas for these two to be considered, in a later section, as candidate neuroendocrine tumours.

Products of neuroendocrine tumours

Amines and metabolites

The principal amine products of apudomas are catecholamines, dopamine, noradrenalin and less often adrenalin, and indolamines, 5-hydroxytryptamine (Fig. 27.1) and N-acetyl-5-methoxytryptamine (melatonin). The latter, formerly regarded as an exclusively pineal hormone, has been demonstrated in the intestinal enterochromaffin cells by Raikhlin et al (1975). It appears not to have been recorded in carcinoid tumours. As shown by Goedert et al (1980), intestinal varieties of the latter can produce substantial amounts of dopamine, noradrenalin and 5-hydroxytryptamine, though not necessarily from one and the same cell. All are identifiable histochemically, either by formaldehyde-induced fluorescence (Fig. 27.1) and spectrofluorometry, with or without ancillary techniques (Björklund et al 1968), or by the more sensitive glyoxylic acid method (Axelsson et al 1973), or by immunocytochemistry using specific antisera. In some cases metabolites of catechol- or indolamines have been identified in neuroendocrine tumours. These include 3,4-dihydroxymandelic and 3-methoxy-4-hydroxymandelic (VMA) acids and, from serotonin, principally 5-hydroxyindole acetic acid (5HIAA).

Prepropeptides and metabolites

Neuroendocrinomas, like the normal cells of the APUD series, synthesise peptides which are larger than the known hormones which are usually their C-terminal fragments. These fragments, technically metabolites, may be broken down by intracellular peptidases into smaller, inactive sequences or oligopeptides. If for any reason the specific peptidases associated with the normal cell are lacking in the cells of a tumour, the prohormone may reach the circulation in measurable amounts. In theory at least, the neoplastic APUD cell retains its capacity to synthesise any of the physiologically active peptides associated with the series. A full search for these, in any given tumour, is clearly beyond the scope of routine diagnostic pathology.

Enzymes

The role of enzymes related to amine metabolism as molecular markers for neuroendocrinomas has already been considered. We are concerned here with two distinct entities, histaminase, and peptidases. The first of these (diamine oxidase) was found in high levels in medullary thyroid carcinomas by Baylin et al (1970, 1972), and regarded at that time as an ectopic placental enzyme and as a biological marker for C cell tumours. Subsequently, however (Baylin 1977), it became clear that the presence of the enzyme in tumour cells of various kinds (particularly MCT and SCC) is not to be regarded as an ectopic manifestation but as due to the expression of a normally repressed mature genome.

Of considerable interest is the problem of the peptidases of neuroendocrine tumours. In normal endocrine and neural tissues many proteolytic enzymes have been recorded, belonging to classes such as esteropeptidases, serine peptidases and trypsin and chymotrypsin-like peptidases. Some of these are clearly responsible for the sequential breakdown of prohormones into biologically effective peptides. As in normal APUD cells, the peptidases of apudomas may be either specific or non-specific and if specific peptidases can be identified, which are directly related to the production of specific active

peptide fragments from inactive large precursor molecules, they may turn out to be cell-specific markers and thus of diagnostic significance (Fig. 27.3).

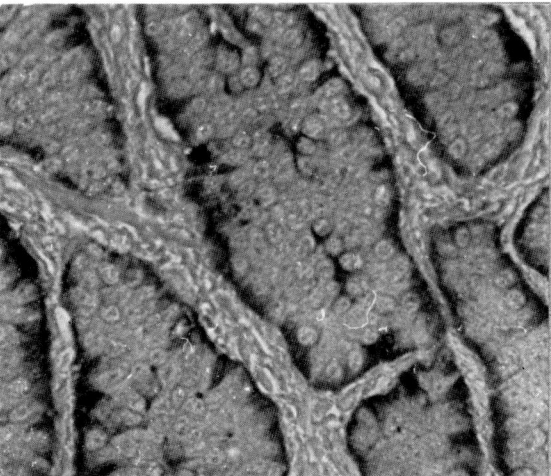

Fig. 27.3 Carcinoid tumour of ileum (male aged 67). Cold formaldehyde-fixed. Shows a strong indoxyl esterase, especially in the region of the basal lamina. This enzyme has the characteristics of an intracellular peptidase. ×750

Multiple hormone production by neuroendocrine tumours

With the increasing availability and more widespread distribution of peptide hormone antibodies it is a matter of no surprise to note the increased frequency with which apudomas are reported to contain more than one hormone, or at least more than one immunologically recognisable peptide, active or not in the physiological or clinical sense. Polyhormonal apudomas were divided by Pearse (1980) into the following five theoretically based classes, each of which derives at least some support from studies of tumours and from experiment.

1. Single cell, single precursor

Production of the peptides of a single family by sequential breakdown of a single protein (i.e. corticolipotropin) yielding ACTH, α-MSH, β-endorphin and β-MSH.

2. Single cell, common precursor

Production of two or more unrelated peptides from a single molecule by the selective action of specific proteases.

3. Single cell, separate precursors

Activation of associated or closely linked genes coding for prepropeptides belonging to more than one family, giving rise to 2 or more unrelated peptides.

4. Multiple cell types, multiple clones

Separate clones, differentiating as distinct cell types, each producing a distinct peptide.

5. Multiple cell types, single clone

Differentiation of a single clone of precursor cells into different cell types producing different peptides.

Certainly more than one member of a single family of hormones can be derived from a single precursor molecule in a single cell type (class 1). Roberts & Herbert (1977), for instance, showed that the 31 kD glycoprotein product of mouse pituitary tumour cell line AtT-20/D-16v contained, at its C-terminal end, the complete sequences of ACTH and β-lipotropin. Another example is the human small cell lung tumour DMS-79 (Bertagna et al 1978) which produces, in culture, ACTH, β-LPH and β-endorphin from a single precursor molecule, and also calcitonin. The authors suggested that two distinct precursors were involved, coded for by closely linked genes.

Morphological and immunocytochemical evidence tends to support classes 4 and possibly 5, rather than classes 1, 2 and 3 for which there is mainly biochemical evidence. Calcitonin, somatostatin and ACTH, the three peptides most commonly found in MCT, were found invariably to be present in distinct and separate cell types by Capella et al (1978) and in 23 of a series of 36 pancreatic endocrine tumours Heitz et al (1976) found that cells producing pancreatic polypeptide could always be distinguished from those producing the other neoplastic pancreatic peptides (insulin, glucagon, VIP, gastrin).

The work of Baylin et al (1976, 1978), already referred to, shows that MCT, at least, is a monoclonal tumour. Since it produces separate cell types, and distinct peptides, the mechanism suggested for tumours of class 5 must be considered operative.

Candidate neuroendocrine tumours

The most important of these, in terms both of human suffering and of academic interest, is the small cell carcinoma of the lung. A number of morphological observations, at light and electron microscopic levels, strongly support an origin for this tumour from the common (endodermal) stem cell of the respiratory tract which, like the crypt base cell of the intestine (Cheng & Leblond 1974) has been considered to give rise to all four types of cell found in the epithelium. Recent studies of SCC carried out by Hashimoto et al (1979) and by Sidhu (1979) can be quoted in support of this view. In respiratory epithelia there is some morphological evidence to support an opposite interpretation and in the case of intestinal neuroendocrine cells recent work by Doyle (1980), on equine small intestine, provides strong evidence for separate origins and hence for separate epithelial and neuroendocrine cell lines. The opposing point of view was also expressed by Bensch et al (1968) who regarded the argyrophil so-called Kultschitsky cells of the respiratory epithelium, with their long slender dendrite-like cytoplasmic processes, as the cells of origin not only of bronchial carcinoids, but also of SCC. Other studies, using mainly EM criteria and granule analysis, have indicated a close relationship between the ultrastructural characteristics of the neuroendocrine cells of the respiratory tract and those of bronchial carcinoids on the one hand (Capella et al 1979) and of SCC on the other (Gould et al 1978). Argyrophilia, and its EM concomitant the presence of neuroendocrine granules, were both demonstrable in half of their cases of SCC in a survey carried out by Tateishi et al (1978). Much other work confirms this finding, as does the observation that SCCs contain NSE and that they produce this isoenzyme when maintained in tissue culture (Tapia et al 1981). The frequent elevation of histaminase activity in SCC, documented by Baylin et al (1975) was regarded by these authors as evidence in favour of an embryological relationship between MCT and SCC, and hence as evidence of a neuroectodermal origin for the latter.

The production of hormonal peptides by SCC presents difficulties in interpretation but the principal facts are not in doubt. SCC is the commonest tumour producing 'ectopic' ACTH and Cushing's syndrome, as first indicated by Meador et al (1962). The figures for ACTH, and for calcitonin, vary widely in different series, depending to some extent on whether they relate to circulating or tumour peptides. Hansen et al (1980) found elevated plasma ACTH in 22/75 cases of SCC (29%) and elevated calcitonin in 48/75 (64%). In studies on tumour and sputum samples Deftos & Burton (1980) found that in 16/31 cases they could show calcitonin cells, in 12/14 β-endorphin and in 11/14 ACTH cells.

Calcitonin was reported to be present in Kultschitsky cells by Becker et al (1980). These peptide statistics cannot *prove* a direct relationship between SCC and neuroendocrine cell precursors in the respiratory tract. But it can be concluded that neuroendocrine cells, whatever their origin, are at least closely related to SCC cells in metabolic terms, and that they could achieve such a transformation without retrograde differentiation.

HYPERPLASIAS

Potential neuroendocrine hyperplasias

There is no reason to suppose that these are confined to a few select members of the APUD cell series. In the central division of the DNES, all the pituitary endocrine cells and the cells of the pineal can certainly exhibit this phenomenon. For obvious reasons they are seldom seen to do so. In the peripheral division of the DNES there can, in theory, be no exceptions to the proposition that all members of the series can be involved in the process of hyperplasia. There is good evidence that they can and do divide, even when fully differentiated, as Jurecka et al (1978) have shown for the adrenomedullary cells in mice. Their normally rather slow rate of division can be accelerated by physiological or other stimuli.

Recorded neuroendocrine hyperplasias

Thyroid gland

The best-documented APUD cell hyperplasia concerns the thyroid C cells. Ample evidence has been provided, from a number of laboratories, to justify the assumption that it is a frequent, if not invariable concomitant of MCT (De Lellis et al 1977). Somatostatin immunoreactive cells may be found in abnormally high numbers in such hyperplasias (O'Briain et al 1979, Deftos et al 1980).

Adrenal gland

Adrenal medullary hyperplasias are a source of some contention. They have been described by Carney et al (1975, 1976), and by De Lellis et al (1976), as accompanying cases of multiple endocrine neoplasia (MEN) type 2. Ljungberg (1972) earlier described similar appearances in two cases of MEN 2. Visser & Axt (1975), on the other hand regarded bilateral adrenomedullary hyperplasia as a distinct and separate entity. There is room for both interpretations.

Respiratory epithelia

'The presence of neuroendocrine cells in the tracheobronchial tree is well-established as is the basic structural and functional relationship between those cells and the enterochromaffin and APUD systems' (Gould et al 1978). In five cases of bronchial dysplasia these authors described accompanying dysplasias of the neuroendocrine cells but no increase in their number. Johnson et al (1980a, b) found hyperplasia of the cells, and their extension into the alveoli, in rats subjected to chronic inhalation of chrysotile or crocidolite (asbestos) for 6–24 months. The use of the neuron-specific enolase marker (Cole et al 1980) should facilitate future studies on respiratory neuroendocrine cell hyperplasias.

Pancreas

Islet cell adenomatosis, as such, was described by Schwartz & Zwiren (1971) and the entity, more usually called nesidioblastosis, has been the subject of a large number of papers, particularly in respect of its relationship to neonatal hypoglycaemia. Focal islet cell adenomatosis as a cause of the latter was described by Klöppel et al (1975) and also by Heitz et al (1977). The hyperplasia in such cases has been supposed chiefly to involve the B cells but D cell hyperplasia was described by Orci et al (1976) in juvenile diabetes and Gepts et al (1977) described PP cell hyperplasia in this condition. Rahier et al (1980), however, showed that D (somatostatin) cells were 20 times more frequent in normal neonatal pancreas, constituting 5% by comparison with the adult level of 0.23% of the total endocrine cell count. The association of islet cell hyperplasias with gastrin-producing tumours was emphasised by Creutzfeldt et al (1975) and by Larsson (1977).

Gastro-intestinal tract

The majority of gastro-intestinal neuroendocrine cell hyperplasias reported in the literature concern the antral G cell. There are accounts also of hyperplasias affecting the ECL and EC cells.

The first description of antral gastrinosis, associated with severe peptic ulcer disease but without gastrinoma, was made by Solcia et al (1970). Later Polak et al (1972) described two types of Zollinger-Ellison syndrome, the classical form, associated with gastrinoma and a second form associated with a pure antral gastrinosis. Further reports of this entity came from Friesen et al (1972) and Cowley et al (1973).

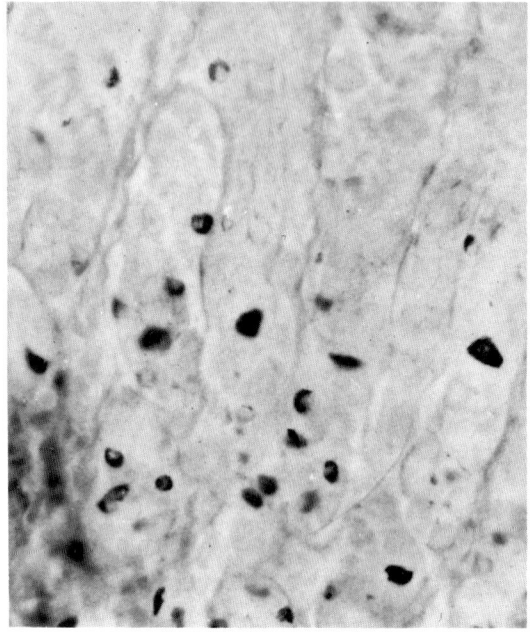

Fig. 27.4 Human pyloric antrum from a case of carcinoma of the stomach. Formaldehyde-fixed. Shows hyperplasia of EC cells in a region adjacent to the tumour. Masson-Fontana. ×550

ECL cell hyperplasia in the fundic mucosa of Z-E patients was described by Bordi et al (1974). It is an invariable component of the gastro-intestinal consequences of pernicious anaemia (Rubin 1973) as is antral G cell hyperplasia.

EC cell hyperplasia was a distinctive feature of many of the cases of gastric carcinoma (Fig. 27.4) recorded by Azzopardi & Pollock (1963), Watanabe (1972), and others, and EC cell hyperplasia was observed in the duodenal mucosa of children with coeliac disease by Challacombe & Robertson (1977). Microproliferation of ileal EC cells was suggested as the source of ileal carcinoids by Sherman et al (1979).

Autonomic nervous system

Strictly speaking, hyperplasias of this system are not classified as neuroendocrine. They cannot be separated, however, from the hyper- and neoplasias of the DNES, in view of a constantly reported association between the two. Several papers have recorded the occurrence of intestinal gangliomatosis (ganglioneuromatosis) in MEN type 2 (Carney et al 1976, Whittle & Goodwin 1976). Paragangliomatosis has been described in man, in association with MEN type 1 (Farhi et al 1976) and also in ageing WAG/Rij rats where there were associated endocrine hyperplasias without endocrinopathy (Van Zwieten et al 1979). Finally, abnormal cutaneous innervation has been observed in MEN 2b cases by Carney et al (1981).

It is obvious that more frequent and more pertinent enquiries, together with the increasing use of molecular markers, will reveal many more of the associations recorded above and many more hyperplasias of neuroendocrine cells and tissues.

REFERENCES

Axelsson S, Björklund A, Falck B, Lindvall O, Svensson L-A 1973 Glyoxylic acid condensation: a new fluorescence method for the histochemical demonstration of biogenic monoamines. Acta physiologica scandinavica 87: 57–62

Azzopardi J G, Pollock D J 1963 Argentaffin and argyrophil cells in gastric carcinoma. Journal of Pathology and Bacteriology 86: 443–451

Baylin S B 1977 Histaminase (diamine oxidase) activity in human tumors: an expression of a mature genome. Proceedings of the National Academy of Sciences of the USA 74: 883–887

Baylin S B, Gann D S, Hsu S H 1976 Clonal origin of inherited medullary thyroid carcinoma and pheochromocytoma. Science 193: 321–323

Baylin S B, Beaven M A, Engelman K, Sjoerdsma A 1970
Elevated histaminase activity in medullary carcinoma of the
thyroid gland. New England Journal of Medicine 283:
1239–1244

Baylin S B, Beaven M A, Buja L M, Keiser H R 1972
Histaminase activity. A biochemical marker for medullary
carcinoma of the thyroid. American Journal of Medicine 53:
723–733

Baylin S B, Abeloff M D, Wieman K C, Tomford J W, Ettinger
D S 1975 Elevated histaminase (diamine oxidase) activity in
small-cell carcinoma of the lung. New England Journal of
Medicine 293: 1286–1290

Baylin S B, Hsu S H, Gann D S, Smallridge R C, Wells S A Jr
1978 Inherited medullary carcinoma: a final monoclonal
mutation in one of multiple clones of susceptible cells.
Science 199: 429–431

Becker K L, Monaghan K G, Silva O L 1980
Immunocytochemical localization of calcitonin in Kulchitsky
cells of human lung. Archives of Pathology and Laboratory
Medicine 104: 196–198

Bensch K G, Gordon G B, Miller L R 1965 Studies on the
bronchial counterpart of the Kultschitsky (argentaffin) cell
and innervation of the bronchial glands. Journal of
Ultrastructure Research 12: 668–686

Bensch K G, Corrin B, Pariente R, Spencer H 1968 Oat-cell
carcinoma of the lung: its origin and relationship to
bronchial carcinoid. Cancer 22: 1163–1172

Bernd P, Gershon M D, Nunez E A, Tamir H 1979
Localization of a highly specific neuronal protein, serotonin-
binding protein, in thyroid parafollicular cells. Anatomical
Record 193: 257–262

Bertagna X Y, Nicholson W E, Sorenson G D, Pettengill O S,
Mount C D, Orth D N 1978a Corticotropin, lipotropin and
β-endorphin production by a human non-pituitary tumor in
culture: evidence for a common precursor. Proceedings of
the National Academy of Sciences of the USA 75:
5160–5164

Bertagna X Y, Nicholson W E, Pettengill O S, Sorenson G D,
Mount C D, Orth D N 1978b Ectopic production of high
molecular weight calcitonin and corticotropin by human
small cell carcinoma cells in tissue culture. Journal of
Clinical Endocrinology and Metabolism 47: 1390–1393

Björklund Λ, Ehinger B, Falck B 1968 A method for
differentiating dopamine from noradrenaline in tissue
sections by microspectrofluorometry. Journal of
Histochemistry and Cytochemistry 16: 263–270

Bordi C, Cocconi G, Togni R, Vezzadini P, Missale G 1974
Gastric endocrine cell proliferation associated with Zollinger-
Ellison syndrome. Archives of Pathology 98: 274–279

Capella C, Bordi C, Mouga G, Buffa R, Fontana P, Bonfasti S
et al 1978 Multiple endocrine cell types in thyroid medullary
carcinoma. Virchows Archiv Abteilung A Pathological
Anatomy and Histology 377: 111–128

Capella C, Gabrielli M, Polak J M, Buffa R, Solcia E, Bordi C
1979 Ultrastructural and histological study of 11 bronchial
carcinoids. Virchows Archiv Abteilung A Pathological
Anatomy and Histology 381: 313–329

Carney J A, Sizemore G W, Tyce G M 1975 Bilateral adrenal
hyperplasia in multiple endocrine neoplasia, type 2, the
precursor of bilateral pheochromocytoma. Mayo Clinic
Proceedings 50: 3–10

Carney J A, Go V L W, Sizemore G W, Hayles A B 1976a
Alimentary tract ganglioneuromatosis. A major component of
the syndrome of multiple endocrine neoplasia type 2b. New
England Journal of Medicine 295: 1287–1291

Carney J A, Sizemore G W, Sheps S G 1976b Adrenal
medullary disease in multiple endocrine neoplasia, type 2,
pheochromocytoma and its precursors. American Journal of
Clinical Pathology 66: 279–290

Carney J A, Heath H III, Perry H O, Pearse A G E, Sizemore
G W 1981 Abnormal cutaneous innervation in multiple
endocrine neoplasia type 2b. Annals of Internal Medicine (in
press)

Challacombe D N, Robertson K 1977 Enterochromaffin cells in
duodenal mucosa of children with coeliac disease. Gut 18:
373–376

Cheng H, Leblond C P 1974 Origin, differentiation and renewal
of the four main epithelial cell types in the mouse small
intestine. V. Unitarian theory of the origin of the four
epithelial cell types. American Journal of Anatomy 141:
537–562

Cole G A, Polak J M, Wharton J, Marangos P, Pearse A G E,
1980 Neuron-specific enolase as a useful histochemical
marker for the neuroendocrine system of the lung. Journal of
Pathology 132: 351–352

Cowley D J, Dymock I W, Boyes B E, Wilson R Y, Stagg B H,
Lewin M R et al 1973 Zollinger-Ellison syndrome type 1:
clinical and pathological correlations in a case. Gut 14:
25–29

Creutzfeldt W 1975 Pancreatic endocrine tumors: the riddle of
their origin and hormone secretion. Israel Journal of Medical
Sciences 11: 762–776

Creutzfeldt W, Arnold R, Creutzfeldt C, Track N S 1975
Pathomorphologic, biochemical and diagnostic aspects of
gastrinomas (Zollinger-Ellison syndrome). Human Pathology
6: 47–76

Deftos L J, Burton D W 1980 Immunohistological studies of
non-thyroidal calcitonin-producing tumors. Journal of
Clinical Endocrinology and Metabolism 50: 1042–1045

Deftos L J, Bone H G III, Parthemore J G, Burton D W 1980
Immunohistological studies of medullary thyroid carcinoma
and C cell hyperplasia. Journal of Clinical Endocrinology
and Metabolism 51: 857–862

DeLellis R A, Nunnemacher G, Wolfe H J 1977 C-cell
hyperplasia: an ultrastructural analysis. Laboratory
Investigation 36: 237–248

DeLellis R A, Wolfe H J, Gagel R F 1976 Adrenal medullary
hyperplasia. American Journal of Pathology 83: 177–190

Doyle D G 1980 The origin of nuclear bodies: a study of the
undifferentiated epithelial cells of the equine small intestine.
American Journal of Anatomy 157: 61–70

Farhi F, Dikman S H, Lawson W, Cobin R H, Zak F G 1976
Paragangliomatosis with multiple endocrine adenomas.
Archives of Pathology and Laboratory Medicine 100:
495–498

Feyrter F 1946 Über die These von peripheren endokrinen
Drüsen. Wiener Zeitschrift für Innere Medizin 27: 9–38

Feyrter F 1953 Über die peripheren endokrinen (parakrinen)
Drüsen des Menschen. Maudrich, Wien

Fletcher L, Rider C C, Taylor C B 1976 Chromatographic and
immunological characteristics of rat brain enolase.
Biochimica et biophysica acta 452: 245–252

Friesen S R, Schimke R N, Pearse A G E 1972 Genetic aspects
of the Z-E syndrome: prospective studies in two kindred;
antral gastrin cell hyperplasia. Annals of Surgery 176:
370–383

Gepts W, De Mey J, Marichal-Pipeleers M 1977 Hyperplasia of
'pancreatic polypeptide'-cells in the pancreas of juvenile
diabetics. Diabetologia 13: 27–34

Goedert M, Otten U, Suda K, Heitz P U, Stalder G A, Obrecht J P et al 1980 Dopamine, norepinephrine and serotonin production by an intestinal carcinoid tumour. Cancer 45: 104–107

Gould V E 1977 Neuroendocrinomas and neuroendocrine carcinomas: APUD cell system neoplasms and their aberrant secretory activities. Pathology Annual 12: 33–62

Gould V E, Aristomenis D, Yannapoulos B S, Summers S C, Terzakis J A 1978 Neuroendocrine cells in dysplastic bronchi. Ultrastructural observations and quantitative analysis of secretory granules and the Golgi complex. American Journal of Pathology 90: 49–56

Gould V E, Valaitis J, Trujillo Y, Chejfec G, Gruhn J G 1980 Neuroendocrinoma of the jejunum: electron microscopic and biochemical analysis. Cancer 46: 713–717

Hansen M, Hansen H H, Hirsch F, Arends J, Christensen J D, Christensen J M et al 1980 Hormonal polypeptides and amine metabolites in small cell carcinoma of the lung, with special reference to stage and subtypes. Cancer 45: 1432–1437

Hashimoto T, Fukuoka M, Nagasawa S, Tamai S, Kusonoki Y, Kawahara M et al 1979 Small cell carcinoma of the lung and its histological origin. American Journal of Surgical Pathology 3: 343–351

Heitz P, Polak J M, Bloom S R, Adrian T E, Pearse A G E 1976 Cellular origin of human pancreatic polypeptide (HPP) in endocrine tumours of the pancreas. Virchows Archiv Abteilung B Cell Pathology 21: 259–265

Heitz P U, Klöppel G, Häcki W H, Polak J M, Pearse A G E 1977 Nesidioblastosis: the pathologic basis of persistent hyperinsulinemic hypoglycemia in infants. Diabetes 26: 632–642

Johnson N F, Wagner J C, Wills H A 1980a Neuroendocrine cell proliferation in the rat lung following asbestos inhalation. Journal of Pathology 131: 261–262

Johnson N F, Wagner J C, Wills H A 1980b Endocrine cell proliferation in the rat lung following asbestos inhalation. Lung 158: 221–228

Jurecka W, Lassmann H, Hörandner H 1978 The proliferation of adrenal medullary cells in newborn and adult mice. Cell and Tissue Research 189: 305–312

Kameya T, Shimosato Y, Adachi I, Abe K, Ebihara S, Ono I 1980 Neuroendocrine carcinoma of the paranasal sinus. A morphological endocrinological study. Cancer 45: 330–339

Klöppel G, Altenähr E, Menke B 1975 The ultrastructure of focal islet cell adenomatosis in the newborn with hypoglycemia and hyperinsulinism. Virchows Archiv A Abteilung Pathological Anatomy and Histology 366: 223–236

Larsson L-I 1977 Two distinct types of islet cell abnormalities associated with endocrine pancreatic tumors. Virchows Archiv Abteilung A Pathological Anatomy 376: 209–219

Larsson L-I 1978 Endocrine pancreatic tumors. Human Pathology 9: 401–416

Leibson L 1979 Endocrinology evolution and evolutionary endocrinology. Perspectives in Biology and Medicine 23: 25–43

Ljungberg O 1972 On medullary carcinoma of the thyroid. Acta pathologica et microbiologica scandinavica A (supplement) 231: 1–57

Marangos P J, Athanasios P, Zis A P, Clark R L, Goodwin F K 1978 Neuronal, non-neuronal and hybrid forms of enolase in brain: structural, immunological and functional comparisons. Brain Research 150: 117–133

Matsuyama M, Inoue T, Ariyoshi Y, Doi M, Suchi T, Sato T et al 1979 Argyrophil cell carcinoma of the uterine cervix with ectopic production of ACTH, β-MSH, serotonin, histamine and amylase. Cancer 44: 1813–1823

Meador C K, Liddle G W, Island D P, Nicholson W E, Lucas C P, Nuckton J G et al 1962 Cause of Cushing's syndrome in patients with tumors arising from 'non-endocrine' tissue. Journal of Clinical Endocrinology and Metabolism 22: 693–703

O'Briain D S, DeLellis R A, Wolfe H J, Reichlin S, Bollinger J, Tashjian A H Jr 1979 Somatostatin immunoreactive cells in C cell hyperplasia and medullary thyroid carcinoma in the rat. Laboratory Investigation 40: 275–276

Orci L, Bactens D, Rufener C, Amherdt M, Ravazzola M, Studer P et al 1976 Hypertrophy and hyperplasia of somatostatin-containing D-cells in diabetes. Proceedings of the National Academy of Sciences of the USA 73: 1338–1342

Pearse A G E 1969 The cytochemistry and ultrastructure of polypeptide hormone-producing cells of the APUD series and the embryologic, physiologic and pathologic implications of the concept. Journal of Histochemistry and Cytochemistry 17: 303–313

Pearse A G E 1980 The APUD concept and hormone production. Clinics in Endocrinology and Metabolism 9: 211–222

Pearse A G E, Polak J M, Heath C M 1974 Polypeptide hormone production by 'carcinoid' apudomas and their relevant cytochemistry. Virchows Archiv B Cell Pathology including Molecular Pathology 16: 95–109

Polak J M, Stagg B, Pearse A G E 1972 Two types of Zollinger-Ellison syndrome: immunofluorescent, cytochemical and ultrastructural studies of the antral and pancreatic gastrin cells in different clinical states. Gut 13: 501–512

Rahier J, Wallon J, Henquin J C 1980 Abundance of somatostatin cells in the human neonatal pancreas. Diabetologia 18: 251–254

Raikhlin N T, Kvetnoy I M, Tolkachev V N 1975 Melatonin may be synthesised in enterochromaffin cells. Nature (London) 255: 344–345

Reid H A S, Richardson W W, Corrin B 1980 Oat cell carcinoma of the esophagus. Cancer 45: 2342–2347

Roberts J L, Herbert E 1977 Characterization of a common precursor to corticotropin and β-lipotropin: cell-free synthesis of the precursor and identification of corticotropin peptides in the molecule. Proceedings of the National Academy of Sciences of the USA 74: 4826–4830

Rubin W 1973 A fine structural characterization of the proliferated endocrine cells in atrophic gastric mucosa. American Journal of Pathology 70: 109–118

Russo A, Buffa R, Grasso G, Giannone G, Sanfilippo G, Sessa F et al 1980 Gastric gastrinoma and diffuse G cell hyperplasia associated with chronic atrophic gastritis. Digestion 20: 416–419

Schmechel D, Marangos P J, Brightman M 1978 Neurone-specific enolase is a molecular marker for peripheral and central neuroendocrine cells. Nature (London) 276: 834–836

Schwartz J F, Zwiren G T 1971 Islet cell adenomatosis. Journal of Pediatrics 41: 646–653

Sherman S P, Li C-Y, Carney J A 1979 Microproliferation of enterochromaffin cells and the origin of carcinoid tumors of the ileum. Archives of Pathology and Laboratory Medicine 103: 636–641

Sidhu G S 1979 The endodermal origin of digestive and respiratory tract APUD cells. American Journal of Pathology 96: 5–20

Sidhu G S, Feiner H, Flotte T J, Mullins J D, Schaefler K, Schultenover S J 1980 Merkel cell neoplasms: histology, electron microscopy, biology and histogenesis. American Journal of Dermatopathology 2: 101–119

Sizemore G W, Heath H III, Carney J A 1980 Multiple endocrine neoplasia type 2. Clinics in Endocrinology and Metabolism 9: 299–315

Solcia E, Capella C, Vassallo G 1970 Endocrine cells of the stomach and pancreas in states of gastric hypersecretion. Rendiconti di gastro-enterologia 2: 147–158

Szijj J I, Csapó Z, Lászlo F A, Kovács K 1969 Medullary cancer of the thyroid gland associated with hypercorticism. Cancer 24: 167–173

Tapia F J, Barbosa A J A, Marangos P J, Polak J M, Bloom S R, Dermody C et al 1981 Neuron-specific enolase is produced by neuroendocrine tumours. Lancet 1: 808–811

Tateishi R, Horai T, Hattori S 1978 Demonstration of argyrophil granules in small cell carcinoma of the lung. Virchows Archiv Abteilung A Pathological Anatomy and Histology 377: 203–210

Tateishi R, Wada A, Hayakawa K, Hongo J, Ishii S, Terakawa N 1975 Argyrophil cell carcinomas (apudomas) of the uterine cervix: light and electron microscopic observations of five cases. Virchows Archiv A Pathologic Anatomy and Histology 366: 257–274

Tateishi R, Taniguchi K, Horai T, Iwanaga T, Taniguchi H, Kabuto T et al 1976 Argyrophil cell carcinoma (apudoma) of the esophagus: a histopathologic entity. Virchows Archiv A Pathologic Anatomy and Histology 371: 283–294

Tischler A S, Dichter M A, Biales B, Greene L A 1977 Neuroendocrine neoplasms and their cells of origin. New England Journal of Medicine 296: 919–925

Van Zwieten M J, Burek J D, Zurcher C, Hollander C F 1979 Aortic body tumours and hyperplasia in the rat. Journal of Pathology 128: 99–112

Visser J W, Axt R 1975 Bilateral adrenal medullary hyperplasia. A clinicopathological entity. Journal of Clinical Pathology 28: 298–304

Watanabe H 1972 Argentaffin cells in adenoma of the stomach. Cancer 30: 1267–1274

Watanabe Y, Yoshida R, Hayaishi O 1979 Immunohistochemical localization of indoleamine 2,3-dioxygenase in the argyrophil cells of the gut and thyroid gland. Acta histochemica et cytochemica 12: 544

Whittle T S Jr, Goodwin M N Jr 1976 Intestinal ganglioneuromatosis with mucosal neuroma, medullary thyroid carcinoma — pheochromocytoma syndrome. American Journal of Gastroenterology 65: 249–257

Wolf-Peeters C, Marien K, Mebis J, Desmet V 1980 A cutaneous APUDoma or Merkel cell tumor. Cancer 46: 1810–1816

Yamaguchi K, Kameya T, Abe K 1980 Multiple endocrine neoplasia type 1. Clinics in Endocrinology and Metabolism 9: 261–284

28

Minerals and pigments

Moshe Wolman

INTRODUCTION

Some minerals and all pigments encountered in pathological specimens can be recognised as such by their colour although their nature cannot be determined in most cases by simple microscopic examination. The presence of others, which are transparent, can sometimes be suspected because of the tissue reaction to them, or clinical history. In many instances transparent minerals might be missed, when the possibility of their occurrence is not considered by the pathologist and special procedures are not instituted.

Mineral and pigment deposits belong to two main categories: endogenous, derived from the body itself in course of metabolic processes and exogenous, introduced into the body from the outside world. The two will be discussed here together. Endogenous minerals which are essential and normal constituents of tissues, cells or active proteins, such as calcium in bones, iron in erythrocytes, copper in caeruloplasmin and zinc in insulin and carbonic anhydrase, are outside the scope of the present volume.

Reactions to various minerals and pigments differ. Some do not elicit any cellular or tissue reactions. This is true for minerals which are essential parts of cell and tissue constituents. For example iron is an essential constituent of haemoglobin and a number of enzymes. Deposition of iron in cells and tissues in haemosiderosis is an indication of increased breakdown of erythrocytes without apparent damage to the storing cells. In haemochromatosis, on the other hand, deposition of iron in tissues is associated with necrosis of some cells and intense reactive fibrosis. It follows that neither the presence, absence, or nature of the reaction can be used at present as a reliable criterion for determining the nature of the iron compound deposited.

This is not the only example in which the type of reaction to a mineral cannot help the pathologist to diagnose the nature of the mineral. It has been shown that the dermal reactions to zirconium lactate and beryllium oxide in humans (whether granulomatous or absent) depend on the presence or absence of hypersensitivity to these compounds (Elias & Epstein 1968). Also starch, used as glove powder, may occasionally elicit a granulomatous reaction with caseous necrosis (Nissim et al 1981), while in most instances only a foreign body giant cell reaction is elicited. With pure carbon particles, however, as in anthracosis, the mineral does not seem to be able to function as a hapten and never elicits a granulomatous reaction.

Thus, the presence of granulomatous reaction or fibrosis indicates that the deposited mineral is not pure carbon, and that the mineral or its derivative has elicited a hypersensitivity reaction, but cannot serve to determine the nature of the mineral. It is clear, therefore, that a definitive diagnosis regarding the nature of the deposit must rely mainly on histochemical and histophysical tests. A possible exception to this are the intranuclear inclusion bodies in renal tubular cells in instances of lead poisoning. As will be shown on page 289 a definitive diagnosis will have to be based also in this case on special tests.

The present chapter will not deal with an extremely useful technique, electron microscopic X-ray microanalysis, which can serve to identify elements in tissue sections, as the apparatus lies outside the reach of most pathology laboratories. A useful review of the subject was written by Chandler (1975).

CARBON PARTICLES AND TATTOOING MATERIALS

Indian ink, a suspension of carbon particles, is often

used in tattooing. Carbon particles are also the main constituent of the pigmented material found in the lungs and lymph nodes in cases of pneumoconiosis. Pure carbon is believed not to elicit any reaction and it can be diagnosed in these organs by a mixture of positive and negative findings. Direct light microscopic examination reveals that it is black and not brown or yellow. Very small carbon particles may appear brown when the microscope illumination is faulty, but this fact does not pose a serious problem to most pathologists. As iron-containing pigments, bile and lipid pigments as well as melanins are not black, the blackness of the deposit rules them out. Black inorganic deposits (for example pieces of shrapnel or dental amalgam) can be recognised as they resist ashing, while no trace of pure carbon remains in a spodogram, prepared by ashing at 600–650°C for 30 min. Ashing is a useful procedure for distinguishing inorganic from organic deposits. Most inorganic material remains while organic substances are burned and volatilised. Spodograms can be prepared from paraffin wax sections or cryosections. Furthermore, a new technique allows incineration to be performed for both light and electron microscopy at nearly room temperature by electrically excited oxygen (Hohman & Schraer 1972).

Graphite

This is a special form of carbon which may be found in pneumoconiosis of miners, or as a result of professional or chance exposure (for example in printers). It may apparently elicit a granulomatous and fibrous reaction. According to Johnson (1980) graphite can be recognised by the birefringence of the periphery of its particles and by the fact that it is not ashed by exposure to 600°C for 5 min, while ordinary carbon particles disappear under this treatment. A number of other compounds, many of which elicit a granulomatous reaction, are used in polychromatic tattoos. They do not usually pose a diagnostic problem to pathologists.

PROSTHETIC AND DENTAL MATERIALS AND TRAUMATICALLY INTRODUCED FOREIGN BODIES

Metal pins, clips and other surgically introduced parts are generally made of metals which are not dissolved in the body. Macrophages containing ferric ions are often found around iron pins and pieces of shrapnel, and can be demonstrated by the Prussian blue reaction described in all standard texts of histological technique. Silver clips, mercury and mercuric amalgam may elicit a granulomatous foreign body reaction. In most instances the hard objects are not included in microscopic sections as they cannot be cut by the microtome knives used. Tiny amalgam granules, liquid mercury and metallic material in soft tissue surrounding the solid masses can be recognized by their black appearance in transmitted light, their persistence in spodograms and the Timm sulphide-silver procedure (Appendix 7).

Silver granules, including those formed by treatment with nitrate and other silver preparations, can be identified by the above procedures and preferably by their solution and disappearance after treatment of the sections in Lugol's iodine for 2 h followed by a 5% sodium thiosulphate rinse.

In injuries caused by explosives, pieces of *clothing* and *building materials* are often found in the tissues. While clothing can mostly be recognised by ordinary and polarised light microscopy, plaster and cement can be recognised by their calcium content as described on page 288.

Different *plastic materials* and *silicones* are used in prostheses. Fine granules are mostly endocytosed by macrophages. Large prostheses often elicit a foreign body reaction with giant cells and occasionally granulomata. The giant cells sometimes contain asteroid inclusions. The polymeric substances and plastics within the macrophages or giant cells are mostly refractile, but they may be birefringent (if the compound has an orderly molecular arrangement). Birefringence may serve to demonstrate also bone cement (methyl methacrylate) as well as polyethylene, PVC and teflon (Józsa & Réffy 1980). Many plastic polymers can be stained by the Congo red or Sirius red procedures used for staining amyloid (Reske-Nielsen et al 1976). It should be noted, however, that foreign inclusions residing for long periods of time in the human body often become encrusted with endogenous constituents. Thus, deposits of polyvinylpyrrolidone (PVP) administered a long time beforehand are often found to contain ferric compounds and lipids. PVP and PVA deposits can be demonstrated by the chlorazol fast pink method.

SILICA, STARCH AND ASBESTOS

Silica and starch

Deposits of silica are mainly found in lungs and lymph nodes of patients who have inhaled silica dust in the course of their work as miners. Silicates were used in surgical glove powders for many years and produced granulomatous reactions on spilling in the operative field. Starch, which replaced silica, produces in most instances a milder foreign body reaction. Both these substances may be found as arterial emboli in drug addicts who have injected themselves intravenously with dissolved drug tablets. Silica is intensely birefringent and can be diagnosed by its persistent birefringence in spodograms heated to 650°C for 1 hour. Starch is also birefringent with a typical Maltese cross appearance and is intensely PAS-positive.

Asbestos

Exposure to asbestos dust is of great importance as it seems to be causally related not only to mesothelioma, but also to a number of other malignant tumours. Light microscopic study of sections reveals in many cases the so-called 'asbestos bodies', mainly in macrophages or giant cells. It has been found, however, that the 'asbestos bodies' (for which the term 'ferruginous bodies' would be preferable) may contain a core of other fibrous materials, such as fibreglass, rather than asbestos. On such a core of asbestos or non-asbestos fibres, protein, calcium and iron are often deposited giving the bodies a yellowish tinge and most bulbous ends. 'Asbestos bodies' are often birefringent, give positive reactions for iron, and withstand incineration. The exact nature of the core can be determined by electron microscopic X-ray microanalysis.

CALCIUM

This element is a normal constituent of all cells and its salts are essential constituents of specialised hard tissues. Deposition of calcium salts is a common occurrence in two pathologic processes. The first is metastatic calcification, where calcium is deposited from hypercalcaemic blood in normal tissues where conditions (e.g. the low pH) favour precipitation of calcium salts. In the second, pathologic calcification, calcium salts are deposited in damaged or dead tissues because local conditions favour precipitation. In both these instances and in normal calcified tissues the deposit consists mainly of apatite. Calcium soaps, which occur at first in fat necrosis, atheroma and tuberculous caseum, probably serve as grains, or crystallisation nuclei, on which apatite is progressively deposited. Many pathologists regard basophilia as evidence for the presence of calcium. This is wrong: basophilia denotes the presence of anionic groups which are free to react with the basic dye. These are chemical moieties which are likely to bind calcium or any other cation present in their neighbourhood. It is true that in most instances basophilic areas in damaged tissue denote the presence of pathologic calcification, but the calcium is sometimes admixed with considerable amounts of iron, as for example in the Gandy-Gamna bodies in the spleen.

Another procedure erroneously believed to indicate calcification is the von Kossa procedure in which silver ions are bound to fixed tissue anionic groups and are then demonstrated by reduction through light. This commonly used procedure is not described in Appendix 7 as it also indicates only the presence of fixed anionic charges (mainly carbonates and phosphates) which are likely to be sites of calcification.

A number of procedures can be used for demonstrating insoluble calcium salts. These are the alizarin S procedure for light microscopy and the Morin procedure for fluorescence microscopy which can be found in most standard texts on technique (cf. Pearse 1968). These procedures demonstrate the deposits but do not prove their nature as numerous other metals are stained by them. The more sensitive GBHA procedure (Appendix 7) is reported to be specific for calcium.

COPPER, ALUMINIUM, ZINC AND TIN

Copper

Copper is deposited in the cornea and the liver in patients suffering from hepatolenticular degeneration (Wilson's disease) and its presence plays a role in the pathogenesis of this disease. In the liver, copper may be occasionally detected within lipid pigment granules. The best and most commonly used method for demonstrating copper is the rubeanic acid procedure (Appendix 7).

Aluminium

This metal seems to play a role in Alzheimer's dementia, which might be similar to the role of copper in Wilson's disease, as well as in a similar process experimentally produced in animals. It may also be found occasionally in humans treated with aluminium salts. It can be demonstrated, but not definitely proven, by fluorescence after staining with Morin (De Boni et al 1974).

Zinc

This is also a constituent of some active proteins (insulin, carbonic anhydrase). It is also a constituent of filling material in dentistry, wound dressing material and smoke bombs. It may be found in wounds, in the gums as well as in the lungs of military personnel exposed to explosion of smoke bombs at close quarters. It can be histochemically demonstrated by the dithizone procedure or less specifically by Timm's method (Appendix 7).

Tin

Tin is occasionally found in the lungs of miners and, like aluminium, can be demonstrated by fluoro-chroming with Morin.

LEAD

This metal is found in tissues mainly in cases of lead poisoning which occurs mostly as an occupational hazard. Deposits are found in various organs, mainly in the gums and bones. In the kidneys chronic lead poisoning is often associated with striking changes which may almost be considered as pathognomonic. Some renal tubular cells are enormously hyper-trophied and their nuclei contain large inclusion bodies which are acid-fast (Goyer & Rhyne 1973). Similar, but less striking, inclusions are produced by bismuth salts. Lead deposits are difficult to diagnose accurately by histochemical means and most authors rely on the sulphide-silver procedure of Timm (cf.

Appendix 7) which demonstrates all the cations which yield black sulphides. For electron microscopy the Timm procedure has been usefully modified by Danscher & Zimmer (1978). However, the light-microscopic appearance of the inclusions and their acid fastness are sufficient for a definite diagnosis.

IRON

Pathologists are likely to encounter mostly haemosiderin (a ferritin-containing breakdown product of haemoglobin) which has to be differentiated from other brownish pigments. Haemosiderin is deposited locally in phagocytic cells wherever haemoglobin breaks down, and diffusely in different cells of various tissues in haemochromatosis and haemosiderosis. Haemosiderin can be easily diagnosed by the Prussian blue method for ferric ions. Figure 28.1 shows a section of liver in a case of haemochromatosis stained by the Prussian blue (Perls') procedure. Intensely stained pigment granules are found in great amounts in macrophages within fibrous septa, but significant amounts of iron are also present in hepatocytes and especially in Kupffer cells. The same procedure can be used for the detection of foreign bodies made of iron which have remained in situ for some time and are surrounded by iron-containing macrophages.

Iron-containing substances found in the body often do not contain ionised iron. Haemoglobin and other haemoproteins, such as myoglobin (which is often found in renal tubules of patients in shock) cannot be demonstrated by the Prussian blue method. They, as well as compounds containing ionised iron, can be shown by micro-incineration followed by examination with either polarised light or dark field microscopy, where they appear red. Figure 28.2 shows a spodogram of a liver of a patient who suffered from transfusion haemosiderosis, photographed under crossed polars. In contrast to Figure 28.1, iron and other inorganic materials are found almost only at the periphery of the lobule. In the lobule birefringent crystals are few and occur practically only in Kupffer cells. Alternatively, in unfixed tissues and often also in tissues fixed for short periods of time in formalin or ethanol, the haemoproteins can be demonstrated by their intrinsic peroxidase activity (Appendix 5).

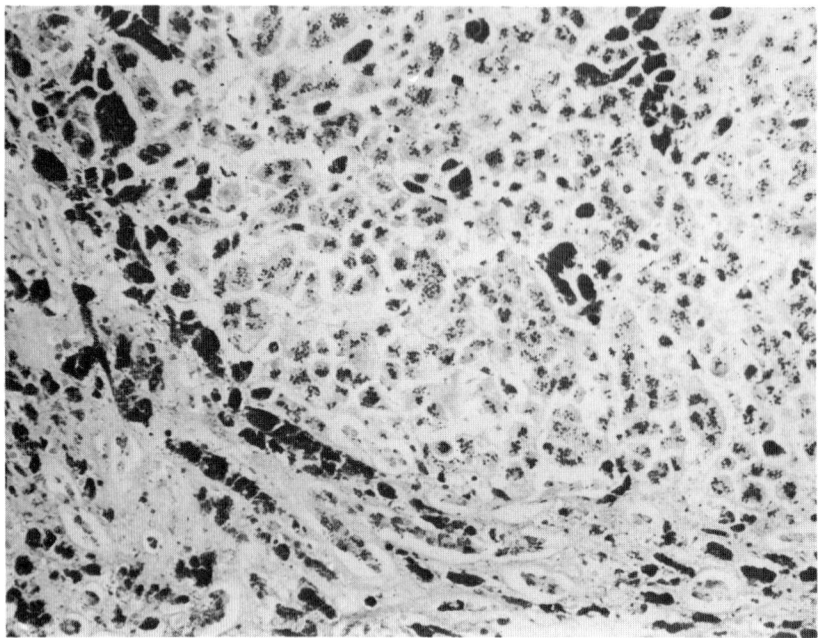

Fig. 28.1 Section of liver in haemochromatosis stained for iron by the Prussian blue procedure. At the left and lower margins are large amounts of intensely stained pigment situated in macrophages lying in a fibrous band delimiting a pseudolobule. In the centre and right-upper part of the figure small discrete granules are situated in hepatocytes, while the solid-stained masses are von Kupffer cells filled with haemosiderin. × 320

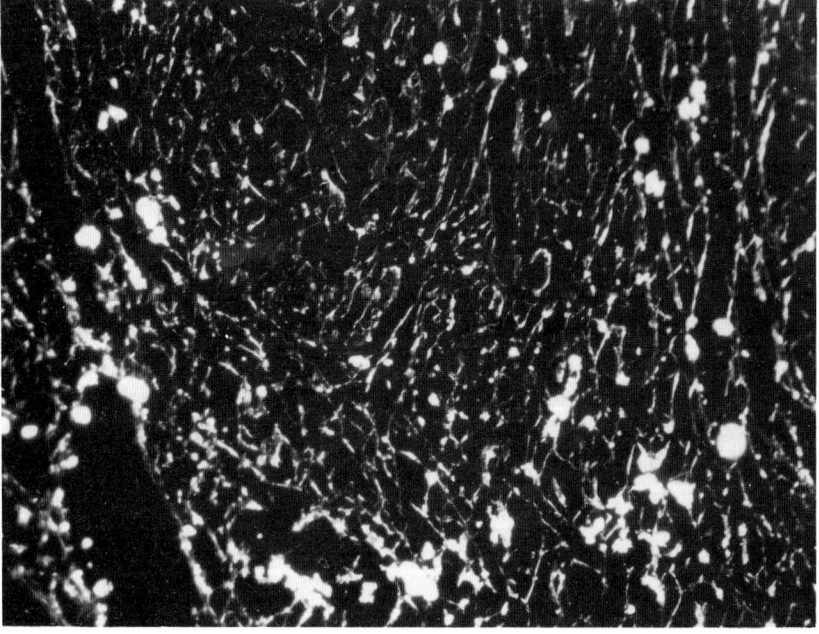

Fig. 28.2 Spodogram of a liver in haemosiderosis photographed under polarised light. Birefringent crystals are seen mainly in the perilobular macrophages at the bottom part of the figure. Few crystals can be seen within the lobule. × 255

IRON-FREE HAEMOGLOBIN DERIVATIVES AND PRECURSORS

Porphyrins

Porphyrins, presumed to be precursors of haem, occur in various organs in porphyrias of different types (acquired and congenital). They can be detected by examination of frozen (preferably cryostat) sections under ultraviolet light, where they emit red or orange-red fluorescence (see p. 157).

Haemoglobin breakdown results in the formation of two pigmented substances: haemosiderin, which has been discussed above, and haematoidin which is identical with bilirubin.

Bilirubin

This can be demonstrated in paraffin wax-embedded or preferably fresh frozen sections by the van den Bergh reaction (Appendix 7), which can demonstrate both the direct (conjugated) and the total (direct and indirect) bilirubin. The procedure is specific and in cases of brownish granules, for example in the liver, may be aided by reactions for iron to exclude haemosiderin and by one of the reactions for lipid pigments (e.g. autofluorescence) to exclude chromolipids. It might be useful to realise that a third pigment, often found together with granules of haemosiderin and haematoidin (for example in haemosiderosis) and called haemofuscin by some authors, is in reality a chromolipid (lipofuscin).

FORMALIN AND MALARIA PIGMENTS

The first is an artefact which forms in tissues congested with blood which were fixed in formalin at low pH levels for long periods. The pigment is dark brown, is not situated within cells but rather on them and its distribution and abundance together with its dark colour mostly allow easy recognition. Confirmation of the nature of this and of the malaria pigment (which is very similar in appearance, but is found mainly in erythrocytes, endothelial cells and phagocytes) can be obtained by their birefringence and by their easy bleaching (within minutes) by concentrated formic acid.

MELANINS, NEUROMELANINS AND LIPOMELANINS

Definite demonstration of melanin granules is often of cardinal importance in diagnostic pathology. While demonstration of 'melanin' in melanosis coli suggests that the patient has probably been using phenolphthalein extensively as a laxative, the finding of melanin in anaplastic tumour cells can clinch the diagnosis of malignant melanoma.

Melanins in the skin and melanocytic tumours are the products of oxidation of aromatic amino acids. They are yellowish-brown in colour and cannot be distinguished from other brownish pigments in routinely stained sections. As in some tumours they are few and difficult to find, intensification of their colour is often of importance. This is best achieved by a silver reduction technique, such as the Masson-Fontana procedure. As reducing activity is present in some lipid pigment granules, the use of complementary tests is advisable. Bleaching, which is obviously useless for the detection of pigments, is a safe procedure for a definite diagnosis of melanin in easily detectable pigment granules. A safe and most useful procedure for diagnosing melanin and premelanin granules is staining for DOPA oxidase (Fig. 28.3) (Appendix 5). There are two drawbacks in the use of this method: (a) it can be used only on unfixed (or shortly fixed in formalin) tissue; (b) it stains mast cell granules well, which might lead to grave errors in diagnosis. Mast cell granules can be easily distinguished from melanin granules, however, as they are intensely metachromatic. The pigment of ochronosis is a melanin and gives the same histochemical reactions as other melanins.

The 'melanin' pigment of melanosis coli exhibits the same histochemical characteristics as skin melanin. However, many of the pigment granules observed in this condition consist of lipid pigment rather than melanin. Mixed melanin and chromolipid granules can also be encountered. In 'pseudomelanosis' of the colon, near tumours and strictures, the pigment is a chromolipid (lipofuscin).

Neuromelanins

These are produced mainly at the expense of catecholamines which serve as synaptic transmitters.

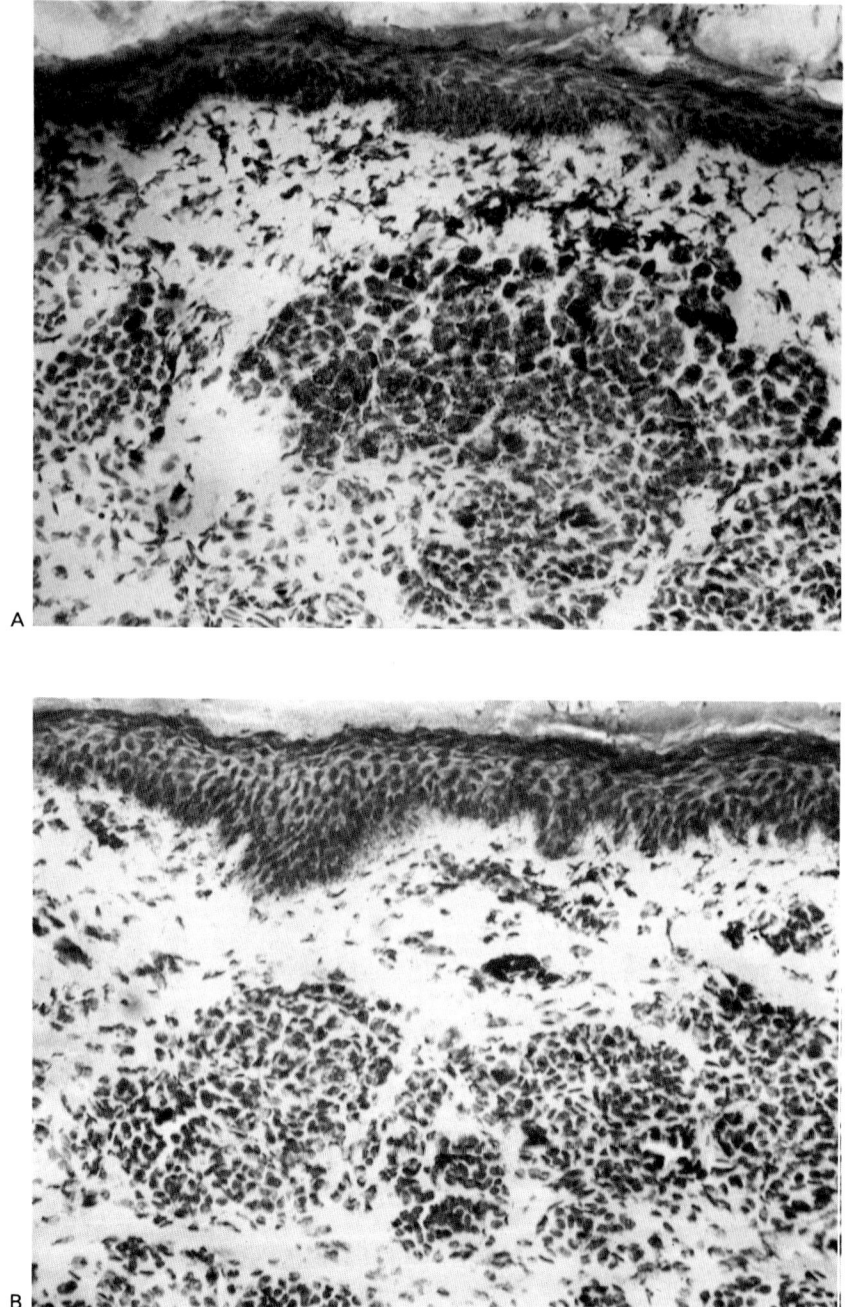

Fig. 28.3 Two unfixed cryostat sections of an intradermal naevus. A is stained by the DOPA-oxidase procedure counterstained with haemalum. B is a control section, incubated in buffer only, showing pre-existent melanin. Pigment present in cell aggregates in A which is absent in B is the product of the enzymic reaction. The cells in A appear larger than those of B, although the magnification in both figures is identical. As in B the melanin-containing cytoplasm shows around the nucleus. ×200

The different pathogenesis from cutaneous melanins is reflected in some histochemical differences between neuromelanin and skin melanin. Both reduce silver diammine and can be bleached by oxidising agents (although the times required might be slightly different in the two cases). The major histochemical difference between the two is, however, the following: neuromelanin is a lipomelanin and not a true melanin. It can be differentiated from true melanin by the fact that frank sudanophilia and autofluorescence can be demonstrated in the granules after bleaching. Without bleaching, fluorescence is quenched and sudanophilia is masked by the melanin moiety. The hepatic pigment of the Dubin-Johnson syndrome and of some similar syndromes is also a lipomelanin. As such it often exhibits autofluorescence (in addition to the typical melanin reactions), the intensity of which can be increased by bleaching.

CHROMOLIPIDS (LIPOFUSCINS, CEROIDS)

Lipid pigments occur in many organs and the intensity of pigmentation is in most instances directly related to age. The pigments are formed by autoxidation of unsaturated lipids with inclusion of different agents, including oxidation catalysts, antioxidants and various proteins. They are therefore increased in different organs also in relation to functional activity, degree of unsaturation of the lipids, presence of oxidation catalysts and relative lack of antioxidants. Chromolipids are a highly heterogeneous group of substances: they differ in their building blocks, in the degree of polymerisation and in the nature and amount of nonlipid materials included in the polymerising mass (Wolman 1980).

The accurate diagnosis of lipid pigment granules is of less practical importance than that of melanins. Chromolipid pigmentation in lipid storage diseases is not known to have any special meaning. Intense pigmentation of the heart and liver mainly indicate age-associated atrophy. Muscle pigmentation (smooth and/or voluntary) in veterinary and fishing practice indicates that the animals were fed a diet excessively rich in unsaturated lipids (such as fish oil), not compensated by adequate amounts of antioxidants. In humans, smooth muscle lipid pigmentation is mostly an indication of pancreatic disease which does not allow ready absorption of fats and fat-soluble vitamins

(vitamin E). In analgesic (mainly phenacetin) abuse, lipid pigment granules occur in the kidneys and liver and can serve as first indication of the nature of the disease. A number of specific diseases, such as Batten's disease (also known as ceroid-lipofuscinosis), chronic granulomatous disease and ceroid storage disease can be diagnosed only after definitive identification of the lipid pigment.

Lipid pigments are yellowish-brown and if present in large amounts they can confer a brownish tinge to the serosal surface of the small intestine (in the 'brown bowel syndrome') or render the liver or myocardium frankly brown (in 'brown atrophy'). Chromolipids cannot be diagnosed as such by simple microscopic examination. They can be diagnosed with certainty by the following tests. As they consist mainly of lipid polymers they can be stained by the Sudan dyes (e.g. Sudan black B, oil red O), while most of them are not dissolved by the lipid solvents used in paraffin wax embedding. Thus, sudanophilia in paraffin wax-embedded sections is diagnostic.

Another constant feature of lipid pigments is their yellow (or orange) fluorescence when illuminated with ultraviolet light. Reducing activity as in the Schmorl reaction or the Masson-Fontana procedure serves well for staining chromolipid granules but is also present in melanin. Chromolipids are often acid-fast, PAS-positive and basophilic. These characteristics, like their insolubility in solvents, depend mainly on the extent of peroxidation and polymerisation. Other characteristics, such as presence of reactions for unsaturated sterols or phospholipids, depend on the nature of the building blocks. Thus, the lipid pigment granules occurring in steroid-secreting organs (adrenal, ovary, testis) often give positive cholesterol reactions and are birefringent.

As has been stated in the preceding sections sudanophilia and autofluorescence cannot always be demonstrated in pigment granules of mixed character. In lipomelanins both these characteristics can be best demonstrated after bleaching of the melanin moiety.

URATES AND OXALATES

In gout articular tophi and deposits in the kidneys mainly consist of monosodium urate. This acid salt, although only slightly soluble in water, is easily lost from sections treated with aqueous solutions except when bound to proteins. Absolute ethanol or methanol, or Carnoy's fixative are therefore recommended. Sodium urate crystals can be easily visualised by polarisation microscopy or alternatively by Gomori's silver methenamine procedure.

Positive identification of sodium urate crystals is of importance in the differential diagnosis of gout v. pseudo-gout. Calcium pyrophosphate dihydrate crystals (which occur in pseudo-gout) exhibit weak birefringence positive in respect to the crystals' length. Sodium urate exhibits negative birefringence in respect to the length. A simple way to differentiate the two consists of examining the crystals under polarised light with a gypsum plate. Crystals lying in the same direction as collagen fibres, which assume the same colour as the collagen, exibit positive birefringence. If they are yellow when collagen is blue or vice versa, they are negative.

Calcium oxalate

CaC_2O_4 is deposited in the kidneys in conditions associated with uraemia, in poisoning by different substances and in endogenous oxalosis. In the last mentioned process oxalates can also be found in other organs. Calcium oxalate is not stained by the commonly used von Kossa method for 'calcium' and can be recognised as such by its birefringence, von Kossa negativity and preferably by the procedure of Pizzolato described in Appendix 7 (see also Fig. 28.4).

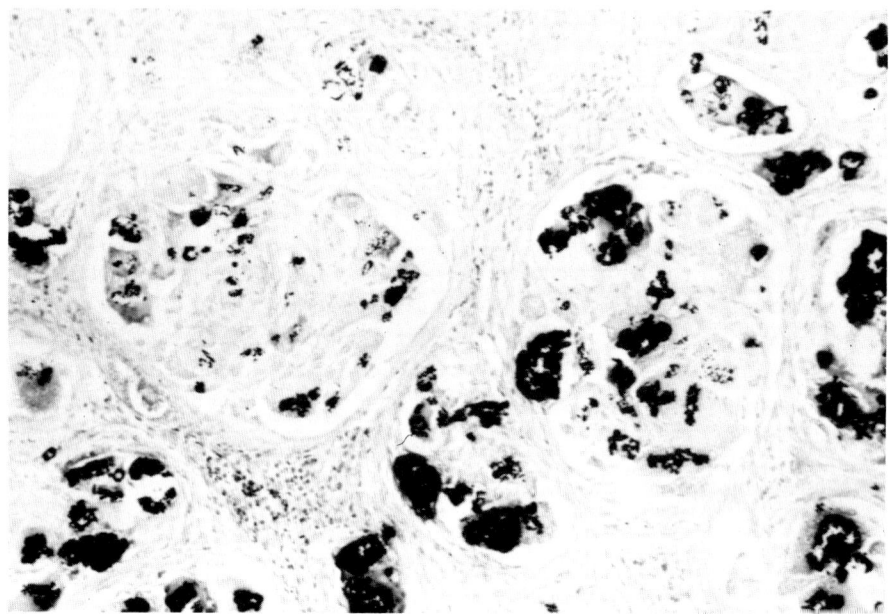

Fig. 28.4 Routine section of an open renal biopsy from a patient with primary oxalosis. Stained by the Pizzolato method for oxalate which appears as black deposits within tubules and glomeruli.

REFERENCES

Chandler J A 1975 Electron probe X-ray microanalysis in cytochemistry. In: Glick D, Rosenbaum R M (eds) Techniques of biochemical and biophysical morphology. Wiley-Interscience, New York, vol 2

Danscher G, Zimmer J 1978 An improved Timm sulphide silver method for light and electron microscopic localization of heavy metals in biological tissues. Histochemistry 55: 27–40

De Boni U, Scott J W, Crapper D R 1974 Intracellular aluminium binding: a histochemical study. Histochemistry 40: 31–37

Elias P M, Epstein W L 1968 Ultrastructural observations on experimentally induced foreign body and organized epithelioid-cell granulomas. American Journal of Pathology 52: 1207–1223

Goyer R A, Rhyne B C 1973 Pathological effects of lead. International Reviews of Experimental Pathology 12: 1–77

Hohman W, Schraer H 1972 Low temperature micro-incineration of thin sectioned tissue. Journal of Cell Biology 55: 328–354

Johnson F B 1980 Identification of graphite in tissue sections. Archives of Pathology and Laboratory Medicine 104: 491–492

Jóźsa L, Réffy A 1980 Histochemical and histophysical detection of wear products resulting from prostheses. Folia histochemica et cytochemica (Krakow) 18: 195–200

Nissim F, Ashkenazy M, Borenstein R, Czernobilsky B 1981 Tuberculoid cornstarch granulomas with caseous necrosis. A diagnostic challenge. Archives of Pathology and Laboratory Medicine 105: 86–88

Pearse A G E 1968 Histochemistry, theoretical and applied, 3rd edn. Churchill Livingstone, Edinburgh

Reske-Nielsen E, Bojsen-Moller M, Velner M, Hansen J C 1976 Polyvinylpyrrolidone-storage disease. Acta pathologica et microbiologica scandinavica A (Pathology) 84: 397–405

Wolman M 1980 Lipid pigments (chromolipids): their origin, nature and significance. Pathobiology Annual 10: 253–267

29

Cytology: preparative techniques

O. A. N. Husain

The use of cytochemistry in the practice of diagnostic cytology mirrors very closely the techniques and their applications to histological sections. The undermentioned schedules may all be applied to cells prepared from mucosal scrapes, brushes or irrigations or fine needle aspirates.

The application of any stain schedule to a smear, as opposed to a section, differs only in the sometimes thinner cell layers of a smear that may require slightly less time in the staining schedule and in the greater liability of the cells to be shed from the smeared slide. In some rare instances the technique may best be carried out on cells in solution, following which they can be layered on to the slide. However, methods now exist to hold cells on the slide.

CELL/SLIDE ATTACHMENT PROCEDURES

1. In *common practice* slides are coated with glycerin albumin (50:50) or pectin (1% aqueous solution) or the spun cell deposit is admixed with a drop of such adhesives and then spread on the slide. Other coatings of a smear by 1% celloidin or 1% carbowax or 1% gelatin are well-established methods possessing some virtue in differing circumstances.

2. *Polylysin technique.* This technique, utilising an ionic bond of the negatively charged cells in solution being layered on to a slide previously coated with a fresh solution of a cationic polymer poly-L-lysine hydrobromide (type 1B, Sigma) with a high molecular weight (the higher the better), has met with success in automated cytology (Husain et al 1978, 1980a) and for cell suspensions (Duguid et al 1979). Details of this procedure are given in Appendix 1.

3. The use of a *cytocentrifuge* improves the cell retention during staining schedules but these

techniques are desdribed in routine bench manuals and need not be detailed here. It is, however, essential to perfect such a method in any particular laboratory by identifying the speed and time required to sediment cells from fluids of differing viscosities so that the cell is optimally laid down and spread for proper interpretation.

STANDARD STAINS IN DIAGNOSTIC CYTOLOGY

The routine stains in diagnostic cytology are the Papanicolau schedule and a Romanowsky stain, usually the May — Grünwald — Giemsa technique. A remarkable amount of morphological behavioural detail can be derived from these two stains which, though not specific, indicate changes that warrant more specific methods to be employed. Toluidine blue is used particularly in touch and scrape preparations with the advantage of speed and simplicity. Haematoxylin and eosin is also widely used after fixation in 3% acetic acid in 95% alcohol or formol saline.

Mucins

Periodic acid-Schiff (PAS) and diastase-PAS, alcian blue and mucicarmine are all standard techniques and the schedules do not differ from those for histological sections. The Hale colloidal iron method before and after hyaluronidase digestion is used for free mesothelial cells in suspension in the diagnosis of mesothelioma (see also Ch. 26).

Mucin stains are also applicable to brush, aspirate and wash smears from the gastric mucosa to identify the presence of sialo- and sulphomucins in sub-types of intestinal metaplasia associated with malignancy as discussed in Chapter 12.

In *alveolar proteinosis* mucin stains are of decisive diagnostic value. The diagnosis of this condition with its rapid waterlogging of the lungs and its characteristic brown sputum can be made by cytology (Vidone et al 1966) prior to the slower but more definitive identification of lamellar inclusions by electron microscopy (Costello et al 1975). Still of uncertain origin, alveolar proteinosis is considered to be caused by excessive surfactant accumulation, possibly as a result of excessive dust or a reaction to it. The peculiar secretions with the accompanying macrophage output threaten to drown the patient if not diagnosed rapidly. This can be done by demonstration of the mucoproteinoceous-like exudate on smear preparations of sputum, lung secretions or washings. The material stains strongly with PAS, is negative to alcian blue, mucicarmine and aldehyde fuchsin stains, while it produces a pale bluish stain background with orange particles with Masson trichrome (Vidone et al 1966) (Table 29.1).

Enzymes

Naphthylamidase

The observation of Sylven & Malmgren (1955) that invasive cells extrude hydrolytic enzymes suggested that the level of such enzymes inside the exfoliated cells might be a useful functional test of malignancy. This has been made use of by the author and his colleagues (Husain et al 1980b) with leucine naphthylamide as the chromogenic substrate and a quantitative assessment of the level of enzyme activity, as described in Appendix 5 (see also Fig. 29.1).

Pentose phosphate shunt

A more recent development is the detection of pentose phosphate shunt enzyme activity in neoplastic cells in the absence of oxygen which may well become a useful functional test in both diagnostic cytology and in automated scanning of cell substrates. The activity of glucose-6-phosphate dehydrogenase shown by the deposition of formazan can then be measured by microdensiometry (Ibrahim et al 1981).

Acid and *alkaline phosphatases* are used with identical schedules to those for sections in the cytological diagnosis of tumours of bone (Ch. 22), prostate (Ch. 24) and other enzyme techniques used in the diagnosis of leukaemia and lymphomas are all applicable to the free and individual cells with identical schedules (see also Ch. 19 and 20).

IMMUNOCYTOCHEMICAL TECHNIQUES

These techniques can ideally be applied to cell samples with minimal modification from the schedules for histological sections (Nadji 1980) (Appendix 6). A wide range of antigens, hormones

Table 29.1 Cytochemical characteristics in sputum diagnosis of pulmonary alveolar proteinosis

Procedure	Sputum in alveolar proteinosis	Sputum in other pulmonary diseases
Periodic acid-Schiff	Positive	Positive
Alcian blue	Negative (many have foci of positive mucinous material along with proteinaceous exudate)	Positive
Mucicarmine	Negative	Positive
Aldehyde fuchsin (pH 1.7)	Negative (yellow)	Usually positive (pink)
Colloidal iron	Pale blue or green with faint pink floccular surround	Bright dark blue with occasional pink staining areas
Masson's trichrome	Green with orange particles. Laminated bodies stain green	Green with other material staining variably, depending on disease

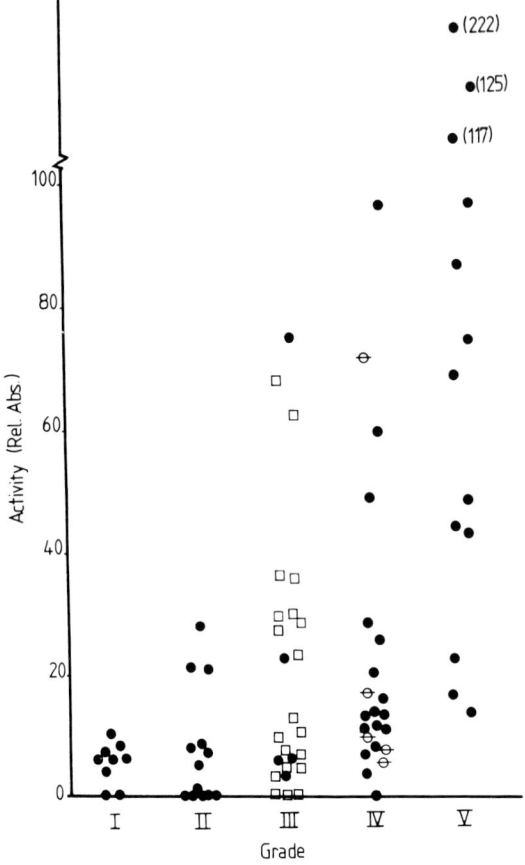

Fig. 29.1 'Cervical intra-epithelial neoplasia and cancers of the cervix'. Scattergram showing the mean value of naphthylamidase activity for each sample.
Grade I — Normal
Grade II — Inflammation
Grade III — Dysplasias and possible carcinoma in situ with insufficient cellular evidence
Grade IV — Carcinoma in situ
Grade V — Invasive cancers, squamous and glandular.
● = cases in grades III, IV and V where the grade has been confirmed histologically in specimens obtained by cone biopsy or hysterectomy; in grades I and II in smears from patients with no known gynaecological abnormality. ⊖ = microinvasion. ☐ = samples from grade III cases with histological confirmation in specimens from punch biopsy only

and various tumour markers now available are being used to identify the origin of cells and tumours from needle aspirates or by brush or washes from the pulmonary and gastro-intestinal tracts and in differentiating the reactive from the neoplastic cells of the lymphoreticular system (Taylor & Burns 1974) (see corresponding earlier chapters).

More recently Coleman et al (1981) Dearnaley et al (1981) and To et al (1981) reported the use of

epithelial membrane antigen (EMA) following on work by Heyderman et al (1979) in the detection of malignant cells in serous effusions and marrow smears.

Another area of importance is the cytodiagnosis of mesotheliomas and their distinction from metastatic tumours using an antimesothelial cell serum (Ch 26 and Singh et al 1979).

THE DNA IN NORMAL AND MALIGNANT NUCLEI

The Feulgen reaction for deoxyribonucleic acid (DNA) in the chromatin of nuclei depends on the liberation of purines from the DNA under the influence of acid.

The rate at which the DNA is hydrolysed will depend on the degree of binding of the DNA within the chromatin. This chromatin is a complex of various proteins with the DNA and the genic effect of DNA molecules may be suppressed when the DNA is fully bound by such protein. De-repression can occur by steroid hormones which is associated with a form of disjunction of protein from the DNA (Paul & Gilmore 1968). Also, it would appear that newly synthesised DNA produced during the S phase of interphase will be more susceptible to hydrolysis than the older DNA. Consequently, there may be no single optimal hydrolysis timing for all species of DNA, old DNA, newly synthesised molecules and derepressed DNA in any one nucleus. This is not made apparent at the usual temperature and acid strength in the routine Feulgen reaction, but if this is slowed down by carrying out the reaction at room temperature and with 5M HCl the hydrolysis curves can be plotted against time by taking cell spreads through the Feulgen hydrolysis for varying periods of time and then measuring the stain reaction in a series of nuclei and plotting the average of, say, 20 cells per time. By doing so, it has been shown that there are two activity curves for normal benign cells maximal at around 20 and 60 min but in the neoplastic and preneoplastic nuclei a more labile form of DNA exists, giving rise to a more rapid production of pigment within the first 3 to 5 min. Interestingly enough the 'normal' cells surrounding the neoplastic cell demonstrate a similar, though not so obvious, phenomenon suggesting that there is a genetic change in the neighbouring tissue (Fig. 29.2).

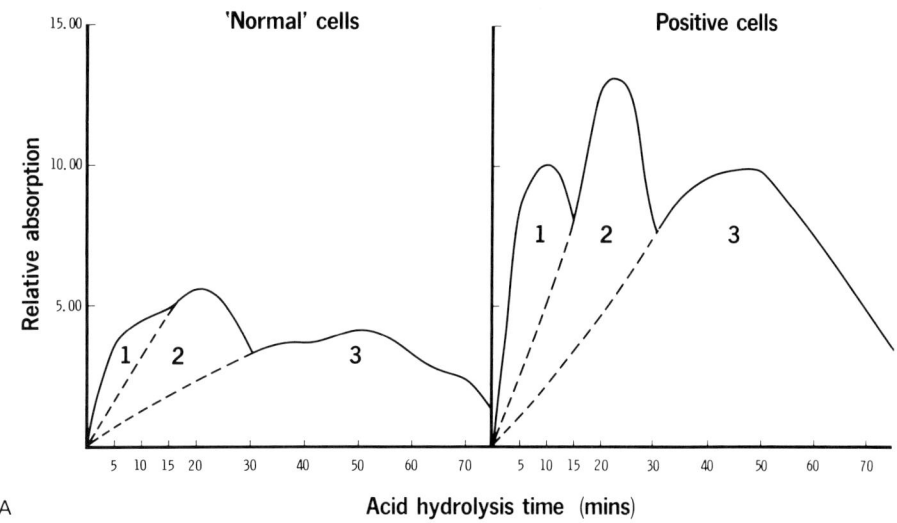

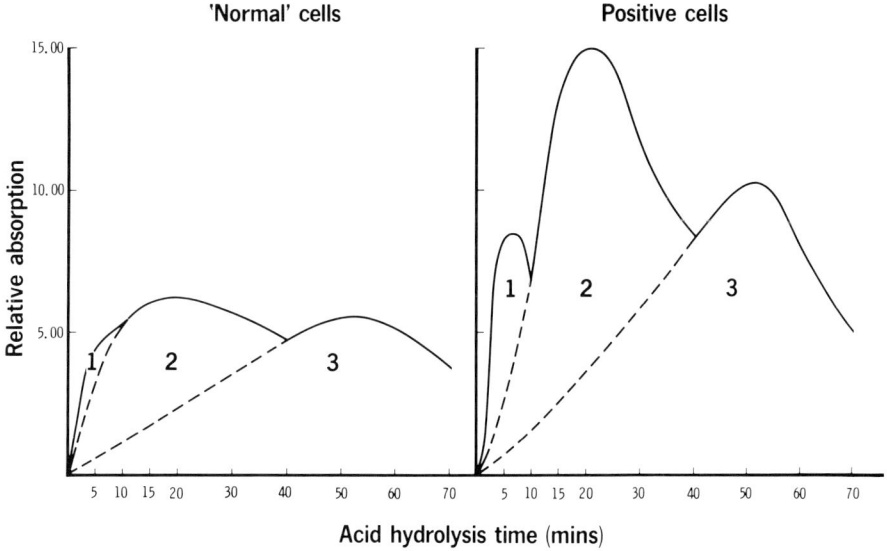

Fig. 29.2 Feulgen hydrolysis curves for (A) invasive squamous carcinoma and (B) carcinoma in situ of the cervix.
'Normal' cells are those adjacent to neoplastic cells. The more readily hydrolysed DNA (area 1) present in the neoplastic cells is also seen in the 'normal' cells. In B the labile component (area 1) appears more substantial than in A. Normal cells from a benign cervix do not show this peak (see text)

REFERENCES

Coleman D V, To A, Ormerod M G, Dearnaley D P 1981 Immunoperoxidase staining in tumour marker distribution studies in cytologic specimens. Acta cytologica 25: 205–206

Costello J F, Moriarty D C, Braithwaite M A, Turner-Warwick M, Corrin B 1975 Diagnosis and management of alveolar proteinosis; the role of electron microscopy. Thorax 30: 121–132

Dearnaley D P, Sloane J P, Ormerod M G et al 1981 Increased detection of mammary carcinoma cells in marrow smears using antisera to epithelial cell membrane antigen. British Journal of Cancer 44: 85–90

Duguid H L D, Wood R A B, Irvine A D, Preece P E, Cuschieri A 1979 Needle aspiration of the breast with immediate reporting of material. British Medical Journal 2: 185–187

Heyderman E, Steele K, Ormerod M G 1979 A new antigen on the epithelial membrane; its immunoperoxidase localisation in normal and neoplastic tissues. Journal of Clinical Pathology 32: 35–39

Husain O A N, Page-Roberts B A, Millett J A 1978 A sample preparation for automated cervical cancer screening. Acta cytologica 22: 15–21

Husain O A N, Millett J A, Grainger J M 1980a Use of polylysin-coated slides in preparation of cell samples for diagnostic cytology with special reference to urine sample. Journal of Clinical Pathology 33: 309–311

Husain O A N et al 1980b Cytodiagnosis of gastric cancer. In: Wright R (ed) Recent advances in gastro-intestinal pathology. W B Saunders, London, 241–254

Ibrahim I T S, Husain O A N, Bitensky L, Chayen J 1981 A potential test for detecting malignant cells in gastric cancer (using glucose-6-phosphate dehydrogenase activity). Medical Research Society, London

Nadji M 1980 The potential value of immunoperoxidase techniques in diagnostic cytology. Acta cytologica 24: 442–447

Paul J, Gilmore R S 1968 Organ-specific restriction of transcription in mammalian chromatin. Journal of Molecular Biology 34: 305–316

Singh G, Whiteside J L, Dekker A 1979 Immunodiagnosis of mesothelioma. Use of antimesothelial cell serum in an indirect immunofluorescence assay. Cancer 43: 2288–2296

Sylven B, Malmgren H 1955 Topical distribution of proteolytic activities in some transplanted mice tumours. Experimental Cell Research 8: 575–577

Taylor C R, Burns J 1974 The demonstration of plasma cells and other immunoglobulin-containing cells in formalin-fixed paraffin-embedded tissues using peroxidase-labelled antibody. Journal of Clinical Pathology 27: 14–20

To A, Coleman D V, Dearnaley D, Ormerod M G, Steele K, Neville A M 1981 The use of antisera to epithelial membrane antigen for the cytodiagnosis of malignancy in serous effusions. Journal of Clinical Pathology 34: 1326-1332

Vidone R A, Hoffman L, Hukill P B, Nesbitt K A, McMahon F J 1966 The diagnosis of pulmonary alveolar proteinosis by sputum examination. Diseases of the Chest 49: 326–332

Appendices

The appendices contain methods mentioned in the
text which the authors and editors have found to be
useful and reliable in the situations described.
Methods and techniques not included in the
appendices can be found in any of the standard
histochemical and histological texts.

Routine sections — sections of fixed, wax-
embedded tissue

Frozen sections — sections of fixed tissue cut
on the freezing microtome
or cryostat

Cryostat sections — sections of unfixed tissue cut
in a cryostat.

Appendix 1

General preparative methods and fixatives

METHODS FOR FREEZING TISSUE

1. Place tissue in container and put into deep freeze. Suitable only for some biochemical examinations. *Useless for histochemistry and morphology.*

2. Freeze tissue on solid carbon dioxide (first place aluminium cooking foil on the carbon dioxide). Suitable for some tissues if small enough samples are frozen (no thicker than 3–4 mm). Useless for muscle or brain.

3. Place tissue on cryostat chuck and freeze with CO_2 jet or in tissue freezing attachments of some cryostats. (Remarks as for 2.)

4. Drop tissue directly into liquid nitrogen. Suitable for some tissues. Layer of gaseous nitrogen around the tissue prevents rapid freezing.

5.* Dust tissue (no larger than 1 cm³) with starch powder and drop into liquid nitrogen. Suitable for all tissues. Good histochemistry and good morphology can be obtained.

6. Drop tissue (1 cm³ maximum) into isopentane or Arcton at $-170°C$ in a liquid nitrogen bath. Suitable for all tissues. The temperature of the isopentane or Arcton is critical and if too cold or too warm, ice crystal artefacts may occur.

7.* Drop tissue (1 cm³ maximum) into hexane maintained at the temperature of solid carbon dioxide in a methanol/CO_2 bath, or acetone/CO_2 bath. Suitable for all tissues.

Notes:
The coating of the tissue with starch powder (as provided with surgical gloves) gives good heat conducting conditions and allows very rapid freezing.

  * Methods 5 and 7 are recommended because they give consistently good results even in inexperienced hands.

  Method 6 requires skill and experience.

  Method 2 is just acceptable if no other method is available (except for muscle where unacceptable results are obtained).

  In extreme circumstances tissue may be frozen on aluminium cooking foil on the freezing surface of the freezer compartment of a domestic fridge. Results will be variable but may be preferable to no frozen tissue.

ORIENTATION AND SUPPORT OF SMALL SPECIMENS FOR CRYOSTAT SECTIONING

Small specimens (portions of needle biopsies, suction rectal biopsies etc) can be supported and oriented
a. in OCT compound (Tissue Tek)
b. on small blocks of gelatine (12%)
c. on small blocks of animal liver or kidney

  The specimen and supporting medium is then frozen as described above, and can be sectioned without damage to the knife.

  OCT compound is inert but should not be used for tissue which might be needed later for biochemical analysis.

GENERAL FIXATIVES

Buffered formaldehyde

40% formaldehyde	100 ml
$NaH_2PO_4.H_2O$	4 g
Na_2HPO_4 (anhydrous)	6.5 g
Distilled water	900 ml

Formol-calcium fixative

40% formaldehyde	100 ml
dried calcium acetate	15.8 g
water to 1 litre	

The pH is approximately 6.8 as made, and requires no adjustment and no marble chips are needed. This formula is more convenient than using calcium chloride.

Zamboni's fixative

Solution A (Stefanini et al 1967, Nature 216: 173–174)
  Saturated picric acid — store at 4°C
Solution B

Paraformaldehyde	100 g
Distilled water	400 ml

Heat to 60°C. Add slowly 1–3 drops of 1M NaOH with stirring until solution is clear
Solution C

$NaH_2PO_4.H_2O$	3.31 g
$Na_2HPO_4 .7H_2O$	33.7 g
Distilled water	1000 ml

Mix just before use 150 ml Solution A, 100 ml Solution B and 750 ml Solution C. The pH should be 7.3–7.4 Osmolality 770–880 mmol.

One-half strength Karnovsky fixative (100 ml)

25% glutaraldehyde (histochemical grade)	10 ml (2.5%)
40% formaldehyde (EM grade)	5 ml (2%)
0.2 M sodium cacodylate buffer	49 ml
anhydrous calcium chloride	50 mg
distilled water	36 ml
	pH 7.3

Mixed aldehyde fixative for immunohistochemistry (100 ml)

25% glutaraldehyde	0.5 ml (0.125%)
40% formaldehyde	5 ml (2%)
0.2 M sodium cacodylate buffer pH 7.3	49 ml
anhydrous calcium chloride	50 mg
distilled water	45.5 ml
	pH 7.3

FIXATIVE TREATMENT FOR LYSOSOMAL ENZYME DEMONSTRATION

Formol-calcium: gum-sucrose

1. Fix small (2 mm thick) blocks of tissue in formol-calcium for 6–18 h at 4°C.

2. Transfer to gum sucrose solution (1 g gum acacia, 30 g sucrose, water to 100 ml) at 4°C with several changes.

3. Keep at 4°C in gum sucrose until needed *or* snap-freeze the tissue and keep frozen.
Note: For some lysosomal enzymes this procedure affords the only means for their preservation and subsequent demonstration.

FIXATIVE TREATMENT FOR BIOGENIC AMINES

Chromaffin reaction

Formaldehyde (40%)	12 ml
5% aqueous potassium dichromate	50 ml
1 M sodium acetate	20 ml
Distilled water	18 ml

Make up immediately before use. The pH should be 5.8.

FIXATIVE TREATMENT FOR POLYPEPTIDE HORMONES

Benzoquinone fixative

This should be prepared immediately before use:

Phosphate buffered saline pH 7.1–7.4	99.6 ml
Benzoquinone (recrystallised)	0.4 g

Bubble nitrogen through container for at least 5 min per 100 ml solution. Do not shake the container. Keep the fixative in the dark at 4°C. It is rendered useless by oxidation. Fixation time 30 min to 4 h for intestine or overnight for 1 cm cubes of brain.

FIXATIVES FOR HAEMATOLOGY

Formaldehyde vapour fixation for lipid staining

Fresh 40% formaldehyde: a few drops are added to a piece of filter paper, which is placed in a Coplin jar. Fixation in the vapour is for 10 min and washed briefly in isopropanol.

Buffered-formalin-acetone for nonspecific esterases

(store at 4°C)

Na_2HPO_4	100 mg
KH_2PO_4	500 mg
Acetone	225 ml
40% formaldehyde	125 ml
Distilled water	150 ml
	pH 6.6

Fix fresh films 30 sec at 4–10°C and wash in distilled water × 3. Allow to dry 10–30 min.

Neutral red uptake

Preparation of tissue
Bone marrow aspirate or peripheral blood in medium in proportion of 1 to at least 2; medium = RPMI 1640 diluted 1/3 in calcium, magnesium-free Hanks balanced salt solution, with added heparin, preservative-free, 100 units/ml; separate mononuclear cells by standard Ficoll

procedure (Aluti et al 1974), and adjust cell concentration to 2×10^7/ml with calcium, magnesium-free Hanks.

Preparation of solution
Neutral red: 1% aqueous, filtered immediately before use.

Method
1. Add 10 μl neutral red to 1×10^6 mononuclear cells
2. Incubate at 37°C for 15 min
3. Wash once in Hanks
4. Examine in a haemocytometer counting chamber.

Results
Myeloid cells: numerous fine granules with occasional coarse aggregates; monocytes: numerous scattered chunks in otherwise clear cytoplasm; lymphocytes: none to a few fine to medium-sized granules (Collins 1981).

REFERENCES

Aluti F et al 1974 Identification, enumeration, and isolation of B and T lymphocytes from human peripheral blood. Scandinavian Journal of Immunology 3: 521–532
Collins R J 1981 Personal communication

RENAL BIOPSIES

If possible divide biopsy
a. for routine histology
b. snap-frozen for IF and
c. for EM.

Direct immunofluorescence microscopy

For transport to the laboratory the tissue should be kept cool (+4°C) and resting on a NaCl (0.9%) moistened compress. The tissue should then be carefully orientated on a chuck prior to snap-freezing.

Cryostat sections (2–4 μm) are cut.

For IF
1. One slide containing 2–3 sections should be prepared for each immune serum (anti-IgA, IgG etc.)
2. Sections are air-dried, fixed in acetone 10–15 min and washed in phosphate buffered saline (PBS) at pH 7.2
3. Sections are stained with appropriate fluorescein-labelled anti-human anti-serum. Anti-sera may be used undiluted or diluted in order to have less background staining*
4. Sections are washed for 30 min in 2 changes of PBS and mounted in glycerol-PBS.
5. Additional sections for routine light microscopy are fixed in formalin and stained i.e. Masson's trichrome.

*Monospecificity has to be confirmed through immuno-electrophoresis.

STORAGE AND TRANSPORT OF SKIN SPECIMENS FOR IMMUNOFLUORESCENCE STUDIES

Skin specimens after snap-freezing can be preserved in a −20°C deep freezer, if available, for several days before transport to the laboratory.

Transport medium (Michael) is used in some centres as a means of transfer of unfrozen tissue to the laboratory at the prevailing temperature. On reception, wash in the buffer through three changes, allow to dry by draining, snap-freeze and then store in a deep freeze until the tests are performed. Test results comparable with those using frozen material may be obtained by the use of the transport medium* so long as the buffer is carefully prepared *with special attention to its pH*.

* Transport medium for IF specimens
Buffer 2.5 ml of 1 M potassium citrate buffer pH 7.0
          5   ml of 0.1 M magnesium sulphate
          5   ml of 0.1 M N-ethyl maleimide
       87.5 ml distilled water
Adjust final solution with 1 M KOH to pH 7.0

Fixative transport medium
55 g $(NH_4)_2SO_4$ in 100 ml buffer.

CYTOLOGY

Cell/slide adhesion by the polylysine technique

Slide-coating solution
Poly-L-lysine hydrobromide (Sigma Type 1B, molecular weight 70 000 to 500 000) is made up 1 mg/ml distilled water. This working solution will keep fairly well at +4°C for up to a month but it is preferable to use it within a few days.

Spread the polylysine solution evenly over a slide with the side of the glass pipette, and allow to dry on sloping racks with the wet side downwards to avoid dust settling on it.

The higher the molecular weight the better the cells stick.

Laying the cells
1. The cell sample is collected either by scraping the mucosal surface (such as the cervix or buccal cavity) with a spatula and shaking the sample in about 10 ml of buffered normal saline in 40% ethanol (6 ml PBS, 4 ml ethanol) or by an irrigation of the same organs by 10% alcohol normal saline solution. Fluid samples from urine or serous fluid or

cyst fluid require little preparation other than shaking vigorously before proceeding.

2. For samples for automation a syringing should be performed using a 2 ml syringe and a 19 gauge needle, aspirating 20 times in about 100 sec. This produces maximal cluster disruption of the secondary clusters (i.e. cells that have aggregated secondarily) without over-much destruction of the cells themselves. Primary sheets and clusters are too difficult to disrupt without creating much cell debris.

3. Centrifuge the cell suspension and then resuspend the cells in 4 volumes buffered normal saline.

4. Using an Eppendorf pipette, a 100 μl aliquot is dropped onto a polylysine-coated slide and covered by a Petri dish to avoid evaporation or dust settling. Leave for 3–5 min then gently drain the fluid from the slide. Fix slide immediately in ethanol-acetic acid (99:1 v/v) for 2 min, followed by alcohol alone for 2 min and then air-dry. Stain with whichever schedule is desired. The cells appear firmly and uniformly fixed to the glass slide.

Appendix 2

Proteins, neurosecretory granules and biogenic amines

SHIKATA ORCEIN METHOD
ALCOHOLIC BASIC FUCHSIN FOR CYSTINE
WOOLLASTON TEST
ARGYROPHILIA (GRIMELIUS)
MASKED METACHROMASIA
LEAD HAEMATOXYLIN
FORMALDEHYDE-INDUCED FLUORESCENCE (FIF)
APUD-FIF
GLYOXYLIC ACID METHOD FOR FIF
   (TISSUE BLOCKS)
GLYOXYLIC ACID METHOD FOR FIF (SECTIONS)

ORCEIN STAIN FOR HEPATITIS B SURFACE ANTIGEN AND COPPER-ASSOCIATED PROTEIN (SHIKATA) (Deodhar et al 1975)

Formalin-fixed. Routine sections.

Solutions required

Acidified potassium permanganate (oxidizing solution)

Potassium permanganate	0.15 g
Dist. water	100 ml
Conc. sulphuric acid	0.15 ml

Orcein solution

Orcein	1.0 g
70% ethanol	100 ml
Conc. HCl	2.0 ml

The pH of the solution should be 1.0 to 2.0

Technique

1. Bring sections to water
2. Treat with acidified potassium permanganate for 15 min
3. Rinse in water and decolorise in 2% oxalic acid for 10 min
4. Rinse in dist. water, then wash in tap water for 3 min
5. Stain in orcein solution for 2–4 h at room temperature

6. Rinse in water, then differentiate in 1% HCl in 70% ethanol
7. Dehydrate, clear and mount.

Results

HBsAg and copper-associated protein stain brown.

Note:

The concentrations recommended are for orcein from BDH. Other orceins may require twice the conc. and twice the amount of HCl (Scheuer 1980).

The orcein solution should be freshly prepared every two weeks.

Source

Orcein (natural). BDH (British Drug Houses).

ALCOHOLIC BASIC FUCHSIN (FOR CYSTINE)

Preparation of tissue: air-dried cryostat sections (5–8 μm) of snap-frozen tissue; air-dried bone marrow films.

Method

1. Flood slides with alcohol
2. Stain for 2–5 min in 0.7% basic fuchsin in 70% alcohol (7 ml 1% basic fuchsin in ethanol, 3 ml water)
3. Rinse in alcohol, clear and mount.

Result

Nuclei: red; cystine: unstained.

Notes

View in polarized light (partial or full). The cystine crystals are birefringent and can be seen without the nuclear stain. However the presence of nuclei is a help in orientation.

WOOLLASTON TEST FOR CYSTINE

Preparation of tissue: air-dried cryostat sections (10 μm) of snap-frozen liver, spleen or lymph node containing high concentrations of cystine. Bone marrow films have insufficient cystine for this test.

Method

Place a cover-slip over the dry section. Choose an area with abundant crystals and observe with polarized light while concentrated hydrochloric acid is introduced under the cover-slip. The 'brick'-shaped crystal aggregates of cystine will dissolve and recrystallize in a fan-shaped cluster of needle-shaped crystals of cystine hydrochloride.

Notes

Beware of fumes and corrosion.

Low concentrations of cystine will fail to recrystallize. Inorganic crystals (phosphates mainly) will disappear completely.

ARGYROPHILIA FOR NEUROSECRETORY GRANULES IN THE DNES

(Grimelius L 1968 Acta Societis medicorum upsaliensis 73: 243–270)

Preparation of tissue: Fix in Bouin or formalin. Tissues fixed otherwise can be post-fixed in Bouin for several hours. Routine sections.

Preparation of solutions

Buffered silver

Dissolve 50 mg AgNO$_3$ in 100 ml of 0.02M acetate buffer at pH 5.6. Make up fresh buffer with freshly distilled water. Discard if solution is cloudy.

Reducing solution

Hydroquinone (quinol) 1 g; anhydrous sodium sulphite 5 g; glass distilled water 100 ml.

Method

1. Bring sections to distilled water
2. Treat for 3 h at 60°C
3. Transfer directly to reducing solution at 60°C
4. If reaction is weak rinse slides in distilled water and return them to fresh silver solution at room temperature for 10–15 min
5. Reduce again as required
6. Wash in distilled water
7. Dehydrate, clear, mount in Canada balsam.

Results

Brown to black staining indicates argyrophilia.

MASKED METACHROMASIA FOR APUD CELLS

Preparation of tissue: Fix preferably in glutaraldehyde-picric acid (25% glutaraldehyde 25 ml, saturated aqueous picric 75 ml, and anhydrous sodium acetate 1g) or in 6% glutaraldehyde (NaH$_2$PO$_4$, 0.2 M 48.75 ml; Na$_2$HPO$_4$, 0.2 M, 76.25 ml; 25% glutaraldehyde 72.5 ml). Check pH to 7.0 and make up to 250 ml with distilled water). Routine sections.

Method

1. Bring sections to water
2. Hydrolyse for 5–20 min in 5M HCl at 60°C
3. Wash in distilled water
4. Stain 2–3 min in 0.1% toluidine blue in 0.1 M acetate buffer at pH 5.0
5. Examine wet, under coverslip.

Results

A reddish stain shows proteins with high levels of side chain carboxyls or carboxamides.

LEAD-HAEMATOXYLIN FOR APUD CELL GRANULES

Preparation of tissue: Glutaraldehyde or formalin-fixed paraffin wax-embedded tissue; fixed frozen sections; cryostat sections.

Preparation of solutions

Stabilised lead solution

Add equal parts of 5% lead nitrate in distilled water and saturated aqueous ammonium acetate; filter. To each 100 ml of filtrate add 2 ml of concentrated (40%) formalin. This stock solution keeps indefinitely in a refrigerator at 0–4°C.

Haematoxylin solution

Dissolve 0.2 g haematoxylin in 1.5 ml of 95% ethanol. Add 10 ml stock lead solution and dilute with 10 ml of distilled water. Stir repeatedly for 30 min, filter and make up to 50–75 ml with distilled water.

Method

1. Bring sections to water
2. Stain in the lead-haematoxylin solution for 1–2 h at 45°C or 2–3 h at 37°C
3. Rinse in distilled water, examine and stain further if necessary
4. Rinse in distilled water
5. Dehydrate, clear and mount.

Results

Blue-black staining of APUD cell granules, nuclear chromatin, nucleoli, nerve fibres, muscle fibres, keratohyalin granules and calcium deposits.

FORMALDEHYDE-INDUCED FLUORESCENCE (FIF) (FREEZE-DRIED FORMALDEHYDE VAPOUR (FDFV) METHOD)

Preparation of tissue: Quench small pieces of tissue in melting Arcton (Freon) 22, precooled in liquid nitrogen. Transfer with cold forceps to freeze dryer and dry at −40°C for 8 h. Transfer to closed paraformaldehyde chamber or vessel and incubate for 3 h at 60–80°C. Impregnate in vacuo with wax containing a high proportion of plastic polymers, and embed in the same wax.

Method

1. Cut sections 2–5 μm and pick up on dry albuminised pre-warmed slides.
2. Mount in Styrolite or Fluorolite and examine by fluorescence microscopy.

Results

Catecholamines and indolamines fluoresce in distinct colours if correct filter combinations are used (dopamine and noradrenalin, greenish; 5-HT yellow).

Notes

For catecholamines, excitation using the 405 nm mercury line is convenient. For 5-HT excitation may be with either the 365 or 405 nm mercury lines. Excitation at 365 nm requires excitation filters UG1 and BG38.

APUD-FIF METHOD

Preparation of tissue

Chop the tissues into small pieces (less than 1 mm³) in Tyrode's solution. Incubate at 37°C in continuously oxygenated Tyrode's solution containing 1 mg/ml 5–HTP or L–DOPA for 1–4 h. Wash in Tyrode's solution at 4°C overnight or for several hours with changes. Blot dry with care (fragility) and subject to the FDFV routine.

Results

Appropriate fluorescence, absent from unincubated controls indicates uptake and decarboxylation of either precursor.

GLYOXYLIC ACID METHOD FOR CATECHOLAMINES (FOR TISSUE BLOCKS)

Preparation of fixation medium
Make a 2% solution of glyoxylic acid monohydrate in 0.1 M phosphate buffer, pH 7.2. Adjust pH to 7.0 to 7.2 with 1 M NaOH.

Method

1. Immerse small pieces of tissue in the fixative for up to 20 min at 4°C
2. Quench tissues in liquid nitrogen-cooled propane or Freon (Arcton) 12, and mount on chucks for cryotomy
3. Cut sections 5–10 μm thick and mount on glass slides
4. Immerse in fixative medium at 4°C for 2–10 min
5. Blot dry and expose to 37°C heat for 3–4 h
6. Mount in Entellan, or other non-fluorescent medium
7. Examine by fluorescence microscopy. Use HBO 200 lamp, primary filters BG38 and BG3, plus interference filter (405 nm). Secondary filter K470 (Leitz).

MODIFIED GLYOXYLIC ACID TECHNIQUE FOR SECTIONS (de la Torre et al 1976)

1. Cut cryostat sections, pick up on slides
2. Dip ×3 in the following solution at room temperature:

sucrose	10.2 g
potassium dihydrogen phosphate	4.8g
glyoxylic acid monohydrate	1.5g
distilled water	100 ml

Bring solution to pH 7.4 with 1M NaOH (about 35 ml). Make up to 150 ml with distilled water. Prepare and use at room temperature the same day. Remove excess fluid quickly with absorbent paper.
3. Dry with hair dryer (cool)
4. Heat dry sections in oven at 80 ± 1°C for exactly 5 min
5. Cover slip with mineral oil (BP) and place on hot plate at 80°C for 1½ min to remove air bubbles
6. Examine with a fluorescence microscope where catecholamines fluoresce.

Appendix 3

Carbohydrates, mucins, mucopolysaccharides

PAS
PB/KOH/PAS
ALCIAN BLUE — PAS
ALCIAN BLUE — (CEC METHOD)
HIGH IRON DIAMINE — ALCIAN BLUE
COLLOIDAL IRON
FEYRTER'S THIONIN
TOLUIDINE BLUE (FOR LYMPHOCYTES)
TOLUIDINE BLUE (HAUST & LANDING) FOR
   SOLUBLE MUCOPOLYSACCHARIDES
HYALURONIDASE DIGESTION
CHONDROITINASE DIGESTION

PERIODIC ACID-SCHIFF (PAS)

Preparation of tissue
Air-dried films of blood or marrow; fresh films preferred, but satisfactory staining has been obtained with material methanol-fixed or Romanowsky-stained months to years before. Sections of fixed embedded tissue, post-fixed cryostat sections, fixed frozen sections, celloidin protected cryostat sections (cell PAS).

Celloidin protection for cell PAS Technique
Air-dried cryostat sections, blood films or cytology smears are dipped in 0.25% celloidin in absolute alcohol and rapidly dried. Follow with the PAS method from step 1.

Preparation of solutions

Schiff's reagent
Add 1 g pararosaniline (or purified basic fuchsin) to 100 ml of distilled water at 95°C. Stir well to dissolve and cool to 60°C. Add 2 g potassium metabisulphite and 20 ml 1 M HCl. Stopper closely, and leave overnight. Add 1 teaspoonful (1–2 g) of activated charcoal, shake and filter. Store at 4°C in stoppered bottles to which xylene is added to maintain the sulphur dioxide concentration.

Method

1. Treat sections with 0.5% aqueous periodic acid, 8 min
2. Wash in running tap water 2 min
3. Rinse in distilled water
4. Cover with Schiff's reagent 15 min
5. Rinse in distilled water and wash in tap water
6. Counterstain with suitable nuclear stain (Mayer H; Carrazzi H. *NOT* Harris H)
7. Wash, dehydrate, clear and mount.

Results
Carbohydrates containing 1:2 glycol groups (*vic*-glycols) are stained red. This includes glycogen, glycolipids, neutral mucosubstances and some non-sulphated acid mucosubstances.

Notes
Glycogen can be removed from sections by treating with salivary amylase (10 min) or 0.1% diastase in 0.02 M phosphate buffer pH 6 containing 0.9% sodium chloride for 1 h at room temperature. Salivary amylase will digest glycogen in celloidin protected sections *through* the celloidin protective layer. Celloidin protection retains glycogen and other water-soluble oligosaccharides.

A delicate nuclear counterstain is required. Harris' haematoxylin overlies the PAS reaction and gives a 'muddy' appearance. Carrazzi's haematoxylin is preferred.

PERIODIC ACID-SODIUM BOROHYDRIDE/SAPONIFICATION/PAS (PB/KOH/PAS)
(Culling et al 1974 Journal of Histochemistry and Cytochemistry 22: 826–831; Reid et al 1973 Journal of Histochemistry and Cytochemistry 21: 473–482)

Preparation of tissue
Routine sections of fixed embedded tissue (sections on cleaned gelatinised slides — see below).

Solutions required

1. 1% aqueous periodic acid
2. 100 ml 2.45% aqueous boric acid
3. 167 ml sodium borohydride solution (dissolve 1.89 g borohydride in 167 ml H_2O)
4. 0.5% potassium hydroxide alcoholic (70% alcohol)
5. Schiff's reagent
6. Carrazzi haematoxylin (or other nuclear stain).

Method

1. 1% aqueous periodic acid — 1 h
2. Wash in running water — 10 min
3. Place the sections in 100 ml 2.45% boric acid and add drop by drop 167 ml borohydride solution over half an hour and leave for a further hour. The container should be surrounded by ice cubes to slow down the reaction (sections come off if at room temp.) $1\frac{1}{2}$ h
4. Wash in running water — 10 min
5. Rinse in 70% alcohol
6. Treat with 0.5% alcoholic KOH — 30 min
7. Rinse in 70% alcohol
8. Wash in running water — 10 min
9. Treat with 1% periodic acid — 5 min
10. Wash in running water — 5 min
11. Treat with Schiff's reagent — 15 min
12. Wash in running water — 10 min
Optional steps 13–14:
13. Counterstain lightly with Carrazzi's haematoxylin
14. Blue in running water — 5 min
15. Dehydrate, clear and mount in DPX.

Result

PAS activity after PB/KOH may indicate presence of o-acylated sialic acids.

Notes

Slide cleaning solution

Potassium dichromate	100 g
Distilled water	850 ml
Concentrated sulphuric acid	80 ml

Carefully add the concentrated sulphuric acid very slowly with constant stirring. (Use a pair of goggles while preparing chromic/sulphuric acid mixture.)

One by one add new glass slides to the chromic/sulphuric acid mixture and stir them with a glass rod. Leave them there for at least 12 h.

The next day, clean the slides in running hot water for 10 to 15 min. Wash in distilled water and then in absolute alcohol. Wipe dry with a clean piece of cloth.

To gelatinise the clean slides

Gelatine	2 g
Distilled water	100 ml
Formaldehyde (40%)	2 ml

Dissolve 2 g of gelatine in 100 ml of distilled water. Add 2 ml of formaldehyde.

Dip the clean slides one by one in a container filled with formaldehyde gelatine solution, drain and dry in 37°C oven for 2 to 3 h.

Sections are mounted on these clean gelatinised slides and dried for at least 2 to 3 h in a 56°C oven.

Poly-L-lysine coated slides may also be used (Appendix 1).

ALCIAN BLUE — PAS

Preparation of tissue
Routine sections of fixed embedded tissue.

Method

1. Dewax and bring sections to water
2. Stain in freshly filtered 1% alcian blue 8GX in 3% aqueous acetic acid at room temperature for 30 min
3. Rinse briefly in distilled water
4. Proceed with PAS method

Result

Acid mucosubstances: blue
Neutral mucosubstances: red

Notes

Some acid mucosubstances containing *vic*-glycol groups may stain purple.

Alcian blue solutions should be freshly prepared, or kept for a maximum of 1 week.

ALCIAN BLUE — CEC METHOD
(Scott & Dorling 1965 Histochemie 5: 221–223)

Preparation of tissue
Post-fixed cryostat sections; frozen sections of fixed tissue; routine sections of fixed embedded tissue.

Solutions required

0.1% alcian blue 8GX in 0.025 M sodium acetate buffer at pH 5.8, containing 0.1 M, 0.2 M, 0.4 M, 0.6 M, 0.8 M and 1.0 M $MgCl_2$.

Method

1. Bring to water
2. Stain overnight in each solution, at room temperature
3. Rinse each section individually in distilled water and then transfer to a fresh distilled water bath for 5 min
4. Dehydrate quickly in alcohols, clear and mount.

Results

Carboxyl groups lose basophilia at low salt concentrations; basophilia of sulphated groups persists at higher salt concentrations.

HIGH IRON DIAMINE/ALCIAN BLUE (HID + AB)
(Spicer 1965 Journal of Histochemistry and Cytochemistry 13: 211–234)

Preparation of tissue
Routine sections of fixed embedded tissue.

Solutions required

1. Diamine solution: Dissolve 120 mg N-N′-dimethyl-*m*-phenylene diamine and 20 mg N-N′-dimethyl-*p*-phenylene diamine simultaneously in 50 ml of distilled water. Add 1.4 ml of 60% ferric chloride. Prepare the diamine solution fresh each time.
2. 1% alcian blue in 3% acetic acid (freshly prepared).

Method

1. Dewax and bring sections to water
2. Treat sections with diamine solution — at room temperature — 18 h
3. Rinse very briefly in distilled water
4. Treat with 1% alcian blue in 3% acetic acid — 30 min
5. Wash well in 80% alcohol
6. Dehydrate, clear and mount in DPX.

Result

Carboxylic groups in mucosubstances stain blue. Sulphated groups stain black or dark brown.

Note
The phenylene diamines are carcinogenic. *Handle with care.* Eastman-Kodak reagents are preferred.

COLLOIDAL IRON

Preparation of tissue
Routine sections of fixed embedded tissue.

Preparation of reagents

Add 2 ml of 60% ferric chloride solution (AR quality) slowly, drop by drop to 250 ml boiling distilled water. Allow to cool. This solution of colloidal iron is stable for years.

Method

1. Bring sections to water
2. Flood with 3% acetic acid for 2 min

3. Stain for one hour at room temperature in:
   4 parts colloidal iron solution
   3 parts distilled water
   1 part acetic acid
4. Wash in several changes of 3% acetic acid
5. Flood with freshly prepared mixture of equal volumes of 2% potassium ferrocyanide and 2% hydrochloric acid (6 ml conc. HCl, 94 ml water) for 10 min
6. Wash well in tap water
7. Counterstain with neutral red
8. Wash, dehydrate, clear and mount in DPX.

Result
Acid groups (carboxyl, phosphate, sulphate, sialic acid residues) stain blue.

Note
This is a relatively non-specific method which can be controlled with hyaluronidase, chondroitin sulphatase and sialidase digestions. Pre-existing ferric iron deposits will also be shown. A control Perls' reaction is necessary for critical studies.

FEYRTER'S THIONIN FOR SIALIC ACID AND OTHER ACIDIC SUBSTANCES

Preparation of tissue
Cryostat sections cut at 5–8 μm from snap-frozen tissue, lightly fixed in formol-calcium for 5–10 min, frozen sections of formalin-fixed tissue mounted on slides.

Method

1. Briefly rinse sections in tap water, wipe around section and allow to come to the point of drying
2. Cover section with 2–3 drops of 1% thionin in 0.5% aqueous tartaric acid
3. Add cover-slip and remove excess dye
4. Seal edges with nail varnish, glyceel etc.

Result

Nuclei: blue; acidic substances: red.

Notes
Observe at intervals of up to 24 h after mounting. Although this is a non-specific method it appears to be remarkably selective for sialic acid residues. Thus gangliosides and other sialic acid-containing substances show a delicate rose-red metachromasia very soon after mounting.

TOLUIDINE BLUE (MUIR, MITTWOCH AND BITTER) FOR LYMPHOCYTIC INCLUSIONS

Preparation of tissue
Air-dried blood films.

Method

1. Fix in methanol for 10 min at room temperature

2. Stain in 0.1% toluidine blue in 30% methanol (1 ml 1% toluidine blue, 3 ml methanol, 6 ml water) for 30 min at room temperature

3. Rinse in acetone, clear and mount in DPX.

Result

Nuclei: blue

Inclusions in lymphocytes: red

Basophils: red granules.

Note

Under oil immersion objective (× 100) count 100 lymphocytes and record the number of them containing metachromatic inclusions.

TOLUIDINE BLUE (HAUST AND LANDING) FOR SOLUBLE MUCOPOLYSACCHARIDES

Preparation of tissue

Cryostat sections (5–8 μm) of snap-frozen tissue; air-dried blood films; air-dried bone marrow films.

Method

1. Fix sections or films for 20 min in an acetone:tetrahydrofuran mixture (1:1 by volume) at room temperature

2. Rinse in acetone

3. Immerse and agitate in:

0.5% toluidine blue in 25% acetone (5 ml 1% toluidine blue, 2.5 ml acetone, 2.5 ml water) for 2 min

4. Rinse in acetone, clear in xylene and mount DPX.

Result

Nuclei: blue

Mucopolysaccharides: red to purple (can appear almost black).

HYALURONIDASE DIGESTION
(Kindblom & Karlsson 1977)

1. Bring dewaxed sections to water

2. Incubate in 0.1 M phosphate buffer pH 6.0 containing 0.3–0.6 mg/ml testicular hyaluronidase (bovine testicular, Sigma type IV, 750 units/mg) for 2 h at 37°C *or* 5 h at room temperature

3. Wash in distilled water

4. Stain with the appropriate technique (alcian blue, colloidal iron etc.)

CHONDROITINASE DIGESTION
(Kindblom & Karlsson 1977)

1. Bring dewaxed sections to water

2. Allow sections to dry

3. Incubate in 0.02 M Tris-HCl buffer pH 8.0 containing 1.25 units chondroitinase ABC/ml for 2 h at 37°C.

Chondroitinase ABC can be obtained from Seikagaku Chemical Industry Ltd, Tokyo, Japan.

Appendix 4

Lipids

OIL RED O
SUDAN BLACK B
BROMINE SUDAN BLACK
SUDAN BLACK FOR HAEMATOLOGY
CHOLESTEROL (SCHULTZ AND PAN)
NILE BLUE SULPHATE
LUXOL FAST BLUE — NEUTRAL RED
ACID HAEMATEIN
TOLUIDINE BLUE FOR SULPHATIDES
OTAN (OSMIUM TETROXIDE α-NAPHTHYLAMINE)
PERIODIC ACID SILVER DIAMINE — OIL RED O
(PASDORO)

OIL RED O METHOD (after Lillie & Ashburn 1943)

Preparation of tissue
Cryostat sections unfixed or post-fixed in formol calcium; frozen sections of fixed tissue; air-dried films of blood, bone marrow or touch preparations fixed in formalin vapour for 10 min.

Preparation of stain
The working solution is prepared an hour in advance by mixing three parts of a stock solution of oil red O (saturated in 99% isopropanol) with two parts of distilled water and filtering the mixture before use.

Method
1. Rinse sections in 60% isopropanol
2. Stain for 15 min in oil red O
3. Differentiate in 60% isopropanol
4. Wash in distilled water and counterstain nuclei with Mayer's or Carrazzi's haematoxylin for 3 min
5. Blue sections in tap water, rinse in distilled and mount in glycerol gelatin.

Results
Cholesteryl esters and triglycerides: red
Some phospholipids: pale pink
Nuclei: blue

Notes
1. Unstained crystalline cholesterol and its esters can be distinguished from fats when the stained section is viewed in polarised light with partly crossed polars. Crystals appear birefringent.
2. For gross staining of arteries with oil red O, the unfixed tissue is cleaned of frank adventitial fat and processed as described for sections, with an extended period (30 min) in the dye followed by differentiation for 1 h.
3. Alternatively, any of the usual oil red O methods with triethyl phosphate or propylene glycol as solvent may be used with identical results.

SUDAN BLACK B (SBB) METHOD FOR GENERAL USE

Preparation of tissue
Cryostat sections unfixed or post-fixed in formol-calcium; frozen sections of fixed tissue.

Method
1. Rinse sections in 70% ethanol
2. Stain in saturated Sudan Black B in 70% ethanol, filtered just before use, for 15 min (2h for Batten's disease)
3. Differentiate with 70% ethanol
4. Counterstain nuclei with 2% carmalum in 1% acetic acid 10 min
5. Wash in water and mount sections in glycerol-gelatin.

Result
Unsaturated cholesterol and triglyceride esters: blue-black
Some phospholipids appear grey
Nuclei are red.

Note
If sections are viewed in polarised light, unstained free and ester cholesterol will appear birefringent whilst the stained phospholipids in myelin exhibit a bronze dichroism. Propylene glycol may be used as solvent for the Sudan black.

BROMINE SUDAN BLACK B (BSBB) METHOD
(Bayliss & Adams 1972)

Preparation of tissue
Cryostat sections post-fixed in formol-calcium 1 h; short fixed frozen sections.

Method
1. Treat sections with 2.5% aqueous bromine at room temperature, inside a fume cupboard, 30 min
2. Wash in distilled water and remove excess bromine with 0.5% sodium metabisulphite, 2 min
3. Wash thoroughly in distilled water and proceed with the Sudan black B method outlined above.

Results
Bromination enhances the reaction of all sudanophilic lipids. In addition, lecithin, free fatty acids and free cholesterol are stained.

SUDAN BLACK B FOR HAEMATOLOGY

Preparation of tissue
Air-dried films of blood or marrow.

Preparation of solutions
Formol-ethanol: one part to 9.
Stock solution: Mix together
    phenol 12 g in ethanol 25 ml,
    disodium hydrogen phosphate ($Na_2HPO_4.12H_2O$)
        0.225 g (or anhydrous 0.089 g) in 75 ml water
    Sudan black B (Gurr or BDH) 0.45 g in ethanol 150 ml.
Filter just before use.

Method
1. Fix slides in formol-ethanol for 30–45 sec
2. Rinse in distilled water
3. Rinse in 70% ethanol
4. Stain in stock solution for 1 h
5. Rinse in 70% ethanol
6. Counterstain with May-Grünwald-Giemsa or Leishman
7. Rinse in running tap water 1 min
8. Mount with aqueous mounting medium (Gurr).

Positive result
Dark brown to black staining; occasionally, especially in pathological states, red metachromatic staining occurs.

REFERENCE

Dacie J V, Lewis S M 1975 Practical haematology, 5th edn. Churchill Livingstone, Edinburgh

CHOLESTEROL (AND ITS ESTERS)

Preparation of tissue
Post-fixed cryostat sections of snap-frozen tissue; frozen sections of formalin-fixed tissue.

Schultz Method
1. Treat sections in 2.5% ferric ammonium sulphate in 0.2 M acetate buffer pH 3.0 for up to 7 days at 37°C
2. Wash 3 × 1 h each in 0.2 M acetate buffer pH 3.0
3. Rinse in distilled water
4. Treat with 5% formalin in tap water for 10 min
5. Drain and allow sections to almost dry
6. Apply reaction medium.

Reaction medium
Add 0.5 ml acetic acid drop by drop slowly to concentrated sulphuric acid with much agitation and cooling. The solution becomes hot and viscous but must remain colourless. Add 1–2 drops of this mixture to the section and cover with a cover slip. Observe immediately. A blue-green colour indicates cholesterol.

Adams PAN method
1. Treat sections in 1% ferric chloride for 4–18 h.
2. Wash well in distilled water 10 min.
3. Air dry.
4. Paint the sections sparingly with the following solution using a soft camel-hair brush:
10 mg 1:2-naphthoquinone-4-sulphonic acid
5 ml ethanol
2.5 ml 60% perchloric acid
0.25 ml 40% formaldehyde
2.25 ml distilled water

Heat painted sections on a hot plate (60–70°C) for 1–2 min. The sections are kept moist by gently repainting. Mount in 60% perchloric acid. A blue colour indicates cholesterol.

Notes
Oxidation is an essential requirement for the successful demonstration of cholesterol.

NILE BLUE SULPHATE (Cain's method)

Preparation of tissue
Post-fixed cryostat sections of snap-frozen tissue; frozen sections of formalin-fixed tissue; air-dried bone marrow films post-fixed in formol-calcium.

Preparation of solution

10 g Nile blue

100 ml 1% sulphuric acid

Reflux gently for 4 h. Cool. Check suitability by adding xylene to 1 ml of dye solution with shaking. A red colour in the xylene layer shows the solution is suitable for staining.

Method

1. Stain sections at 60°C for 5 min
2. Wash (dip only) in distilled water at 60°C
3. Differentiate for 30 sec only, in 1% acetic acid at 60°C
4. Wash in cold water
5. Mount in glycerine jelly.

Result

Neutral fat: red

Free fatty acids: blue

Nuclei: pale blue.

Notes

Many authors refer to 'positive with Nile blue' without specifying the colour — or the method used. Both are important in interpretation. The method is only of pathological significance in Wolman's disease where the storage cells stain a deep purple (the red of neutral fat plus the blue of free fatty acids).

LUXOL FAST BLUE — NEUTRAL RED

Preparation of tissue

Post-fixed cryostat sections of snap-frozen tissue; frozen sections of formalin-fixed tissue; routine sections.

Method

1. Bring sections to 95% alcohol
2. Stain for 16–18 h at 60°C in 0.1% luxol fast blue in 95% alcohol (in a closed container)
3. Rinse in 70% alcohol to remove excess surface dye
4. Wash in tap water
5. Differentiate in 0.05% aqueous lithium carbonate until no further dye comes away (over-differentiation is not possible)
6. Wash well
7. Counterstain with 1% aqueous neutral red
8. Rinse, dehydrate, clear and mount.

Result

Nuclei: red

Myelin and several substances in various neuronal storage disorders are stained a deep blue-black

Connective tissue and muscle are pale blue/green.

Note

The differentiation step 5 is not critical as in some other methods for luxol fast blue, and over-differentiation is not a problem.

ACID HAEMATEIN REACTION

Preparation of tissue

Post-fixed cryostat sections; frozen sections of formalin-fixed tissue.

Preparation of reagents

1. Calcium-dichromate solution:
   5 g potassium dichromate
   2 g calcium chloride (hexahydrate)
   (or 1 g anhydrous)
   Water to 100 ml
2. Haematoxylin solution:
   50 mg haematoxylin
   48 ml distilled water
   1 ml 1% sodium iodate
   Heat to boiling, cool to 37°C and
   add 1 ml acetic acid.
   Use immediately.
3. Differentiating solution:
   250 mg potassium ferricyanide
   250 mg sodium tetraborate (decahydrate)
   100 ml distilled water

Method

1. Wash sections in distilled water
2. Transfer to calcium-dichromate solution, at 37°C for 12–16 h
3. Wash for 30 min in many changes of distilled water
4. Stain for 2 h at 37°C in freshly prepared haematoxylin solution
5. Rinse in distilled water
6. Differentiate for 2 h at 37°C in freshly prepared borax-ferricyanide solution
7. Wash in distilled water and mount in glycerin jelly.

Result

Lecithin and sphingomyelin are blue.

Note

The method can be made more specific by alkaline hydrolysis (2 M NaOH, 1 h 37°C) of the sections prior to the calcium-dichromate treatment. Lecithin is susceptible to alkaline hydrolysis while sphingomyelin is unaffected. The method is more reliable in the CNS than in the viscera.

There is some disagreement about the exact nature of what is stained but for the purposes of this book, sphingomyelin and lecithin stain blue. Some sphingo-myelins stain yellow-brown.

A control section delipidised in chloroform:methanol (2:1), should be included to assess the extent of any non-lipid staining.

Several batches of haematoxylin have proved unsatisfactory for this method. That supplied by Sigma Chemical Co. Ltd. is recommended.

Cryostat sections tend to become detached from their slides during alkaline hydrolysis, but can be remounted onto slides after the acid rinse if handled gently.

In the study of Wallerian degeneration of peripheral nerves, and in the study of demyelinating disorders, sections may be counterstained with oil red O at stage 7. Normal myelin stains blue, degenerating myelin stains red.

TOLUIDINE BLUE FOR SULPHATIDES IN METACHROMATIC LEUCODYSTROPHY

Preparation of tissue
Cryostat sections cut at 5–8 μm of snap-frozen tissue, fixed for 10 min in formol-calcium; frozen sections of formalin-fixed tissue; smears of urinary deposits fixed in formaldehyde vapour.

Method
1. Rinse fixed sections and smears in tap water.
2. Stain overnight at room temperature in 0.01% toluidine blue in McIlvaine buffer pH 5.0
3. Rinse in tap water, dehydrate in *acetone*, clear and mount in DPX.

Result
Sulphatides are stained yellow, brown or purple. Nuclei are blue.

Note
Dehydration must be in acetone. Alcohol will destroy the metachromasia. Mast cells are also stained.

OSMIUM TETROXIDE-α-NAPHTHYLAMINE (OTAN) METHOD (Adams 1959)

Preparation of tissue
Cryostat sections of calcium-formal fixed material, mounted onto chrome-gelatin-coated slides.

Prepararation of solutions

Osmium reagent

1% osmium tetroxide	10 ml
1% potassium chlorate	30 ml

α-naphthylamine reagent
Add a few crystals of α-naphthylamine to 40 ml distilled water at 40°C. Filter and use at 37°C.

Method
1. Treat sections for 18 h in osmium-chlorate reagent in a well-stoppered container in a fume cupboard
2. Wash well for 10 min
3. Treat with α-naphthylamine at 37°C 20 min
4. Wash well in distilled water
5. Counterstain sections in 2% alcian blue in 5% acetic acid for 15 sec
6. Rinse and mount sections in glycerol-gelatin.

Results
Cholesterol esters in degenerating myelin are black. Phospholipids in normal myelin are orange.

Note
Osmium vapour is toxic; α-naphthylamine may be slightly contaminated with the carcinogen β-naphthylamine, therefore both these reagents must be handled with due care.

PERIODIC ACID SILVER DIAMINE-OIL RED O (PASDORO) METHOD (Bayliss et al 1970)

Preparation of tissues
Frozen sections of formol-calcium fixed material (not too long fixed) or post-fixed cryostat sections of fresh brain.

Preparation of silver diamine solution
Rapidly add 35 ml 10% silver nitrate solution to 5 ml 28% ammonia water in a conical flask under agitation. Cautiously add a further 5–10 ml silver nitrate until the brown precipitate has dissolved and a faint opalescence remains. Dilute with an equal volume of water.

Method
1. Treat sections with 0.5% aqueous periodic acid, 10 min
2. Wash well in distilled water
3. Immerse in silver diamine at 25°C for 3 min
4. Wash quickly in three changes of distilled water
5. Treat with 10% formaldehyde, 10 min
6. Wash well in distilled water
7. Tone with 0.2% gold chloride, 2 min
8. Rinse in distilled water
9. Remove any unreduced silver with 5% sodium thiosulphate, 2 min
10. Wash sections
11. Stain with the oil red O technique (see p. 314).
12. Counterstain nuclei with 0.2% light green in 0.2% acetic acid, 12 min.

Results
Characteristic nuclei of glia and neurons and basement membranes: black. Fats, including cholesterol esters in degenerating myelin: red. 'Background' structures: green.

Appendix 5

Enzymes

HEXAZOTIZED PARAROSANILINE (HPR)

Stock solution
4% pararosaniline hydrochloride or basic fuchsin in 2 M HCl (heat gently, cool and filter).

Add equal volumes of stock 4% pararosaniline to freshly prepared 4% aqueous sodium nitrite; wait 1–2 min until the solution becomes a pale straw colour. For critical studies add about 5 mg solid sulphamic acid per ml of hexazotized pararosaniline to destroy excess nitrous acid.

ACID PHOSPHATASE (GOMORI) for general use

Preparation of tissue
Post-fixed cryostat sections, fixed frozen sections.

Preparation of solutions

Incubating medium

Sodium β-glycerophosphate	31.5 mg
dissolved in	
0.1 M acetate buffer pH 5.0	5 ml
0.008 M lead acetate	5 ml

Dilute ammonium sulphide
1 ml of 10% ammonium sulphide diluted to 200 ml with distilled water.

Method
1. Incubate at 37°C for 30 min
2. Wash in tap water, 30 sec
3. Immerse in dilute ammonium sulphide
4. Wash well, counterstain nuclei, mount in glycerine jelly.

Result
Activity is indicated by a brown/black reaction product.

Notes
Use medium immediately, do not filter, do not adjust pH.

ACID PHOSPHATASE (simultaneous coupling technique) for general use

Preparation of tissue

Cryostat or frozen sections of tissue fixed in formaldehyde, or less preferably, glutaraldehyde; or freeze-dried cryostat sections coated with celloidin; Fixed tissue must be rinsed in buffer thoroughly before sections are cut to remove excess fixative.

Preparation of incubation medium:

Naphthol AS-BI phosphate	5 mg

dissolved in 0.5 ml N, N-dimethylformamide
    Add to

0.1 M acetate buffer, pH 5.2, containing:	50 ml
Hexazotised pararosaniline (p. 318)	2 ml

or

Fast Blue B, BB, RR or Fast Red Violet	30 mg

Mix thoroughly, adjust pH to 5.2 and filter before use.

Method

1. Incubate sections for 30–120 min at 37°C
2. Wash thoroughly in tap water
3. Place in 4% formaldehyde for several hours (to prevent the formation of gas bubbles in the sections)
4. Rinse in tap water
5. If desired, counterstain nuclei with haematoxylin or nuclear fast red or methyl green
6. Mount in Kaiser's glycerine jelly or dehydrate clear and mount in DPX

Results

Enzyme-active sites are red (if hexazotised pararosaniline is used) or blue or violet (if fast blue salts are used).

Note

The formaldehyde post-incubation (Step 3) can be replaced with 70% ethanol for 10–30 min at room temperature.

ACID PHOSPHATASE (for haematology)

Preparation of tissue
Air-dried films of blood or marrow.

Preparation of solutions

Veronal-acetate buffer: sodium acetate trihydrate 9.75 g, sodium barbitone 14.71 g; make up to 500 ml with distilled water.

Incubation medium
10 mg Naphthol AS-BI phosphate in 1 ml N, N-dimethylformamide

5.0 ml veronal acetate buffer	
13.0 ml distilled water	
1.6 ml hexazotized pararosaniline (p. 318)	

Adjust pH to 5.0.

For tartrate inhibition studies, add 75 mg L-tartaric acid to each 10 ml medium; readjust pH to 5.0.

Method

1. Fix in formol-calcium at 4°C for 10 min
2. Rinse with distilled water 3 times and allow to drain
3. Incubate for 1 h at 37°C; for tartrate resistance, place one slide in tartrate-containing medium
4. Rinse with distilled water
5. Counterstain with methyl green 15 min
6. Rinse with distilled water and mount in glycerine jelly.

Positive result

Orange-red reaction product.

ACID PHOSPHATASE for smears or touch preparations of bone tumours

Preparation of tissue
Unfixed air-dried touch preparations or smears of bone tumours.

Preparation of incubating medium

0.1 M acetate buffer pH 5.4	10 ml
Naphthol AS-TR phosphate solution (40 mg/ml N, N-dimethylformamide)	0.02 ml
Fast Ponceau salt L	6–10 mg

Method

1. Incubate at 37°C, 20 min
2. Wash in running water
3. Carrazzi haematoxylin, 2–3 min
4. Wash in running water till blue
5. Mount in PVP or glycerine jelly.

Result

Site of acid phosphatase activity: red.

ALKALINE PHOSPHATASE for smears or touch preparations of bone tumours

Stock solutions

0.2 M Tris buffer pH 8.3:

Tris (hydroxymethyl) aminomethane	24.2 g
1M hydrochloric acid	100 ml

Make up to 1 litre with distilled water.

Preparation of incubating medium

0.2 M Tris buffer pH 8.3	10 ml
Naphthol AS-TR phosphate solution	0.02 ml
(40 mg/ml N, N-dimethylformamide)	
Fast Red salt TR	6–10 mg

Method

Use unfixed smears, air-dried
1. Incubate at 37°C, 20 min
2. Wash in running water
3. Carrazzi haematoxylin, 2–3 min
4. Wash in running water till blue
5. Mount in PVP or glycerin jelly.

Result

Site of alkaline phosphatase activity: red.

'NEUTRAL' PHOSPHATASE for smears or touch preparations of bone tumours (Jeffree 1970)

Stock solutions

Veronal acetate buffer pH 7.5:

Sodium acetate (trihydrate)	9.714 g
Barbitone sodium	14.714 g
Make up to 500 ml with distilled water	

Take 50 ml of this solution, add 49 ml 0.1 M hydrochloric acid. Dilute to 250 ml with distilled water to give a final pH of 7.5.

Preparation of incubating media

A. Veronal acetate buffer pH 7.5	10 ml
Naphthol AS-TR phosphate solution	0.02 ml
(40 mg/ml N, N-dimethylformamide)	
Fast Red salt TR	6–10 mg
B. Veronal acetate buffer pH 7.5	10 ml
Naphthol AS-TR phosphate solution	0.02 ml
(40 mg/ml N, N-dimethylformamide)	
Fast Ponceau salt L	6–10 mg

Method

Two unfixed smears, air-dried, are required.
1. Incubate one smear in each of the incubating media (A and B above) at 37°C for 20 min
2. Wash in running water
3. Carrazzi's haematoxylin, 2–3 min
4. Wash in running water till blue
5. Mount in PVP or glycerin jelly.
A staining reaction present in A only indicates the presence of 'neutral' phosphatase.

A staining reaction present in both A and B indicates the presence of alkaline phosphatase.

PHOSPHORYLCHOLINE PROCEDURE FOR PROSTATIC ACID PHOSPHATASE (PRAP) ACTIVITY (after Serrano et al 1976)

Unfixed and formalin-fixed tissue

Solutions

Acetate buffer:

sodium acetate .3H$_2$O	19 g
acetic acid	3.6 ml
distilled water	1 l
after mixing, dissolve 4.8 g lead nitrate	
adjust pH to 5.0	

Incubation medium:

acetate buffer	11.4 ml
distilled water	38.6 ml
phosphorylcholine chloride	129 mg

Formalin-Macrodex solution:

1 volume formaldehyde	40% v/v
1 volume CaCl$_2$	1% w/v

8 volumes Macrodex dextran poviet 70 dissolved in 0.9% NaCl (Povite Production)

Mix and add about 2 g CaCO$_3$ to neutralize formic acid.

L(+) tartrate stock solution 1 M:

The 1 M stock solution contains 150 mg of L(+) tartrate per ml distilled water. Add 0.5 ml of this solution to 50 ml incubation medium (final concentration 10 mmol/l).

Method

1. Fix cryostat sections (6 μm) in formalin-Macrodex solution at 4°C for 20 min
2. Wash in distilled water for about 2 min
3. Incubate in medium at 37°C for 30–60 min
4. Wash in distilled water for about 2 min
5. Treat with ammonium sulphide 1% v/v for 1 min
6. Wash in distilled water
7. Mount in glycerin-gelatin.

Result

The cytoplasm of the prostatic carcinoma cells is (dark) brown or black. The benign prostatic epithelial cells stain black. Controls should consist of sections with prostatic benign or malignant epithelial cells and preferably with bladder mucosa. By adding L(+) tartrate (10 mmol/l) or omitting the formalin fixation (step 1) the sensitivity aspects of the acid phosphatase isoenzymes (see Ch. 24) can be tested.

For fixed tissue step 1 can be omitted.

LEUCOCYTE ALKALINE PHOSPHATASE

Preparation of tissue

Air-dried films of blood, preferably unanticoagulated, or marrow; fix without delay; if staining cannot be done within 6 hours, fixed dried slides may be stored at −20°C.

Preparation of solutions

Fixative
Formol-ethanol: neutral formalin (40% formaldehyde) to ethanol, 1 part to 9 parts. Keep refrigerated; may be used for up to 3 weeks.

Substrate
Naphthol AS phosphate: dissolve 30 mg in 0.5 ml N, N-dimethyl formamide; add 100 ml 0.05 M tris buffer pH 9.0. Freeze in 10 ml aliquots; thaw at room temperature and check pH just before use.

Incubation medium
To a thawed aliquot of Naphthol AS phosphate solution add 10 mg Fast Blue BB and agitate.

Method
1. Fix for 30 sec at 0–5°C
2. Wash in running tap water for 10–15 sec and allow to dry
3. Filter incubation medium immediately onto slide and leave for 10 min
4. Rinse gently in running tap water and air-dry
5. Counterstain with 0.1% aqueous neutral red 10 min
6. Rinse in tap water
7. Air-dry, do not coverslip, and examine immediately under oil-immersion.

Positive result
Blue granules; reaction may be scored:
  0: no granules
  1: few granules
  2: few to moderate number granules
  3: numerous granules
  4: cytoplasm crowded with granules
Normal range for 100 neutrophils 35–100.

REFERENCE

Dacie J V, Lewis S M 1975 Practical haematology, 5th edn. Churchill Livingstone, Edinburgh

MYOFIBRILLAR ATPase (1) (after Hayashi & Freiman 1966) for muscle

Preparation of tissue
Cryostat sections (10 μm) post-fixed in methanol-free formalin.

Preparation of solutions

Methanol-free formalin (MFF) fixative
Dissolve 20 g paraformaldehyde in 100 ml distilled water by heating to 60°C. Add 10% NaOH gradually until formalin solution clears. Dissolve sucrose 62.5 g, sodium cacodylate 10.7 g, sodium pyrophosphate 0.625 g and make total volume up to 500 ml. Adjust pH to 7.4; store at 4°C.

Incubating medium

ATP (disodium salt)	30 mg
0.1 M Tris-HCl buffer containing 0.018 M calcium chloride	20 ml
Adjust pH to 9.5.	

Method
1. Air dry sections 15 min
2. Fix in MFF fixative 4°C for $1\frac{1}{2}$ h
3. Wash in running tap water 2 min; rinse in distilled water
4. Incubate at 37°C for 1 h
5. Wash in two changes of 1% calcium chloride
6. Immerse in 2% cobalt nitrate 3 min
7. Rinse rapidly in 3 changes tap water, 3 changes distilled water
8. Immerse in 0.5% ammonium sulphide
9. Wash in tap water; rinse in distilled water
10. Counterstain nuclei in methyl green (optional)
11. Mount in aqueous mountant.

Results
Type 1 fibres: pale
Type 2A fibres: intermediate
Type 2B fibres: dark

MYOFIBRILLAR ATPase(2) (after Brooke & Kaiser 1970) for muscle

Preparation of tissue
Air-dried cryostat sections (10 μm)

Preparation of solutions:

Pre-incubating solutions
0.2 M acetate buffer at pH 4.3 and pH 4.6.

Incubating medium
As for myofibrillar ATPase method (1)

Method
1. Air-dry sections 1 h
2. Immerse sections in 0.2 M acetate buffers at pH 4.3 and pH 4.6 at room temperature for 30 min

3. Wash in running tap water 2 min
4. Rinse 3 times in distilled water
5. Incubate at 37°C for 40 min
6. Follow steps 5–11 in myofibrillar ATPase method (1).

Results

pre-incubation at pH 4.3
Type 1 fibres: dark
Type 2A fibres: pale
Type 2B fibres: pale
Type '2C' fibres: intermediate

pre-incubation at pH 4.6
Type 1 fibres: dark
Type 2A fibres: pale
Type 2B fibres: intermediate
Type '2C' fibres: intermediate

GLUCOSE-6-PHOSPHATASE

Preparation of tissue
Air-dried cryostat sections (5–8 μm) of snap-frozen tissue.

Incubation medium

Glucose-6-phosphate (dipotassium salt)	5 mg
Distilled water	5.5 ml
0.05M Tris buffer pH 7.0	4.0 ml
2.29% lead acetate	0.6 ml

Mix well and filter.

Method

1. Incubate sections for 15 min at 37°C
2. Wash in tap water
3. Treat with dilute ammonium sulphide (1 ml of 10% solution diluted to 200 ml with tap water)
4. Wash well in tap water
5. Lightly counterstain with Carrazzi's haematoxylin
6. Wash and mount in glycerine jelly.

Result

Sites of enzyme activity are stained brown.

Notes
  Sections of very fatty tissue may float off.
  The dilute ammonium sulphide should be only a pale yellow.
  Specificity of the reaction may be checked by pre-incubation of the sections in 0.05 M acetate buffer pH 5.0 for 10 min at 37°C. This procedure inactivates glucose-6-phosphatase but has no effect on other phosphatases.

The tissue should be *fresh* and the delay between excision and freezing should be less than 5 min. Post mortem tissue may be suitable but negative results cannot be considered conclusive.

Tris buffer at pH 7.0 is used because the coefficient of temperature variation for this buffer reduces the pH during incubation to 6.5, which is optimal for glucose-6-phosphatase.

ACETYLCHOLINESTERASE

Preparation of tissue

Cryostat sections of snap-frozen tissue cut at 10 μm are air-dried and fixed for 30 sec in 4% formaldehyde in 0.1 M calcium acetate (formol-calcium). Frozen sections of formol-calcium-gum sucrose treated blocks of tissue.

Incubation medium

A. 5 mg   acetylthiocholine iodide
   6.5 ml 0.1M acetate buffer pH 6.0
   0.5 ml 0.1M sodium citrate
   1.0 ml 30 mmol/l copper sulphate
   1.0 ml distilled water
   0.2 ml 4 mmol/l iso octamethyl pyrophosphoramide (iso OMPA)

B. Add 1.0 ml 5 mmol/l potassium ferricyanide just before use

A and B can be made in bulk and stored in aliquots at − 20°C.

Method

1. Rinse the fixed sections for 10 sec in tap water
2. Incubate at 37°C for 1 h in the above medium
3. Wash briefly in tap water
4. Treat with 0.05% *p*-phenylene diamine dihydrochloride in 0.05 M phosphate buffer pH 6.8 for 45 min at room temperature
5. Wash in tap water
6. Treat with 1% osmium tetroxide for 10 min at room temperature
7. Wash well in tap water, counterstain lightly (10 sec) in Carrazzi haematoxylin (or Mayer's haemalum), wash, dehydrate, clear and mount in DPX.

Result

Nerve fibres and cells containing acetylcholinesterase are stained dark brown to black.

Notes

  Do not counterstain too heavily.
  Thoroughly dehydrate because areas with strong activity are resistant to dehydration.

Beware of areas of RBCs which appear to have nerve fibres lying over. This is due to the acetylcholinesterase of the RBC membrane.

NADH-TETRAZOLIUM REDUCTASE (Pearse 1972)

Preparation of tissue
Fresh cryostat sections (10 μm).

Preparation of solutions

Incubating medium

MTT (1 mg/ml)	2.5 ml
0.1 M Tris-HCl buffer pH 7.4	2.5 ml
0.5 M cobalt chloride	0.5 ml
Distilled water	3.5 ml

Adjust pH to 7.0; store at $-20°$C. NADH added just before use, 2 mg per 1 ml of stock incubating solution.

Method
1. Incubate at 37°C for 40 min
2. Drain off incubating medium and post-fix in formol-calcium at room temperature for 15 min
3. Wash in tap water 2 min
4. Rinse in distilled water
5. Counterstain nuclei in methyl green
6. Wash in distilled water
7. Mount in glycerine jelly.

Results
Black formazan deposits indicate sites of activity (mainly mitochondrial). In muscle higher activity in Type 1 fibres is due to larger number of mitochondria per unit area.

NITROBLUE TETRAZOLIUM TEST (Derias & Adams 1978)

Preparation of tissue
Slices of fresh myocardium rinsed in cold running water to remove the blood.

Preparation of incubating medium

Nitro blue tetrazolium	50 mg
dissolved in 0.2 M Tris- HCl buffer pH 7.4	100 ml
Sodium cyanide	50 mg
β-nicotinamide adenine dinucleotide (NAD)	100 mg

Adjust to pH 7.1 with Tris.

Method
1. Incubate slices at 37°C for 30 min
2. Wash in running water
3. Store in formol saline.

Result
Normal myocardium: dark blue

Areas of infarcted muscle show diminished staining, the degree of which depends upon the age of the infarct.

Notes
Loss of staining can be detected as early as 1 to 5 h after the clinically determined onset of the infarct. Material can be tested for 3 days after death and even longer if stored at 4°C.

It is particularly important to add NAD, as this coenzyme is rapidly lost post-mortem. The superior results obtained with this modification can be attributed to adding NAD (Derias & Adams, 1982).

SUCCINATE DEHYDROGENASE (Pearse 1972)

Preparation of tissue
Fresh cryostat sections (10 μm).

Preparation of solutions

Prepare stock solution containing MTT and cobalt chloride in Tris-HCl buffer as in method for NADH-TR reductase.

Succinate solution (0.6 M)

Sodium succinate	1.62 g
Distilled water	8.0 ml
1 M HCl	0.05 ml

Adjust pH to 7.0 and make up total volume to 10 ml. Store at $-20°$C.

Incubating medium
Add 0.1 ml succinate solution per 0.9 ml of stock incubating solution just before use.

Method
As for NADH-tetrazolium reductase.

CYTOCHROME OXIDASE (Seligman et al 1968)

Preparation of tissue
Cryostat sections (10 μm), air-dried.

Preparation of solutions

3, 3' diaminobenzidine tetrahydrochloride (DAB)	2.5 mg
Catalase (20 μg/ml)	0.5 ml
Cytochrome c (Type 2)	5 mg
Sucrose (optional)	375 mg
0.1 M phosphate buffer pH 7.4	4.5 ml

Adjust pH to 7.4 before use.

Method
1. Incubate at 22°C for 1 h
2. Rinse in distilled water
3. Post-fix in formol-calcium 15 min
4. Dehydrate, clear xylene and mount in DPX.

Results

Brown reaction product indicates sites of cytochrome oxidase activity.

Distribution is mitochondrial with higher activity in Type 1 muscle fibres.

Note: DAB is carcinogenic — **handle with care.**

MYOPHOSPHORYLASE (1) (Meijer 1968)

Preparation of tissues

Air-dried cryostat sections (10 µm) of snap frozen skeletal muscle

Preparation of solutions

Incubating medium

Glucose-1-phosphate (dipotassium salt)	50 mg
Adenosine 5′ monophosphate	15 mg
EDTA	25 mg
Sodium fluoride	20 mg
Dextran	1 g
0.2 M acetate buffer pH 5.9	20 ml
Absolute ethanol	5 ml

Adjust pH to 5.9 before use.

Method
1. Air-dry fresh frozen sections 15 min
2. Incubate 37°C for 1 h
3. Wash rapidly in distilled water
4. Immerse in Gram's iodine 1 min
5. Wash in distilled water
6. Blot gently, allow to air dry completely
7. Clear in xylene and mount in DPX.

Results

Deep purple reaction in sarcoplasm due to presence of newly synthesised amylose-like straight-chain polysaccharide. Higher activity in Type 2 muscle fibres.

Skeletal muscle shows no reaction (i.e. stains yellow) in the McArdle's disease, while the smooth muscle enzyme in blood vessels walls is unaffected and acts as an internal control for the method.

MYOPHOSPHORYLASE (2)

Preparation of tissue

Cryostat sections (10 µm) of snap-frozen skeletal muscle. Air-dried.

Preparation of incubation medium

0.1 M acetate buffer pH 5.9	10 ml

To this add in the order given with stirring between additions:

Glucose-1-phosphate (dipotassium salt)	100 mg
Sodium fluoride	180 mg
Glycogen (rabbit liver *or* oyster)	2 mg
AMP	5 mg
ATP	5 mg
0.1 M $MgCl_2$	1 ml
Ethanol	2 ml
PVP (MW 44 000)	900 mg

Store at −20°C in 2 ml aliquots.

Method
1. Incubate for 30 min at 37°C
2. Rinse (2 sec) in 40% ethanol and rapidly air-dry
3. Fix in ethanol 3 min
4. Rapidly air-dry
5. Treat sections in large volume (10 ml) of Lugol's iodine diluted 1–10 with water, for 5 min
6. Mount in glycerol-Lugol's iodine (9 vols–1 vol)

Result

Newly synthesized glycogen in skeletal and smooth muscle stains brown — purple — blue.

Note

Skeletal muscle shows no reaction (i.e. stains yellow) in McArdle's disease, while the smooth muscle enzyme in blood vessel walls is unaffected and acts as an internal control for the method.

Lugol's iodine as used here is 1 g iodine, 2 g potassium iodide, 100 ml distilled water.

PHOSPHOFRUCTOKINASE (Bonilla & Schotland 1970)

Preparation of tissues

Cryostat sections (10 µm) air-dried.

Preparation of solutions

Incubating medium

Fructose-6-phosphate	80 mg
Nitro blue tetrazolium (5 mg/ml)	1 ml
NAD	20 mg
ATP	10 mg
0.05 M arsenate-HCl buffer pH 7.0	9 ml

Adjust pH to 7.0 before using.

Method
1. Air-dry sections at room temperature for 1h
2. Fix in 80% ethanol at 4°C for 20 min
3. Drain slides and air-dry for a further 30 min
4. Incubate at 37°C for 1 h
5. Rinse in distilled water
6. Post-fix in formal-calcium, 10 min
7. Counterstain nuclei with methyl green
8. Rinse in distilled water
9. Mount in glycerine jelly.

Results
This is a multistep method in which the final reaction product reflects the localisation of the final enzyme in the pathway (NADH–TR).

Notes
If this method is to be used to investigate PFK deficiencies, the integrity of the metabolic pathway via fructose-1,6-diphosphate aldolase must be demonstrated. To do this, substitute fructose-1,6-diphosphate for fructose-6-phosphate in the incubating medium. Magnesium ions (0.01 mmol/l $MgCl_2$) and one drop of a glyceraldehyde-3-phosphate dehydrogenase (Sigma, from rabbit muscle 40 – 80 units/mg protein) enzyme preparation should improve the method. A control without fructose phosphate is also necessary.

MYOADENYLATE DEAMINASE (Fishbein et al 1978)

Preparation of tissues
Cryostat sections (10 μm), air-dried.

Preparation of solutions

Incubating medium
Adenosine-5'-monophosphate	4 mg
Distilled water	7.0 ml
Nitro blue tetrazolium (5 mg/ml)	2.0 ml
Potassium chloride 3 M	7.0 ml

Add potassium chloride slowly while stirring. Adjust pH to 6.1 and add dropwise 5 mg dithiothreitol dissolved in 0.3 ml distilled water just before using.

Method
1. Incubate at room temperature for 1h
2. Rinse briefly in distilled water
3. Mount in glycerine jelly.

Results
Deep blue reaction end-product more intense in Type 1 muscle fibres.

Note
For control use 4 mg inosine-5'-monophosphate instead of AMP.

LACTASE

Preparation of tissue
Cryostat sections unfixed or fixed in cold chloroform-acetone (1:1) 5 min.

A. Indigogenic procedure according to Lojda and Kraml (cf. Lojda et al 1979a)

Incubation medium
5-Bromo-4-Chloro-3-indoxyl-β-D-fucoside	3 mg
Dissolve in N,N-dimethylformamide	0.3 ml
0.1M citric acid phosphate buffer, pH 6	6 ml
1.65% potassium ferricyanide (50 mmol/l)	0.5 ml
2.11% potassium ferrocyanide (50 mmol/l)	0.5 ml
Mix thoroughly.

Method
1. Incubate sections in the medium at 37°C 2 h
2. Rinse in distilled water
3. Fix in 4% formaldehyde for 5 min at room temperature
4. Rinse in distilled water
5. Counterstain with Kernechtrot nuclear fast red
6. Wash, dehydrate, clear and mount.

Results
Enzyme-active sites are stained turquoise.

Notes
The medium can be used repeatedly. After incubation it is filtered and stored frozen in closed vessels. This can be repeated at least 10 times and is sufficient for 50 biopsies.

B. Azo-coupling procedure according to Lojda (cf. Lojda et al 1979a)

Incubation medium
1-naphthyl-β-d-glucoside	5 mg
Dissolve in N,N-dimethylformamide	0.25 ml
Buffered hexazotized pararosaniline	10 ml
(made up of 0.3 ml hexazotized pararosaniline and 9.7 ml 0.1 M citric acid-phosphate buffer, pH 6.5; if the pH is between 5.5 and 6 correction is unnecessary, otherwise adjust with 1 M and 0.1 M NaOH)
Mix well and filter.

Method
1. Incubate sections in the medium in a refrigerator 15 h
2. Rinse in distilled water

3. Place in 4% formaldehyde for several hours at room temperature
4. Rinse in tap water
5. Nuclei can be counterstained with haematoxylin
6. Wash, dehydrate, clear and mount.

Results
Enzyme-active sites are stained brown.

Notes
This method is recommended only in cases when indoxyl-fucoside is not available.

SUCRASE AND TREHALASE WITH NATURAL SUBSTRATES

Preparation of tissue
Unfixed cryostat sections.

Glucose oxidase-peroxidase-diaminobenzidine (GO-PO-DAB) method according to Lojda (cf. Lojda et al 1979a)

Preparation of solutions
1. 1% aqueous solution of agar (dissolve in a water bath at 80–90°C or by repeated boiling over a flame); check pH with indicator paper and adjust to 6.0). Store at 4°C
2. 4% solutions of sucrose and trehalose (chromatographically pure). Store at 4°C
3. Diaminobenzidine solution: dissolve 2–4 mg diaminobenzidine tetrahydrochloride in a few drops of N,N-dimethylformamide, add 0.3 ml distilled water and 2.1 ml 0.1 M citric acid-phosphate buffer, pH 6.5. Prepare freshly
4. Glucose oxidase – peroxidase solution: dissolve 0.5–1 mg glucose oxidase (degree of purity I, Boehringer, Mannheim, FRG, about 70–140 units) and 0.1 mg horseradish peroxidase, (type VI, Sigma, St. Louis, USA, about 30 purpurogallin units) in 0.9 ml 0.1M citric acid – phosphate buffer, pH 6.5. Prepare freshly.

Incubation medium

4% sucrose or trehalose	0.5 ml
Diaminobenzidine solution	0.8 ml
Glucose oxidase-peroxidase solution	0.3 ml
Mix	
1% aqueous agar (60°C)	1.6 ml
Mix thoroughly.	

Method
1. Cover sections mounted on slides with gel medium and allow to set; the solution is enough for three slides containing several sections each
2. Incubate 2 h in a wet chamber at 37°C

3. Wash slides carefully (substrate gel must not peel off) in 5% acetic acid for 5 min
4. Rinse in distilled water
5. Cover slides with wet blotting paper and allow to dry at room temperature.

Results
When active disaccharidase is present a brown staining develops in sections and in the agar overlay. Its intensity is the measure of sucrase or trehalase activity.

Notes
If the activity of disaccharidases is low, controls are carried out in which the substrates are replaced by an aliquot of distilled water. Most of the reaction product (diaminobenzidine brown) is formed in the gel overlay. Therefore the gel must not be removed. It is left to dry into a film and an assessment of total activity is possible.

BRUSH BORDER α-GLUCOSIDASES (SUCRASE, ISO MALTASE AND GLUCOAMYLASE)

Preparation of tissue
Cryostat sections fixed in cold chloroform-acetone (1:1) 5 min

Azo-coupling procedure according to Lojda and Gossrau
(cf. Lojda et al 1979a)

Incubation medium

6-Br-2-naphthyl-α-D-glucoside	2 mg
or	
2-naphthyl-α-D-glucoside	5 mg
Dissolve in N,N-dimethylformamide	0.5 ml
Buffered hexazotized pararosaniline	10 ml

(made up of 9.4–9.7 ml 0.1 M citric acid-phosphate buffer, pH 7, and 0.3–0.6 ml hexazotized pararosaniline; adjust pH to 6.5 with 1M and 0.1M NaOH)
Mix well and filter.

Method
1. Incubate sections in the medium 2 h at room temperature or 15 h in a refrigerator at 4°C
2. Rinse in distilled water
3. Place in 4% formaldehyde for several hours at room temperature
4. Rinse in tap water
5. If desired, counterstain nuclei with haematoxylin
6. Mount in glycerin jelly, Apathy's syrup or similar medium.

Results
Enzyme-active sites are stained orange-red.

Notes

The 2-naphthyl-derivative is cleaved more rapidly than the bromo-derivative, is less expensive and more soluble. However, the bromo-derivative is more easily obtainable.

ENTEROPEPTIDASE (ENTEROKINASE)

Preparation of tissue
Unfixed cryostat sections.

Indirect procedure of Lojda and Mališ (cf. Lojda et al 1979a)

Preparation of solutions
1. 1% aqueous solution of agar (dissolve in a water-bath at 80–90°C or by repeated boiling over a flame); check pH with indicator paper and adjust to 6.5). Store at 4°C.

2. *Incubation medium*

Benzyloxycarbonyl-glycyl-glycyl-arginine-4-methoxy-2-naphthylamide	2 mg
or	
Benzoyl-L-arginine-2-naphthylamide (BANA)	4 mg
Dissolve in N,N-dimethylformamide	0.15 ml
0.1M Tris-maleate or Tris-HCl buffer, pH 6.5, with 0.1% calcium chloride	2 ml
Mix well	
Fast Blue B	5 mg

Mix thoroughly and filter
Add 2 mg trypsinogen and 2 ml 1% warm (60°C) agar
Mix well.

Method
1. Cover sections mounted on slides with gel medium and allow to set; the solution is enough for 3 slides containing sections of several biopsies each
2. Incubate 2 h in a wet chamber at 37°C
3. Carefully place the gel-covered slides bearing the sections for 5 min in Petri dishes with 2% copper sulphate at room temperature (optional)
4. Wash 5 min in distilled water
5. Dry the underside of each slide, examine under the microscope with 10 × magnification, and where appropriate take pictures
6. For storage let agar gels dry into a film.

Results
Sites of reaction product in the section and in the overlying agar appear reddish or blue-violet.

Notes

It is necessary to perform controls with media without trypsinogen. Because generated trypsin can activate trypsinogen it is impossible to assess the enteropeptidase activity correctly. However, a similar condition is also in the gut 'in vivo'. The method reflects the activation of trypsinogen.

For better localization but less sensitivity use gly-(asp)₄-lys-2-naphthylamide as substrate.

DIPEPTIDYL (AMINO) PEPTIDASE IV

Preparation of tissue
Cryostat sections fixed in cold chlorform-acetone (1:1) 5 min

Azo-coupling method according to Lojda (cf. Lojda et al 1979a)

Incubation medium

Glycyl-proline-4-methoxy-2-naphthylamine	4 mg
Dissolve in N,N-dimethylformamide	0.5 ml
Buffered Fast Blue B	10 ml

(made up of 10 mg Fast Blue B dissolved in 10 ml 0.1 M phosphate or cacodylate buffer, pH 7.2–7.4
Mix well and filter.

Method
1. Incubate sections in the medium at room temperature 45 min
2. Rinse in distilled water
3. Place in 2% copper sulphate for 5 min (can be omitted)
4. Rinse in distilled water
5. Place in 4% formaldehyde for several hours
6. Rinse in distilled water
7. Mount in glycerin jelly, Apathy's gum syrup or similar medium.

Results
Enzyme-active sites are stained deep red.

Notes
It is advisable to evaluate the reaction as soon as possible because of fading of the reaction product after several days. Sites with low activity become less discernible or cannot be seen at all.

NON-SPECIFIC ESTERASE (α-NAPHTHYL ACETATE METHOD) FOR GENERAL USE

Preparation of tissue
Fresh-frozen cryostat sections (5–10 μm) post-fixed in acetone or formol-calcium at 4°C for 1 h. Frozen sections of formol-calcium/gum sucrose-treated small blocks.

Incubation medium

α-naphthyl acetate	2.5 mg
(dissolved in 2-methoxyethanol)	0.2 ml
0.2M disodium hydrogen phosphate	7.5 ml
distilled water	2.5 ml
hexazotized pararosaniline	0.8 ml

Adjust pH to 6.5 with 0.2M disodium hydrogen phosphate or 1M HCl.

Method

1. Incubate sections at room temperature for 1–20 min
2. Wash well in tap water
3. Counterstain nuclei with methyl green or Carrazzi's haematoxylin
4. Wash
5. Mount in glycerine jelly or dehydrate, clear and mount in DPX.

Results

Sites of esterase activity are coloured brown.

Note

After step 2 the sections may be stained with Sudan black technique shown on page 314. The combination esterase-Sudan black method is useful in the study of lipid degenerative disorders.

α-NAPHTHYL ACETATE ESTERASE (ANAE) (FOR HAEMATOLOGY)

Preparation of tissue

Air-dried films of blood or marrow; unfixed films may be stored at room temperature for up to 2 weeks without appreciable loss of enzyme activity.

Incubation medium

α-naphthyl acetate 50 mg dissolved in 2.5 ml 2-methoxyethanol	
67 mmol/l phosphate buffer pH 7.6	44.5 ml
Hexazotized pararosaniline	3.0 ml

Adjust pH to 6.1, filter.

Method

1. Fix in buffered formalin acetone (p. 304) at 4–10°C for 30 sec
2. Wash in distilled water 3 times
3. Incubate for 45 min at room temperature
4. Rinse with distilled water 3 times
5. Counterstain with methyl green for 15 min
6. Wash with tap water, dry and mount with DPX.

Positive result

Dark brown-red reaction product.

Note

For testing fluoride sensitivity, add 1.5 mg sodium fluoride per ml of phosphate buffer.

ACID ESTERASE IN BLOOD FILMS (FOR WOLMAN'S DISEASE)

Preparation of tissue

Air-dried blood films. Fix for 10 min in formol-calcium.

Incubation medium

α-naphthyl acetate 12.5 mg (dissolved in 0.2 ml 2-methoxy ethanol).	
0.05M phosphate buffer pH 5.0	50 ml
Hexazotized pararosaniline (p. 318)	3 ml

Adjust to pH 5.8 with 1M NaOH

Method

1. Rinse fixed films in tap water
2. Incubate for up to 4 h
3. Wash in tap water
4. Counterstain with Carrazzi's haematoxylin
5. Wash, dehydrate, clear and mount in DPX.

Result

Enzyme activity is shown by a dark reddish-brown colour.

Notes

Lymphocytes (T lymphocytes) usually have one or two small dense spots of activity. No activity is seen in neutrophils. Monocytes (non-specific esterase) are densely stained.

The inhibitor E600 is not necessary for blood film studies because no non-specific esterase activity is present in lymphocytes, and monocytes are easily recognizable by their intense activity.

ACID ESTERASE IN TISSUES

Preparation of tissue

Frozen sections (5–10 μm) of formol-calcium gum-sucrose treated tissue are essential.

Incubation medium

α-naphthyl acetate 2.5 mg in 0.2 ml 2-methoxyethanol.	
0.05 M phosphate buffer pH 5.0	10 ml
Hexazotized pararosaniline (p. 318)	0.6 ml

Adjust to pH 5.0 with 1 M NaOH and add 0.1 ml 10^{-2}M E600 in 0.1 M phosphate buffer pH 5.0

Method

1. Treat free-floating sections in 10^{-4}M E600 in 0.1 M phosphate buffer pH 5.0 at 37°C for 30 min

2. Transfer sections to the incubating medium for 30 min at 37°C

3. Wash, mount on slides, counterstain with Carrazzi's haematoxylin, wash, dehydrate, clear and mount in DPX.

Result
Acid esterase activity is shown by a reddish-brown stain.

Notes
The inhibitor E600 is used to inhibit the non-specific esterases which are very active in liver and other tissues and which would mask acid esterase activity unless inhibited. Fixed tissue is essential because acid esterase is labile and its activity cannot be demonstrated in cryostat sections of snap-frozen tissue.

CHLOROACETATE ESTERASE (CAE) FOR HAEMATOLOGY AND GENERAL USE

Preparation of tissue
Air-dried films of blood or marrow; unfixed films may be stored at room temperature for up to 2 weeks without appreciable loss of enzyme activity. Routine sections of fixed wax-embedded tissue; frozen sections of fixed tissue; post-fixed cryostat sections.

Incubation medium
67 mmol/l phosphate buffer pH 7.6	47.5 ml
Hexazotized pararosaniline (p. 318)	0.24 ml
Naphthol AS-D chloroacetate	5 mg

(dissolved in 2.5 ml N,N-dimethylformamide)
  Adjust pH to 7.6.

Method
1. Fix films in buffered formalin acetone (p. 304) at 4–10°C for 30 sec
2. Wash in distilled water 3 times
3. Incubate away from direct light, at room temperature for 10 min
4. Wash with tap water
5. Counterstain with methyl green for 15 min
6. Wash with tap water, dry and mount with DPX.

Positive result
Bright orange-red reaction product.

Notes
This medium can also be used on routine sections of fixed, wax-embedded tissue to show mast cells and granulocytes. Incubate for up to 2 h; counterstain with Carrazzi's haematoxylin.

DUAL ESTERASE (α-NAPHTHYL ACETATE ESTERASE, ANAE: CHLOROACETATE ESTERASE, CAE) FOR HAEMATOLOGY

Preparation of tissue
Air-dried films of blood or marrow; unfixed films may be stored at room temperature for up to 2 weeks without appreciable loss of enzyme activity.

Preparation of solutions
CAE medium:
M/15 phosphate buffer pH 7.6	9.5 ml
Naphthol AS-D chloroacetate	1 mg
in N,N-dimethylformamide	0.5 ml
Fast Blue BB	5 mg

Method
1. Fix, wash and incubate in ANAE medium for haematology (p. 328)
2. Wash with distilled water 3 times
3. Filter CAE medium (above) onto slides and leave 20 min at room temperature
4. Wash with distilled water 3 times
5. Counterstain with methyl green 15 min
6. Wash with tap water, dry and mount with DPX.

Positive result
ANAE activity dark red-brown; CAE activity blue.

REFERENCE

Yam L T, Li C Y, Crosby W H 1971 Cytochemical identification of monocytes and granulocytes. American Journal of Clinical Pathology 55: 283–291

β-GALACTOSIDASE

Preparation of tissue
Air-dried cryostat sections (5–8 μm) of snap-frozen tissue; frozen sections of formol-calcium gum-sucrose treated tissue; air dried bone marrow films.

Incubation medium
5-bromo-4-chloro-3-indoxyl-β-galactoside	3 mg
dissolved in 2 drops 2-methoxyethanol	
McIlvaine buffer pH 4.0	7 ml
50 mmol/l (2.11%) potassium ferrocyanide	0.5 ml
50 mmol/l (1.65%) potassium ferricyanide	0.5 ml
Sodium chloride	47 mg

Method
1. Cover sections with medium (use small volumes and prevent drying out) and incubate at 37°C for 5–18 h
2. Rinse in water
3. Counterstain with neutral red
4. Rinse, dehydrate, clear and mount in DPX.

Result

Enzyme activity is blue. Nuclei are red.

Notes

Blood films often have a layer of uncoloured precipitate over them after incubation. This can be removed, without damaging the film, by a jet of water from the wash bottle.

The enzyme is stable in dried films and its activity can be shown in films several years old. Neutrophils have diffuse cytoplasmic activity; lymphocytes have one or two dense spots of activity; eosinophils have granular activity; platelets are also positive.

The sections can be stained by the oil red O technique for neutral fat (p. 314) after Stage 2. The combination of β-galactosidase and oil red O methods is useful in the study of lipid degenerative disorders.

HEXOSAMINIDASE (β-GLUCOSAMINIDASE)

Preparation of tissue

Air-dried blood films; post-fixed cryostat sections of snap-frozen tissue; frozen sections of formol calcium-gum sucrose treated tissue; air-dried monolayers of cultured fibroblasts on cover slips.

Incubation medium

2.5 mg Naphthol AS-BI-β-D-glucosaminide dissolved in
   0.1 ml 2-methoxyethanol
5 ml distilled water
2 ml McIlvaine buffer pH 5.0
0.5 ml hexazotized pararosaniline
   Adjust pH to 4.1 (for maximum activity)
   Adjust pH to 5.0 (for best localization).

Method

1. Incubate lightly fixed films (5 min in formol-calcium) or sections for 1–4 h at 37°C
2. Wash in tap water
3. Counterstain in Carrazzi's haematoxylin
4. Wash (mount), dehydrate, clear and mount in DPX.

Result

A red colour indicates enzyme activity.

Notes

This method is of use only in testing for the total deficiency in Sandhoff's disease and cannot be used to detect Tay-Sachs disease.

β-GLUCURONIDASE FOR HAEMATOLOGY

Preparation of tissue

Air-dried films of blood or marrow; treat preferably within 4 h. Heparin and citrate are inhibitory. EDTA should be less than 5 mg/ml.

Incubation medium

Naphthol AS-BI β-D-glucuronic acid	1.4 mg
(dissolved in 1–2 drops N,N-dimethyl formamide)	
0.1 M acetate buffer pH 5.0	10 ml
hexazotized pararosaniline (p. 318)	0. 3ml
Adjust to pH 5.2; filter.	

Method

1. Fix in buffered formalin acetone (p. 304) at 4–10°C for 30 sec; or in formol-calcium for 1–2 min
2. Wash in distilled water 3 times
3. Incubate for 90 min at 37°C
4. Wash well in distilled water
5. Counterstain with methyl green for 15 min or Carrazzi haematoxylin for 2 min
6. Rinse in tap water, air-dry and mount with DPX.

Positive result

Red reaction product.

α-MANNOSIDASE FOR LANGERHANS CELLS, HISTIOCYTES X AND MELANOMA

Preparation of tissue

Cryostat sections (10 μm) of snap frozen skin biopsy mounted on a semipermeable membrane (see Histochemical Journal 1982 14: 697)

Preparation of medium

2.5 mg α-naphthyl—α-mannoside dissolved in 0.1 ml
   2-methoxyethanol
5 ml distilled water
5 ml McIlvaine (citrate-phosphate) buffer pH 5.0
0.5 ml hexazotized pararosaniline
500 mg polyethylene glycol 6000
   Adjust pH to 5.5

Incubate for 4 hours at 37°C or overnight at room temperature, cut out membrane, counterstain with Carrazzi's haematoxylin, dehydrate, clear and mount in DPX.

Activity is shown by a reddish brown colour with yellow background.

Note

Similar, but less intense activity can be shown in unfixed cryostat sections on cover slips incubated in the above medium.

MYELOPEROXIDASE

Preparation of tissue

Air-dried films of blood or marrow, preferably unanticoagulated; films may be made from anticoagulated material (heparin, EDTA, oxalate) which has not stood for more than 6 h at room temperature; once made, films should be fixed as soon as possible but staining may be postponed for up to 2 days.

Preparation of solutions

Benzidine: Because benzidine is a potential carcinogen, use gloves and fume cupboard. Add reagents in order listed, mixing well after each addition: ethanol 30% 100 ml, benzidine dihydrochloride (Koch-Light) 0.3g, zinc sulphate ($ZnSO_4.7H_2O$) 3.8% 1.0 ml; (a precipitate will form after addition of the zinc sulphate, but this will dissolve when the other reagents are added); sodium acetate ($NaC_2H_3O_2.3H_2O$) 1 g (or anhydrous 0.6 g), sodium hydroxide 1M 1.5 ml. Filter and store at room temperature in dark bottle.

Buffer for Giemsa: pH 6.8 — A: potassium dihydrogen phosphate (KH_2PO_4) 9.1g/l; B: sodium dihydrogen phosphate (NaH_2PO_4) 9.5g/l (or $NaH_2PO_4.2H_2O$ 11.9 g/l) in proportion of A 50.8 to B 49.2. Alternatively use buffer tablets (Hopkin and Williams).

Fixative: Formol-ethanol 1 part to 6.

Giemsa: Add 50 g powder to 2 l glycerol, mix well; leave in 56°C water-bath for 24 h with occasional shaking; add 2.5 l methanol, shake well, leave to stand 7 days at room temperature before filtering; dilute 1 to 9 with buffer immediately before use; do test runs for optimal staining time for each batch.

Incubation medium: Filter 5 ml stock benzidine solution into 10 ml cylinder; add 1 drop hydrogen peroxide 6% to 2 drops distilled water; add 1 drop diluted hydrogen peroxide to the stock solution immediately before use.

Method

1. Fix slides 60 sec
2. Wash with distilled water
3. Cover with incubating medium for 20 sec
4. Rinse in distilled water
5. Counterstain with buffered Giemsa for appropriate time
6. Rinse with tap water
7. Air-dry and mount in DPX (Gurr).

Positive result

Green-black in neutrophils and monocytes, brown-black in eosinophils.

REFERENCES

Dacie J V, Lewis S M 1975 Practical haematology, 5th edn. Churchill Livingstone, Edinburgh
Kaplan L S 1965 Simplified myeloperoxidase stain using benzidine dihydrochloride. Blood 26: 215–219

BENZIDINE-PEROXIDASE PROCEDURE (After Van Duijn)

Preparation of tissue
Unfixed cryostat or frozen sections. Smears. Better preservation of structure in sections or smears fixed in formalin at 4°C for 30–60 min, followed by rinse in water.

Preparation of solutions

Incubating medium
To 9 ml of saturated benzidine solution (prepared by dissolving 100 mg of benzidine in 20 ml of distilled water at 80–90°C, filter after the solution reaches room temperature and use fresh) add 1 ml of saturated NH_4Cl solution, 1 ml of 5% ethylene diamine tetra-acetic acid buffered to pH 6.0 with NaOH, and 1 drop of 3% H_2O_2 (1:10 dilution from the regular 30% H_2O_2 solution). To be used fresh.

Method

1. Incubate for 5–10 min at room temperature
2. Rinse in water
3. Mount in glycerogel, Apathy, or fructose syrup.

Results

Peroxidase activity sites — red blood cells and granules in leucocytes are stained blue. Colour fades within a few days or weeks.

DOPA-OXIDASE REACTION (after Okun)

Preparation of tissue
Cryostat or frozen sections of unfixed tissue or of tissue fixed in cold formalin for 12–24 h.

Method

1. Incubate each of 4 consecutive serial sections for 3 h at 37°C in one of the following:
   a. 0.1M phosphate buffer, pH 7.4
   b. 0.1M phosphate buffer, pH 7.4 containing 1–2 mg of DL-DOPA
   c. As in b, but also containing 1 mmol/l diethyldithiocarbamate (= 38.7 mg/100 ml).
2. Rinse in water, counterstain with Mayer's haemalum or Kernechtrot (nuclear fast red), dehydrate, clear and mount.

Results

Pigment formed as a result of the enzymic activity appears brown-black in section b. By comparing with sections a and c this pigment can be distinguished from pre-existing melanin.

Note

Presence of the reaction indicates melanogenetic capacity of cells and is useful, when present, in identifying naevus and melanoma cells. Mast cells are also stained by this procedure and are often difficult to distinguish from melanocytes. The perivascular location of mast cells and metachromasia of their granules can be used for definitive diagnosis.

Diethyldithiocarbamate is added (1c) to chelate the copper on which DOPA-oxidase is dependent.

GLUCOSE-6-PHOSPHATE DEHYDROGENASE (for cytology)

Preparation of slides

Fresh gastric or bronchial brush smears are air-dried.

Reaction medium

The reaction medium contains:

5 mmol/l glucose-6-phosphate	1.5 mg/ml
3 mmol/l NADP	2.4 mg/ml
0.67 mmol/l phenazine methosulphate	0.02 mg/ml
5 mmol/l neotetrazolium chloride	3.3 mg/ml

In 0.05M glycylglycine buffer pH 8.0 containing 34 mmol/l calcium chloride and 40% w/v Polypep 5115 (Sigma) to stabilize the section.

Method

1. Bubble the reaction medium with N_2 for at least 2 min
2. Incubate in the reaction medium in rings, in an atmosphere of nitrogen, watching for the intensity of the colour
3. Wash in distilled water
4. Mount in Farrant's medium.

Result

The activity due to glucose-6-phosphate dehydrogenase is shown by the deposition of formazan (red-purple stain).

Control

Incubate as for the test but omit either the substrate or the NADP from the reaction medium.

NAPHTHYLAMIDASE (for cytology)

Preparation of slides

Fresh air-dried smears.

Reaction medium

0.1M acetate buffer pH 6.1	10 ml
0.85% solution of sodium chloride	8 ml
0.2M potassium cyanide	1 ml
leucyl-2-naphthylamide 8 mg and	in 1 ml distilled
leucine amide hydrochloride 32 mg	water
Just before use, add Fast Blue B	10 mg

Method

1. Incubate the smears in the reaction medium in a Coplin jar at 37°C for 15 min
2. At 15-min intervals the reaction medium should be replaced with fresh medium. Repeat the process until sufficient reaction is obtained
3. Rinse well in 0.85% solution of sodium chloride at room temperature
4. Transfer to an 0.1M solution of copper sulphate at room temperature for 1 min (to chelate and so intensify the dye that has been formed)
5. Mount in Farrant's medium.

Result

Blue or purple particulate stain

Control

Smears can be incubated in the absence of substrate or the full incubation medium can be used on sections in which the naphthylamidase activity has been inhibited by treatment with absolute ethyl alcohol at 37°C for 30 min.

Note

Leucine amide hydrochloride is used in 4 times concentration of leucyl-β-naphthylamides to divert aminopeptidases which act preferentially on leucine amide hydrochloride, releasing a non-chromogenic reaction product which will not therefore contribute to the final colour.

Appendix 6

Immunohistochemical methods

IMMUNOHISTOCHEMICAL TECHNIQUES — GENERAL

Preparation of buffered saline

Tris-buffered saline (TBS) is used for diluting antibodies and for washing sections. This is prepared by adding a tenth volume of pH 7.6 Tris-HCl buffer (0.5M) to 0.15M saline.

Alternatively phosphate buffered saline (PBS) (p. 336) may be used in some procedures.

Removal of mercury pigment from formal sublimate-fixed tissues

Immerse dewaxed slides in iodine solution (0.5% iodine in 70% alcohol) for 5 min. Rinse in tap water. Immerse in 'hypo' solution (2.5% sodium thiosulphate in distilled water) for 1 min. Wash in tap water.

Trypsinisation

The solution for this procedure should be prepared immediately before use and not re-used. Dissolve trypsin and $CaCl_2$ in distilled water to give a final concentration for each constituent of 0.1%. Adjust pH to 7.8 with 0.1 M NaOH. Pre-warm trypsin solution in a staining dish to 37°C in a water-bath (this takes 20–30 min). Simultaneously place a rack of de-waxed slides for trypsinisation in TBS and warm to 37°C. When both solutions are at this temperature transfer slides from TBS to trypsin and incubate for appropriate time. This varies from section to section but is usually in the range 10–30 min. Wash and then proceed to stage 3 of the immunoperoxidase procedure.

Trypsin (or pronase) may be used at lower concentrations (from 0.005%).

Storage of reagents

Reagents

Antisera or peroxidase-labelled reagents should not be kept for any length of time after they have been diluted to their working concentrations. However if stored undiluted at 4°C there should be little or no loss of activity over a period of months or even years. A bacteriostatic agent is usually present in all commercial reagents. If in doubt, however, sodium azide may be added (at 0.1%) to prevent the growth of micro-organisms.

H_2O_2 should be obtained as a 30% solution (also referred to as '100 volume' H_2O_2) and stored in the dark at 4°C. It is stable for many months under these conditions.

Diaminobenzidine should be stored at $-20°C$ and warmed to room temperature before opening the container and weighing out. This precaution avoids the condensation of atmospheric vapour on the cold reagent.

Trypsin is stable in powder form at $-20°C$ for many months. It should be noted that numerous different preparations of this enzyme are available and that the same type (and preferably batch) should always be used.

IMMUNOPEROXIDASE STAINING OF ROUTINE SECTIONS AND SMEARS BY THE PAP METHOD

This technique is shown schematically in Figs. 11.1, 20.3 & 23.1. The full details of the technique are as follows:

1. De-wax
Xylene 60 sec
Xylene 60 sec
Ethanol 60 sec
Ethanol 60 sec
70% alcohol 60 sec
Wash in tap water.

2. Remove mercury pigment
Tissues which have been fixed in formol sublimate should be treated to remove residual mercury pigment (see above for details). If formol saline or buffered formalin has been used as fixative it may be necessary to digest the sections with trypsin before immunocytochemical staining (see text and section above for details of trypsinisation).

3. Block endogenous peroxidase
0.5% H_2O_2 in methanol 15 min
Wash in tap water
Rinse in distilled water
Transfer to TBS.

4. Reduce non-specific staining
Apply normal swine serum (1/10) for 10 min
Drain off but do not wash
Dry slides around sections.

5. Apply first layer antibody
Apply rabbit antibody at appropriate dilution (see below)
Allow reagent to remain on section/smear for at least 30 min
Wash slides in TBS for 5 min
Dry slides around section.

6. Apply second layer antibody
Apply swine anti-rabbit immunoglobulin (the optimal dilution is usually approx. 1/25)
Leave this reagent on slides for at least 30 min
Wash slides in fresh TBS for 2 min
Dry slides around sections.

7. Apply third layer antibody
Apply rabbit PAP (optimal dilution is usually approx. 1/100)
Leave this reagent on the section for at least 30 min
Wash slides in fresh TBS for 3–5 min

8. Development of peroxidase reaction
Prepare a solution of diaminobenzidine tetrahydrochloride (0.6 mg/ml) in TBS. Prepare aliquots in advance to reduce risk of any carcinogenic effect. 5 ml of solution is sufficient for staining 12–80 sections. Immediately before use add H_2O_2 to give a final concentration of 0.01% (see above for details of storage of H_2O_2). Remove slides from the TBS in

which they are being washed, tip off excess TBS and flood the slides with enzyme substrate. Allow reaction to develop for 7–10 min. The development of the peroxidase reaction may be followed under the microscope using a low-power objective. However, slides examined in this way should be placed in a suitable container, e.g. a Petri dish, to avoid spilling substrate onto the microscope stage. When the reaction has developed wash slides in fresh TBS for 2 min.

9. Counterstain
Sections may be counterstained with haematoxylin (Carrazzi or Mayer).

10. Dehydration and mounting
70% ethanol 60 sec
100% ethanol 60 sec
100% ethanol 60 sec
Xylene 60 sec
Xylene 60 sec
Mount slides in DPX or other xylene-miscible mountant.

IMMUNOPEROXIDASE STAINING OF ROUTINE SECTIONS OR SMEARS* BY THE INDIRECT TECHNIQUE

This technique is shown schematically in Figs 11.1, 20.3 & 23.1. The method follows the procedure given above for PAP staining up to and including stage 5 (application of primary antibodies). Stage 6 however consists of applying peroxidase conjugated swine anti-rabbit Ig (1/25). This reagent is left on the slides for at least 30 min. They are then washed in fresh TBS for 3–5 min before developing the peroxidase reaction, counterstaining and mounting (see stages 8–10 in PAP method).

It is often necessary to add normal human serum (at a final concentration of 1/25 to the peroxidase conjugate to block any cross-reactivity against human immunoglobulin. At the completion of incubation wash slides in fresh TBS for 3 to 5 min.

* Smear preparations would benefit from several saline washes to remove excess protein and eliminate background staining.

Immunoperoxidase staining of cryostat sections
Cryostat sections (5 μm) are mounted on gelatin-coated slides, air dried for 30 min and desiccated by placing them in the vacuum chamber of a freezer drier for 4–18 h. If this apparatus is not available dry the slides for a number of hours at room temperature or leave them (wrapped in aluminium foil or cling film) at 4°C overnight. Following drying sections are fixed in acetone at room temperature for 10 min and air-dried. They may at this stage either be stained immediately or wrapped in aluminium foil and stored at −20°C for future staining.

Immunoperoxidase staining
Staining may be performed by either a two-stage indirect method or by the PAP procedure, both of which are detailed above. The following technique describes immunoperoxidase staining of tissue antigens using monoclonal antibodies and a two-stage immunoperoxidase procedure.

1. *Apply first layer antibody*
Monoclonal antibody is applied at the appropriate dilution (see below) to the dry cryostat sections or smears. Incubation is carried out for at least 30 min. Wash slides in TBS for 5 min. Dry slides around sections/smears.

2. *Apply immunoperoxidase conjugate*
Apply peroxidase-conjugated rabbit anti-mouse immunoglobulin and leave this reagent on the slides for at least 30 min. The appropriate concentration is usually approximately 1/50 and it is usually necessary to add normal human serum at a final concentration of 1/25, in order to block cross-reactivity against human Ig. Wash slides in fresh TBS for 3–5 min.

3. *Development of peroxidase reaction*
See Section 8 of PAP procedure for staining of paraffin wax sections.

4. *Counterstain*
See Section 9 of PAP procedure for staining paraffin wax sections.

5. *Dehydration and mounting*
See Section 10 of PAP procedure for staining paraffin wax sections.

Establishment of optimal dilutions for immunoperoxidase reagents
At first sight the fact that immunoperoxidase techniques involve incubation with either two (in the indirect method) or three (in the PAP technique) different reagents before developing the enzyme reaction presents a daunting prospect of multiple chequerboard titrations to establish optimal working dilutions for each reagent. However by following the scheme detailed below this task can be simplified considerably.

1. *Control tissue*
When establishing a new immunoperoxidase staining method a normal tissue sample containing the relevant antigen should be obtained for use as a positive control. Tonsil tissue is widely used for this purpose. If the tissue is to be embedded in paraffin it should be fixed by a fixation schedule which optimally preserves antigenic reactivity. In the case of immunoglobulin Bouin's or formol sublimate fulfils this requirement.

2. *Choice of dilutions for second and third stage reagents*
Indirect immunoperoxidase technique. When using this technique the first step is to make dilutions of the second stage peroxidase-conjugated anti-Ig antibody (e.g. 1:10, 1:30, 1:90 and 1:270) and to carry out incubations on tissue sections for 30 min before washing and developing the peroxidase reaction. The lowest dilution at which no non-specific staining occurs should then be used to establish the optimal dilution of the primary antiserum (see Section 3 below).

PAP method. When using this technique it is important to realise that (owing to the peculiar geometry of the PAP sandwich — see Fig. 20.3) it is almost impossible to use the second stage linking antibody or the third stage PAP complexes at too high a concentration. Hence in initial testing these reagents should be kept at low dilutions (e.g. 1:10 for the bridging antiserum, 1:25 for PAP) and the primary antiserum titrated out (see below).

3. *Dilution of primary antiserum*
Having chosen working concentrations of the second and third stage reagents the primary antiserum should be tested at a range of different dilutions. When using polyclonal antisera it is usually appropriate to test the following dilutions: 1:20, 1:60, 1:180, 1:540 and 1:1620. From this titration it should be possible to select a concentration of primary antiserum at which there is no non-specific background labelling, but which is well above the concentration at which the specific staining begins to diminish in intensity.

Having chosen a working dilution for the primary antiserum in this way the peroxidase-conjugated anti-Ig reagent or the second and third stage reagents of the PAP technique can then be titrated out in order to establish the most economical concentration for their use. However it should be re-emphasised (see above) that in the PAP technique there is virtually no upper limit to the concentration at which the second and third stage reagents can be used (i.e. they will not cause non-specific background staining however low their dilution). In consequence there is an argument for erring on the side of over-concentration in order to avoid any possibility of sub-optimal staining.

IMMUNOFLUORESCENCE (Indirect technique)

Preparation of tissues
Routine sections after various modes of fixation (formol-calcium, Bouin, cold ethanol, etc.) are cut at 5 μm and floated on a warm (not hot) water-bath. They are picked up on clean, non-albuminised slides and dried at 37°C overnight.

Cryostat sections from fixed or unfixed blocks are picked up on formol-gelatine coated slides and dried overnight at 37°C or for 15 min or more at room temperature.

Preparation of solutions

Phosphate buffered saline
Dissolve 435 g NaCl in 3–4 l of distilled water with continuous stirring. Add 13.6 g KH_2PO_4 (anhydrous) and make up to 5 l. For use, make a 1/10 dilution with distilled water. Check pH and adjust to 7.1 to 7.4.

Method

1. Place sections on racks in covered Petri dishes containing wet cotton-wool
2. Apply to each section a drop of appropriately diluted antiserum in phosphate buffered saline. Leave for 45–60 min at room temperature.
3. Rinse in 3 changes of PBS for 5 min in each
4. Wipe dry (except section region of the slide) and replace in the Petri dish
5. Apply a drop of the appropriate fluorescein- or rhodamine-labelled anti-globulin. Leave for 45–60 min and again wash in PBS
6. Wipe dry and mount in buffered glycerine (9 parts glycerine to 1 part of PBS)
7. Examine by fluorescence microscopy.

Notes
Controls must be employed.
1. Normal rabbit or guinea-pig serum as first layer
2. Inappropriate serum as first layer
3. An iserum pre-absorbed with specific antigen. With high dilutions of antiserum (1/1000 and above) the first layer procedure is carried out for 24–48 h at 4°C.

INDIRECT ALKALINE PHOSPHATASE METHOD FOR IMMUNOCYTOCHEMISTRY

Solutions

A. Veronal acetate buffer pH 9.2
Boil 500 ml distilled water and allow to cool (to remove CO_2)

Sodium acetate (trihydrate)	0.9715 g
Sodium barbitone	1.4715 g
Distilled water (CO_2-free)	247.5 ml
0.1M hydrochloric acid	2.5 ml

B. Substrate solution
Suspend 5 mg Naphthol AS: BI phosphoric acid (sodium salt) in 1 drop of dimethylformamide. Mix 10 ml veronal acetate buffer pH 9.2 and 5 mg Brentamine Fast Red TR. Mix the two solutions together, avoiding contact with skin.

Method

1. Dewax, take to water through xylene and alcohols
2. Treat with 15% acetic acid to block endogenous alkaline phosphatase — 5 min

3. Rinse in PBS (pH 7.4), then wipe excess from the tissues so that the antiserum is not diluted too much on the slide
4. Incubate in a moist chamber, with *1st antibody* (e.g. anti-human carcinoembryonic antigen) diluted in 0.5% BSA in PBS or in 1:20 goat serum: PBS (60 min)
5. Wash off with 0.5% BSA in PBS
6. Wash several times in PBS (pH 7.4) containing 2–3 drops detergent (BRIJ 1:100 or TWEEN 80).
7. Wash off with PBS, and wipe excess from tissue. Incubate in a moist chamber with *2nd antibody* (anti-rabbit immunoglobulin alkaline phosphatase conjugate) diluted in 0.5% BSA in PBS or in 1:20 goat serum: PBS (60 min) Repeat steps 5 and 6
8. Wash in distilled water
9. Prepare substrate solution B just before use, and put 1 ml onto each slide
10. Leave at room temperature (60 min)
11. Rinse in distilled water, then wash in tap water
12. Counterstain with Mayer's haemalum (optional)
13. Blue in running tap water and sat. lithium carbonate (optional)
14. Mount in glycerin jelly (water base mounting medium)
15. If slides are to be kept permanently the cover-slip should be ringed with nail varnish or sealant when dry.

Note
Incubations at room temperature, unless otherwise stated. Steps 4 and 7, dilutions of 1st and 2nd antibody vary and have to be worked out for each source of antibody and tissue.

Results
Sites of immunocytochemical activity: red.
Nuclei: blue if counterstained.

IMMUNOHISTOCHEMICAL METHODS — SKIN

Direct IF tests

Fresh-frozen tissue
10 cryostat sections at 6–8 μm.

Antibodies
Anti-human IgA, IgG, IgM and complement (C3), fibrinogen, fluorescein-labelled.

1. The slides are laid out in a shallow plastic container in which a layer of moistened paper towelling has been placed
2. Each section is covered with the appropriate conjugate, and incubated in the covered container at 37°C for 30 min
3. The sections are then washed with several changes of phosphate-buffered saline (PBS) pH 7.2 in Coplin jars or staining jars, for 30 min

4. The excess fluid is blotted from the slides, care being taken not to touch the specimens, and the sections are mounted in 60% glycerine on PBS or Hydramount

5. Examine by fluorescence microscopy.

Indirect IF tests

Patient's serum

1. Each serum sample is absorbed with washed concentrated A and B red blood cells by incubating 1–2 ml of serum with an equal quantity of washed concentrated A and B cells for 30 min at 37°C

2. The mixtures are spun down, and the absorbed serum removed. It may be stored in a domestic refrigerator until required for use, preferably no longer than the following day

3. The serum sample is submitted to a series of doubling dilutions in phosphate buffered saline (PBS) to give dilutions of 1/10, 1/20, 1/40, 1/80, 1/160.*

Substrate

Fresh frozen rabbit oesophagus
  4. a. Five cryostat sections for each sample
      b. in addition 5 sections are cut for positive and
      c. 5 for negative controls.

5. The diluted serum is placed on the sections which are incubated at 37°C for 30 min

6. They are then washed in PBS pH 7.2 with several changes for 30 min

7. The area about the substrate is now dried and IgG conjugate is applied. Incubate at 37°C for 30 min

8. Wash in PBS as before, and mount in 60% glycerine in PBS. The slides can be stored in a refrigerator until they are read (within 72 h).

A positive control is usually obtained from a known high titre case of pemphigus, the serum being aliquoted into tubes each containing 0.1 ml and stored at −40°C. A negative control is also prepared by absorbing blood group antigens from a known normal serum.

---

* These dilutions may be extended to 1/2560, if e.g. in pemphigus, it is required to 'titre out' in order to monitor response to therapy.

STAINING OF METASTATIC MAMMARY CARCINOMA CELLS IN BONE MARROW
(see Dearnaley et al 1981)

Carcinoma cells do not stain with anti-EMA in routinely processed smears of bone marrow cells. Erythrocytes should be removed and the remaining cells washed prior to smearing.

Preparation of smears

1. Collect marrow aspirate into 2 ml Hepes buffered tissue culture medium (TC 199) containing 100 μl heparin

2. Layer cell suspension onto 10 ml of a mixture of Ficoll/sodium metrizoate, density = 1.077 ('Lymphoprep': Nyegaards Co.) in a 30 ml Sterilin universal container

3. Centrifuge at 400 g for 20 min

4. Remove supernatant together with white cell layer at the medium-Lymphoprep interface with a Pasteur pipette into suitable centrifuge tube. Make up to 20 ml with TC 199

5. Centrifuge at 200 g for 4 min. Aspirate supernatant, resuspend pellet in 20 ml TC 199. Re-centrifuge, aspirate supernatant leaving approx. 0.3 ml behind

6. Resuspend pellet and transfer to 0.5 ml conical centrifuge tube (Sarstedt). Centrifuge at 300 g for 3 min

7. Aspirate supernatant leaving volume approximately equal to that of cell pellet

8. Gently pipette the cells up and down to resuspend cell pellet and to disaggregate clumps of loosely cohesive cells

9. Use 3 μl cell suspension for each smear

10. Prepare thin smears and immediately fix in 100% ethanol for at least 30 min

11. Store smears dry, in dark, at −20°C

Treatment of smears prior to immunohistochemical stain

1. Immerse in 23 mg/ml periodic acid (0.1 M). Wash in tap water.

2. Immerse in 200 mg/l potassium borohydride (0.05 M) for 2 min. Wash in tap water.

3. Immerse in 20% acetic acid for 10 min. Wash in tap water. Finally rinse in PBS.

Proceed as for sections using alkaline phosphatase conjugate (p. 336) except that 250 μl of diluted antiserum or conjugate diluted in 5% normal serum in PBS should be used at Steps 4 and 7.

Note

Steps 1 and 2 abolish non-specific staining of some haemopoietic cells by the conjugated second antibody.

Step 3 destroys endogenous alkaline phosphatase present in osteoblasts and leucocytes.

Normal serum should be of the same species as that of the second antibody.

We use a rabbit primary antiserum prepared as described by Heyderman et al (1979). A goat anti-serum is available from Sera-Lab Ltd. (Crawley Down, Sussex, U.K.). Alkaline phosphatase conjugates can be purchased from Sigma Ltd.

The correct dilution of primary antiserum and conjugate must be determined beforehand. This can be done using sections of human kidney which will show a strong stain of the distal and collecting tubules.

A few myeloid cells from a few patients may give a weak stain on the cell membrane. These can be distinguished from tumour cells, which will be visualised by this stain, by their cytological detail.

REFERENCES

Dearnaley D P, Sloane J P, Ormerod M G, Steele K et al 1981 Increased detection of mammary carcinoma cells in marrow smears using antisera to epithelial membrane antigen. British Journal of Cancer 44: 85–90

Heyderman E, Steele K, Ormerod M G 1979 A new antigen on the epithelial membrane: its immunoperoxidase localisation in normal and neoplastic tissue. Journal of Clinical Pathology 32: 35–39

SPECIFIC RED CELL ADHERENCE TEST (SRCA)

Five-micron-thick *histological sections* are dewaxed in several changes of xylene, passed through alcohol and rinsed in three changes of 0.05M tris buffered saline (TBS), pH 7.4, for 5 min each. The slides are then incubated with isoantibody (anti-A or anti-B blood grouping human sera) in moist Petri dishes for 15 min. The slides are then washed in three changes of TBS for 15 min each and incubated with isologous indicator erythrocytes for 15 min.

The *erythrocytes* are outdated blood bank cells washed three times in TBS and resuspended to give a 3 to 5% suspension. The slides are carefully inverted on wood applicator sticks in Petri dishes filled with just enough TBS to cover the inverted surface of the slide. This allows the unreacted erythrocytes to settle off the slide. The slides are then gently moved to a clear area of the Petri dish and are read and photographed while inverted in TBS. For blood group O patients H antigen is assayed for by using a *Ulex europaeus* extract in place of an antiserum as previously described (DeCenzo et al 1975). Twenty grams of *Ulex europaeus* seeds are ground up and homogenized in 100 ml of TBS at 0°C. This preparation is centrifuged on the Beckman preparatory ultracentrifuge at 15 000 RPM for 120 minutes and the supernatant is recentrifuged at 36 000 RPM for 120 minutes at 0°C. The preparation is used in place of the human blood grouping antisera and is incubated with the slides for 1 h. The O erythrocytes are treated with 0.01% papain for 30 min at 37°C and then washed in TBS. This is done to increase their reactivity. The O erythrocytes are incubated with the slides for 30 min prior to inverting the slides in saline.

It should be stated that particular care is needed to ensure accurate results in the O blood group patients. The delicate nature of the antiserum and/or the surface antigen itself requires careful attention to the preparation of the *Ulex europaeus*, avoiding its use a second time after thawing and gentle handling of the slides while inverting them and flooding the Petri slides. On the other hand tissue from patients in the A, B, or AB groups shows consistent results under routine techniques.

Specificity controls should include positive and negative controls and sections from inflammatory lesions. Quality controls should include (a) fresh antisera and fresh indicator red cells (RC), (b) incubations in Ulex extract and (c) pH of TBS (7.4).

Appendix 7

Pigments, metals etc.

BILIRUBIN
SILVER SULPHIDE METHOD FOR HEAVY METALS
   (TIMM)
PEROXIDE-SILVER METHOD FOR CALCIUM OXALATE
   (PIZZOLATO)
CALCIUM (GLYOXAL-BIS-2-HYDROXYANIL) METHOD
COPPER (RUBEANIC ACID)
COPPER (RHODANINE)
SCHMORL REACTION
AUTOFLUORESCENT PIGMENTS
CHLORAZOL FAST PINK FOR PVA AND PVP

BILIRUBIN REACTION (Van den Bergh, after Raia)

Preparation of tissue
For staining all bilirubin, routine sections of formalin-fixed tissue may also be used. For differential staining of conjugated bilirubin fresh cryostat sections obtained after 18–24 h fixation in cold CaCl$_2$ (1%)-formalin should be used. The fixed tissue should be immersed for 1–2 days in 0.88M sucrose containing 1% gum acacia at 4°C.

Preparation of solution
Stock diazo solution: To 100 ml of distilled water add 2 ml of concentrated HCl, then 200 mg of 2,4 dichloroaniline, shake and filter. Keeps for 2 weeks at 4°C
   Sodium nitrite: 1%. Keeps for 2 weeks at 4°C
   Accelerator solution: Dissolve 6 g caffeine, 10 g sodium benzoate and 10 g urea in 100 ml water. To 35 ml of this solution add 25 ml concentrated (40%) formaldehyde and 30 ml water.

Method for bilirubin (direct procedure)
1. Immerse sections for 30 min in the staining solution (prepared by adding 0.5 ml of the sodium nitrite to 25 ml of the stock diazo solution and leaving in ice water for 20 min before use)
2. Wash for 3 min in running water
3. Counterstain with haematoxylin
4. Dehydrate rapidly in ethanol, clear and mount.

Method for all bilirubin (the indirect procedure)
1. Prepare the solution as above (0.5 ml of sodium nitrite and 25 ml of the stock diazo solution, left in ice water for 20 min), add to 50 ml of the accelerator solution. If a precipitate forms, dissolve by heating to 38–40°C. Immerse sections in this for 30 min
2. Wash, counterstain, dehydrate and mount as in the direct bilirubin procedure.

Results
The method for all bilirubin stains bilirubin everywhere yellow-brown. The direct procedure stains hepatic and renal bilirubin but not that found in neurons in kernicterus.

SILVER SULPHIDE METHOD FOR HEAVY METALS (after Timm)

Preparation of tissue
For paraffin wax embedding: Fix small pieces of tissue for 10–24 h in 70% ethanol containing 1–2 ml of concentrated (about 10%) ammonium sulphide per 100 ml. For cryostat sections: Suspend thin (less than 2 mm) pieces of tissue in a covered container containing 1–2 ml of concentrated ammonium sulphide at room temperature for 20 min.

Preparation of solution
To 100 ml of 20% gum arabic, prepared with daily stirring 14 days ahead of time, add 1 ml of 10% AgNO$_3$ and stir well. Then add 10 ml of a solution containing 0.5 g citric acid and 0.2 g hydroquinone. All reagents should be dissolved in double distilled water and the solution used within seconds of its preparation.

Method
1. Dewax and bring to water. Cryostat sections should be mounted on slides or cover-slips
2. Treat in the solution for variable lengths of time (20 min to 4–6 h), until sections are light brown
3. Wash well in distilled water
4. Counterstain with a suitable nuclear stain (e.g. haemalum, or Nuclear Fast Red)
5. Rinse, dehydrate, clear and mount.

Results

Heavy metals (Pb, Au, Ag, Fe, Cd, Cu, Co, Ni, Zn, Hg): brown-black.

Notes

The procedure does not allow identification of the metal. Metals present in protein complexes will be demonstrated only if they are ionized. Thus, haemoglobin is not stained while haemosiderin (ferritin) is easily demonstrable by the technique.

PEROXIDE-SILVER METHOD FOR CALCIUM OXALATE (after Pizzolato)

Preparation of tissue

Routine, frozen or cryostat sections of formalin-fixed tissue.

Method

1. Dewax and bring to water
2. Treat for 15 min with 12% (2M) acetic acid
3. Rinse well with distilled water
4. Put sections face-up on two glass rods over a bowl and pour on each 2 ml of a solution made by mixing equal volumes of concentrated H_2O_2 (30%) and 2% $AgNO_3$ in distilled water. If bench area is not well illuminated place a 60-Watt tungsten electric lamp 15 cm above the slides. Stain for 15–30 min
5. Wash thoroughly with distilled water, counter-stain for 2–3 min in 0.1% safranin in 1% acetic acid
6. Dehydrate and mount.

Results

Calcium oxalate deposits are stained black. Calcium phosphate and carbonate crystals are dissolved out by the acetic acid rinse.

CALCIUM: GBHA METHOD (after Wolters et al 1979)

Preparation of tissue

Cryostat sections of fresh tissue or sections of unfixed freeze-dried tissue preferred. Routine sections of fixed tissue can also be used, although much of the original calcium content may not be present in such tissue.

Preparation of solutions:

GBHA solution

Glyoxal-bis-(2-hydroxyanil) (GBHA) dissolved in:	2.5 g
75% ethanol containing:	50 ml
NaOH	1.7 g

Na_2CO_3-KCN solution:

Add solid anhydrous Na_2CO_3 to 50 ml 96% ethanol and shake until no more carbonate appears to dissolve. Filter, and add solid KCN. Shake again until the solution is saturated. Filter, and store solution in stoppered bottle.

Method

1. Mount sections on albuminized slides
2. Dewax sections in chloroform
3. Immerse sections in 0.1% celloidin (dissolved in 50:50 v/v absolute ethanol:diethyl ether) for about 30 sec, and dry in air
4. Flood sections with GBHA solution at room temperature and decant after 10 min
5. Rinse in 75% ethanol and subsequently in 96% ethanol
6. Immerse for 15 min in the ethanolic Na_2CO_3-KCN solution at room temperature
7. Rinse in 96% ethanol and subsequently in three changes of absolute ethanol
8. Clear in xylene and mount in a synthetic resin (e.g. Entallan or DPX)
9. Examine the sections with and without a green filter in the substage of microscope.

Result

Calcium deposits are coloured red, and black when viewed with a green filter.

Notes

Both ionic and insoluble calcium appears to be visualised by this method. In early versions of the method (e.g. Kashiwa & Atkinson 1963, as given in Pearse 1972), only some forms of calcium, probably insoluble salts, are stained.

RUBEANIC ACID METHOD FOR COPPER (after Uzman)

Preparation of tissue

Thin (1 mm or less) slices of fresh or formalin-fixed tissue; fixed or unfixed cryostat or frozen sections, or formalin-fixed routine sections after dewaxing and hydration. Metal-containing fixatives, such as Zenker or Susa and fixatives kept in metal containers should be avoided.

Method

1. Immerse for 10 min in 0.1% rubeanic acid in 70% ethanol
2. Add to the alcoholic solution sodium acetate to reach a concentration of 0.2%, stir and leave at room temperature for 2 days
3. Rinse in 70% ethanol. Leave in 70% ethanol for 1 day
4. Slices of tissue: dehydrate, clear, embed in paraffin wax, cut sections, deparaffinize, hydrate. Cryostat and other sections: hydrate
5. Counterstain with Mayer's haemalum or Nuclear Fast Red
6. Wash, dehydrate, clear and mount.

Results

Copper salts are stained greenish-black.

Notes
Other metal salts (for example: iron and lead) also react with rubeanic acid, but they do not exhibit the greenish tinge.

RHODANINE STAIN FOR COPPER (Lindquist 1969)

Preparation of tissue
Post-fixed cryostat sections; frozen sections; routine sections.

Solutions required

Rhodanine stock solution
p-Dimethylaminobenzylidene rhodanine   0.2 g
Ethanol   100 ml
   To prepare working solution, dilute 3 ml of stock solution (well shaken) with 47 ml distilled water.

Borax solution
Disodium tetraborate   0.5 g
Distilled water   100 ml

Technique
   1. Bring sections to water
   2. Incubate in rhodanine working solution at 37°C (or 56°C) 18 h (or 3 h)
   3. Rinse in several changes of distilled water and stain with Carrazzi's haematoxylin, 1 min
   4. Rinse in distilled water and then quickly in borax solution. Rinse well in distilled water
   5. Dehydrate, clear and mount.

Results
Copper deposits stain bright red. Bile stains green.

Note
If fading occurs, it can be reduced by staining at 56°C and by using certain mounting media (e.g. DPX).

REFERENCE

Lindquist R S 1969 Studies on the pathogenesis of hepatolenticular degeneration. II. Cytochemical methods for the localization of copper. Archives of Pathology 87: 370–379

SCHMÖRL REACTION

Preparation of tissues
Cryostat sections (10 μm); frozen sections; routine sections.

Preparation of solutions

Schmörl reagent
Add 18 ml of 1% ferric chloride to 6 ml of 1% potassium ferricyanide. Mix and use immediately.

Method
1. Air-dry sections 15 min
2. Cover with reagent for 15 min
3. Rinse in 1% acetic acid
4. Wash in distilled water and blot gently
5. Allow to air-dry completely
6. Clear in xylene and mount in DPX.

Results
Blue granular reaction indicates the presence of lipofuscin pigment, copper associated protein etc.

AUTOFLUORESCENT PIGMENTS

Preparation of tissue
Air-dried cryostat sections (5–8 μm) of snap-frozen tissue; routine sections of wax-embedded fixed tissue.

Method
1. Mount cryostat sections in DPX
   (dewax and mount routine sections in DPX)
2. Examine by fluorescence microscopy with dark-ground illumination (excitation filter 300–370 nm (UG5); barrier filter 410 nm.)

Result
Lipofuscin (wear and tear pigment) has orange-yellow fluorescence. Pigment in Batten's disease has a yellowish fluorescence.

Notes
The background tissue is deep blue on initial mounting. This colour becomes brighter on storage.
Cryostat sections have much less background than routine sections.

CHLORAZOL FAST PINK FOR PVA AND PVP

Preparation of tissue
Routine sections of formalin-fixed tissue.

Method
1. Dewax and bring sections to water
2. Stain in 1% chlorazol fast pink in 50% ethanol at 60°C for 10–30 min
3. Rinse in water
4. Counterstain with Carrazzi's haematoxylin
5. Wash, dehydrate, clear and mount.

Result
PVP and PVA stain pink-red.

Index